Marijuana / Cannabinoids

Neurobiology and Neurophysiology

Edited by

Laura Murphy

Andrzej Bartke

Department of Physiology
Southern Illinois University School of Medicine
Carbondale, Illinois

CRC Press
Boca Raton Ann Arbor London Tokyo

Library of Congress Cataloging-in-Publication Data

Marijuana/cannabinoids : neurobiology and neurophysiology / [edited by] Laura Murphy, Andrzej Bartke.
p. cm. — (CRC series in physiology of drug abuse)
Includes bibliographical references and index.
ISBN 0-8493-7931-8
1. Cannabinoids—Physiological effect. 2. Marijuana—Physiological effect. 3. Neuropharmacology. I. Murphy, Laura, 1955– . II. Bartke, Andrzej. III. Series.
[DNLM: 1. Cannabinoids—pharmacology. 2. Marijuana Smoking—adverse effect. 3. Nervous System—drug effects. 4. Neurobiology. 5. Neurophysiology. QV 77.7 M3352]
QP801.C27M36 1992
615′.7827—dc20
DNLM/DLC
for Library of Congress

92-20119
CIP

Direct all inquiries to CRC Press, Inc., 2000 Corporate Blvd., N. W., Boca Raton, Florida, 33431.

International Standard Book Number 0-8493-7931-8

Library of Congress Card Number 92-20119
Printed in the United States 1 2 3 4 5 6 7 8 9 0
Printed on acid-free paper

PREFACE

Within the last few years, marijuana research has progressed in leaps and bounds, due primarily to the development of potent cannabinoid-type compounds and demonstration of the presence of cannabinoid receptors in the brain. As the central nervous system is a primary site of marijuana action, much research, both basic and clinical, has focused on the neural consequences of cannabinoid action. Thus, it seems both timely and appropriate that this 1992 volume in the *Physiology of Drug Abuse* series deals with marijuana, one of the most widely used illicit drugs worldwide, and the effects of its principal components, the cannabinoids, on the brain. Topics ranging from cannabinoid structure/activity, and sites and mechanisms of cannabinoid action in the brain, to the effects of marijuana and cannabinoids on neurophysiology and behavior are presented.

Particular thanks go to the authors for making the time in their schedules to participate, for meeting deadlines, and for maintaining high standards of quality.

CRC Series in PHYSIOLOGY OF DRUG ABUSE
Edited by Ronald R. Watson

Published Titles:

Alcohol and Cancer

Alcohol and Neurobiology: Brain Development and Hormone Regulation

Alcohol and Neurobiology: Receptors, Membranes, and Channels

Drugs of Abuse and Neurobiology

Marijuana/Cannabinoids: Neurobiology and Neurophysiology

Forthcoming Titles:

Neurobiology of Opiates

Nutrition and Alcohol

Opioid Peptides in Substance Abuse

THE EDITORS

Dr. Laura L. Murphy received her Ph.D. degree in Reproductive Endocrinology in 1983 from the Medical College of Georgia in Augusta and completed her postdoctoral training at Duke University in 1986 where she studied the effects of cannabinoids on the female neuroendocrine system. Dr. Murphy held an assistant scientist position in the Department of Physiology at Southern Illinois University School of Medicine in Carbondale (1987–1989) before assuming her present position of research Assistant Professor. Her interests are primarily in female reproductive endocrinology and include the effect of cannabinoids on reproductive and neuroendocrine development and function. Dr. Murphy is the recipient of a research grant from the National Institute on Drug Abuse. She is the author of over 50 publications and is a member of several national and international endocrine, neuroscience, and cannabinoid research societies.

Dr. Andrzej Bartke received his Ph.D. degree in Zoology from the University of Kansas in 1965 and had postdoctoral training at the Institute for Cancer Research in Philadelphia, and at the Worcester Foundation for Experimental Biology in Shrewsbury, Massachesetts. He held faculty positions at the Jagiellonian University in Krakow, Poland (1965–1967), Worcester Foundation for Experimental Biology (1969–1978), and University of Texas Health Science Center in San Antonio, Texas (1978–1984) before assuming his present position of Professor and Chairman of the Department of Physiology at Southern Illinois University in 1984. His interests are primarily in male reproductive endocrinology, including the roles of prolactin and growth hormone, seasonal breeding, and effects of cannabinoids on reproductive and neuroendocrine development and functions. He has published over 330 research papers and book chapters. He is a recipient of a Research Career Development Award from NIH, is founding editor of the *Journal of Andrology*, and served as Chairman of the Reproductive Biology Study Section at NIH and President of the American Society of Andrology and the Society for the Study of Reproduction.

CONTRIBUTORS

Syed F. Ali
Division of Neurotoxicology
National Center for Toxicological Research
Jefferson, Arkansas

John R. Baily
Rocky Mountain Labs
Hamilton, Montana

N. K. Barkov
Science Production Centre of Medical Biotechnology
Ministry of Health
USSR

Sumner Burstein
Department of Biochemistry and Molecular Biology
University of Massachusetts Medical School
Worcester, Massachusetts

L. D. Chait
Drug Abuse Research Center
Department of Psychiatry
The Pritzker School of Medicine
The University of Chicago
Chicago, Illinois

Paul Consroe
Department of Pharmacology and Toxicology
College of Pharmacy
University of Arizona
Tuscon, Arizona

Dominique Croix
INSERM U-156
Lille-Cedex, France

W. A. Devane
Department of Natural Products
Hebrew University
Medical Faculty
Jerusalem, Israel

S. M. Dudkin
Science Production Centre of Medical Biotechnology
Ministry of Health
USSR

J. Charles Eldridge
Department of Physiology and Pharmacology, and
Center for Neurobiological Investigation of Drug Abuse
Bowman Gray School of Medicine
Wake Forest University
Winston-Salem, North Carolina

David M. Evans
Saint Louis University School of Medicine
Saint Louis, Missouri

J. J. Fernández-Ruiz
Department of Biochemistry
Faculty of Medicine
Complutense University
Madrid, Spain

Eliot L. Gardner
Departments of Psychiatry and Neuroscience
Program in Brain and Behavior
Albert Einstein College of Medicine
New York, New York

R. Glaser
Department of Chemistry
Ben Gurion University
Beersheva, Israel

Devin B. Houston
Saint Louis University School of Medicine
Saint Louis, Missouri

Allyn C. Howlett
Saint Louis University School of Medicine
Saint Louis, Missouri

Torbjörn U. C. Järbe
Department of Psychology
University of Uppsala
Uppsala, Sweden
Division of Addiction Research and Treatment
Department of Mental Health Sciences
Hahnemann University
Philadelphia, Pennsylvania

Philip W. Landfield
Department of Pharmacology
College of Medicine
University of Kentucky Chandler Medical Center
Lexington, Kentucky

Jean Leonardelli
Laboratoire d'Histologie
Faculté de Médecine
Université Lille II
Lille-Cedex, France

A. I. Machula
Science Production Centre of Medical Biotechnology
Ministry of Health
USSR

Roy J. Mathew
Department of Psychiatry
Duke University Medical Center
Durham, North Carolina

Diane A. Mathis
Division of Addiction Research and Treatment
Department of Mental Health Sciences
Hahnemann University
Philadelphia, Pennsylvania

Raphael Mechoulam
Department of Natural Products
Hebrew University
Medical Faculty
Jerusalem, Israel

M. Navarro
Department of Psychobiology
Faculty of Psychology
Complutense University
Madrid, Spain

Merle G. Paule
Division of Neurotoxicology
National Center for Toxicological Research
Jefferson, Arkansas

R. G. Pertwee
Department of Biomedical Sciences
Marischal College
University of Aberdeen
Aberdeen, Scotland

J. Pierri
Drug Abuse Research Center
Department of Psychiatry
The Pritzker School of Medicine
The University of Chicago
Chicago, Illinois

J. A. Ramos
Department of Biochemistry
Faculty of Medicine
Complutense University
Madrid, Spain

F. Rodríguez de Fonseca
Department of Biochemistry
Faculty of Medicine
Complutense University
Madrid, Spain

Reuven Sandyk
Department of Psychiatry
Albert Einstein College of Medicine
New York, New York

Andrew C. Scallet
Division of Neurotoxicology
National Center for Toxicological Research
Jefferson, Arkansas

William Slikker, Jr.
Division of Neurotoxicology
National Center for Toxicological Research
Jefferson, Arkansas

Gerard Tramu
Laboratoire de Neurocytochimie
Université de Bordeaux I
Talence-Cedex, France

Tibor Wenger
2nd Department of Anatomy
Semmelweis University School of Medicine
Budapest, Hungary

William H. Wilson
Associate Professor of Psychiatry
Duke University Medical Center
Durham, North Carolina

TABLE OF CONTENTS

1 Cannabinoid Geometry and Biological Activity

INTRODUCTION

Almost all molecules of pharmacodynamic importance in biology—enzymes, receptors, and transmitters—are chiral (asymmetric). Although they possess one or more chiral centers, generally only one enantiomer is formed in the body. This well-defined molecular geometry puts severe restrictions on the structure of compounds (synthetic or natural) interacting with these natural substances. Hence, both structural and stereochemical requirements have to be defined for every group of biologically active products, be they drugs or molecular tools for research investigations. The cannabinoids are no exception. In this review we shall try to (1) summarize the structural and stereochemical requirements for cannabinoid activity, (2) compare the metabolism, biological activity, and receptor binding of those few cannabinoids whose activity has been investigated with both enantiomers, and (3) analyze the geometry of cannabimimetically active compounds.

STRUCTURAL AND STEREOCHEMICAL REQUIREMENTS FOR CANNABINOID ACTIVITY

While the structural requirements for cannabimimetic activity are reasonably well established, the stereochemical ones are still being debated.[1-4] The main reason for this discrepancy is technical: while the chemistry of cannabinoids is not complicated by comparison to many other natural products, most cannabinoids are oily materials that are difficult to purify in large amounts; hence, one can reasonably assume that some of the synthetic products of the early syntheses were not completely pure (in particular, as regards the presence of enantiomers). Most of the early cannabinoid tests were performed *in vivo*. It is quite possible that such tests frequently detected nonspecific activity[5] in addition to activity caused by to the presence of enantiomeric impurities.

Nearly 20 years ago we formulated tentative rules for cannabimimetic structure-activity relationships (SAR).[1] Most of these rules have thus far withstood the erosion of time. Over the years, exceptions were noted and refinements were made. In the last few years, synthetic compounds with structures apparently unrelated to those of the classical cannabinoids have shown cannabimimetic activity. Hence, it is possible that in the near future our understanding of cannabimimetic SAR may undergo a revision.

Methods for Measuring Cannabimimetic Activity

The *in vivo* and *in vitro* methods for measuring cannabimimetic activity have been reviewed;[5-7] hence, only a short list of methods currently used are presented here. Gratifyingly, the levels of activity measured by any one of the listed methods are reasonably comparable to most of the others.

Overt Behavior in Rhesus Monkeys or Baboons[1,8]

Active cannabinoids injected i.v. cause a pattern of behavior resembling that of humans under the influence of cannabinoids. The test is semiquantitative. At low doses (0.05 mg/kg), Δ^9-THC (1) causes tranquility, drowsiness, decreased motor activity, occasional partial ptosis, and occasional head drop; at higher doses (0.1 to 0.25 mg/kg), the drug causes stupor, ataxia, suppression of motor activity, full ptosis, and typical crouched posture (thinker position) maintained for up to 3 h (the animal may, however, regain normal behavior for short periods of time if external sensorial stimuli are applied); at doses above 0.5 mg/kg, the drug causes severe stupor and ataxia, full ptosis, immobility, crouched posture lasting for more than 3 h, and absence of reaction to external stimuli. A compound is not considered to be cannabimimetic if, at 5 mg/kg i.v., it fails to induce the above characteristic syndromes.

Dog Ataxia[9,10]

This simple, though very sensitive test is also semiquantitative in nature. Δ^9-THC (1) at 0.2 mg/kg injected i.v. causes static ataxia and depressed activity and the tails are typically tucked. This test has been widely used since the beginning of this century. Walton, in his classical book on marijuana published more than 50 years ago,[9] presents numerous photographs of dogs under the influence of the crude drug. Walton[9] defined six stages of intensity of effects: (1) a slight depression, (2) a barely recognizable ataxia, (3) an obvious ataxia, (4) a marked ataxia in which the animal frequently pitches forward and barely catches itself, (5) inability to stand alone, and (6) inability to rise and plunge about. The ataxia is chiefly manifested by swaying movements. More recently, Martin's group in Virginia has used the dog ataxia test as one of several routine tests for determining cannabinoid activity.[10]

The Mouse Ring Test[11]

In this simple assay, introduced by Pertwee, a mouse injected s.c. with the drug is placed on a horizontal ring and the percentage of time spent immobile during a 5-min exposure is determined.

Spontaneous Activity in Rats and Mice[10]

Mice injected i.v. with the drug are placed in a photocell activity chamber. Interruptions of the photocell beams are recorded to quantify the locomotor activity. In a standard variation, rats are used in an open field test with ambulation, rearing, defecation, grooming, and urination recorded.

Drug Discrimination[12-14]

Assays based on drug discrimination are at present the most specific *in vivo* tests available. Rats or pigeons are trained to emit one response when trained with Δ^9-THC and an alternative response when trained without drug. There is no generalization to Δ^9-THC (1) from various CNS-active drugs. However, some exceptions have been noted, and it has been suggested that "the discriminative stimulus properties of THC may have some commonality with the effect of diazepam in a subpopulation of rats trained to discriminate THC".[14]

Binding to the Cannabinoid Receptor[15,16]

The recent discovery of a cannabinoid receptor and the availability of labeled ligands has made possible the efficient screening of compounds for cannabinoid activity. Thus far, however, the binding of only a limited number of cannabinoids has been reported.

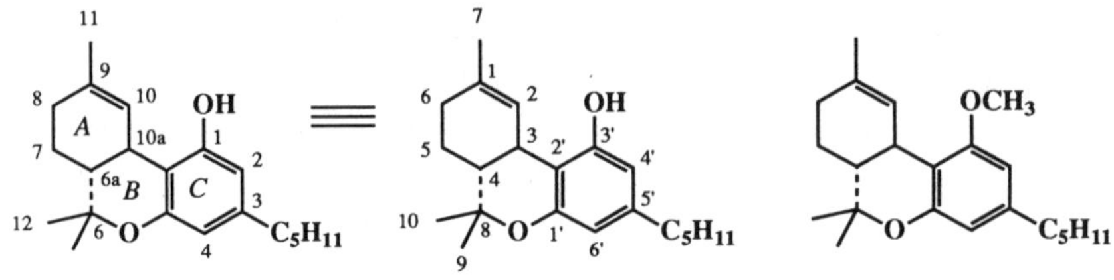

Δ-9-THC (1) Δ-1-THC (terpene nemenclature) (2)

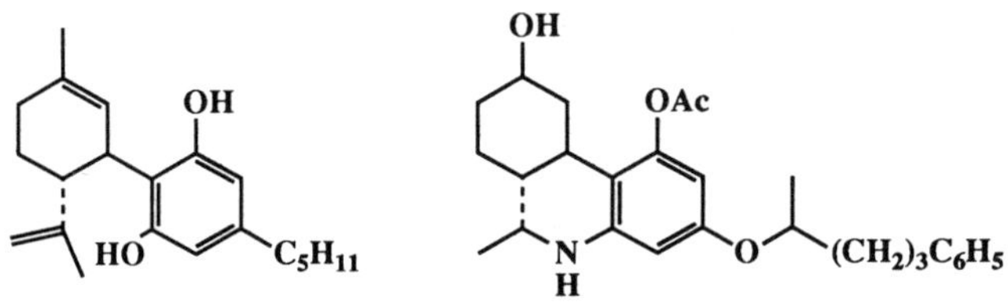

Cannabidiol (3) (CBD) Levonantradol (4)

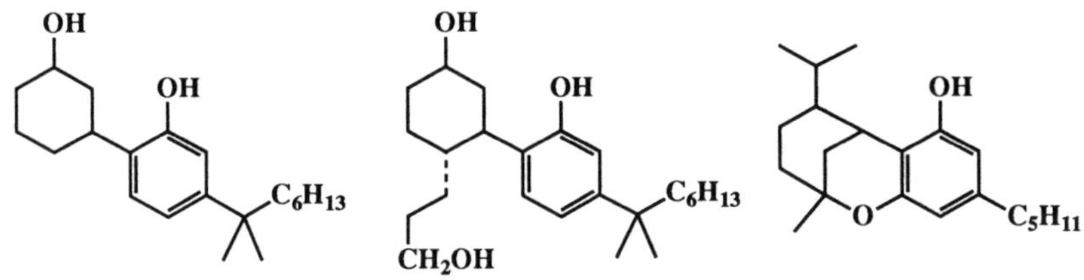

CP-47497 (5) CP-55940 (6) (7)

SCHEME 1.

Structural Requirements for Cannabimimetic Activity

In the following discussion, we shall use the benzopyran numbering in order to conform to the rest of the articles in this monograph. For comparative purposes, the terpenoid numbering is indicated in Scheme 1. While we use statements such as "retains activity", "eliminates activity", etc., it should

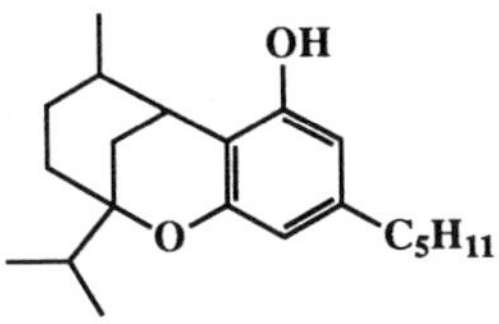

(8)

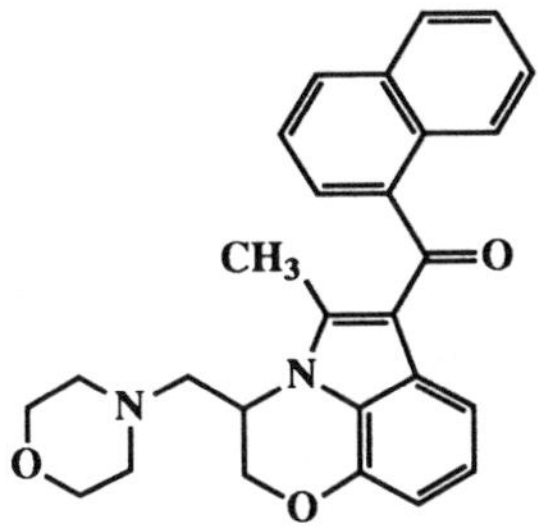

Win-55212-2 (9)

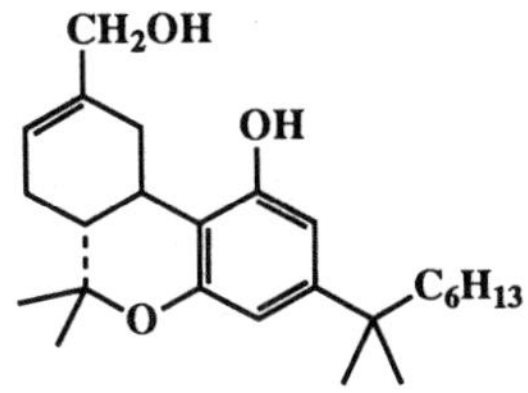

HU-210 (10)

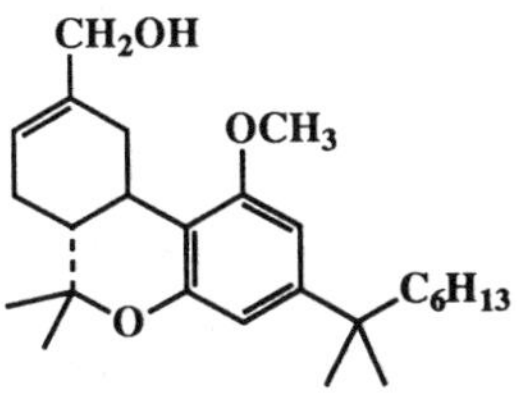

(11)

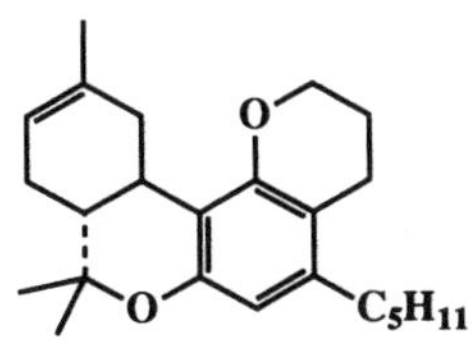

(12)

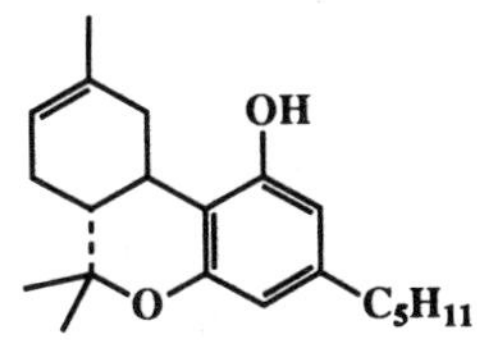

Δ-8-THC (13)

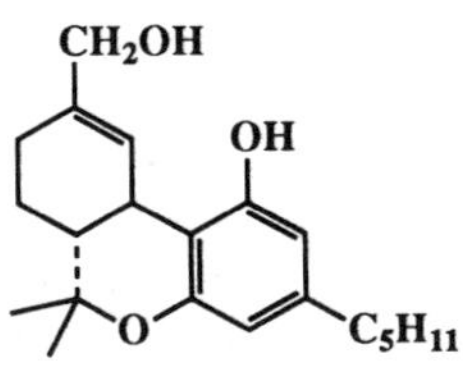

11-OH-Δ-9-THC (14)

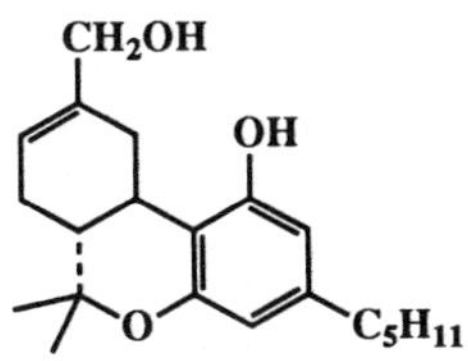

11-OH-Δ-8-THC (15)

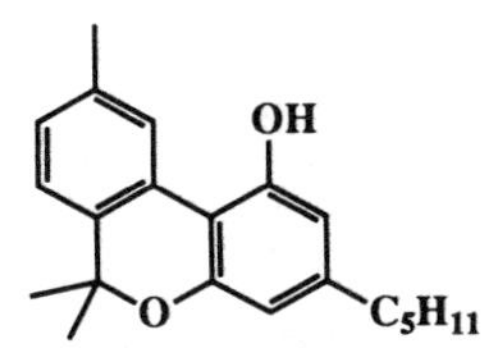

Cannabinol (16)

SCHEME 1. (continued)

be understood that such determinations depend on the sensitivity and specificity of the test. For example, while methoxy Δ^9-THC (2) is found to be inactive in most tests,[1,3] highly sensitive ones have determined a level of activity of approximately 4% compared to that of Δ^9-THC itself.[3]

1. A dihydrobenzopyran-type structure, with a hydroxyl group on the C-1 aromatic position and an alkyl group on the C-3 aromatic position, is

present in most active cannabinoids, including Δ^9-THC (1). Opening of the pyran ring forming cannabidiol (CBD) (3) type compounds leads to complete loss of activity.

However, several major exceptions have been found. (1) Levonantradol (4) and some of its derivatives:[17] this modification indicates that the benzopyran ring structure may be exchanged with a corresponding tetrahydroquinoline-type structure. (2) Compounds synthesized at Pfizer, numbered CP-47497 (5), CP-55940 (6), and some of their derivatives:[17] this is a major simplification of the cannabinoid molecule, which indicates that a heterocyclic B ring may not be an absolute requirement for activity.[18,19] However, as cannabidiol (a diphenol) has no cannabimimetic activity, and while CP-47497 (a monophenol) is cannabimimetic, the exact requirements need further elaboration. It seems that only monophenolic, but not resorcinolic, compounds are active when ring B is missing. The synthesis and testing of additional compounds should be achieved before strict rules can be formulated on the point. (3) Compounds of type 7 and 8 that are not planar; hence, planarity is not an absolute requirement.[18,19] (4) Recently, several amino alkyl indoles (AAI) (such as structure 9) structurally unrelated to the cannabinoids, were found to have a partial cannabimimetic pharmacological profile and bind to the cannabinoid receptor. The SAR in this series remains to be clarified.[20,21]

2. The C-1 phenolic group has to be free. While phenol esters exhibit activity *in vivo*,[1,2] this is probably due to hydrolysis in the body, as no activity has been detected *in vitro*.[22]

Etherification of the phenolic group has been considered to lead to elimination of the activity throughout the cannabinoid series. Recently we observed that etherification (on the phenolic group alone) of the highly potent (–)-11-hydroxy-Δ^8-THC-dimethylheptyl (10) leading to compound 11 does not completely abolish binding to the receptor or activity in drug discrimination in the pigeon. The ether (11) is considerably less cannabimimetic than the starting material in both tests. The activity is obviously not due to ether removal which, while possible *in vivo*, will not be expected in the binding assay. As mentioned above, low levels of activity (approximately 4% of the activity of Δ^9-THC) have been observed for the methyl ether of Δ^9-THC.[3]

Seltzman et al.[23] have reported that O,2-propano-Δ^8-THC (12) depressed rectal temperature and produced antinociception and ring immobility in mice. However, it differed from Δ^9-THC (1) and Δ^8-THC (13) in that it only

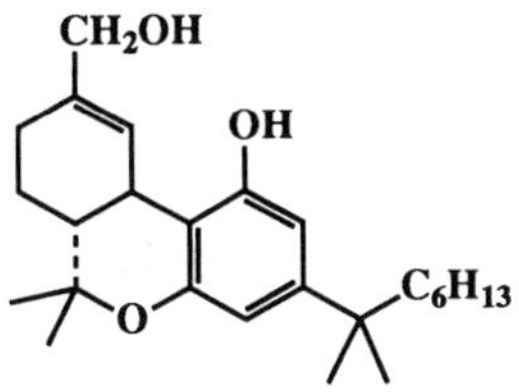

(17)

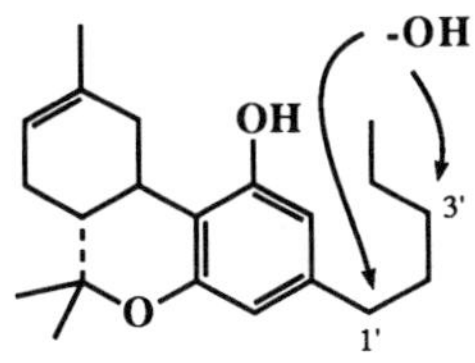

1'-OH-Δ-8-THC (18)
3'-OH-Δ-8-THC (19)

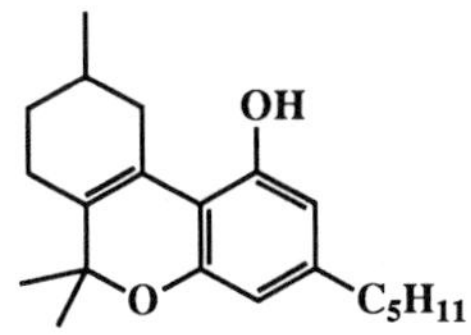

Δ-6a,10a-THC (20)

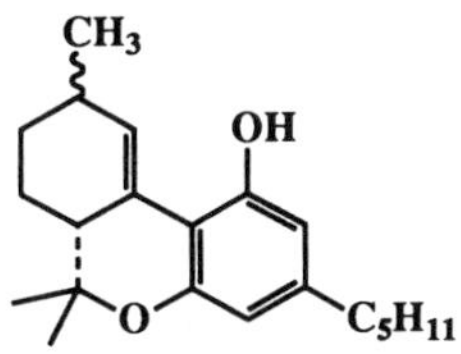

Δ-10,10a-THC (21)

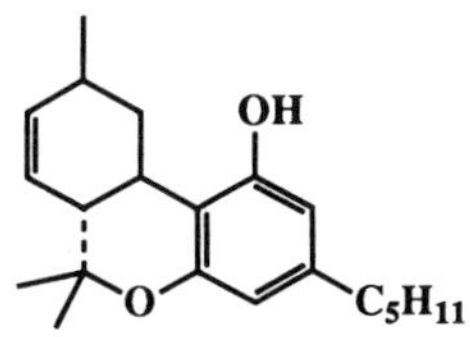

Δ-7-THC (22)

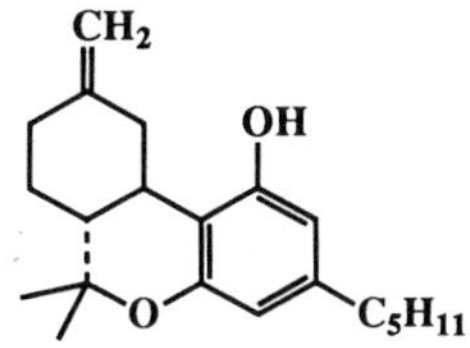

Δ-9,11-THC (23)

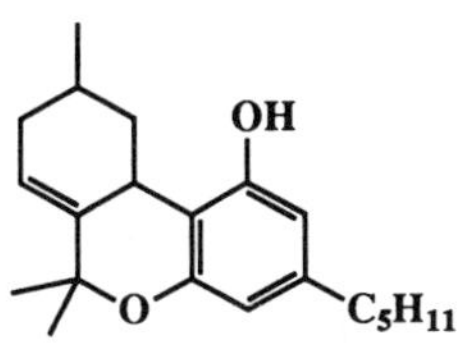

Δ-6a,7-THC (24)

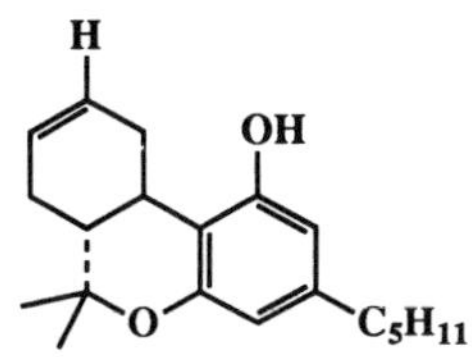

11-Nor-Δ-8-THC (25)

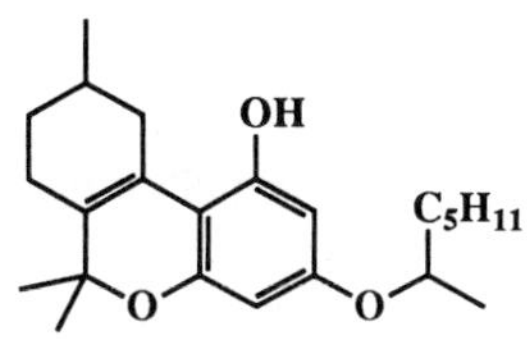

(26)

SCHEME 2.

weakly depressed locomotor activity and failed to substitute for Δ^9-THC in the drug discrimination paradigm. A similar separation of cannabinoid pharmacological effects has not been possible heretofore. These results are of considerable theoretical interest and will be further discussed in the section on "The Geometry of Cannabimimetically Active Cannabinoids". Replacement of the phenolic group by an amino group in some cannabinoids retains activity; replacement by a thiol group in all cases tested thus far eliminates activity.[24]

3. A major metabolic path in the cannabinoid series involves hydroxylation at C-11.[25,26] Both 11-hydroxy-Δ^9-THC (14) and 11-hydroxy-Δ^8-THC (15) are cannabimimetic[27,28] and somewhat more potent in tests than the parent substances. Further oxidation to the 11-aldehyde or the corresponding carboxylic acid leads to loss of cannabimimetic activity,[28] although analgesic activity may be retained.[29] Cannabinol (16) is weakly cannabimimetic; hydroxylation at C-11 causes an increase in activity.[3] 11-hydroxy-Δ^8-THC-DMH {HU-210, (10)} is apparently the most potent cannabimimetic known thus far (see the section on "Biological Activity").[31-35] The enhancement of activity is due to the simultaneous introduction of a hydroxyl group at C-11 and the substitution of the pentyl side chain with a 1,1-dimethylheptyl (DMH) one. Surprisingly 11-hydroxy-Δ^9-THC-DMH (17) is less active than HU-210 (10),[35,36] although Δ^9-THC (1) is more active than Δ^8-THC (13). While monohydroxylation on the terpene ring usually leads to active derivatives, dihydroxylation generally causes loss of activity.[1,3]

Hydroxylation on the side chain has been investigated thoroughly as such compounds are cannabinoid metabolites. While hydroxylation at the C-1′ position leading to (18) eliminates activity, a substitution of this type on any one of the other four side chain positions retains activity. Hydroxylation at the C-3′ position leading to (19) potentiates activity.[25,37]

4. It is now well established that the Δ^9 isomer is the main, or perhaps the only, cannabimimetic principle in cannabis.[38,39] Occasional reports mention the presence of the Δ^8 isomer (13), but it is possible that it is actually formed by isomerization.[39,40] Other double-bond THC isomers have been prepared by synthesis.[41,42] In discrimination tests,[43] the activity is as follows: Δ^9-THC (1) >Δ^8-THC (13) >$\Delta^{6a,10a}$-THC (20) >$\Delta^{10,10a}$-THC (21). Δ^7-THC (22) and $\Delta^{9,11}$-THC (23) are inactive in animal tests.[41,44] However, these compounds have not been extensively tested. $\Delta^{6a,7}$-THC (24) has not been tested at all.

5. The double bond in the terpene ring is not essential for activity. Reduction of Δ^9-THC leads to the C-9 hexahydrocannabinol epimers which are both active. Their activity is discussed in the section on "Stereochemical Requirements for Cannabimimetic Activity".

6. The methyl group on position C-9 is not an absolute requirement for activity. 11-nor-Δ^8-THC (25) is active in dogs.[45]

7. The alkyl side chain on the aromatic ring is of considerable importance for cannabinoid activity. A side chain of up to three or four carbon atoms apparently is not sufficient for robust activity. The C_5 chain, present in most of the natural cannabinoids, obviously leads to potent compounds. Elonga-

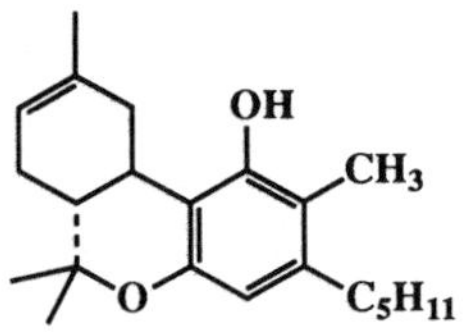

(27)

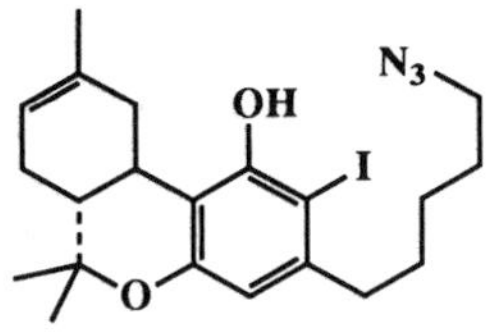

(28)

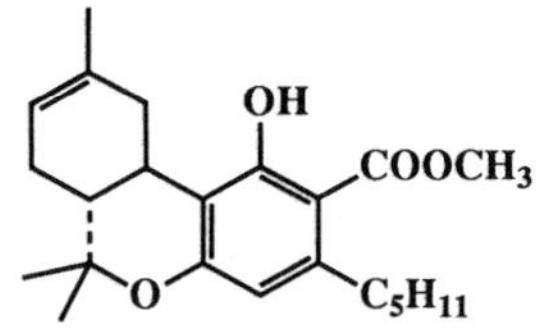

(29)

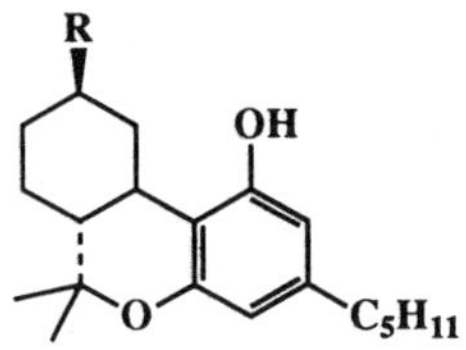

R = CH_3 (30a)
R = CH_2OH (30b)

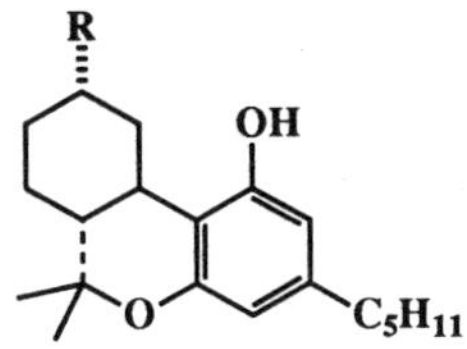

R = CH_3 (31a)
R = CH_2OH (31b)

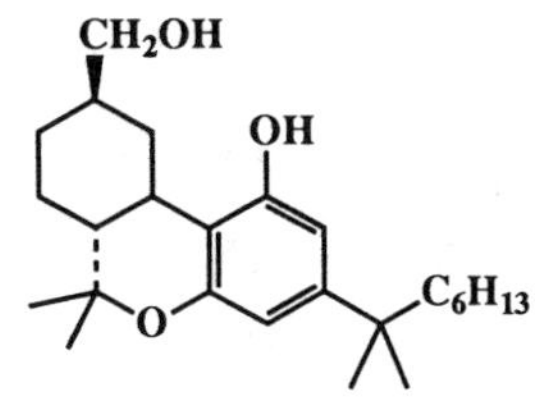

(32)

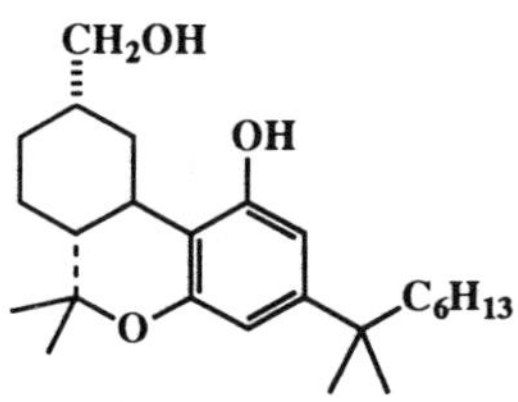

(33)

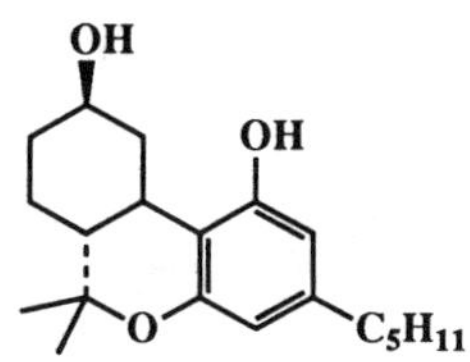

(34)

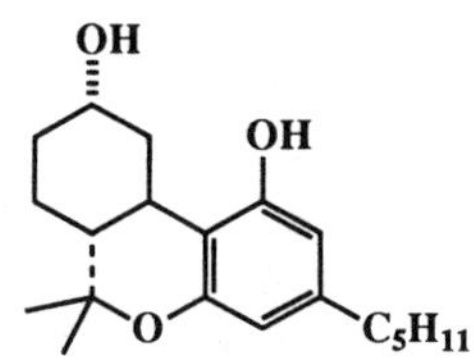

(35)

SCHEME 3.

tion and branching of the side chain confer potentiation. The most active side chains seem to be the 1,1-dimethylheptyl and 1,2-dimethylheptyl ones.[1,3]

An all-carbon side chain may be replaced by one containing an oxygen as an ether; for an example, see compound 26. These modifications do not seem to cause qualitative changes in activity. However, only a few examples have been studied.[17,46,47]

8. Substitution by alkyl groups on C-2 (e.g., compound 27) retains cannabinoid activity, while substitution at C-4 eliminates it.[48] Recently, in connection with research on radioactive labeling with iodine, it was reported

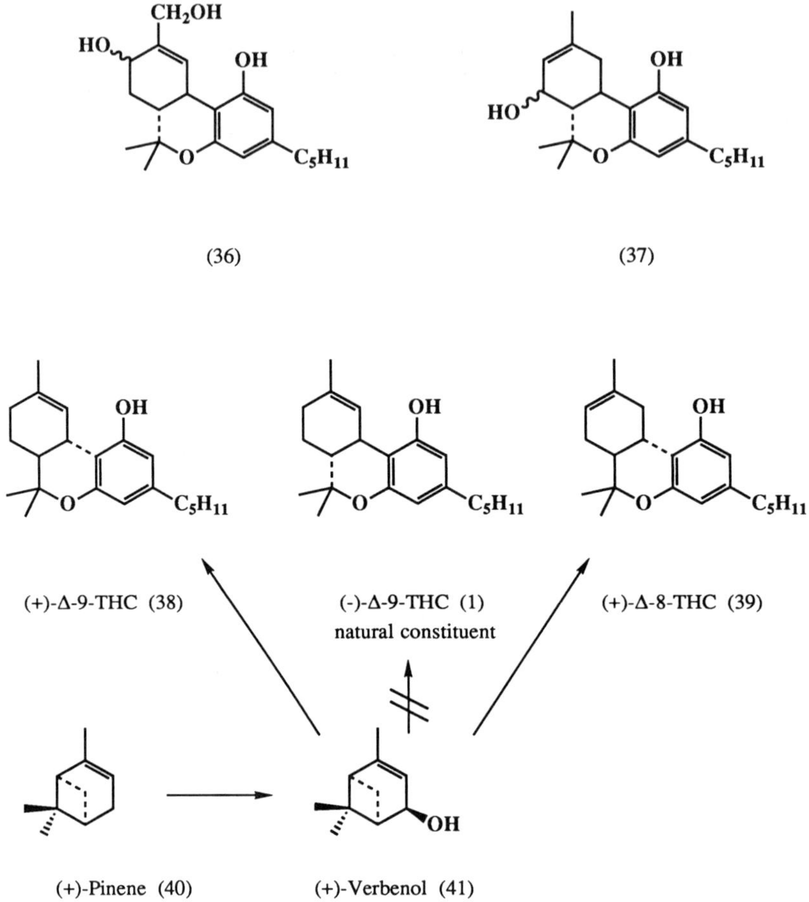

SCHEME 4.

that 2-iodo-5[1]-azido-Δ^8-THC (28) showed cannabinoid activity.[49] Electronegative groups such as carboxyl (e.g., compound 29), methoxycarbonyl or acetyl at either C-2 or C-4 eliminate activity.[48]

A considerable number of additional modifications of the cannabinoid molecule have been prepared and tested, such as replacement of the terpene ring with a heterocyclic one, amino and sulfur substitutions at C-11 and C-12, various esters on the C-1 phenolic group (in order to confer solubility in an aqueous medium), expansion or contraction of the heterocyclic ring, etc. Many of these compounds are cannabimimetic, but none of them contribute significantly to our understanding of cannabinoid activity. For a detailed listing, see the review by Razdan.[3]

The above, somewhat sketchy, recapitulation of structural requirements points out that three parts of the cannabinoid molecule are of central importance: the phenolic group, the side chain, and possibly the molecular arrangement around C-9. These requirements fit the classical cannabinoids, represented by the THCs as well as the nonclassical cannabinoids mentioned above (such as CP-47497 (5)), but obviously do not fit the novel cannabimimetic amino alkyl indoles (9).

Stereochemical Requirements for Cannabimimetic Activity

In this part of our review, we shall look at those cannabinoids in which the configurations of only some of the asymmetric centers have been modified. Pairs of enantiomeric cannabinoids that have been investigated thus far will be discussed in the next section. The reason for this arbitrary separation is that while enantiospecificity (i.e., the specific biological activity of only one of the mirror images of a molecule or enantiomer) is of basic importance to binding to an enzyme or a receptor, the change of configuration at a single asymmetric center and a concomitant change of biological activity may only give us evidence as to the SAR in a particular series of modifications.

1. The stereochemistry at the 6a,10a in the natural active cannabinoids is *trans* (6a*R*,10a*R*).[50] A few *cis* isomers have been tested and have shown very low activity.[2] However, *cis* compounds have not been studied over a wide range of tests.

2. As mentioned in the previous section, the double bond in the terpene ring is not essential for activity. Reduction of Δ^9-THC leads to hexahydrocannabinol epimers that are both active, and the equatorial epimer (30a) is considerably more active than the axial one (31a).[1,5] The same relationship is observed with the 11-hydroxyhexahydrocannabinols (30b, 31b).[52] Recently, the corresponding 11-hydroxy compounds with a 1,1-dimethyl heptyl side chain were prepared. The equatorial epimer (32) was found to bind to the cannabinoid receptor even better than the highly potent starting material HU-210 (10), while the axial one (33) binds considerably less.[53] The equatorial epimer has also been tested in a discrimination test and was found to be equipotent to HU-210. It seems that an equatorial substitution (i.e., one in which the methyl or hydroxymethyl group is in the plane of the cyclohexane ring) is preferable to an axial one.

3. In two additional series, an equatorial substituent on C-9 was found to confer greater activity than an axial one. While (9*R*,6a*R*)-$\Delta^{10,10a}$-THC ac-

etate (21) (in which the 11-methyl group is equatorial) is active, the corresponding axial epimer is inactive in discrimination tests in pigeons.[43] When the methyl group in the hexahydrocannabinols is replaced by a hydroxyl group, activity is retained.[54] Again, the equatorial epimer (34) is much more active than the axial one (35) in animal tests.

4. Several hydroxylated metabolites of Δ^9-THC and Δ^8-THC are known in both epimeric forms. Thus 8α and 8β-hydroxy-Δ^9-THC (36) and 7α and 7β-hydroxy-Δ^8-THC (37) have been identified as relatively minor metabolites. They have been synthesized and tested. Some relatively minor differences in activity were observed.[1,3]

CANNABINOID ENANTIOMERS: A COMPARISON OF THEIR METABOLISM, BIOLOGICAL ACTIVITY, AND RECEPTOR BINDING

METABOLISM

The metabolism of (–) Δ^9-THC (1), (–) Δ^8-THC (13), (–) CBD (3), and several of the other major natural cannabinoids has been investigated in considerable detail, and numerous reviews are available.[25,26] The major primary metabolic pathways are: (1) hydroxylation of the C-11 (in THC), followed by oxidation to the corresponding carboxylic acid; (2) hydroxylation at C-8 (in Δ^9-THC) or C-7 (in Δ^8-THC); and (3) hydroxylation on the side chain, followed by cleavage and oxidation to the corresponding acid. Comparable metabolic pathways are followed for CBD and other natural cannabinoids. The secondary metabolism involves glucuronidation of the carboxylic and phenolic groups. These glucuronides are stored in the body for a relatively long period of time and can be detected frequently in urine *several weeks* after cannabis consumption.

The unnatural, synthetic (+) enantiomers of the cannabinoids have received much less attention than the natural (–) enantiomers. Only one synthetic route has been reported for these compounds.[55] Further routes to the unnatural cannabinoids have been explored and the enantiomeric forms of nabilone, a synthetic antiemetic cannabinoid, have been prepared. However the (+)-THCs have not yet been reported by these routes.[56,57]

In (+)-Δ^9-THC (38), initial hydroxylation at C-11 is the major *in vivo* metabolic pathway, followed by oxidation to the carboxylic acid.[58] Addi-

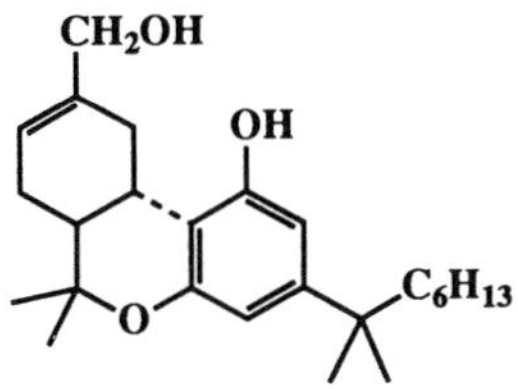

(+)-HU-211 (42)

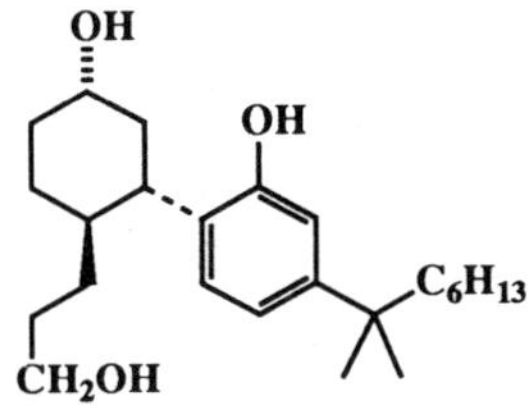

(+)-CP-56667 (43)

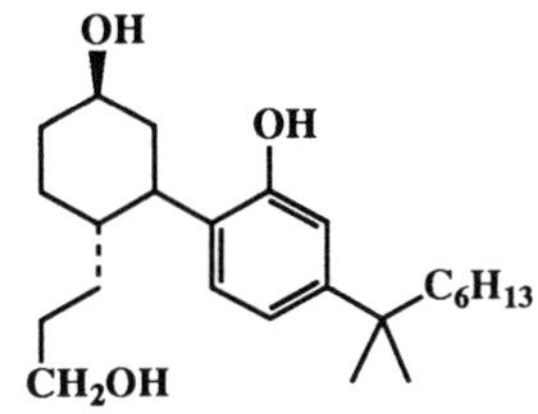

(-)-CP-55940 (6)

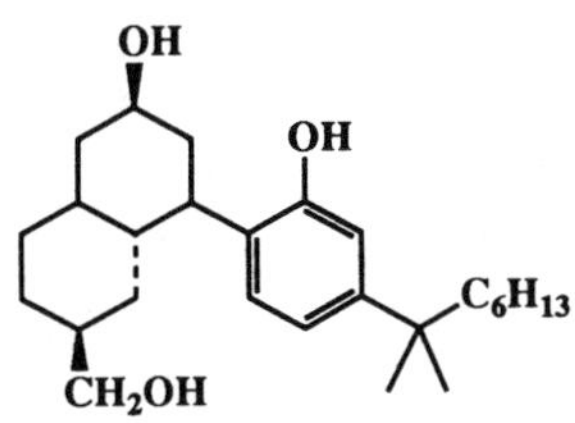

(-)-CP-55244 (44)

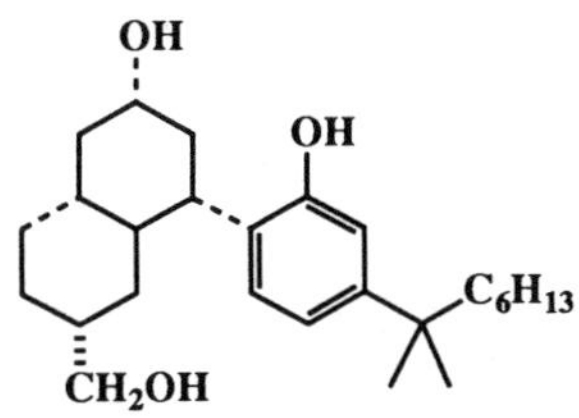

(+)-CP-55243 (45)

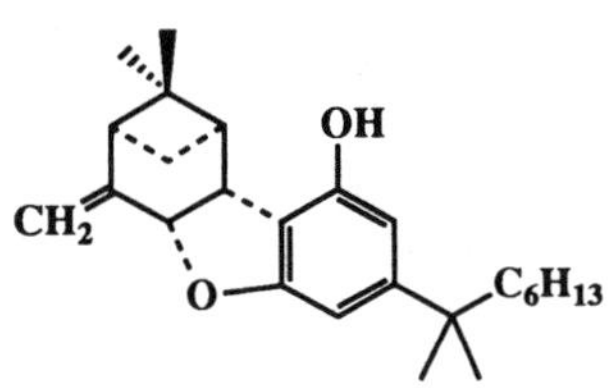

(46)

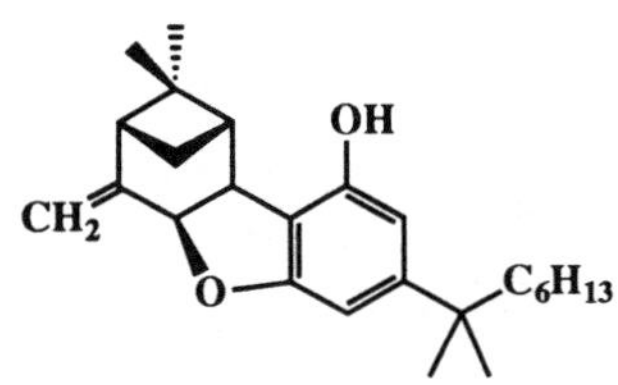

(47)

SCHEME 5.

tional hydroxylation on the side chain occurs. Hence metabolism is very similar to that of the naturally occurring (–) isomers. The main difference observed was the position of the side chain hydroxylation. 1′-Hydroxylation was preferred in (+)-Δ^9-THC, 2′-hydroxylation was mostly observed in (–)-Δ^9-THC. Likewise, less conversion to the carboxylic derivative was noted in the (+) series.[58]

A somewhat related picture was observed with (+)-Δ^8-THC (39). *In vivo,* the major metabolic route was again hydroxylation at C-11, followed by side chain hydroxylation, with the C-1′ position preferred over the C-2′. In addition, dihydroxylation at positions C-7 and C-11 is much more prominent in (–)-Δ^8-THC than in (+)-Δ^8-THC.[59]

The conclusion from these observations was that "the differences in metabolism between the (+) and (–) isomers would not be expected to contribute significantly to the absence of psychoactivity shown by the (+)-isomer."[58]

Biological Activity

Dewey et al.[2] have published a review on the pharmacological effects of cannabinoid stereoisomers that covers most of the work done until the early 1980s. Their review compares (+)- with (–)-Δ^9-THC as regards schedule-controlled responding and decreased overt behavior in the monkey, static ataxia in dogs, hypothermia, decreased spontaneous activity and the ring test in the mouse, and action on the isolated ileum of the guinea pig. It also compares the potency of (+)-Δ^8-THC (39) with that of (–)-Δ^8-THC (13) as regards static ataxia in the dog, analgesia and hypothermia in the mouse, depressed overt behavior in the monkey, drug discrimination, and acetylcholine turnover in the rat. The main conclusion reached was that the (3*S*,4*S*) (+) compounds exhibit some cannabimimetic activity, approximately 5 to 10% of those of the (3*R*,4*R*) (–) compounds. Both types of compounds were found to bind to plasma proteins equally well.

We assume that most of the (+)-THC enantiomers used in the above reviewed publications were prepared by the same synthesis.[55] Commercial (+)-α-pinene (40), the starting material for this synthesis, has about 90 to 95% optical purity and, if the intermediates or the final products are not further purified, THCs with a corresponding level of enantiomeric purity will be obtained. We are not aware that such purifications were indeed performed before the pharmacological tests. Hence, we have to conclude that much, if not all, of the cannabimimetic activity observed may have been due to contamination by the natural (–)-enantiomers.

We have recently repeated the above synthesis.[60] The intermediate crystalline *cis*-verbenol (41) was recrystallized to 99% optical purity and was further employed in the above synthesis. (+)-Δ^8-THC (39) and (+)-Δ^9-THC (38) thus obtained were tested in the mouse ring test, and compared with the corresponding (–)-enantiomers. The natural (–)-THCs showed significant activity at 1 mg/kg. No activity was observed with the (+)-enantiomers at 50 mg/kg. However, at 100 mg/kg, definite activity was noted, as would be expected with compounds that are 99% optically pure.

Until further work is reported, we have to assume that complete stereospecificity exists as regards cannabimimetic activity within the THC series. Most of the work done in the past with the (+)-enantiomers has to be reevaluated.

An example of absolute stereospecificity is the case of the enantiomers of the dimethylheptyl homologs of 11-hydroxy-Δ^8-THC (10 and 42). The synthesis of these enantiomers has been published.[32] It should be pointed out that in this synthesis, both an intermediate and the 11-OH-Δ^8-THC-DMHs themselves are crystalline compounds that can be obtained with absolute enantiomeric purity.

The pharmacological work with these compounds indeed points to the enormous difference in activity between the two enantiomers. As mentioned above, the (–)-enantiomer (HU-210) (10) is probably the most active cannabimimetic compound synthesized thus far.[35] In drug stimulus generalization tests, it was found that in the pigeon (–)-11-OH-Δ^8-THC-DMH (HU-210) was about 87 times more active than Δ^9-THC.[33] The (+)-11-OH-Δ^8-THC-DMH enantiomer (HU-211) (42) was inactive in the pigeon at doses about 4500 higher than the ED_{50} of the (–)-enantiomer; the respective dose in the rat was almost 1000 times higher. Parallel results were obtained in the mouse ring test and in the rotarod test in rats.[31]

The effects of the enantiomers of 11-OH-Δ^8-THC-DMH on spontaneous activity, rectal temperature, tail-flick latency, and catalepsy were also studied in mice and in the dog ataxia model.[34] The (–)-enantiomer (HU-210) (10) was active in all tests at doses between 3 and 100 μg/kg, while the (+)-enantiomer (HU-211) was inactive at 30 mg/kg in the mouse and 1 mg/kg in the dog. Thus, the (–)-enantiomer (HU-210) was 100 to 800 times more potent than Δ^9-THC in the mouse.

On the basis of the above results, the authors conclude that "the degree of stereoselectivity demonstrated with the enantiomers of 11-OH-Δ^8-THC-DMH greatly exceeded that which had been demonstrated previously for the cannabinoids. The degree of stereoselectivity is on the order of that demonstrated with the opiates, which have a well-defined receptor system. It is likely that the extreme potency of (–)-11-OH-Δ^8-THC-DMH unmasks the true stereoselectivity of the cannabinoids. The stereoselectivity and potency of the (–)-enantiomer is certainly consistent with the existence of a specific receptor for the cannabinoids which is involved in the mediation of some of the pharmacological effects of the cannabinoids."[34] Parallel results were obtained on inhibition of adenylate cyclase and binding to the cannabinoid receptor (see the section on "Noncannabimimetic Effects").[61]

Other enantiomeric pairs have also been investigated. For example, the nitrogen-containing cannabinoid levonantradol (4), whose stereochemistry parallels that of natural Δ^9-THC, has been compared with its enantiomer dextronantradol.[17,35] In a series of analgetic tests, the levo compound was active (MPE_{50}) at about 0.2 mg/kg, while the dextro compound was not active up to 10 mg/kg. In several tests for cannabimimetic activity (spontaneous activity, hypothermia, and catalepsy), levonantradol was active between 0.123 and 1.5 mg/kg, whereas dextronantradol was inactive.

The same overall picture was seen with the "nonclassical" cannabinoid pairs (–)CP-55940 (6)—(+)CP-56667 (43), and (–)CP55244(44)—(+)CP-55243 (45).[17,35,36] In both series, the (–) enantiomers were found to be highly cannabimimetic in a battery of *in vivo* tests. The (+)-enantiomers were essentially inactive. The results are presented in Table 1. As the relative ED_{50} (compared to that of Δ^9-THC) observed for a given drug was similar for all assays, results were averaged to give an estimate of their overall potency, with Δ^9-THC having a potency of 1.0 and HU-210 a potency of 100.0.

The degree of potency and particularly the high enantioselectivity within the four groups of compounds presented above and in Table 1 are strong indications of a molecular mechanism that is highly specific, presumably involving a receptor.

A few other cannabinoid enantiomeric pairs have also been investigated. Less stereospecificity was observed. If the reason for the low specificity is not trivial (the presence of an impurity for example), one can presume that such molecules only partially fit the cannabinoid receptor and the stereochemistry of the active materials is therefore less demanding. The enantiomeric benzofuran cannabinoids 46 and 47 will be further discussed in the section on "The Geometry of Cannabimimetically Active Cannabinoids". The ratio of the activity of the more active cannabinoid (47) to the less active one (46) (in a drug discrimination assay in the pigeon) was only about 20:1.[62]

$\Delta^{6a,10a}$-THC (20) (as a racemate) was synthesized about 50 years ago by the groups of Roger Adams (in the U.S.) and Lord Todd (in the U.K.) and was the first identified compound to show hashish-type activity.[63] Surprisingly, its pure enantiomers were not prepared and tested until a few years ago.[42] As the racemate had already been administered to humans, the enantiomers were also directly tested i.v. in human volunteers. The cannabimimetic effects observed with one of the enantiomers, (9S)-$\Delta^{6a,10a}$-

TABLE 1
Relative Potency of Cannabinoid Enantiomeric Pairs

Compound	Potency
(–)Δ^9-THC (1)	1.0
(–)HU-210 (10)	100.0
(+)HU-211 (42) (estimated)	<0.05
(–)-Levonantradol (4)	2.0
(+)-Dextronantradol	<0.13
(–)-CP-55940 (6)	10.0
(+)-CP-56667 (43)	<0.2
(–)-CP-55244 (44)	33.33
(+)-CP-55243 (45)	<0.14

Note: The potency numbers represent an average of various *in vivo* tests, as compared to that of Δ^9-THC (activity 1.0) (also see text).

THC, were indistinguishable from those caused by Δ^9-THC. The adjectives used by the volunteers to describe experiences with either drug included "peaceful", "pleasant", "relaxed", "dreamy", "dry mouth", "satisfied", "carefree", "happy", "hungry", "time slowed", etc. The potency of the synthetic drug however was one third to one sixth that of the natural Δ^9-THC. No effects were noted with the other enantiomer, (9*R*)-$\Delta^{6a,10a}$-THC.[64] These enantiomers were later assayed in discrimination tests in pigeons.[43] In these tests the (9*S*)-enantiomer was found to be only two to three times more active than the (9*R*)-enantiomer. Presumably, the differences between the results in man and in pigeons are due to the limited range of material administered i.v. to the volunteers. It is quite possible that at higher doses the volunteers on the (9*R*)-enantiomer would have also reported to be "peaceful", "relaxed", and "dreamy".

Noncannabimimetic Effects

The synthesis of pure cannabinoid enantiomeric pairs has made possible investigations on and comparison of noncannabimimetic compounds and their biological and pharmacological properties.

CBD

The pharmacological effects of cannabidiol (CBD) (3) (and a few of its derivatives) have been well investigated, mostly as regards movement and seizure disorders.[65] Recently, two groups reported that CBD has an antianxiety effect in rats and mice in the elevated plus-maze at surprisingly low dose levels.[66,67] These results give support to previous reports that CBD has anxiolytic effects in human volunteers,[65] as well as in animals.[68] The (+)-enantiomer of CBD (48) has not yet been investigated in these tests. In view of the activity of both (–)-CBD (48) and (+)-CBD (49) in anticonvulsion tests,[69] work along these lines may be of importance.

Administration of Δ^9-THC to proestrous rats has been shown to suppress the preovulatory LH surge and ovulation.[70] (–)-Cannabidiol (3) was inactive, but (+)-cannabidiol (48) had significant activity at 10 mg/kg. The (+)-CBD-DMH homolog (49) was considerably more potent than (–)-CBD-DMH (50).[71] These observations have not been followed up. It should be of considerable interest to investigate other members of the (+)-cannabinoid series.

HU-211

As mentioned above, the (+)-enantiomer of 11-hydroxy-Δ^8-THC-DMH (HU-211) (42) is nonpsychotropic.[31-34] However, HU-211 was found to induce a highly significant inhibition of cisplatin-induced emesis in pigeons, particularly at doses of 2.5 and 3.0 mg/kg ($p < 0.001$).[72] In contrast to these results, Δ^9-THC failed to cause a significant antiemetic effect at any of the doses injected. This is in agreement with clinical findings showing only a low inhibition of *cis*platin emesis in cancer patients concurrently treated with Δ^9-THC. Cupric salts enhanced the activity of both HU-211 and of Δ^9-THC. These results indicate, for the first time, that separation between the antiemetic and psychotropic activity is possible in the cannabinoid series and may open a new route to antiemetic drugs.

HU-211 induces stereotypy and locomotor hyperactivity in mice and mild tachycardia in rats. These properties are consistent with those of noncompetitive antagonists of the NMDA subclass of glutamate receptors, suggesting that HU-211 might function as an NMDA-receptor antagonist. Indeed, HU-211 protects against the tremorogenic, convulsive, and lethal effects of NMDA in mice. Binding studies indicate that HU-211 blocks NMDA receptors.[73] The (–)-enantiomer (HU-210) was inactive.

Recent evidence has indicated that brain damage induced by ischemia is mediated, *inter alia*, by NMDA receptors. In view of the absence of psychotropic or other untoward side effects seen by HU-211 administration at doses that block NMDA effects, this cannabinoid seems to be a suitable candidate for drug development.

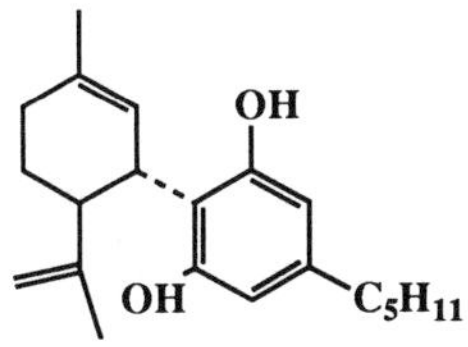

(+)-CBD (48)

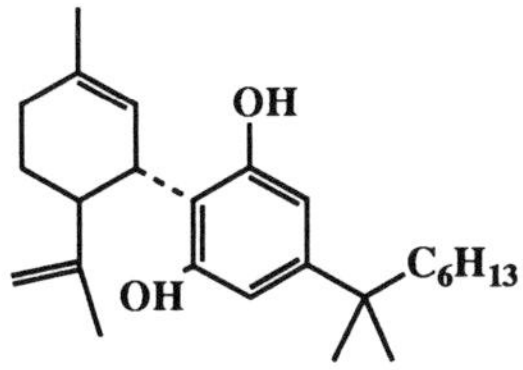

(49)

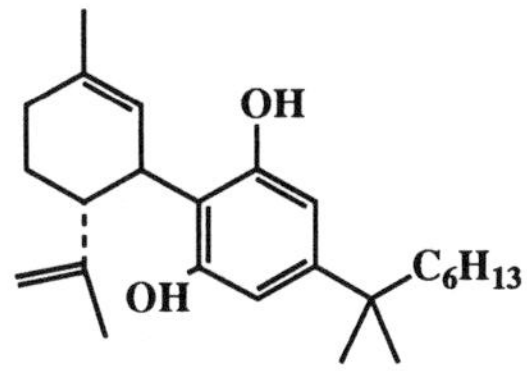

(50)

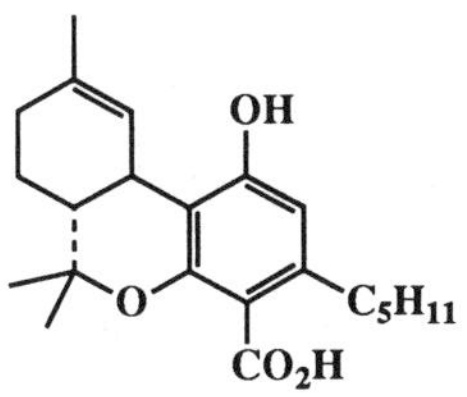

Δ-9-THC acid B (51)

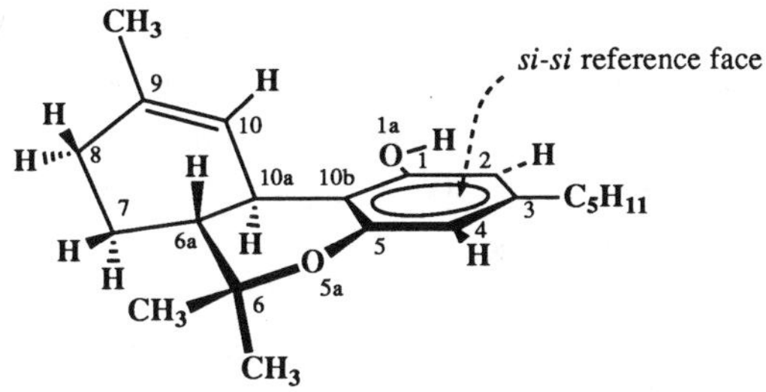

(52)

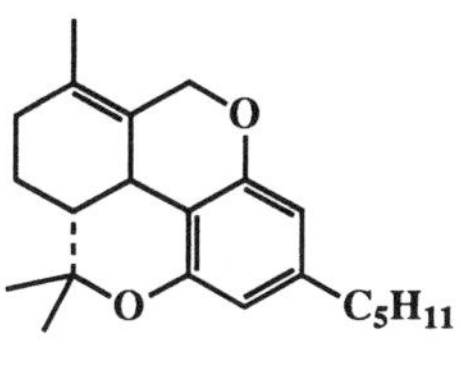

(53)

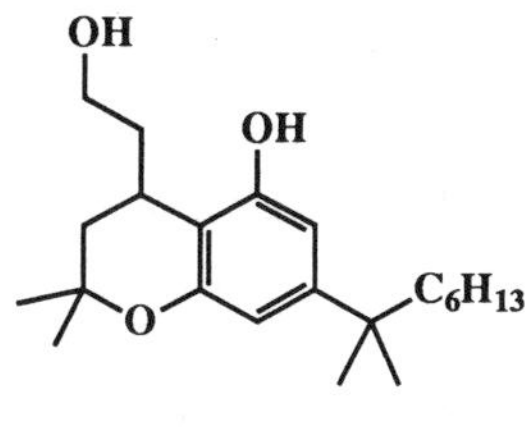

(54)

SCHEME 6.

Dextronantradol and (+)-CP-56667

These nonclassical, nonpsychotropic cannabinoids whose (–)-enantiomers, (4) and (6), respectively, are highly potent cannabimimetics (see above) seem to have been only marginally investigated for noncannabinoid activity. However, they have been found to cause changes in the effects of Δ^9-THC. Thus, CP-56667 reduced the effect of THC on spontaneous activity in mice, but enhanced the catalepsy.[35,36]

Adenylate Cyclase Inhibition and Receptor Binding

The discovery by Howlett that cannabinoid drugs can inhibit the adenylate cyclase activity of a neuroblastoma cell line led to an extensive biochemical and pharmacological characterization of this response (reviewed in Reference 16 and this book). In assessing the ability of cannabinoid drugs to inhibit the secretin-stimulated adenylate cyclase activity of an N18TG2 membrane preparation, a marked enantioselectivity was observed for several compounds (Table 2). The potency ratio of compounds with (–)-THC-like stereochemistry to their corresponding (+)-enantiomers ranges from less than twofold for the simple bicyclic CP-47497 (5) (K_{inh} = 79 n*M*) to greater than 1000-fold for the highly potent HU-210 (K_{inh} = 7.2 n*M*) and CP-55244 (K_{inh} = 5 n*M*). The great enantioselectivity, as well as the involvement of a G-protein in the inhibitory response, suggested that cannabinoid drugs were working through distinct receptors.

Using [^{3}H]CP-55940 (^{3}H-6), a centrifugation binding assay was developed that identified and characterized a cannabinoid receptor in rat brain.[15] Since the receptor is coupled to a G-protein, its affinity for cannabinoid drugs can be modulated by guanine nucleotides and various ions, similar to the behavior of other G-protein-linked receptors. The assay conditions employed were those that maximized the specific binding of [^{3}H]CP-55940 (^{3}H-6) to a rat cortical membrane preparation. With this centrifugation assay, CP-55940 (6) displayed K_i = 68 p*M* and its (+)-enantiomer was 50-fold less potent, with K_i = 3.4 n*M*. HU-210 (10) was able to fully compete for the specific binding of [^{3}H]CP-55940 in heterologous displacement studies with K_i = 234 p*M*; the (+)-enantiomer was 1500-fold less potent. Though CP-55940 (6) and HU-210 (10) are similarly potent in both the binding studies and the inhibition of adenylate cyclase, it appears that the more rigid structure of the 3-ring HU-210 (10) confers a greater enantioselectivity than the bicyclic CP-55940 (6).

The identification of a cannabinoid receptor in rat brain was quickly followed by the elegant autoradiographic analyses of Herkenham et al.,[74] and the cloning of a cannabinoid receptor by Matsuda et al.[75] In addition to detailing the distribution of the cannabinoid receptor in the brains of several species using classical autoradiography, Herkenham et al. have used slide-mounted “sausage” sections of whole rat brain to pharmacologically characterize the receptor. The brain preparation and assay conditions are quite different from those used in the previously described centrifugation assay

TABLE 2
Adenylate Cyclase Inhibition and Receptor Binding by Cannabinoid Enantiomeric Pairs

	Inhibition of adenylate cyclase		Binding to cannabinoid receptor	
	K_{inh} (n*M*)[a]	IC_{50} (n*M*)[b]	K_i (n*M*)[c]	K_i (n*M*)[d]
(–)-CP-55940 (6)	25	0.87	0.068	15
(+)-Enantiomer	>5000	96.3	3.4	470
HU-210 (10)	7.2		0.234	
HU-211 (42)	[e]		360	
(–)-CP-55244 (44)	5			1.4
(+)-CP-55243 (45)	[e]			18,000
(–)-Levonantrodal (4)	100			14
(+)-Dextronantrodal	>5000			26,000
(–)-CP-47497 (5)	79			79
(+)-Enantiomer	135			86
(–)-Δ^9-THC (1)		13.5		420
(+)-Δ^9-THC (38)		773		7,700

[a] References 61 and 77; N18TG2 neuroblastoma cell membranes.
[b] Reference 75; inhibition of cAMP accumulation in transfected CHO-K1 cells.
[c] References 15 and 61; centrifugation assay with rat P2 cortical membranes.
[d] References 74 and 76; "sausage" sections from whole rat brain.
[e] No significant inhibition at 10 μ*M*.

that may account for the 200-fold increase in observed K_i values. The K_i value of CP-55940 (6) derived from competitive inhibition studies using the "sausage" technique was 15 n*M* vs. a K_i = 68 p*M* using the centrifugation assay. However, the enantiospecificity was similar for both assays: the (+)-enantiomer was 30-fold less potent than CP-55940 using the "sausage" technique, compared with 50-fold less potent using the centrifugation assay. They tested several pairs of enantiomers for their ability to inhibit the binding of [^{3}H]CP-55940 to the "sausage" sections and the results correlate well with the ability of these drugs to inhibit adenylate cyclase activity in neuroblastoma membranes (Table 2). CP-47497 (5) and its (+)-enantiomer, which exhibited less than a twofold potency ratio in the adenylate cyclase assay, were almost equipotent, with K_i values of 79 and 86 n*M*, respectively. The most potent compound, CP-55244 (44), had a K_i = 1.4 n*M*, with its (+)-enantiomer (CP-55243) (45) being 12,000-fold less potent. The potency ratio of these (–)- and (+)-enantiomers was greater than 1000-fold in the adenylate cyclase assay.

Using Chinese hamster ovary K1 cells stably transfected with the cDNA for a rat cortical cannabinoid receptor, Matsuda et al.[75] were also able to demonstrate the inhibition of forskolin-stimulated cyclic AMP production by cannabinoid drugs. In this system, CP-55940 exhibited an IC_{50} of 0.87 n*M* and its (+)-enantiomer was 110-fold less potent. (–)-Δ^9-THC (1) exhibited an IC_{50} of 13.5 n*M*, while (+)-Δ^9-THC (38) was 57-fold less potent.

In both the binding studies and the adenylate cyclase assays, the conformationally more rigid 3-ring, highly potent cannabinoids, CP-55244 (44) and HU-210 (10), exhibit the greatest enantioselectivity, followed by the potent but more flexible bicyclic CP-55940 (6). The more simple bicyclic, CP-47497 (5), exhibited little enantioselectivity in either test.

THE GEOMETRY OF CANNABIMIMETICALLY ACTIVE CANNABINOIDS

X-ray crystallography has yet to provide us with a solid-state structure of Δ^9-THC (1). However, analogs of 1 have been studied by this method; for example, the natural Δ^9-THC acid B (51).[78,79]

Initial ^{1}H NMR 1D studies provided partial information on the structure of Δ^9-THC in solution.[80] Recently, ^{1}H and ^{13}C NMR 1D and 2D techniques have been used by Kriwacki and Makriyannis to extensively study the conformation in $CDCl_3$ solution.[81] Their analysis clearly shows that the solution-state conformation of Δ^9-THC is very similar to the solid-state conformation of the crystalline carboxylated analog (51). Furthermore, both structures are very similar to those calculated by molecular mechanics modeling.[62,80,82] This agreement of structural features is really quite reasonable since the Δ^9-THC ring-skeleton is fairly rigid.

The orientation and electronegativity of substituents (e.g., X and Y in a XCH_2CH_2Y fragment) influence the magnitude of NMR vicinal coupling constants.[83] Using Lambert's R-factor method in which the orientation and electronegativity effects on $^3J_{HH}$ are cancelled out,[84] we have calculated a $45 \pm 1°$ value for dihedral angle C(6a)-C(7)-C(8)-C(9) based on the relevant coupling constants of Kriwacki and Makriyannis[81] (and estimated standard deviations of 0.2 Hz). This solution-state value compares quite favorably with the 42° value determined for crystalline 51.

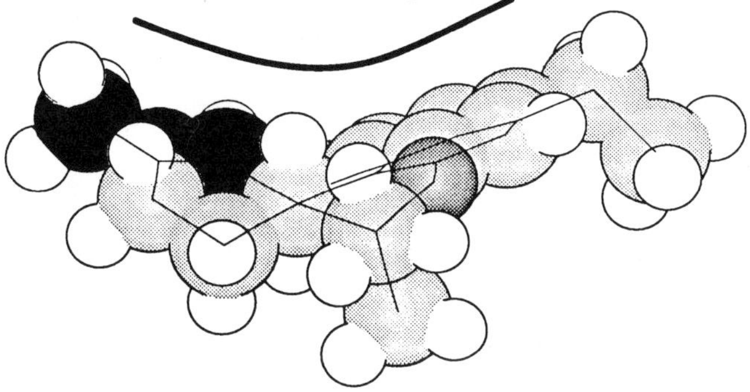

FIGURE 1. Space-filling MM2 calculated model of Δ^9-THC (1) [with ethyl group instead of *n*-amyl group] looking down the C(6a)-C(10a) bond showing the "butterfly"-type arrangement of rings A and B. C(9)-methyl, C(9), and C(10) colored black.

The conformation of ring A is that of a chair containing a *synperiplanar* (i.e., circa 0°) torsion angle [C(8)-C(9)-C(10)-C(10a)]. In this conformation, H(6a) is mutually *antiperiplanar* (i.e. circa 180°) to its H(7α) and H(10a) neighbors, while H(8β) and H(6a) have a pseudo *cis*-1,3-diaxial relationship. The equatorial positions at C(6a) and C(10a) provide points of fusion for ring B, thus resulting in a *trans* A,B-ring junction. The conformation of ring B is also that of a chair containing a synperiplanar (circa 0°) torsion angle [O(5a)-C(5)-C(10b)-C(10a)]. In ring B of the natural (–)-(6a*R*,10a*R*)-Δ^9-THC (1), H(6a) is mutually antiperiplanar (circa 180°) to both its H(10a) and pro-R [axial]methyl C(6)-CH_3 (α) neighbors as depicted in structure 52. As a result of *trans*-fusion, the two flattened-chair rings (A, B) are disposed in a "butterfly-type" arrangement in which fragment C(10a)-C(10)-C(9)-CH_3 is closer to one of the two *diastereotopic* (i.e., symmetry unrelated) faces of the resorcinol ring (ring C) (see Figure 1). *re,si*-Descriptors[85] may be used to define the particular face of the resorcinol ring that is closer to this ring A fragment. Using descriptors for trigonal carbons C(5) and C(10b), the above-mentioned fragment is oriented closer to the *si-si* face of the C(5,10b) double bond reference plane in (–)-Δ^9-THC. Irradiation of the OH afforded only one NOE intensity enhancement [7% into H(2)] and none to H(10), thus testifying to a preferred orientation in $CDCl_3$ solution as shown in 52.[81] In the solid-state, the *n*-amyl side chain of 51 is arranged in an extended staggered-type conformation so that the second carbon of the chain is approximately perpendicular to ring-C.[78] Long-range benzylic NMR coupling constants are consistent with such an arrangement, but do not rule out dynamic averaging of other conformations.[81]

The importance of the OH group lone pair and its spatial orientation vs. the phenoxy proton and its orientation has been the subject of much recent work.[23,86,87] Molecular mechanics has shown two low-energy conformations for the phenoxy proton.[86] In one of them, it is almost synperiplanar to H(2); the other is almost antiperiplanar to H(2). The lack of cannabimimetic activity of 1-methoxy-Δ^9-THC (2) (approximately 20 times less potent than 1), may be rationalized in terms of either (1) the absence of a phenoxy proton or (2) the inability of the lone pairs of the 1-methoxy oxygen to become oriented into the pharmacologically important conformation. Using the Van der Waal's radii of constituent atoms, volume maps of energy-minimized 1 and 2 were calculated by Semus and Martin.[88] Models were made in which the phenoxy proton or the methoxy methyl were located either *syn* or *anti* to C(2). An electropositive atom (hydrogen) 2.8 Å from the oxygen atom probed for a hypothetical H bond in which the 1-substituent is the acceptor. A lone pair 2.8 Å from the phenoxy proton probed for a hypothetical H bond in which the phenolic proton is the donor. Overlap of the cannabinoid volume with the volume of the probing unit (H or lone pair) was considered to be evidence that a hydrogen-bonding interaction between the 1-substituent on the cannabinoid and a receptor/membrane was not possible. Hydrogen bonding involving oxygen lone pairs (using H probe) when either OMe or OH is in an *anti* orientation to C(2) was ruled out since (1) overlap was observed for the OH model and (2) the OMe model does not represent an energy minimum for the cannabinoid. On the other hand, no volume overlap was noted for hydrogen bonding involving oxygen lone pairs when either O<u>Me</u> or O<u>H</u> is in a *syn* orientation to C(2). Furthermore, no volume overlap was noted for hydrogen bonding involving the phenyl proton (using the lone pair probe) when O<u>H</u> is either *anti* or *syn* to C(2).[88]

Another approach to this problem was taken by preparing rotationally restricted THC ethers 53 and 12.[23] Behavioral and pharmacological evaluations were made for these analogs. Compound (53) was completely inactive in all mouse behavioral tests, as well as the rat drug discrimination model.[87] Compound 12 was found to be similar to Δ^8-THC in that it depressed rectal temperature and produced antinociception and ring immobility in mice. However, it differed from Δ^8-THC in that it only weakly depressed locomotor activity and failed to substitute for Δ^9-THC in the drug

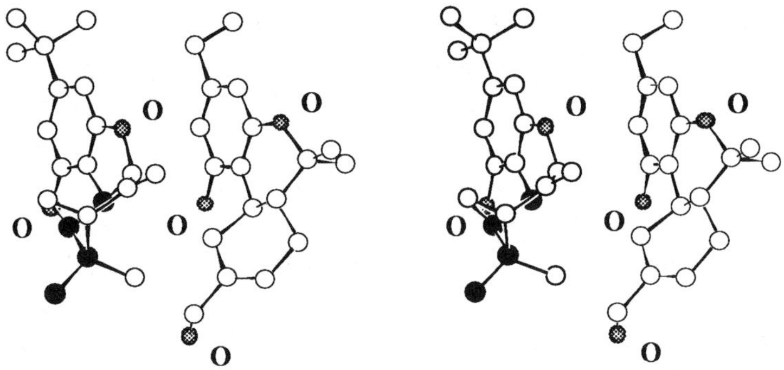

(47) (1) (47) (1)

FIGURE 2. Stereo view comparison of MM2 calculated models for 1 [on the right], and 47 [on the left]. For the purpose of simplification in the molecular mechanics calculations, a *t*-butyl group has replaced the dimethylheptyl side chain in the model for 47, and an ethyl group has replaced the *n*-amyl group in the model for 1. Hydrogen atoms have been omitted for clarity. The four atoms in 47 corresponding to C(10b), C(10), C(9), and CH_3 in (1) are colored in black.

discrimination paradigm. On this basis, it was stated that the orientation of the lone pairs of electrons on the phenolic hydroxyl plays a role in the mediation of some, but not all, behavioral effects of the cannabinoids. Molecular mechanics studies and molecular electrostatic potential maps were calculated for compounds 12 and 53. Two conformations of the more active 12 were found. These differ primarily in the twist of the O,2-propano moiety, whereby the new pyran ring assumes a chair conformation with a synperiplanar torsion angle. Inversion of the ring torsion angles is possible for this new ring and, thus, two diastereomeric conformations exist.

Recently, enantiomeric bicyclic cannabinoids having a furanyl B ring (46) and (47) were prepared and tested[62] (see the section on "Biological Activity"). NMR spectroscopy together with molecular mechanics modeling studies have shown that the bicyclo[3.1.1]heptyl moiety in the more active (–)-enantiomer of compound 47 contains four carbon atoms (colored black in the stereo view, Figure 2) located in positions similar to those in the C(10a)-C(10)-C(9)-CH_3 fragment of (–)-1. The geminal dimethyl groups on the disubstituted methano bridge in 47 reside in different environments. The methyl group *syn* (on the same side as) to the unsubstituted methano bridge in 47 corresponds spatially to the C(9)-methyl in (–)-1 and to the equatorial

C(9)-methyl in hexahydrocannabinol compound 30a. In saturated ring-A analogs of THC (30a, 31a), the two C(9) epimers show different cannabimimetic activity (see above). Molecular mechanics calculations of models for 30a and 31a were recently presented by Reggio et al.[86] They discussed the orientation of ring A relative to the resorcinol ring. One of their conclusions was that protrusion of a C(9)-substituent (e.g., the axial C(9)-methyl in 31a towards the "bottom" face [i.e., towards the *re-re* reference plane as in 52]) might hamper or prevent cannabinoid molecules from binding at their site of action. The methyl group *anti* to (on the opposite side of) the unsubstituted methano bridge in the active (–)-47 molecule corresponds spatially to the axial C(9)-methyl in 31a and both are located relatively close to the resorcinol ring reference plane. Since 47 shows high cannabimimetic activity, it is probable that the weak activity of 31a results from an absence of the properly oriented C(9)-substituent (i.e., an equatorial methyl), rather than the presence of the improperly axially oriented C(9)-methyl. However, paradoxically the absence of a substituent at C(9) (i.e., 9-nor-Δ^9-THC, 25) does not remove all activity.[45,88] The overlap of the benzofuran (47) and Δ^9-THC (1) is presented in Figure 3. The structures were calculated by molecular mechanics (MM2 program).[62] For simplicity, the *n*-amyl group of 1 was replaced by an ethyl group, and the 1,1-dimethylheptyl group of 47 was replaced by a *t*-butyl group. In this comparison of structures 1 and 47, the carbon and hydrogen atoms common to the CH_3-C(9)-C(10)-C(10a) fragment are highlighted in both molecules. In addition, carbon and hydrogen atoms in 47 that are spatially different than those in rings-A,B of 1 are also highlighted.

The importance of the disposition of C(9)-substituents relative to the resorcinol ring is also demonstrated by compounds 34 and 5.[17] Johnson and Melvin have shown that the relative orientation of the C(9)-OH to the phenolic-OH is similar in molecular mechanics energy minimized models for these two compounds. Compounds 5, 34, and 54 are all active analgesic agents,[17] and all share three common structural features: the aromatic ring, the alkyl chain at C-3, and four atoms corresponding to the CH_3-C(9)-C(10)-C(10a) fragment in Δ^9-THC. As analgesic activity in this series parallels cannabimimetic activity, one can presume that these features are also structurally significant for the latter.

The above results indicate that a simple correlation seems to exist between the relative geometry of rings A and C and cannabimimetic activity. The C(10a), C(10), C(9), C(11) carbon chains attached to the resorcinol ring in compounds 1, 47, and 10 are reasonably equivalent in terms of the respec-

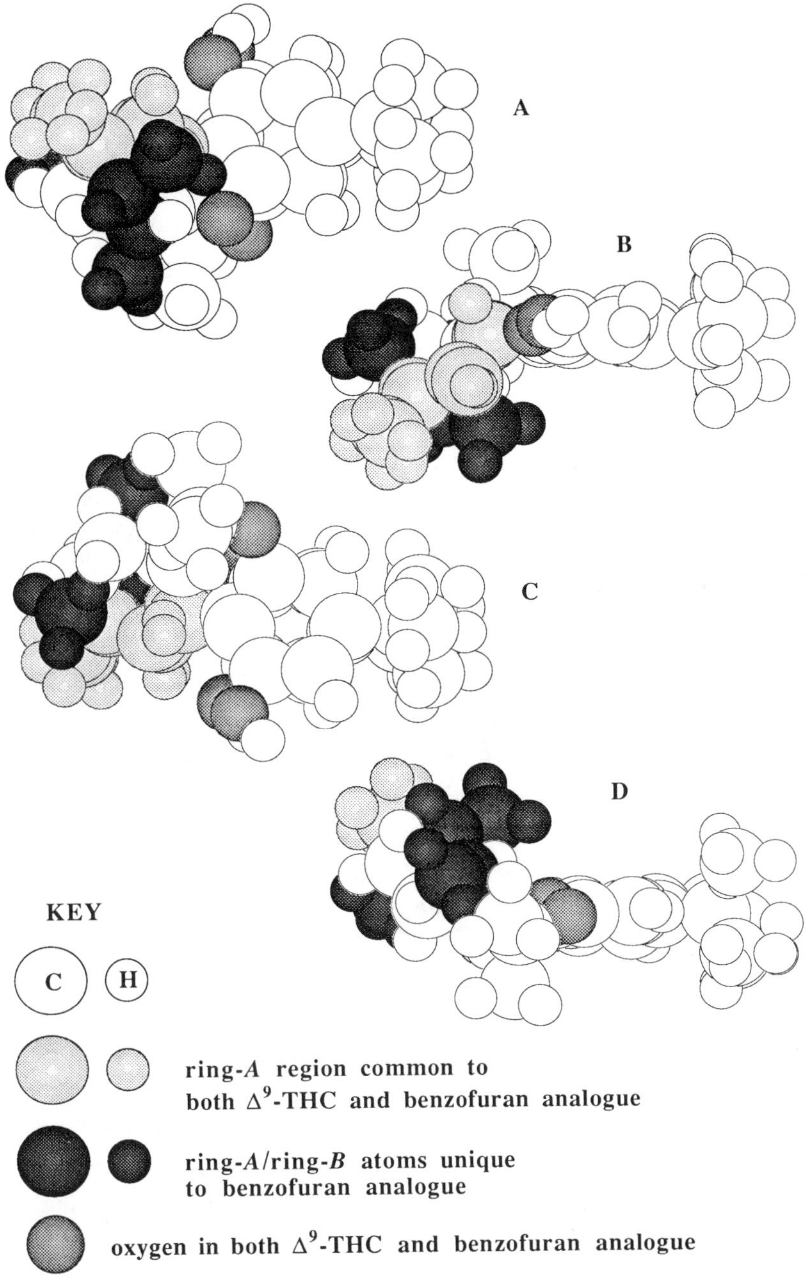

FIGURE 3. Overlay of molecular calculated structures 1 and 47. View A is looking down upon the *si-si* face of the resorcinol ring. Views A to D represent rotations of 90° in a direction parallel to the C(10a)—C(10b) bond. The CH_3-C(9)-C(10)-C(10a) fragment common to both structures is shown. Ring A/ring B atoms unique to 47 are also highlighted.

tive distances of the corresponding atoms to the midpoint of the aromatic ring. In addition, these chains of carbon atoms in models for the same compounds (all of which are cannabimimetic) show equivalent signs and magnitudes for their respective angles of deviation relative to the *si-si* ("top face of the molecule") face of the plane. The corresponding carbon atoms in the respective enantiomers (which either lack cannabimimetic activity or are much weaker than their enantiomers) show equivalent magnitudes, but opposite signs for the respective angles of deviation.

QUO VADIMUS?

The recent identification of a cannabinoid receptor has been a major step forward in our understanding of cannabinoid action. The realization that cannabinoid activity is very highly enantiospecific has contributed much to this advance. We now believe that four major goals are to be accomplished and that their achievement is possible on the basis of the knowledge acquired and the techniques available.

1. *Identification of a specific endogenous cannabinoid transmitter.* Preliminary work in our laboratory has indicated that such a transmitter(s) exists. It seems to be a lipid.
2. With the help of the natural endogenous transmitter and the accumulated knowledge on SAR (in particular, as regards stereochemical requirements), *the tertiary structure of the cannabinoid receptor* and the *mode of binding of cannabinoids to the receptor* could be clarified.
3. We still know very little about the *functions of the cannabinoid system* in the body. Is it involved in the regulation of movement and in short-term memory as presently presumed? Does it play a role in the modulation of emotions?
4. A major disappointment thus far has been the lack of success in introducing a cannabinoid (synthetic or natural) as a major *new drug*. While Δ^9-THC and nabilone are very helpful in preventing vomiting during cancer chemotherapy, these compounds do not represent a major therapeutic breakthrough. Let us hope that recent reports on the separation of cannabimimetic from therapeutic activity in synthetic molecules, in particular enantiomers, may revitalize this area and lead to new drugs in neurology (particularly stroke and muscular dystrophy) and against glaucoma.

REFERENCES

1. **Mechoulam, R. and Edery, H.,** Structure-activity relationships in the cannabinoid series, in *Marijuana: Chemistry, Pharmacology, Metabolism and Clinical Effects*, Mechoulam, R., Ed., Academic Press, New York, 1973.
2. **Dewey, W. L., Martin, B. R., and May, E. L.,** Cannabinoid stereoisomers: pharmacological effects, in *Handbook of Stereoisomers: Drugs in Psychopharmacology*, Smith, D. F., Ed., CRC Press, Boca Raton, FL, 1984, 317.
3. **Razdan, R. K.,** Structure-activity relationships in cannabinoids, *Pharmacol. Rev.*, 38, 75, 1986.
4. **Mechoulam, R., Lander, N., Srebnik, M., Breuer, A., Segal, M., Feigenbaum, J. J., Järbe, T. U. C., and Consroe, P.,** Stereochemical requirements for cannabimimetic activity, in *Structure-Activity Relationships of the Cannabinoids*, Rapaka, R. S. and Makriyannis, A., Eds., NIDA Research Monograph 79, Rockville, MD, 1987, 15.
5. **Martin, B. R.,** Cellular effects of cannabinoids, *Pharmacol. Rev.*, 38, 45, 1986.
6. **Dewey, W. L.,** Cannabinoid pharmacology, *Pharmacol. Rev.*, 38, 151, 1986.
7. **Paton, W. D. M. and Pertwee, R. G.,** The pharmacology of Cannabis in animals, in *Marijuana: Chemistry, Pharmacology, Metabolism and Clinical Effects*, Mechoulam, R., Ed., Academic Press, New York, 1973.
8. **Grunfeld, Y. and Edery, H.,** Psychopharmacological activity of the active constituents of hashish and some related cannabinoids, *Psychopharmacologia*, 14, 200, 1969.
9. **Walton, R. P.,** *Marihuana, America's New Drug Problem*, J.B. Lippincott, Philadelphia, 1938, 168.
10. **Martin, B. R., Balster, R. L., Razdan, R. K., Harris, L. S., and Dewey, W. L.,** Behavior comparisons of the stereoisomers of tetrahydrocannabinols, *Life Sci.*, 29, 565, 1981.
11. **Pertwee, R. G.,** The ring test: a quantitative method for assessing the cataleptic effect of Cannabis in mice, *Br. J. Pharmacol.*, 46, 753, 1972.
12. **Järbe, T. U. C., Johansson, J. O., and Henriksson, B. G.,** Characteristics of tetrahydrocannabinol produced discrimination in rats, *Psychopharmacologia*, 48, 181, 1976.
13. **Järbe, T. U. C., Swedberg, M. D. B., and Mechoulam, R.,** A repeated tests procedure to assess onset and duration of the cue properties of (–)-delta-9-THC, (–)-delta-8-THC and (+)-delta-8-THC, *Psychopharmacologia*, 75, 152, 1981.
14. **Mokler, D. J., Nelson, D. B., Harris, L. S., and Rosencrans, J. A.,** The role of benzodiazepine receptors in the discriminative stimulus properties of delta-9-THC, *Life Sci.*, 38, 1581, 1986.
15. **Devane, W. A., Dysarz, F. A., Johnson, M. R., Melvin, L. S., and Howlett, A. C.,** Determination and characterization of a cannabinoid receptor in rat brain, *Mol. Pharmacol.*, 34, 605, 1988.
16. **Howlett, A. C., Bidaut-Russell, M., Devane, W. A., Melvin, L. S., Johnson, M. R., and Herkenham, M.,** The cannabinoid receptor: biochemical, anatomical and behavioral characterization, *Trends Neurosci.*, 13, 420, 1990.
17. **Johnson, M. R. and Melvin, L. S.,** The discovery of nonclassical cannabinoid analgetics, in *Cannabinoids as Therapeutic Agents*, Mechoulam, R., Ed., CRC Press, Boca Raton, FL, 1986.
18. **Houry, S., Mechoulam, R., Fowler, P. J., Macko, E., and Loev, B.,** Benzoxocin and benzoxonin derivatives: novel groups of terpenophenols with CNS activity, *J. Med. Chem.*, 17, 287, 1974.

19. **Houry, S., Mechoulam, R., and Loev, B.,** Benzoxocin and benzoxonin derivatives: novel groups of terpenophenols with CNS activity, a correction, *J. Med. Chem.*, 18, 951, 1975.
20. **Bell, M. R. et al.,** Antinociceptive (aminoalkyl) indoles, *J. Med. Chem.*, 34, 1099, 1991.
21. **Haubrich, D. R. et al.,** Pharmacology of pravadoline: a new analgesic agent, *J. Pharmacol. Exp. Ther.*, 255, 511, 1990.
22. **Banerjee, S. P., Mechoulam, R., and Snyder, S. H.,** Cannabinoids: influence on neurotransmitter uptake in rat brain synaptosomes, *J. Pharmacol. Exp. Ther.*, 194, 74, 1975.
23. **Seltzman, H. H., Hsieh, Y.-A., Pitt, C. G., and Reggio, P. H.,** Synthesis of rotationally restricted THC ethers, *J. Org. Chem.*, 56, 1549, 1991.
24. **Matsumoto, K., Stark, P., and Meister, R. G.,** 1-Amino- and 1-mercapto-7,8,9,10-tetrahydro-6H-dibenzo[b,d]pyrans, *J. Med. Chem.*, 20, 17, 1977.
25. **Agurell, S.,** Pharmacokinetics and metabolism of delta-1-THC and other cannabinoids with emphasis on man, *Pharmacol. Rev.*, 38, 21, 1986.
26. **Harvey, D. J. and Paton, W. D. M.,** Metabolism of the cannabinoids, *Rev. Biochem. Toxicol.*, 6, 221, 1984.
27. **Lemberger, L., Crabtree, R. E., and Rowe, H.M.,** 11-Hydroxy-delta-9-THC: pharmacology, disposition and metabolism of a major metabolite of marihuana in man, *Science*, 177, 62, 1972.
28. **Mechoulam, R., Ben-Zvi, Z., Agurell, S., Nilsson, I. M., Nilsson, J. L. G., Edery, H., and Grunfeld, Y.,** Delta-6-THC-7-oic-acid, an urinary delta-6-THC metabolite: isolation and synthesis, *Experientia*, 29, 1193, 1973.
29. **Burstein, S. H., Hull, K., Hunter, S. A., and Latham, V.,** Cannabinoids and pain responses: a possible role for prostaglandins, *FASEB*, 2, 3022, 1988.
30. **Hollister, L. E.,** Structure-activity relationships in man of cannabis constituents, and homologs and metabolites of delta-9-THC, *Pharmacol. Rev.*, 11, 3, 1974.
31. **Mechoulam, R., Feigenbaum, J. J., Lander, N., Segal, M., Järbe, T. U. C., Hiltunen, A. J., and Consroe, P.,** Enantiomeric cannabinoids: stereospecificity of psychotropic activity, *Experientia*, 44, 762, 1988.
32. **Mechoulam, R., Lander, N., Breuer, A., and Zahalka, J.,** Synthesis of the individual, pharmacologically distinct, enantiomers of a tetrahydrocannabinol derivative, *Tetrahedron: Asymmetry*, 1, 315, 1990.
33. **Järbe, T. U. C., Hiltunen, A. J., and Mechoulam, R.,** Stereospecificity of the discriminative stimulus functions of the dimethylheptyl homologs of 11-OH-delta-8-tetrahydrocannabinol in rats and pigeons, *J. Pharmacol. Exp. Ther.*, 250, 1000, 1989.
34. **Little, P. J., Compton, D. R., Mechoulam, R., and Martin, B.,** Stereochemical effects of 11-OH-delta-8-THC-dimethylheptyl in mice and dogs, *Pharmacol. Biochem. Behav.*, 32, 661, 1989.
35. **Thomas, B. F., Compton, D. R., and Martin, B. R.,** Characterization of the lipophilicity of natural and synthetic analogs of delta-9-THC and its relationship to pharmacological potency, *J. Pharmacol. Exp. Ther.*, 255, 624, 1990.
36. **Little, P. J., Compton, D. R., Johnson, M. R., Melvin, L. S., and Martin, B. R.,** Pharmacology and stereoselectivity of structurally novel cannabinoids in mice, *J. Pharmacol. Exp. Ther.*, 247, 1046, 1988.
37. **Ohlsson, A., Agurell, S., Leander, K., Dahmen, J., Edery, H., Porath, G., Levy, S., and Mechoulam, R.,** Synthesis and psychotropic activity of side chain hydroxylated delta-6-THC metabolites, *Acta Pharm. Suec.*, 16, 21, 1979.
38. **Gaoni, Y. and Mechoulam, R.,** Isolation, structure and partial synthesis of an active constituent of hashish, *J. Am. Chem. Soc.*, 86, 1646, 1964.

39. **Turner, C. E., Elsohly, M. A., and Boeren, E. G.,** Constituents of *Cannabis sativa* L. XVII. A review of the natural constituents, *J. Nat. Prod.,* 43, 169, 1980.
40. **Gaoni, Y. and Mechoulam, R.,** Concerning the isomerization of delta-1- to delta-6-tetrahydrocannabinol, *J. Am. Chem. Soc.,* 88, 5673, 1966.
41. **Mechoulam, R., Ben-Zvi, Z., Varconi, H., and Samuelov, Y.,** Cannabinoid rearrangements: synthesis of delta-5-tetrahydrocannabinol, *Tetrahedron,* 29, 1615, 1973.
42. **Srebnik, M., Lander, N., Breuer, A., and Mechoulam, R.,** Base catalysed double bond isomerizations of cannabinoids: structural and stereochemical aspects., *J. Chem. Soc., Perkin Trans. I,* 2881, 1984.
43. **Järbe, T. U. C., Hiltunen, A. J., Mechoulam, R., Srebnik, M., and Breuer, A.,** Separation of the discriminative stimulus effects in stereoisomers of delta-2 and delta-3-tetrahydrocannabinols in pigeons, *Eur. J. Pharmacol.,* 156, 361, 1988.
44. **Binder, M., Edery, H., and Porath, G.,** Delta-7-THC, a nonpsychotropic cannabinoid: structure activity considerations in the cannabinoid series, in *Marihuana: Biological Effect,* Nahas, G. G. and Paton, W. D. M., Eds., Pergamon Press, Oxford, 1979, 71.
45. **Wilson, R. S. and May, E. L.,** 9-Nor-delta-8-THC, a cannabinoid of metabolic interest, *J. Med. Chem.,* 17, 475, 1974.
46. **Loev, B., Bender, P. E., Dowalo, F., Macko, E., and Fowler, P. J.,** Cannabinoids. Structure-activity studies, related to 1,2-dimethylheptyl derivatives., *J. Med. Chem.,* 16, 2100, 1973.
47. **Korbinits, D. et al.,** Cannabinoids with an ether side chain. Synthesis, solubilization and analgesic properties, *Eur. J. Med. Chem.,* 20, 492, 1985.
48. **Edery, H., Grunfeld, Y., Porath, G., Ben-Zvi, Z., Shani, A., and Mechoulam, R.,** Structure-activity relationships in the THC series. Modifications on the aromatic ring and on the side chain, *Arzneim. Forsch.,* 22, 1995, 1972.
49. **Burstein, S. H., Audette, C. A., Charalambous, A., Doyle, S. A., Guo, Y., Hunter, S. A., and Makriyannis, A.,** Detection of cannabinoid receptors by photoaffinity labelling, *Biochem. Biophys. Res. Commun.,* 176, 492, 1991.
50. **Mechoulam, R. and Gaoni Y.,** The absolute configuration of delta-1-tetrahydrocannabinol, the major active constituent of hashish, *Tetrahedron Lett.,* 1109, 1967.
51. **Gaoni, Y. and Mechoulam, R.,** The isomerization of cannabidiol to tetrahydrocannabinols, *Tetrahedron,* 22, 1481, 1966.
52. **Mechoulam, R., Lander, N., Varkony, T. H., Kimmel, I., Becker, O., Ben-Zvi, Z., Edery, H., and Porath, G.,** Stereochemical requirements for cannabinoid activity, *J. Med. Chem.,* 23, 1068, 1980.
53. **Mechoulam, R., Devane, W. A., Breuer, A., and Zahalka, J.,** A random walk through a cannabis field, *Pharmacol. Biochem. Behav.,* 40, 461, 1991.
54. **Wilson, R. S., May, E. L., Martin, B. R., and Dewey, W. L.,** 9-Nor-9-hydroxyhexahydrocannabinols. Synthesis. Some behavioral and analgesic properties and comparison with the tetrahydrocannabinols, *J. Med. Chem.,* 19, 1165, 1976.
55. **Mechoulam, R., Braun, P., and Gaoni, Y.,** Syntheses of delta-1-THC and related cannabinoids, *J. Am. Chem. Soc.,* 94, 6159, 1972.
56. **Archer, R. A., Stark, P., and Lemberger, L.,** Nabilone, in *Cannabinoids as Therapeutic Agents,* Mechoulam, R., Ed., CRC Press, Boca Raton, FL, 1986.
57. **Huffman, J. W., Joyner, H. H., Lee, M. D., Jordan, R. D., and Pennington, W. T.,** Synthesis of both enantiomers of nabilone from a common intermediate. Enantiodivergent synthesis of cannabinoids, *J. Org. Chem.,* 56, 2081, 1991.
58. **Harvey, D. J.,** *In vivo* metabolism of (+)-*trans*-delta-9-THC in the mouse, *Biomed. Environ. Mass Spectrosc.,* 16, 117, 1988.

59. **Harvey, D. J. and Marriage, H. J.,** Metabolism of (+)-*trans*-delta-8-THC in the mouse *in vitro* and *in vivo*, *Drug Metab. Dispos.,* 15, 914, 1987.
60. **Mechoulam, R., Breuer, A., Feigenbaum, J. J., and Devane, W. A.,** Nonpsychotropic synthetic cannabinoids as therapeutic agents, *Farmaco,* 46 (Suppl. 1), 267, 1991, and unpublished results.
61. **Howlett, A. C., Champion, T. M., Wilken, G. H., and Mechoulam, R.,** Stereochemical effects of 11-OH-delta-8-tetrahydrocannabinol-dimethylheptyl to inhibit adenylate cyclase and bind to the cannabinoid receptor, *Neuropharmacology,* 29, 161, 1990.
62. **Mechoulam, R., Breuer, A., Järbe, T. U. C., Hiltunen, A. J., and Glaser R.,** Cannabimimetic activity of novel enantiomeric benzofuran cannabinoids, *J. Med. Chem.,* 33, 1037, 1990.
63. **Mechoulam, R.,** The pharmacohistory of *Cannabis sativa*, in *Cannabinoids as Therapeutic Agents*, Mechoulam, R., Ed., CRC Press, Boca Raton, FL, 1986.
64. **Hollister, L. E., Gillespie, H. K., Mechoulam, R., and Srebnik, M.,** Human pharmacology of 1*S* and 1*R* enantiomers of delta-3-THC, *Psychopharmacology,* 92, 505, 1987.
65. **Consroe, P. and Snider, S. R.,** Therapeutic potential of cannabinoids in neurological disorders, in *Cannabinoids as Therapeutic Agents*, Mechoulam, R., Ed., CRC Press, Boca Raton, FL, 1986.
66. **Guimaraes, F., Chiaretti, T. M., Graeff, F. G., and Zuardi, A. W.,** Antianxiety effect of cannabidiol in the elevated plus-maze, *Psychopharmacologia,* 100, 558, 1990.
67. **Onaivi, E. S., Green, M. R., and Martin, B. R.,** Pharmacological characterization of cannabinoids in the elevated plus maze, *J. Pharmacol. Exp. Ther.,* 252, 1002, 1990.
68. **Musty, R. E., Conti, L. H., and Mechoulam, R.,** Anxiolytic properties of cannabidiol, in *Marihuana '84,* Harvey, D. J., Ed., IRL Press, Oxford, 1985, 1.
69. **Leite, J. R., Carlini, E. A., Lander, N., and Mechoulam, R.,** Anticonvulsant effect of (–) and (+) isomers of CBD and their dimethyl heptyl homologs, *Pharmacology,* 124, 141, 1982.
70. **Nir, I., Ayalon, D., Tsafriri, A., Cordova, T., and Lindner, H. R.,** Suppression of the cyclic surge of LH secretion and of ovulation in the rat by delta-1-THC, *Nature (London),* 243, 470, 1973.
71. **Cordova, T., Ayalon, D., Lander, N., Mechoulam, R., Nir, I., Puder, M., and Lindner, H. R.,** The ovulation blocking effect of cannabinoids: structure-activity relationships, *Psychoneuroendocrinology,* 5, 53, 1980.
72. **Feigenbaum, J. J., Richmond, S. A., Weissman, Y., and Mechoulam, R.,** Inhibition of *cis*platin induced emesis in the pigeon by a non-psychotropic synthetic cannabinoid, *Eur. J. Pharmacol.,* 169, 159, 1989.
73. **Feigenbaum, J. J., Bergmann, F., Richmond, S. A., Mechoulam, R., Nadler, V., Kloog, Y., and Sokolovsky, M.,** A non-psychotropic cannabinoid acts as a functional *N*-methyl-D-asparate (NMDA) receptor blocker, *Proc. Natl. Acad. Sci. U.S.A.,* 86, 9584, 1989.
74. **Herkenham, M. et al.,** Cannabinoid receptor localization in brain, *Proc. Natl. Acad. Sci. U.S.A.,* 87, 1932, 1990.
75. **Matsuda, L. A., Lolait, S. J., Brownstein, M. J., Young, A. C., and Bonner, T. I.,** Structure of a cannabinoid receptor and functional expression of the cloned cDNA, *Nature (London),* 346, 561, 1990.

76. **Herkenham, M., Lynn, A. B., Johnson, M. R., Melvin, L. S., de Costa, B. R., and Rice, K. C.,** Characterization and localization of cannabinoid receptors in rat brain: a quantitative *in vitro* autoradiographic study, *J. Neurosci.,* 11, 563, 1991.
77. **Howlett, A. C., Johnson, M. R., Melvin, L. S., and Milne, G. M.,** Nonclassical cannabinoid analgetics inhibit adenylate cyclase: development of a cannabinoid receptor model, *Mol. Pharmacol.,* 33, 297, 1988.
78. **Rosenqvist, E. and Ottersen, T.,** The crystal and molecular structure of delta-9-THC acid B, *Acta Chem. Scand. B.,* 29, 379, 1975.
79. **Mechoulam, R., Ben-Zvi, Z., Yagnitinsky, B., and Shani, A.,** A new tetrahydrocannabinolic acid, *Tetrahedron Lett.,* 2339, 1969.
80. **Archer, R. A., Boyd, D. B., Demarco, P. V., Tyminski, I. J., and Allinger, N. L.,** Structural studies of cannabinoids. A theoretical and proton magnetic resonance study, *J. Am. Chem. Soc.,* 92, 5200, 1970.
81. **Kriwacki, W. and Makriyannis, A.,** The conformational analysis of delta-9- and delta-9,11-THCs in solution using high resolution NMR spectroscopy, *Mol. Pharmacol.,* 35, 495, 1989.
82. **Reggio, P. H. and Mazurek, A. P.,** A molecular reactivity template for cannabinoid activity, *J. Mol. Struct.,* 149, 331, 1987.
83. **Haasnoot, C. A. G., de Leeuw, F. A. A. M., and Altona, C.,** The relationship between proton-proton NMR coupling constants and substituent electronegativities, *Tetrahedron,* 36, 2783, 1980.
84. **Lambert, J. B.,** Structural chemistry in solution. The R value, *Acct. Chem. Res.,* 4, 87, 1971.
85. **Hanson, K. R.,** Applications of the sequence rule. Naming the paired ligands gg at a tetrahedral atom, *J. Am. Chem. Soc.,* 88, 2731, 1966.
86. **Reggio, P. H., Greer, K. V., and Cox, S. M.,** The importance of the orientation of the C9 substituent to cannabinoid activity, *J. Med. Chem.,* 32, 1630, 1989.
87. **Reggio, P. H., Seltzmann, H. H., Compton, D. R., Prescott, W. R., Jr., and Martin, B. R.,** Investigation of the role of the phenolic hydroxyl in cannabinoid activity, *Mol. Pharmacol.,* 38, 854, 1990.
88. **Semus, S. F. and Martin, B. R.,** A computergraphic investigation into the pharmacological role of the THC-cannabinoid phenolic moiety, *Life Sci.,* 46, 1781, 1990.
89. **Martin, B. R., Dewey, W. L., Harris, L. S., Bechner, J., Wilson, R. S., and May, E. L.,** Marihuana-like activity of new synthetic THCs, *Pharmacol. Biochem. Behav.,* 3, 849, 1975.

2 The Cannabinoid Receptor

THE CANNABINOID RECEPTOR AS A G PROTEIN-COUPLED RECEPTOR

What Are G Protein-Coupled Receptors?

A broad range of neuromodulators and hormones transduce their signals to target cells via receptors coupled to their effector systems via G proteins. The evolution of our current understanding of G protein-coupled receptors has been reviewed in the recent literature,[1-11] and follows from the development of the research literature in second messenger signaling systems. Notably, the Nobel Prize-winning discovery by Rall and Sutherland that the effects of glucagon and epinephrine on liver metabolism were transduced by cyclic AMP[12] led to the subsequent recognition that many hormones and neurotransmitters could stimulate the production of cyclic AMP via the enzyme adenylate cyclase (see Table 2 in the review by Birnbaumer[2]). The increased intracellular cyclic AMP subsequently stimulates the rate of phosphorylation of key enzymes and proteins via cyclic AMP-dependent protein kinases in target cells (reviewed by Shenolikar[13]). The alterations in substrate enzyme or protein activity within the target cells may be maintained under conditions of phosphorylation, or reversed by dephosphorylation resulting from the activity of phosphatases in the cells (see Reference 14 for review).

The concept that the receptor protein was genetically and structurally distinct from the catalytic protein of adenylate cyclase[15-19] explained how multiple receptors could regulate cyclic AMP synthesis to a limited maximum within a cell population.[20,21] This concept was also essential to the recognition that proteins other than the receptor for hormones may be involved in the regulation of adenylate cyclase activity. The ability of GTP to regulate adenylate cyclase was initially described by Rodbell and colleagues in a series of papers that characterized the response to glucagon in a purified membrane preparation from rat liver.[22,23] It was noted that GTP (being supplied as a contaminant of the substrate ATP) was essential for glucagon-stimulated adenylate cyclase activity.[23] In addition, the affinity of the agonist glucagon for its receptor was decreased in the presence of GTP and its analogs.[22] Rodbell and colleagues demonstrated that nonhydrolyzable GTP analogs were sufficient to produce both effects, thereby demonstrating that phosphorylation using GTP as the substrate was not essential to the mechanism. Subsequent studies in other laboratories demonstrated that β-adrenergic agonists could stimulate a hormone-specific GTPase activity in membrane preparations,[24] and that inhibition of the hormone-regulated GTPase activity by cholera toxin was the mechanism by which this toxin irreversibly activates adenylate cyclase activity.[25] The recognition that a protein other than the receptor and the catalytic protein was the target for the action of GTP was the result of solubilization of membrane proteins using detergents, and reintroduction of these detergent extracts into membranes.[26]

The *in vitro* detergent extract reconstitution assay system was used to characterize the GTP- and NaF-regulation conferring subunit, initially referred to as G/F or N, but now established in the literature as G_s.[26-28] It was determined that a protein that could be radiolabeled by GTP affinity analogs and could activate adenylate cyclase exhibited an apparent molecular weight of 42 kDa on denaturing SDS polyacrylamide gel electrophoresis.[29] However, the apparent molecular weight of the reconstituting G_s activity (after adjustment for the detergent mass) using nondenaturing separation techniques was considerably greater than 42 kDa, but could be reduced by conditions that activated G_s.[30] This finding suggested that G_s exhibited subunit structure. Subsequent purification of G_s activity demonstrated the presence of α (45 and 52 kDa), β (35 and 36 kDa),[31] and γ (10 kDa)[32] subunits.

This subunit structure of G_s resembled that characterized for the GTP-binding protein component (transducin or Gt) associated with rhodopsin to stimulate a cyclic GMP-specific phosphodiesterase activity in retinal rod cells.[33] Subsequent homologies between the rhodopsin-transducin and the adenylate cyclase-G_s systems have been supported by studies in the literature (reviewed by Stryer[34]). It became apparent that the regulation of adenylate cyclase by hormones and neuromodulators that inhibit the enzyme and decrease the intracellular cyclic AMP concentration also required GTP (discussed by Cooper[35]). The signal transduction protein complex referred to as G_i was discovered and characterized after pertussis toxin was shown to block the biological responses to hormones that inhibited adenylate cyclase.[36] The mechanism of pertussis toxin was to ADP-ribosylate the G_i heterotrimer and thereby render it unable to interact with receptors, that is, to inhibit G_i. The similarities within these three signal transduction pathways led to the concept that stimulation of the production of other second messengers resulting from hormone or neuromodulator interaction with receptors may also involve a GTP regulatory step. Substantial evidence has accumulated to demonstrate that receptors regulating polyphosphoinositide hydrolysis via a specific phospholipase C,[5-7,9] arachidonic acid hydrolysis from phospholipids via a specific hormone-sensitive phospholipase A_2,[1] and phosphatidic acid formation via hormone-regulated phospholipase D[10] can interact with G proteins homologous to those described for the adenylate cyclase and rhodopsin systems. Recent studies using patch clamp technology have demonstrated that certain ion channels can also be the effector proteins coupled to receptors via G proteins.[2-4,8,9]

How Does the Receptor-G Protein Interaction Produce Its Response?

The receptor, G protein heterotrimer complex, and the effector enzyme exhibit a cycle of reversible interactions in the presence of GTP plus hormone or neuromodulator. The design of a sequence of events in this cycle and the original evidence supporting it have been reviewed by Birnbaumer[2,3] and Gilman.[6] To summarize from their discussions: after the hormone agonist interacts with the receptor, the hormone-receptor complex interacts with a G protein heterotrimer. As the result of this interaction, the α subunit of the G protein is able to bind GTP and assume an activated state. This transiently activated G protein rapidly dissociates to an activated α subunit having a bound GTP, and the remaining β-γ dimer. The hormone-receptor complex associated with the α subunit exhibits a low affinity for the agonist in the presence of GTP, which promotes dissociation of the ligand and dissociation of the receptor from the activated GTP-α protein. The activated GTP-α protein is able to transiently interact with one or more effector proteins to result in their activation. The ability of the α protein to remain activated depends upon the presence of bound GTP. A GTPase activity is intrinsic to this protein, and the hydrolysis of GTP to GDP ultimately causes conversion to the inactive state. The α subunit bearing

GDP has a high affinity for the β-γ dimers, and reassociation can occur. The subsequent release of GDP allows the G protein to interact with the receptor and, in the absence of GDP or GTP, the receptor assumes a state having a high affinity for the agonist ligand.[2] Since Mg^{2+} is required for GTP hydrolysis and modifies the affinity of the α subunits for guanine nucleotides, divalent cations are important modifiers of hormone-receptor affinity and receptor-effector coupling.

Regulation of adenylate cyclase by inhibitory receptors that couple to G_i follows a GTP-driven cycle as described for stimulatory receptors that couple to G_s. Following dissociation of the G_i heterotrimer, it is unclear how the adenylate cyclase activity is curtailed. Evidence from reconstitution studies demonstrates that the released β-γ dimers are able to interact with free α_s subunits and form complete heterotrimers. Such an interaction would prevent α_s monomers from interacting with the catalytic protein of adenylate cyclase. In addition, inhibition of adenylate cyclase occurs in the absence of G_s, suggesting that perhaps the α_i subunit can inhibit the catalytic protein directly. A tool that has been instrumental in determining G_i-mediated activity is pertussis toxin, which prevents the interaction of G_i with inhibitory receptors. Thus, subsequent activation of G_i by GTP and the dissociation step are not able to progress. G proteins other than G_i, such as G_o and G_t can be ADP-ribosylated by pertussis toxin under appropriate conditions;[2,6,8] however, these G proteins do not inhibit adenylate cyclase. For this reason, experimental results following pertussis toxin treatment of intact cells or tissues must be interpreted carefully.

The importance to the target cell of G protein cycles for receptor-effector signal transduction is several-fold. First, selectivity is provided: each target cell expresses a select set of pharmacologically distinct receptors with which to respond to the environmental milieu, plus a complement of different G proteins and distinct effector enzymes. Second, because the stoichiometry of the components can change due to cellular regulation of gene expression, translation, posttranslational modification, and protein degradation, the nature of the response to the receptor-hormone complex may potentially be altered. Finally, the cycle provides amplification of the response: a single hormone-receptor interaction can potentially activate several G proteins, which in turn can activate numerous effector proteins prior to inactivation.

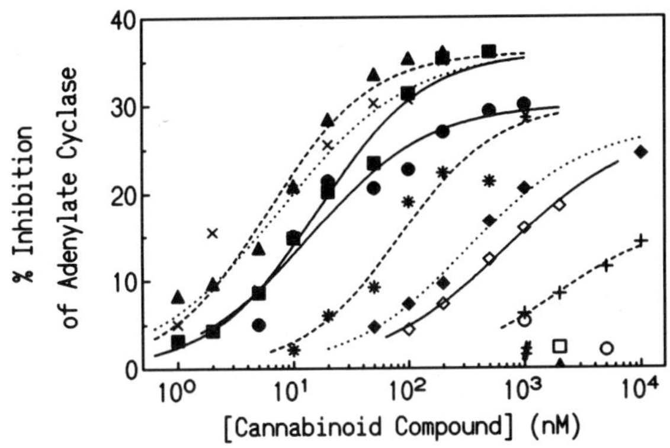

FIGURE 1. Inhibition of adenylate cyclase by cannabinoid compounds. Log dose–response curves for cannabinoid compounds to inhibit adenylate cyclase in membranes from N18TG2 neuroblastoma cells are shown for: (–)CP-55244 (▲), (–)CP-55940 (■) DALN (×), (–)11-OH-Δ^8-THC-DMH (●), (–)11-OH-Δ^9-THC (✳), Δ^9-THC (◆), Δ^8-THC (◇), Cannabinol (+), cannabidiol (#), (+)11-OH-Δ^8-THC-DMH (○), (+)CP-55940 (□), and (+)CP-55244 (Δ). Data are from References 38, 41, 43, and 49.

Cannabinoid Receptor Interactions with G_i to Inhibit Adenylate Cyclase

Studies from our laboratory have demonstrated that inhibition of adenylate cyclase is a cellular action that can be pharmacologically correlated with human effects and a number of animal behaviors (see below). Initial studies were performed using a cloned line (N18TG2) of neuroblastoma cells in culture. These studies determined that cannabimimetic drugs decrease the cyclic AMP content of neuronal cells.[37] Adenylate cyclase in plasma membranes from these cells was inhibited by centrally active cannabinoid drugs.[38,39] Several lines of evidence suggested that a cannabinoid receptor was mediating the effects on adenylate cyclase. The response was rapid and reversible both in intact cells and in membrane preparations, indicating that an irreversible cytopathological change was not responsible for the effects.[38] The drug concentrations that were required were in a range that would be expected to be present at the site of action *in vivo*.[39] The inhibition of cyclic AMP synthesis by Δ^9-THC occurred at nanomolar concentrations of drugs in intact cells[40] and with a K_{inh} of 360 n*M* for

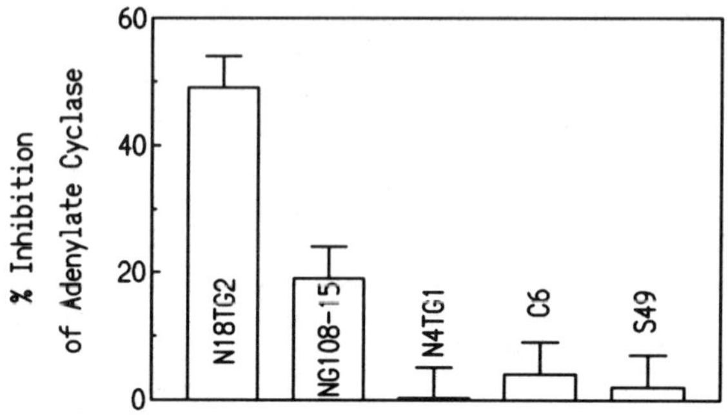

FIGURE 2. Cell specificity of the inhibition of adenylate cyclase. The percentage of inhibition of hormone-stimulated adenylate cyclase activity by 1 μ*M* DALN is shown in purified plasma membranes from the indicated cultured cell lines. Error bars represent the average coefficient of variation for the experiments. Adenylate cyclase was stimulated by 1 μ*M* PGE_1 in the neuroblastoma clones N18TG2 and N4TG1 and the hybrid cell NG108-15, and stimulated by 1 μ*M* isoproterenol in the glioma C6 and the lymphoma S49 cells. Data are from References 40 and 42.

inhibition of adenylate cyclase in membrane preparations[41] (Figure 1). The relative potency of natural and synthetic cannabinoid compounds was consistent with that reported for a number of biological responses *in vivo*.[41] The cannabinoid inhibition of adenylate cyclase was exhibited only in certain cell types[40,42] (Figure 2). Even within neuroblastoma cells, a clone was found that failed to respond to cannabinoid compounds by inhibiting adenylate cyclase.[42] The glioma cell line C6 failed to respond, as did the lymphoma cell line S49. Since the effects on the enzyme are not universally observed, this cellular selectivity would argue for a phenotypic difference between responding and nonresponding cells such as would be expected for the expression of cannabinoid receptors. The final argument favoring a cannabinoid receptor is that the regulation of adenylate cyclase required a functional G_i protein.[39,40]

The requirement for G_i for the cannabinoid inhibition of adenylate cyclase was initially based on the biochemical evidence that GTP was required at the same concentration (EC_{50} = 100 μ*M*) as was required for inhibition of adenylate cyclase by carbachol acting at the muscarinic receptor.[39] This concentration is considerably greater than that required for G_s to transduce the response to stimulatory hormones. It could also be shown that desacetyllevonantradol (DALN) could induce a state of irreversible inhibition when the nonhydrolyzable analog GppNHp was substituted for GTP under the appropriate conditions. This would result from cannabinoid-stimulated dissociation of the G_i subunits and the failure of the heterotrimer to reassociate because hydrolysis to GDP could not occur. The requirement for divalent cations was also identical to that of the muscarinic response. The maximum inhibition occurred at lower concentrations of Mg^{2+} than would be required for hormone-stimulated adenylate cyclase activity. These properties characterized the cannabinoid response as being typical of that expected for hormones and neuromodulators that act via G_i.

Data to support the requirement for G_i in the cannabinoid response came from the use of pertussis toxin as a tool to block the interaction of receptors with the G_i heterotrimer.[40] Treatment of intact cells with pertussis toxin blocked the ability of DALN to inhibit cyclic AMP accumulation. Inasmuch as the ADP-ribosylation is irreversible, the regulation of adenylate cyclase by cannabinoid drugs was blocked in membranes that were purified from pertussis toxin-treated cells. It was also possible to treat the membranes directly with pertussis toxin and block the cannabinoid response. These studies confirmed that the regulation of adenylate cyclase by cannabinoid drugs required the interaction with a functional G_i.

Having the knowledge that the cannabinoid response was associated with a second messenger system for which we could define a functional receptor,[43] our laboratory undertook the development of a ligand binding assay for the cannabinoid receptor based on the tritium-labeled, biologically potent bicyclic cannabinoid analgetic CP-55940.[44] Binding to rat brain P_2 membranes was linear within the range of 10 to 50 μg of protein per milliliter. The pH optimum was between 7 and 8, and specific binding could be eliminated by heating the membranes to 60°C, characteristic of ligand-protein interactions. Specific binding (defined as total binding displaced by 1 μ*M* Δ^9-THC or 100 n*M* DALN) was saturable; the B_{max} for rat cortical P_2 membranes was 1.85 pmol/mg of protein. The K_d was 133 p*M* when determined by Scatchard analysis of binding using a P_2 membrane preparation that had been preincubated at 37°C and extensively washed to remove endogenous guanine nucleotides. However, using less severe conditions of membrane preparation devoid of the 37°C-incubation, the affinity is somewhat lower. This may be a reflection of the receptor's interaction with G_i. This is supported by kinetic studies that uncovered two affinity states. The binding $K_{on} \approx 2.6 \times 10^{-4}$ pM^{-1} min^{-1} and the off rate was biphasic with a K_{off} ≈ 0.016 min^{-1} and $K_{off'}$ >0.06 min^{-1}. The two K_d values estimated from the kinetic constants approximated 55 p*M* and exceeded 200 p*M*, respectively. The binding of the agonist ligand [^{3}H]CP-55940 was decreased by a nonhydrolyzable GTP analog. The guanine nucleotide induced a more rapid dissociation of the ligand from the binding site, consistent with an allosteric regulation of the receptor by a G protein.

No evidence for receptor subtypes was found from the log dose-response curves determined for cannabinoid compounds to displace [^{3}H]CP-55940 binding.[44] The Hill coefficient for CP-55940 approximated 1, indicating that under the conditions of assay, a single class of binding sites was determined that did not exhibit cooperativity. Binding of [^{3}H]CP-55940 was displaced by cannabinoid drugs in the following order of potency: CP-55940 $\geq$ DALN > 11-OH-Δ^9-THC = Δ^9-THC > cannabinol. Cannabidiol and cannabigerol displaced [^{3}H]CP-55940 by less than 50% at 1 μ*M* concentrations. This order of potency is comparable to both the inhibition of adenylate cyclase *in vitro* and the analgetic activity of these compounds *in vivo*.[43] With these studies, the criteria for a high affinity, stereoselective, pharmacologically distinct cannabinoid receptor in brain tissue were fulfilled.

BIOLOGICAL RESPONSES REGULATED BY THE CANNABINOID RECEPTOR

Structure-Activity Relationship Studies

The definition of a receptor as a pharmacologically distinct entity resides in the potency ratios for a series of structurally diverse agonist or antagonist congeners. In the case of the cannabinoid receptor, the lack of potent antagonists has been a limiting factor in defining cannabinoid receptor-mediated responses. However, a large number of cannabinoid agonists, active and inactive metabolites, and related structures are available, and many of these have been used to describe the *in vivo* responses to cannabinoid compounds. The activities of over 300 cannabinoid compounds in typical animal models have been compiled by Razdan in a comprehensive review.[45] These studies have included the tachycardia and subjective rating of human experience with cannabinoid compounds, behaviors observed in monkeys, static ataxia measured in dogs, and various behaviors determined in rodents. The most extensive structure-activity relationship investigation of cannabinoid compounds in mice has been developed recently by Martin and co-workers.[46-48] These investigators were able to take advantage of the differing time courses to determine within the same series of mice four parameters comprising spontaneous activity in an open field, rectal temperature, antinociception determined by tail flick latency, and the ring stand test for catalepsy.

The inhibition of adenylate cyclase by a series of marijuana constituents and cannabinoid drug metabolites paralleled the ability of these compounds to produce a subjective response and tachycardia in humans and to alter various behaviors in animals[41] (see Figure 1). Δ^9-THC was more potent than Δ^8-THC, and cannabinol was only partially active at its maximally effective concentration. Cannabidiol, cannabigerol, cannabichromene, and olivetol were inactive *in vitro* as well as *in vivo*. Hydroxylation at the C(11) position resulted in compounds having approximately one order of magnitude greater potency to inhibit adenylate cyclase as well as to produce behavioral responses in humans and animals. Hydroxylation at other positions reduced the potency or eliminated activity. Displacement of binding of [^{3}H]CP-55940 in rat brain membranes paralleled the ability of these compounds to inhibit adenylate cyclase and evoke these behaviors.

Stereochemical selectivity both *in vivo* and *in vitro* could be demonstrated using the pure enantiomers of 11-OH-Δ^8-THC-dimethylheptyl. Within the Martin four-parameter studies of cannabinoid behavior in mice, the (–)-enantiomer exhibited ED_{50} values between 4 and 21 μg/kg, whereas the (+)-enantiomer was inactive at 30 mg/kg.[48] In dogs, static ataxia was pronounced at 10 μg/kg of the (–)-enantiomer, but could not be discerned at 1 mg/kg of the (+)-enantiomer.[48] Similarly, enantioselectivity was noted in the ability of this pair to bind to the cannabinoid receptor and inhibit cyclic AMP production in intact neuroblastoma cells and adenylate cyclase activity in membranes.[49]

Extensive structure-activity relationship studies by Pfizer, Inc. have demonstrated that specific requirements within the cannabinoid structure are necessary to produce potent analgesia.[50,51] A three-point association between the agonist and the receptor mediating analgesia was hypothesized to consist of: (1) the C ring hydroxyl, (2) the phenolic A ring hydroxyl, and (3) the A ring alkyl hydrophobic side chain. Potent tricyclic and bicyclic structures were synthesized as cannabinoid analgetics that conform to this agonist-receptor three-point interaction model. The structure-activity relationship profile for inhibition of adenylate cyclase *in vitro* was consistent with this same three-point association of agonists with the receptor *in vivo*.[43] Within the Pfizer series, the order of potency for both analgesia and inhibition of adenylate cyclase was CP-55244 ≥ CP-55940 ≥ desacetyllevonantradol

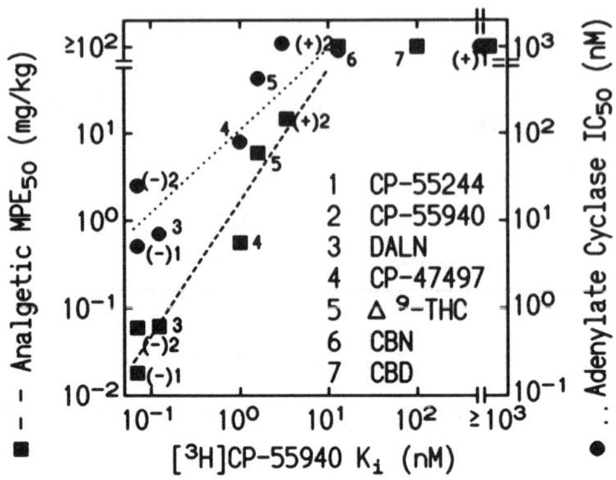

FIGURE 3. Structure-activity relationship data for binding to the cannabinoid receptor correlated with inhibition of adenylate cyclase *in vitro* and antinociception *in vivo*. Inhibition of secretin (500 n*M*)-stimulated adenylate cyclase was determined using membranes from N18TG2 cells, and correlated with binding to the receptor with a slope of 0.94 ± 0.19, $r = 0.93$ ($n = 6$). Antinociceptive activity was determined using the phenylbenzylquinone stretching test in mice, and correlated with binding to the receptor with a slope of 1.50 ± 0.20, $r = 0.97$ ($n = 6$). Data are from References 38, 41, 43, 44, and 49 through 51.

> Δ^9-THC > CBN >> CBD (Figure 3). Further investigation of the Pfizer series of compounds in the Martin four-parameter mouse model demonstrated this same parallel order of potency for each of the behaviors tested.[47] Enantio- and stereoselectivity were exhibited by the bicyclic and tricyclic cannabinoid compounds for the behavioral responses,[47,48] the inhibition of adenylate cyclase,[43] and binding to the [^{3}H]CP-55940 site in brain.[44,52] These findings support the pharmacological definition of the [^{3}H]CP-55940 binding site as the receptor that is associated with the regulation of adenylate cyclase *in vitro* and that mediates cannabinoid induced analgesia, hypothermia, spontaneous activity, and catalepsy *in vivo*.

Effects of Cannabinoid Drugs on Biological Responses

The interpretation of *in vivo* and *in vitro* responses as being attributable to the above-described cannabinoid receptor must be made based on pharmacological and biochemical criteria. Natural and synthetic cannabinoid ligands have been used to characterize the pharmacology of this receptor for both regulation of adenylate cyclase and certain biological responses in animals. The determination of a similar ratio of potencies must be applied to the studies of other cellular phenomenon and biological responses to be assured that they result from cannabinoid interactions with the cannabinoid receptor. In order to define cannabinoid receptor "subtypes", a different structure-activity relationship profile would have to be described for the given biological response.

A usual criterion for the pharmacological definition of a receptor includes the specificity of antagonists to block the response. Potential antagonists developed by Sterling Research Group have been described as competitive blockers of the regulation of adenylate cyclase and the action of cannabinoids in the mouse vas deferens[53,54] (see below). However, the relatively poor potency has precluded the ability to behave as useful antagonists in *in vivo* studies.[53] Until a high potency competitive cannabinoid antagonist can be developed for general use in *in vivo* and *in vitro* assays, the determination of receptor-mediated vs. nonreceptor mechanisms of cannabinoid actions will have to rely on the relative activities of agonist compounds.

Considerable research literature has demonstrated that lipophilic cannabinoid compounds are able to interact with biological membranes.[55-60] This property may be responsible for altered responses of membrane-associated enzymes and proteins, and subsequent cellular effects. It is clear, however, that membrane fluidity changes are not responsible for the inhibition of adenylate cyclase by cannabimimetic agonists.[61] In contrast, in more rigid membranes in which cannabinoid compounds increase membrane fluidity, the response is to decrease the thermal transition temperature, thereby increasing adenylate cyclase activity.[62] The role that alterations in membrane fluidity may play in responses observed in the intact animal has yet to be resolved.[63] Using a series of cannabinoid compounds having a wide range of potencies *in vivo*, it was demonstrated that lipophilicity defined by the octanol/water partition coefficient determined by HPLC did not correlate with biological activity in the Martin four-parameter mouse model.[64]

CANNABINOID RECEPTORS IN THE BRAIN

Anatomical Localization of Cannabinoid Receptors

Cannabinoid receptors, quantitated by [^{3}H]CP-55940 binding to rat brain membranes, were found in greatest abundance in the rat cortex, cerebellum, hippocampus, and striatum, with smaller but significant binding also found in the hypothalamus, brain stem, and spinal cord.[65] Herkenham and colleagues have used quantitative autoradiography of [^{3}H]CP-55940 binding to the cannabinoid receptor in rat brain slices to discern greater detail in regional localization.[66-67] Since this preparation is unfixed tissue, a receptor-G protein interaction is observed, i.e., a decrease in the relative affinity of the receptor for agonist ligands in the presence of analogs of GDP or GTP.[67]

Within the cerebellum, cannabinoid receptors were located predominantly within the molecular layer and were extremely sparse within the deep nuclei.[66,67] Evidence supports the localization of these receptors on the axons of the granule cells.[67] The static ataxia which has been described as a measure of cannabimimetic activity in larger animals such as the dog[46,48] may be the result of cannabinoid interactions with the receptors present in the cerebellum.[66]

Moderately dense binding was found throughout the rat cortex, particularly in layers I and VI.[67] One might expect that cognitive and perceptual effects noted by humans administered cannabinoid compounds[68] could be attributed to information processing at various cortical regions. Areas of the hippocampal formation exhibiting the densest receptor binding were the CA3 field of Ammon's horn and the molecular layer of the dentate gyrus; however, the remaining areas also exhibited fairly dense binding.[67] In humans, cannabimimetic compounds disrupt the consolidation of short-term memory,[69,70] suggesting hippocampal involvement. In chronically instrumented behaving rats, Deadwyler and colleagues have demonstrated that Δ^9-THC impairs a tone discrimination behavior concurrent with a diminished response of the dentate granule cells to incoming information from the perforant path.[71,72]

Within the olfactory tubercle, cannabinoid receptors were most dense in the ependymal and subependymal layers, and were moderately dense in the internal granular and plexiform layers.[67] Cannabinoid receptors were of moderate density within the amygdalar nuclei, mammillary bodies, and lateral hypothalamus.[66,67] Binding density was fairly low within several other hypothalamic nuclei, particularly the median eminence. Various endocrine effects of cannabinoid compounds have been reported,[73] but the sites of action and mechanisms have yet to be discerned. Hypothermia is an action of cannabinoid drugs that may be regulated at the level of the hypothalamus or autonomic structures. The brain stem and spinal cord were of note due to the paucity of cannabinoid receptors in these regions, with a few exceptions.[66,67] These regions may nevertheless be involved in the cannabinoid effects to decrease body temperature,[74-76] produce changes in heart rate in man and animals,[77-79] suppress nausea and vomiting,[80,81] produce antinociception,[50,51,82-84] or decrease intraocular pressure.[85]

Regions of the basal ganglia that exhibited a very high density of cannabinoid receptors included the lateral striatum, a region that receives inputs from the cortex involved in sensorimotor functions.[66,67] Dense binding was also observed in the striatal outflow nuclei comprising the globus pallidus, entopeduncular nucleus, and the substantia nigra pars reticulata. It is of note that the greatest cannabinoid receptor density in the basal ganglia exists in the regions that process motor behaviors and regulate sensorimotor information.[66,67,86] This is in contrast to the moderate receptor density found in the medial and ventral regions of the striatum, nucleus accumbens, and the ventral pallidum, areas subserving limbic functions and believed to be involved in "reward".[86] The cannabinoid-induced catalepsy in mice[87] and the synergistic effects of cannabinoids on immobility in reserpinized rats[88] may be regulated in part at the level of the basal ganglia, as evidenced by stereotaxic implantation of Δ^9-THC or levonantradol. Implantation of Δ^9-THC into the nucleus accumbens resulted in increased spontaneous activity in mice.[89]

In an effort to determine the neuronal localization of cannabinoid receptors within the basal ganglia, Herkenham and colleagues[86] examined [^{3}H]CP-55940 binding following selective lesions of striatal afferent and efferent pathways. In these studies, the nigrostriatal pathway was unilaterally destroyed by infusion of 6-hydroxydopamine into the medial forebrain bundle. Four weeks later, the animals exhibited amphetamine-induced rotation and loss of dopamine uptake sites in the striatum and substantia nigra pars compacta, indicative of loss of dopaminergic inputs to the striatum. No change in cannabinoid receptor density was found to be associated with the lesioned dopaminergic inputs to the striatum from the substantia nigra. In contrast, striatal neurons and efferent projections were selectively destroyed by unilateral infusion of ibotenic acid into the caudate-putamen. The loss of striatal neurons was accompanied by a loss of cannabinoid receptors in addition to dopamine D_1 and D_2 receptors in the striatum. Furthermore, cannabinoid and D_1 receptors were depleted within the globus pallidus, entopeduncular nucleus, and substantia nigra pars reticulata. This suggests that these receptors are present not only on the neuronal cell bodies but also on the axons and terminals of these cells.

Cannabinoid receptor binding localization in higher mammalian species, including guinea pigs, dogs, marmosets, rhesus monkeys, and humans, exhibited a pattern that generally resembled that observed for rats.[66] Cannabinoid receptors were also prevalent in the nervous systems of lower vertebrates including chickens, turtles, and trout.[90] A low concentration of cannabinoid receptors was discernable in fruit flies (unpublished observations, A. C. Howlett).

Biochemical Evidence for Second Messenger Regulation by Cannabimimetic Compounds in the CNS

Our laboratory undertook a determination of the effects of cannabimimetic drugs to alter cyclic AMP production in selected brain regions that would be expected to be targets for their effects. Using rat brain slice preparations, we evaluated the effect of DALN on basal- and forskolin-stimulated cyclic AMP accumulation in the regions exhibiting the greatest cannabinoid receptor density.[65] DALN reduced cyclic AMP levels in the hippocampus, frontal cortex, and striatum. In the cerebellum, however, the response to DALN was biphasic, with cyclic AMP accumulation being decreased at lower and increased at higher concentrations. DALN reduced cyclic AMP accumulation in isoproterenol-stimulated slices in the cortex and cerebellum, but not in the hippocampus. Cells that responded to vasoactive intestinal peptide with an increase in cyclic AMP accumulation in the hippocampus and cortex also responded to DALN. The modulation of cyclic AMP accumulation in the striatum by DALN could be attenuated following stereotaxic implantation of pertussis toxin,[65] supporting the involvement of a G protein in the cannabinoid response in the brain.

In further studies,[91] the inhibition of cyclic AMP accumulation by DALN was compared to that of morphine in rat striatal slices. Both drugs were equally efficacious in reducing the stimulation of cyclic AMP levels by D_1 dopaminergic agonists (SKF-38393) or vasoactive intestinal peptide. Inhibition by DALN, unlike that of morphine, was not reversed by naloxone. The response to maximally effective concentrations of DALN plus morphine was neither additive nor synergistic. Morphine and DALN, while sharing analgetic properties as well as a common transducer (G_i) nevertheless act at different receptors. The evidence suggests that one point of commonality may be that the opioid and cannabinoid receptors may be co-localized on cells responsive to dopamine and to VIP.

PHARMACOLOGICAL SPECIFICITY OF RECEPTORS FOR CANNABINOID COMPOUNDS

FAILURE OF A NUMBER OF NEUROMODULATORS AND THEIR ANALOGS TO BIND TO THE CANNABINOID RECEPTOR

Early reports used radiolabeled ligands for a variety of receptor families to examine the ability of Δ^9-THC to affect the binding of these ligands to their corresponding receptors. Such reports have shown interaction of the cannabinoid drugs with β-adrenergic,[92-94] dopaminergic,[94,95] as well as opioid[94,96] receptors. However, alterations in binding characteristics reported for these pharmacological receptor types might be due to the perturbation of membrane fluidity evoked by high concentrations of cannabinoid compounds.[58] Thus, we studied many of these ligands for their ability to specifically inhibit the binding of [^{3}H]CP-55940.

Addition of Δ^9-THC had been shown to result in increased binding of the β-adrenergic antagonist [^{3}H]dihydroalprenolol to its receptor. This effect was apparently due to an increase in receptor affinity.[93,94] For the converse experiment, in our hands, no effect of adrenergic ligands on binding to the cannabinoid receptor were observed (see Table 1). Perhaps a change in membrane fluidity resulted in the observed effects on [^{3}H]dihydroalprenolol binding, as speculated by the authors.[94]

The opioids and cannabinoids share the behavioral responses of analgesia, inhibition of intestinal motility, and a depression of locomotor activity, and certain of the cannabinoid-induced responses could be partially reversed by the opioid antagonist naloxone.[97,98] Evidence exists for partial cross-tolerance between these two pharmacological systems for certain behaviors.[97,99-103] Nevertheless, the evidence for cannabinoid compounds affecting ligand binding to opioid receptors is conflicting. Bloom and Hillard showed that concentrations of 4 to 50 μM Δ^9-THC decreased [^{3}H]naloxone binding to washed membranes from mouse brains compared to a detergent-vehicle control, but had no effect on the binding of [^{3}H]D-ala-D-leu-enkephalin.[94] The cannabinoid effects on [^{3}H]naloxone binding were the result of an apparent decrease in the opioid antagonist affinity with no change in B_{max}. Because these changes in the binding behavior occurred at concentrations of cannabinoids which also altered the parameters of fluorescence polarization of diphenylhexatriene in mouse synaptic plasma mem-

TABLE 1
Effect of Various Ligands for Specific Receptor Types on Binding of [^{3}H]CP-55940 to Rat Brain Membranes

	Concentration (μM)	% Inhibition of [^{3}H]CP-55940 binding
Adrenergic ligands		
Epinephrine	10	5
Norepinephrine	1	0
Isoproterenol	1	7
DL-Propanolol	1	25
Clonidine	1	0
Yohimbine	1	0
Prazosin	1	0
Dopaminergic ligands		
Apomorphine	1	20
(+)-SKF-38390	1	0
Quinpyrole	1	25
Haloperidol	1	5
Spiroperidol	1	10
Opioid ligands		
[D-Trp^2]-Met Enkephalinamide	1	3
Dermorphin	1	7
β-Endorphin	6	2
Naloxone	1	0
Morphine sulfate	1	0
D-Ala^2-D-Leu^5-enkephaline	1	2
D-Ala^2-Met^5-enkephalinamide	10	0
Sigma ligands		
Ethyl ketocyclazocine	1	19
Phencyclidine	1	0
SKF-10047	1	0
Haloperidol	1	5
Cholinergic ligands		
Nicotine	6	0
Nicotinic acid	5	3
Mecamylamine	5	0
Hexamethonium	6	0
Atropine	6	0
Carbachol	1	0
Oxotremorine	1	0
McN-A343	1	0
Pirenzipine	1	4
Neuroleptics		
Promazine	5	0
Chlorpromazine	5	0
Haloperidol	5	0
Trifluoperazine	5	0

TABLE 1 (continued)
Effect of Various Ligands for Specific Receptor Types on Binding of [^{3}H]CP-55940 to Rat Brain Membranes

	Concentration (μM)	% Inhibition of [^{3}H]CP-55940 binding
GABA receptor ligands		
GABA	10	3
Baclofen	6	0
Muscimol	5	3
β-carboline carboxylate Me	5	0
Bicuculline	5	8
Diazepam	5	7
Flunitrazepam	5	10
Picrotoxin	5	8
Serotonergic ligands		
Serotonin	6	2
8-OH-DPAT	6	0
Ketanserin	6	0
5-Methoxytryptamine	1	0
ICS-205930	6	0
Methiothepin	6	0
Metergoline	6	0
Hormones and neuromodulators		
ACTH	6	2
α-Melanocyte stimulating hormone	6	0
Cholecystokinin	10	0
Follicle stimulating hormone	6	20
Luteinizing hormone	6	11
Luteinizing hormone releasing hormone	6	0
Neuropeptide Y	10	0
Neurotensin	10	21
Oxytocin	6	0
Secretin	1	0
Glucagon	1	0
Somatostatin	10	9
Substance P	10	0
VIP	10	0
Steroid receptor ligands		
Androsterone	10	0
Cortisone	1	2
Corticosterone	1	0
Dehydro-epi-androsterone	10	0
Estradiol	1	0
Pregnenolone sulfate	10	0
Progesterone	1	0
Triiodothyronine	1	0
Thyroxine	1	5

TABLE 1 (continued)
Effect of Various Ligands for Specific Receptor Types on Binding of [^{3}H]CP-55940 to Rat Brain Membranes

	Concentration (μM)	% Inhibition of [^{3}H]CP-55940 binding
Amino acids		
Aspartate	10	0
Glutamate	10	0
Glycine	10	0
Histidine	1	2
Taurine	10	0
Tryptophan	10	0
Peptides		
Bovine adrenal medulla dodecapeptide (BAM-12P)	1	10
Bradykinin (1-5)	1	0
Chick Bursin	5	4
Cycloleu-gly	10	0
Kyotorphin	50	0
Neokyotorphin	50	0
Neuromedin N	5	0
Morphiceptin	10	0
Metorphamide (adrenorphin)	10	0
Val-morphiceptin	10	0
Nonsteroidal antiinflammatory drugs		
4-Aminoantipyrine	10	0
Indomethacin	1	10
Naproxen	3	1
Isopyrin	10	0
Mepirizole	10	0
Arachidonic acid and metabolites		
Arachidonic acid	3.3	0
5(*S*)-HpETE	10	7
5(*S*)-HETE	10	10
8-HETE	5	14
9-HETE	5	14
11-HETE	5	14
12-HETE	5	17
15-HETE	5	17
5(*S*)-12(*S*)-di HETE	5	5
15(*S*)-HpETE	5	0
12(*S*)-HpETE	5	0
Prostaglandins		
PGE_2	7.5	5
$PGF_{2\alpha}$	10	0
PGG_2	4	0
PGH_2	4	5

TABLE 1 (continued)
Effect of Various Ligands for Specific Receptor Types on Binding of [^{3}H]CP-55940 to Rat Brain Membranes

	Concentration (μM)	% Inhibition of [^{3}H]CP-55940 binding
Leukotrienes		
LTB_4	5	15
LTC_4	5	0
LTD_4	5	0
LTE_4	5	0
LTF_4	5	0
6-*trans*-LTB_4	5	6
N-acetyl-LTE_4	5	0
20-OH-LTB_4	5	15
20-COOH-LTB_4	5	15
Miscellaneous agents		
Imipramine	5	1
Pentobarbital	10	0
Retinol	10	10
13-*cis*-Retinol	10	5
Retinal	10	2
13-*cis*-Retinal	10	0
Ergocalciferol	10	0
dl-α-Tocopherol	10	0
Isobutylmethylxanthine	1	35
Phenylisopropyladenosine	1	0
Forskolin	1	39
Harmine	1	0
2-Methylharmine	1	0
Dihydroergotamine	10	5
Bromocryptine	10	0
DL-Carnitine	10	0
L-Carnosine	10	0

branes, these investigators interpreted their results as being due to an action of cannabinoids at the membrane. In contrast, Δ^9-THC in the same concentration range was reported to cause noncompetitive inhibition of ligand binding of [^{3}H]etorphine, [^{3}H]dihydromorphine, and [^{3}H]D-pen^2-D-pen^5-enkephalin to μ- and δ-opioid receptors in rat brain membranes as well as inhibiting binding of the antagonist [^{3}H]naloxone.[96] In this study, Δ^9-THC and cannabidiol were equipotent, and levonantradol and dextronantradol were 10-fold less potent than Δ^9-THC in decreasing the binding of [^{3}H]dihydromorphine. Thus, the pharmacology of the response demonstrates that the cannabinoid receptor is not involved in these effects on the binding assays. When the binding of [^{3}H]D-ala-D-leu-enkephalin was examined in membranes from N18TG2 neuroblastoma cells, no effects of 1 μM Δ^9-THC or DALN were discerned. Furthermore, no evidence was found for cannabinoid-opioid interactions in the effector response of inhibition of cyclic AMP accumulation in the neuroblastoma cells.[42] Finally, opioids did not inhibit the binding of [^{3}H]CP-55940 to the cannabinoid receptor in rat brain membranes (see Table 1).

Previous work *in vivo* implicated Δ^9-THC as intrinsic to a direct or indirect nicotinic cholinergic pathway involved in immobility in the rat because its action could be inhibited by the antagonist mecamylamine.[104] Neither mecamylamine nor nicotine at concentrations above 1 μM were capable of inhibiting the specific binding of [^{3}H]CP-55940, arguing against the possibility that cannabinoids bind to nicotinic receptors. Atropine at these high concentrations was also unable to compete for ligand binding, further dismissing the possibility of a muscarinic action for the cannabinoids.

Binding to the dopamine D_2 (neuroleptic) receptor was shown to be perturbed by concentrations of cannabinoids exceeding 1 μM.[94] Since many of the clinical parameters observed upon exposure to the cannabinoids, namely, their antiemetic properties, production of hypothermia, and their sedative-like effects, are similar to those associated with use of antipsychotic neuroleptic agents, we tested several of the latter compounds in the binding assay. Representatives of the phenothiazines (promazine and chlorpromazine) and the butyrophenones (haloperidol) were unable to displace binding of the ligand to the cannabinoid receptor, even at high concentrations.

The reported involvement of serotonin receptor subtypes in analgesia or emesis led us to examine whether serotonin and many of its analogs were capable of interacting with the cannabinoid receptor. A number of ligands specific for the various serotonin receptor subtypes were unable to compete with [^{3}H]CP-55940 for the cannabinoid receptor, even at concentrations as high as 6 μM (Table 1).

Prostaglandins have been shown to decrease adenylate cyclase activity in fat cells via a G_i-coupled receptor.[35] It was reported that DALN was able to displace the binding of [^{3}H]prostaglandin D_2, albeit with low potency (IC_{50}

> 10 μ*M*).[105] It was therefore of interest to determine whether a metabolite of the arachidonic acid cascade could bind to the cannabinoid receptor. As can be seen from Table 1, an exhaustive list of prostaglandins and leukotrienes were unable to modify ligand binding.

It has been reported that cannabinoid compounds can influence binding to steroid receptors in brain and other tissues. Investigations of Δ^9-THC interaction with the estrogen receptor have yielded conflicting results.[106-108] A recent preliminary report described Δ^9-THC binding to a glucocorticoid receptor in brain tissue.[109] Clearly, estradiol and glucocorticoids failed to alter the binding of [^{3}H]CP-55940 at concentrations expected for binding to their respective receptors (Table 1).

A Series of Aminoalkylindoles Bind to the Cannabinoid Receptor

Recent reports indicate that the synthetic compound pravadoline and its analogs, the aminoalkylindoles, act as analgesics.[110,111] Antinociceptive activity was demonstrated in rodents using tests of chemical, thermal, and mechanical stimuli. Pravadoline also diminished the electrically induced twitch response of mouse vas deferens preparations and inhibited rat gastrointestinal transit.[110,111] Neither the analgesic effects nor the smooth muscle relaxation effects of pravadoline involved opioid receptors because neither could be reversed by naloxone.[110] Pravadoline differed from the nonsteroidal antiinflammatory drugs in that it exhibited no gastrointestinal cytotoxicity and was not antiinflammatory.[110] Certain structural analogs of pravadoline were more potent as antinociceptive agents, but were unable to inhibit prostaglandin formation.[111] Pharmacologic studies of several of these analogs in the mouse vas deferens and guinea pig ileum preparations indicated that their effects were not mediated by muscarinic cholinergic, α-adrenergic, serotonin subtypes 2 and 3, μ, κ, or δ opioid, P1-purinergic, neurokinin-1, bradykinin, or PGE_2 receptors.[111]

Aminoalkylindole analogs also inhibited adenylate cyclase in rat brain membranes.[53] The effects could be discerned using striatal or cerebellar membranes, but not membranes from cortex, amygdala, thalamus, hypothalamus, hippocampus, or brain stem.[53] The inhibition was not observed in the absence of GTP, and was evident for both basal and forskolin-stimulated adenylate cyclase activity.[53] The order of potency for a series of aminoalkylindole compounds to inhibit adenylate cyclase paralleled their ability to inhibit the electrically stimulated mouse vas deferens contractions, and enantioselectivity was demonstrated for both responses.

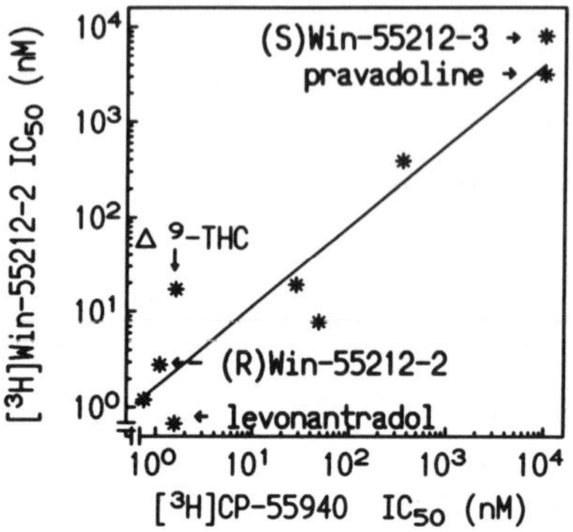

FIGURE 4. Correlation of binding affinities of cannabinoid and aminoalkylinodole compounds to the cannabinoid receptor determined by [^{3}H]CP-55940 in cortex and [^{3}H]Win-55212-2 in cerebellar membranes from rats. The slope of the correlation curve is 0.84 ± 0.11, $r = 0.95$ ($n = 9$).

The most potent of the aminoalkylindole compounds, Win-55212-2, was consequently tritium-labeled and its specific binding was found to be displaced by Δ^9-THC and other CNS-active cannabinoid compounds with potency ratios consistent with their cannabinoid activities reported in animals.[112] Moreover, in the converse experiment, a number of aminoalkylindoles were able to displace [^{3}H]CP-55940 binding in the cannabinoid receptor assay.[112] The order of potency of aminoalkylindole analogs to displace [^{3}H]CP-55940 binding to the cannabinoid receptor paralleled their ability to inhibit cerebellar adenylate cyclase, inhibit mouse vas deferens contractions, and behave as analgesics in several assays[112] (Figure 4). Thus, the evidence is strong that this class of compounds is eliciting these responses as a result of interaction with the cannabinoid receptor.

An intriguing outcome of the development of this series of compounds is the advent of potentially useful antagonists of cannabinoid responses. One analog, Win-56098, competitively antagonized the effects of pravadoline and Win-55212-2 in the mouse vas deferens response.[53] At 10 μM Win-56098, the dose ratio for pravadoline was 40, and for Win-55212-2 was 100. By comparison, the dose ratios for Δ^9-THC and levonantradol were 6 and 4, respectively, which were significant but not as great as for the aminoalkylindoles. Win-56098 was also able to shift the curve of inhibition of adenylate cyclase by Win-55212-2 to the right.

These studies point to the possibility that unique and potent antagonists for the cannabinoid receptor can be developed from this series of compounds.

Speculation on the Endogenous Ligand for the Cannabinoid Receptor

The identification of the presence of a pharmacologically specific receptor for cannabinoid drugs on central neuronal membranes derived from many animal species,[44] together with the characterization of their molecular mechanism of action,[39,40] has led to a search for an endogenous ligand(s) for this receptor. Using the radiolabeled ligand [^{3}H]CP-55940, we eliminated many neuromodulator agonists as putative candidates for the endogenous cannabinoid by their failure to reduce the level of specific binding to rat brain membranes (Table 1). From the above studies it can be seen that a broad range of ligands were unable to compete for binding to the cannabinoid receptor site even at the relatively high concentrations used. To date, the only report of an endogenous compound capable of inhibiting the binding of cannabinoids to a site in brain membranes has implicated myelin basic protein as the "endogenous ligand".[113] However, no evidence has been obtained to suggest that this large (14.5 kDa) molecule has a neurotransmitter-like role. As proposed by the authors, myelin basic protein may interact with its binding site due to its polycationic nature, since the ligand used to detect this binding site was the charged trimethylammonium analog of Δ^8-THC.[113]

With the advent of novel synthetic ligands and their use as pharmacological tools, together with molecular biology techniques, many new receptor subtypes are being discovered. It is therefore naive to ignore the possibility that the cannabinoid receptor may be a novel subtype of one of the receptors listed above. However, as indicated in Table 1, none of the ligands for these "classical" receptors appear capable of interacting with the cannabinoid receptor in a specific manner. It is with this observation in mind that one can speculate on the nature of the endogenous ligand for this receptor. Clearly, the nature of the receptor sets a precedent for a neuromodulatory compound as an agonist ligand. A number of drugs have actions at a synaptic protein or ion channel resembling an allosteric modulation of a response, e.g., benzodiazepines, barbiturates, phencyclidine. For those sites of action, an endogenous substance may not necessarily be present in the brain. However, the interaction of Δ^9-THC and its cannabimimetic analogs with a G protein-coupled receptor to produce a response is typical of the mode by which neuromodulators interact with neurons in the brain. Thus, one would predict that a neuromodulatory substance would interact with such a site.

Determination of a substance as a neuromodulator demands that certain criteria be fulfilled. First, the substance must be shown to be released at a synapse in the manner of a neuromodulator or else be able to reach the receptor site via the circulation in the manner of a hormone. In either case, the neuromodulator must be shown to be synthesized and packaged such that release can occur upon stimulation. Second, the neuromodulator must be shown to produce a response at the target cells. The characterization of the cannabinoid receptor's interaction with adenylate cyclase and its localization in the CNS set up the requirements for testing this criteria. The substance must be shown to bind to the cannabinoid receptor and to inhibit adenylate cyclase in the manner expected of a G_i-linked response. Furthermore, the appearance of the substance and its interaction with the receptor must occur within the structures of the brain known to possess cannabinoid receptors. Last, neuromodulators must not remain within the synapse for extensive lengths of time and therefore a means by which the substance is removed is considered important. Thus, a metabolizing enzyme or reuptake system is generally expected for neuromodulators.

Work in our laboratory is currently aimed at identifying a molecule capable of specific interaction with the cannabinoid receptor. Applying the criterion that a neuromodulatory substance must be synthesized and packaged such that it can be released upon stimulation, we examined the ability of the calcium ionophore A23187 to release a compound that could subsequently interact with the cannabinoid receptor on rat brain membranes. We have reported that the release of receptor-binding activity was dependent on the concentration of A23187 with maximal release occurring at 1 μM.[114] Release was also dependent on the time of exposure to the ionophore. Until the nature of the released substance can be chemically characterized and shown to fulfill all of the criteria of a neuromodulator, we cannot be certain of its physiological significance. It is clear that a number of substances found in extracts of brain tissue can diminish binding to G protein-coupled receptors, including NaCl and GTP. Thus, a conservative approach to interpreting the results of purification of an endogenous ligand for the cannabinoid receptor must be taken.

CELLULAR REGULATION OF CANNABINOID RECEPTORS

Tertiary Structure: Deductions from Amino Acid Sequence Derived from cDNA

The large number of characterized receptors that modulate adenylate cyclase activity through G proteins has revealed a structure common to this receptor type. The visual pigment protein rhodopsin provided the first structural analysis and was quickly followed by cloning of the gene and cDNA for the β_2-adrenergic and other receptors.[115,116] Prediction of the amino acid sequences indicates that these receptors possess 7 regions of 20 to 25 hydrophobic amino acids that span the membrane as α-helical structures. The amino terminus, usually containing at least two potential sites for N-linked glycosylation, is on the extracellular side of the cell membrane. The carboxyl tail extends into the intracellular compartment and is rich in serine and threonine residues, providing putative sites for phosphorylation by cyclic AMP-dependent kinases, protein kinase C, and certain receptor-associated kinases. Those amino acid residues linking the transmembrane helices form loops which also insert into the cytoplasmic space of the cell. The third intracellular loop, which joins the fifth and sixth hydrophobic domain, is thought to be the site of receptor interaction with G proteins. Differences in amino acid composition in this region may determine the particular G protein to which a receptor binds. Although the degree of homology between different G protein-linked receptors varies widely, enough exists such that antibodies produced against a peptide sequence of one receptor will react with another receptor.[117]

Recently, cDNA sequences for the cannabinoid receptor have been derived from rat and human brain.[118,119] The two receptors are 98% identical in the predicted amino acid sequences. The rat cannabinoid receptor cDNA, SKR6, was isolated from a cerebral cortex library using an oligonucleotide probe derived from the substance K family of homologous receptors. Translation of the cDNA sequence yields a 473-amino acid polypeptide with a molecular mass of 52,823. The primary structure of the gene product has seven hydrophobic domains, several potential sites for glycosylation, and many residues which are highly conserved among G protein-coupled receptors. The particular G protein to which the polypeptide can couple is not obvious based on the primary structure of the cytoplasmic loops of the protein.[118]

Posttranslational Modification

Recent studies indicate that certain posttranslational modifications may have a role in the functioning of G protein-associated receptors.[120-122] The focus of most studies on posttranslational processing is on three types of modification: glycosylation, phosphorylation, and attachment of fatty acid residues to cysteine.

Most G protein-linked receptors possess asparagine residues with a consensus sequence of Asn-X-Ser/Thr (where X may be any amino acid except proline or aspartate) near the amino terminus and are potential sites for *N*-glycosylation.[123] The presence of these sequences in the superfamily of G protein-coupled receptors suggests that *N*-glycosylation may play an important role in receptor function and/or expression. Glycosylation of membrane proteins including receptors has been studied primarily as a step in biosynthesis and localization of the protein.[124] Work from several studies indicates that protein glycosylation is required for transport of muscarinic receptors to the cell surface.[125] However, the exact role of receptor glycosylation remains controversial, as some groups have reported that *N*-glycosylation had no effect on receptor expression or function, while others have reported impaired coupling to G protein after enzymatic receptor deglycosylation or inhibition of glycosylation.[120,126-128] Three consensus amino acid sequences for glycosylation are predicted from the cDNA sequence of the cloned cannabinoid receptor.[118,119] Studies are underway to determine if the cannabinoid receptor is indeed glycosylated and, if so, what role this modification may have in the function or processing of the receptor. It would appear that blocking the *N*-linked glycosylation with tunicamycin in N18TG2 neuroblastoma cells has little effect on the ability of DALN to inhibit cyclic AMP accumulation.[129]

Much evidence indicates that phosphorylation may be important in regulating receptor function.[130] Receptor phosphorylation is thought to be the major mediator of the rapid uncoupling of the receptor from the associated G protein.[130-134] This process is the first step contributing to agonist-induced desensitization. The inhibition of adenylate cyclase by cannabimimetic compounds rapidly desensitizes in the cultured cell system, suggesting that phosphorylation may be of importance to cannabinoid receptor regulation.[129,135] Mutant forms of the β_2-adrenergic receptor that lack cAMP-dependent protein kinase (PKA) phosphorylation sites in the intracellular regions were significantly blunted in the extent of agonist-induced down-regulation.[122] In addition to being phosphorylated by PKA, the β_2-adrenergic receptor and rhodopsin are both phosphorylated (at sites distinct from those for PKA) by an associated receptor kinase.[130,136] These kinases, known as β-adrenergic receptor kinase (or βARK) and rhodopsin kinase, respectively, phosphorylate the receptor only upon binding of an agonist ligand. Phosphorylation by this kinase may be involved in regulating the ability of the receptor to interact with the G protein. However, phosphorylation by these kinases apparently has little or no role in receptor down-regulation or sequestration because mutants lacking the putative sites for βARK phosphorylation appear to down-regulate normally.[122] The deduced amino acid sequence of the cannabinoid receptor is abundantly supplied with serine and threonine residues, especially in the carboxyl terminus of the protein, and would appear to be an ideal substrate for phosphorylation.

Recent studies have reported that several membrane proteins, including rhodopsin and the human β_2-adrenergic receptor, are covalently modified by fatty acids.[137,138] Such a modification may represent a general feature of receptors and may have significant functional importance. Rhodopsin has been shown to be acylated by palmitic acid at two cysteine residues located on the inner face of the cell membrane.[138,139] The homologous cysteine in the β-adrenergic receptor is also palmitoylated, and if deleted by site-directed mutagenesis, results in reduced ability to mediate agonist stimulation of adenylate cyclase.[121] Thus, the modified cysteine may be crucial in mediating normal coupling of the receptor to the signal transduction system. Some membrane-bound proteins can be modified by isoprenylation at a cysteine residue four amino acids from the C-terminal. In the case of the cannabinoid receptor, the consensus sequences of CAAX (where A is an amino acid with an aliphatic side chain and X is any amino acid) or CCGG at the C-terminal are missing, indicating that isoprenylation is not possible for the cannabinoid receptor.

Another modification of cysteine residues involves the oxidation of sulfhydryl groups on neighboring cysteine residues to form disulfide bridges. This apparently occurs with rhodopsin, as mutants of this receptor lacking one or more cysteine residues located proximal to the third hydrophobic domain demonstrated decreased agonist binding.[139] In addition, glycosylation was also defective in the mutants.[139] Although this phenomenon has not been noted in other G protein-linked receptors, the homologous cysteine residues are present in this family of receptors. However, this particular cysteine homolog is not present in the reported amino acid sequence of the cannabinoid receptor.[118,119]

Although cloning and cDNA isolation are useful techniques in determining the primary structure of a receptor, they are of limited value in determining the structure of the native protein. To determine what specific covalent modifications have occurred, the protein of interest must be isolated from its naturally occurring source in a form suitable for study. As a first step toward purification, the cannabinoid receptor has been solubilized using low concentrations of the zwitterionic detergent CHAPS.[140] The solubilized form of the receptor retains high-affinity agonist binding, indicative of association with a G protein(s). Determination of posttranslational receptor modification can thus be accomplished through binding to lectin affinity columns to determine the nature and extent of glycosylation as well as aid in purification of the protein. Purification to homogeneity will allow determination of other possible modifications through the use of sequence analysis and site-directed mutagenesis.

THE IMPORTANCE OF THE CANNABINOID RECEPTOR

Preparations of *Cannabis sativa* (e.g., marijuana, hashish, and bhang) have a long history as therapeutic agents. Therapeutic applications include analgesia, sedation, attenuation of the nausea and vomiting of cancer chemotherapy, appetite stimulation, decreasing intestinal motility of diarrhea, antirheumatic and antipyretic actions, decreasing bronchial constriction of asthma, decreasing intraocular pressure of glaucoma, and treatment of convulsant disorders (see reviews by Dewey,[141] Lemberger,[142] Mechoulam,[143] and Hollister.[144]) It is not clear whether all of these actions can be attributed to an interaction of Δ^9-THC with the cannabinoid receptor. Variations in structure-activity profiles of various cannabinoid compounds suggest that certain of these diverse therapeutic actions may be mediated by differing mechanisms of action. With the exception of the antinociceptive effects, structure-activity relationship studies have not been rigorously applied to these potential therapeutic responses in an effort to determine whether the cannabinoid receptor is involved.

Cannabinoid drugs have also been used in this and other cultures for their mood altering properties.[141,144] Reported effects include euphoria, tranquility, difficulty in thinking or remembering, rapid flow of thoughts, and dreamy states.[68] Impairment of short-term memory consolidation has been documented at doses typical of casual users of marijuana.[69,70] Perceptual changes reported include altered perception of visual or auditory stimuli and distortions in duration of time.[68] Somatic effects reported include dizziness, dry mouth, paraesthesias, increased body awareness, weakness, incoordination, fatigue, and sleepiness.[68] These responses have been considered to be unwanted side effects in the use of cannabinoid compounds as therapeutic agents. At our current level of understanding, it is unclear that these effects can all be attributed to an interaction with the cannabinoid receptor.

Various animal models have been developed for the study of cannabinoid drugs (see review of structure-activity profiles by Razdan[45]). In rodents, low to moderate doses of Δ^9-THC result in hypermotility and hyperreactivity, and higher doses result in hypomotility and catalepsy. A decrease in body temperature is also a property of CNS-active cannabinoid drugs as well as other pharmacological classes of compounds. A unique static ataxia is observed in dogs in response to CNS-active cannabinoid compounds. Numerous reports in the literature describe other behavioral responses, endocrine effects, and immune responses associated with cannabinoid compounds. Neither the involvement of specific CNS regions or cell types nor the mechanism(s) of action at the neuronal level for these animal behaviors has been fully elucidated. Again, stringent pharmacological characterization of these and other responses in animals must be applied in order to support the hypothesis that the cannabinoid receptor regulates these responses.

The effects of cannabinoid drugs at the cellular level have been addressed by numerous laboratories using *in vitro* studies. As detailed in the reviews by Martin[145] and Pertwee,[146] effects of cannabinoid drugs have been observed on neurotransmitter release and/or reuptake, activities of enzymes including Na/K ATPase and monoamine oxidase, modification of arachidonic acid metabolism, and alteration of ligand binding constants for several neurotransmitter receptors. It has not been possible to correlate many of these effects with a biological response because effects of non-CNS-active cannabinoid compounds often appeared to be as great or better than for Δ^9-THC. Further, many of these studies utilized concentrations of cannabinoid drugs that may be in excess of the concentration expected to be present in the brain. The perturbation by cannabinoid drugs of membrane fluidity is a factor that must be considered in the interpretation of a vast majority of the literature regarding the *in vitro* effects of cannabinoid drugs.

Cannabinoceptive neurons must be present in regulatory pathways in the brain for a variety of behavioral responses. The development of antagonists will be instrumental in the elucidation of which behaviors can be attributed to these receptors and how the neuronal pathways are regulated. Until a highly potent antagonist becomes available, neuroanatomists, physiologists, and pharmacologists must utilize the precedents established by the studies with potent cannabimimetic agonists to describe these responses.

REFERENCES

1. **Axelrod, J., Burch, R. M., and Jelsema, C. L.,** Receptor-mediated activation of phospholipase A_2 via GTP-binding proteins: arachidonic acid and its metabolites as second messengers, *TINS*, 11, 117, 1988.
2. **Birnbaumer, L.,** G proteins in signal transduction, *Annu. Rev. Pharmacol. Toxicol.*, 30, 675, 1990.
3. **Birnbaumer, L., Abramowitz, J., and Brown, A. M.,** Receptor-effector coupling by G proteins, *Biochim. Biophys. Acta*, 1031, 163, 1990.
4. **Brown, A. M.,** A cellular logic for G protein-coupled ion channel pathways, *FASEB J.*, 5, 2175, 1991.
5. **Cockcroft, S.,** Polyphosphoinositide phosphodiesterase: regulation by a novel guanine nucleotide binding protein G_p, *TIBS*, 12, 75, 1987.
6. **Freissmuth, M., Casey, P. J., and Gilman, A. G.,** G proteins control diverse pathways of transmembrane signaling, *FASEB J.*, 3, 2125, 1989.
7. **Hokin, L. E.,** Receptors and phosphoinositide-generated second messengers, *Annu. Rev. Biochem.*, 54, 205, 1985.
8. **Miller, R. J.,** Receptor-mediated regulation of calcium channels and neurotransmitter release, *FASEB J.*, 4, 3291, 1990.
9. **Neer, E. J. and Clapham, D. E.,** Roles of G protein subunits in transmembrane signalling, *Nature (London)*, 333, 129, 1988.
10. **Shukla, S. D. and Halenda, S. P.,** Phospholipase D in cell signalling and its relationship to phospholipase C, *Life Sci.*, 48, 851, 1991.
11. **Simon, M. I., Strathmann, M. P., and Gautam, N.,** Diversity of G proteins in signal transduction, *Science*, 252, 802, 1991.
12. **Rall, T. W., Sutherland, E. W., and Berthet, J.,** The relationship of epinephrine and glucagon to liver phosphorylase. IV. Effect of epinephrine and glucagon on the reactivation of phosphorylase in liver homogenates, *J. Biol. Chem.*, 224, 463, 1957.
13. **Shenolikar, S.,** Protein phosphorylation: hormones, drugs, and bioregulation, *FASEB J.*, 2, 2753, 1988.
14. **Cohen, P.,** The structure and regulation of protein phosphatases, *Annu. Rev. Biochem.*, 58, 453, 1989.
15. **Brunton, L. L., Maguire, M. E., Anderson, H. J., and Gilman, A. G.,** Expression of genes for metabolism of cyclic adenosine 3′:5′-monophosphate in somatic cells, *J. Biol. Chem.*, 252, 1293, 1977.

16. **Insel, P. A., Maguire, M. E., Gilman, A. G., Bourne, H. R., Coffino, P., and Melmon, K. L.,** Beta adrenergic receptors and adenylate cyclase: products of separate genes?, *Mol. Pharmacol.*, 12, 1062, 1976.
17. **Orly, J. and Schramm, M.,** Coupling of catecholamine receptor from one cell with adenylate cyclase from another cell by cell fusion, *Proc. Natl. Acad. Sci. U.S.A.*, 73, 4410, 1976.
18. **Haga, T., Haga, K., and Gilman, A. G.,** Hydrodynamic properties of the β-adrenergic receptor and adenylate cyclase from wild type and variant S49 lymphoma cells, *J. Biol. Chem.*, 252, 5776, 1977.
19. **Limbird, L. E. and Lefkowitz, R. J.,** Resolution of β-adrenergic receptor binding and adenylate cyclase activity by gel exclusion chromatography, *J. Biol. Chem.*, 252, 799, 1977.
20. **Klainer, L. M., Chi, Y.-M., Freidberg, S. L., Rall, T. W., and Sutherland, E. W.,** Adenyl cyclase. IV. The effects of neurohormones on the formation of adenosine 3′, 5′-phosphate by preparations from brain and other tissues, *J. Biol. Chem.*, 237, 1239, 1962.
21. **Birnbaumer, L. and Rodbell, M.,** Adenyl cyclase in fat cells. II. Hormone receptors, *J. Biol. Chem.*, 244, 3477, 1969.
22. **Rodbell, M., Krans, H. M. J., Pohl, S. L., and Birnbaumer, L.,** The glucagon-sensitive adenyl cyclase system in plasma membranes of rat liver. IV. Binding of glucagon: effect of guanyl nucleotides, *J. Biol. Chem.*, 246, 1872, 1971.
23. **Rodbell, M., Birnbaumer, L., Pohl, S. L., and Krans, H. M. J.,** The glucagon-sensitive adenyl cyclase system in plasma membranes of rat liver. V. An obligatory role of guanyl nucleotides in glucagon action, *J. Biol. Chem.*, 246, 1877, 1971.
24. **Cassel, D. and Selinger, Z.,** Catecholamine-stimulated GTPase activity in turkey erythrocyte membranes, *Biochim. Biophys. Acta*, 252, 538, 1976.
25. **Cassel, D. and Selinger, Z.,** Mechanism of adenylate cyclase activation by cholera toxin: inhibition of GTP hydrolysis at the regulatory site, *Proc. Natl. Acad. Sci. U.S.A.*, 74, 3307, 1977.
26. **Ross, E. M. and Gilman, A. G.,** Resolution of some components of adenylate cyclase necessary for catalytic activity, *J. Biol. Chem.*, 252, 6966, 1977.
27. **Ross, E. M., Howlett, A. C., Ferguson, K. M., and Gilman, A. G.,** Reconstitution of hormone-sensitive adenylate cyclase activity with resolved components of the enzyme, *J. Biol. Chem.*, 253, 6401, 1978.
28. **Howlett, A. C., Sternweis, P. C., Macik, B. A., Van Arsdale, P. M., and Gilman, A. G.,** Reconstitution of catecholamine-sensitive adenylate cyclase. Association of regulatory component of the enzyme with membranes containing the catalytic protein and β-adrenergic receptors, *J. Biol. Chem.*, 254, 2287, 1979.
29. **Pfeuffer, T.,** GTP-binding proteins in membranes and the control of adenylate cyclase activity, *J. Biol. Chem.*, 252, 7224, 1977.
30. **Howlett, A. C. and Gilman, A. G.,** Hydrodynamic properties of the regulatory component of adenylate cyclase, *J. Biol. Chem.*, 255, 2861, 1980.
31. **Northup, J. K., Sternweis, P. C., Smigel, M. D., Schleifer, L. S., Ross, E. M., and Gilman, A. G.,** Purification of the regulatory component of adenylate cyclase, *Proc. Natl. Acad. Sci. U.S.A.*, 77, 6516, 1980.
32. **Hildebrandt, J. D., Codina, J., Risinger, R., and Birnbaumer, L.,** Identification of a gamma subunit associated with the adenylyl cyclase regulatory proteins N_s and N_1, *J. Biol. Chem.*, 259, 2039, 1984.
33. **Kuhn, H.,** Light-induced, reversible binding of proteins to bovine photoreceptor membranes. Influence of nucleotides, *Neurochemistry*, 1, 269, 1980.

34. **Stryer, L.,** Visual excitation and recovery, *J. Biol. Chem.*, 266, 10711, 1991.
35. **Cooper, D. M. F.,** Bimodal regulation of adenylate cyclase, *FEBS Lett.*, 138, 157, 1982.
36. **Katada, T. and Ui, M.,** Direct modification of the membrane adenylate cyclase system by islet-activating protein due to ADP-ribosylation of a membrane protein, *Proc. Natl. Acad. Sci. U.S.A.*, 79, 3129, 1982.
37. **Howlett, A. C.,** Inhibition of neuroblastoma adenylate cyclase by cannabinoid and nantradol compounds, *Life Sci.*, 35, 1803, 1984.
38. **Howlett, A. C. and Fleming, R. M.,** Cannabinoid inhibition of adenylate cyclase. Pharmacology of the response in neuroblastoma cell membranes, *Mol. Pharmacol.*, 26, 532, 1984.
39. **Howlett, A. C.,** Cannabinoid inhibition of adenylate cyclase: biochemistry of the response in neuroblastoma cell membranes, *Mol. Pharmacol.*, 27, 429, 1985.
40. **Howlett, A. C., Qualy, J. M., and Khachatrian, L. L.,** Involvement of G_i in the inhibition of adenylate cyclase by cannabimimetic drugs, *Mol. Pharmacol.*, 29, 307, 1986.
41. **Howlett, A. C.,** Cannabinoid inhibition of adenylate cyclase: relative activities of marihuana constituents and metabolites, *Neuropharmacology*, 26, 507, 1987.
42. **Devane, W. A., Spain, J. W., Coscia, C. J., and Howlett, A. C.,** An assessment of the role of opioid receptors in the response to cannabimimetic drugs, *J. Neurochem.*, 46, 1929, 1986.
43. **Howlett, A. C., Johnson, M. R., Melvin, L. S., and Milne, G. M.,** Nonclassical cannabinoid analgetics inhibit adenylate cyclase: development of a cannabinoid receptor model, *Mol. Pharmacol.*, 33, 297, 1988.
44. **Devane, W. A., Dysarz, F. A., III, Johnson, M. R., Melvin, L. S., and Howlett, A. C.,** Determination and characterization of a cannabinoid receptor in rat brain, *Mol. Pharmacol.*, 34, 605, 1988.
45. **Razdan, R. K.,** Structure-activity relationships in cannabinoids, *Pharmacol. Rev.*, 38, 75, 1986.
46. **Martin, B. R., Balster, R. L., Razdan, R. K., Harris, L. S., and Dewey, W. L.,** Behavioral comparisons of the stereoisomers of tetrahydrocannabinols, *Life Sci.*, 29, 565, 1981.
47. **Little, P. J., Compton, D. R., Johnson, M. R., Melvin, L. S., and Martin, B. R.,** Pharmacology and stereoselectivity of structurally novel cannabinoids in mice, *J. Pharmacol. Exp. Ther.*, 247, 1046, 1988.
48. **Little, P. J., Compton, D. R., Mechoulam, R., and Martin, B. R.,** Stereochemical effects of 11-OH-Δ^8-THC-dimethylheptyl in mice and dogs, *Pharmacol. Biochem. Behavior,* 22, 661, 1989.
49. **Howlett, A. C., Champion, T. M., Wilken, G. H., and Mechoulam, R.,** Stereochemical effects of 11-OH-Δ^8-tetrahydrocannabinol-dimethylheptyl to inhibit adenylate cyclase and bind to the cannabinoid receptor, *Neuropharmacology*, 29, 161, 1989.
50. **Johnson, M. R. and Melvin, L. S.,** The discovery of nonclassical cannabinoid analgetics, in *Cannabinoids as Therapeutic Agents*, Mechoulam, R., Ed., CRC Press, Boca Raton, FL, 1986, 121.
51. **Melvin, L. S. and Johnson, M. R.,** Structure-activity relationships of tricyclic and nonclassical bicyclic cannabinoids, in *Structure-Activity Relationships of the Cannabinoids*, in NIDA Research Monograph 79, Rapaka, R. S. and Makriyannis, A., Eds., NIDA, Bethesda, MD, 1987, 31.

52. **Howlett, A. C.,** Cannabinoid receptor ligands, in *Trends in Medicinal Chemistry '90*, Agranat, I., Sarel, S., and Mechoulam, R., Eds., Blackwell Scientific Publications, in press, 1991.
53. **Pacheco, M., Childers, S. R., Arnold, R., Casiano, F., and Ward, S. J.,** Aminoalkylindoles: actions on specific G-protein-linked receptors, *J. Pharmacol. Exp. Ther.*, 257, 170, 1991.
54. **Casiano, F. M., Haycock, A. D., Kuster, J., and Ward, S. J.,** Putative aminoalkylindoles (AAI) antagonists, *NIDA Res Monogr.*, 105, 295, 1990.
55. **Lawrence, D. K. and Gill, E. W.,** The effects of Δ^1-tetrahydrocannabinol and other cannabinoids on spin-labeled liposomes and their relationship to mechanisms of general anesthesia, *Mol. Pharmacol.*, 11, 595, 1975.
56. **Bruggemann, E. P. and Melchior, D. L.,** Alterations in the organization of phosphatidylcholine/cholesterol bilayers by tetrahydrocannabinol, *J. Biol. Chem.*, 258, 8298, 1983.
57. **Tamir, I. and Lichtenberg, D.,** Correlation between the psychotrophic potency of cannabinoids and their effect on the ^{1}H-NMR spectra of model membranes, *J. Pharm. Sci.*, 72, 458, 1983.
58. **Hillard, C. J., Harris, R. A., and Bloom, A. S.,** Effects of the cannabinoids on physical properties of brain membranes and phospholipid vesicles: fluorescence studies, *J. Pharmacol. Exp. Ther.*, 232, 579, 1985.
59. **Makriyannis, A., Banijamali, A., Jarrell, H. C., and Yang, D. P.,** The orientation of (–)-delta 9-tetrahydrocannabinol in DPPC bilayers as determined from solid-state ^{2}H-NMR, *Biochim. Biophys. Acta*, 986, 141, 1989.
60. **Makriyannis, A., Yang, D. P., Griffin, R. G., and Das Gupta, S.,** The perturbation of model membranes by (–)-delta-9-tetrahydrocannabinol. Studies using solid state ^{2}H and ^{13}C-NMR, *Biochim. Biophys. Acta*, 1028, 31, 1990.
61. **Howlett, A. C., Scott, D. K., and Wilken, G. H.,** Regulation of adenylate cyclase by cannabinoid drugs: insights based on thermodynamic studies, *Biochem. Pharmacol.*, 38, 3297, 1989.
62. **Hillard, C. J., Pounds, J. J., Boyer, D. R., and Bloom, A. S.,** Studies of the role of membrane lipid order in the effects of delta 9-tetrahydrocannabinol on adenylate cyclase activation in the heart, *J. Pharmacol. Exp. Ther.*, 252, 1075, 1990.
63. **Wing, D. R., Leuscher, J. T. A., Brent, G. A., Harvey, D. J., and Paton, W. D. M.,** Quantification of *in vivo* membrane-associated delta-1-tetrahydrocannabinol and its effect on membrane fluidity, in *Marihuana '84*, Harvey, D. J., Ed., IRL Press, Oxford, 1985, 85.
64. **Thomas, B. F., Compton, D. R., and Martin, B. R.,** Characterization of the lipophilicity of natural and synthetic analogs of Δ^9-tetrahydrocannabinol and its relationship to pharmacological potency, *J. Pharmacol. Exp. Ther.*, 255, 624, 1990.
65. **Bidaut-Russell, M., Devane, W. A., and Howlett, A. C.,** Cannabinoid receptors and the modulation of cyclic AMP accumulation in the rat brain, *J. Neurochem.*, 55, 21, 1990.
66. **Herkenham, M., Lynn, A. B., Little, M. D., Johnson, M. R., Melvin, L. S., De Costa, B. R., and Rice, K. C.,** Cannabinoid receptor localization in brain, *Proc. Natl. Acad. Sci. U.S.A.*, 87, 1932, 1990.
67. **Herkenham, M., Lynn, A. B., Johnson, M. R., Melvin, L. S., De Costa, B. R., and Rice, K. C.,** Characterization and localization of cannabinoid receptors in rat brain: a quantitative *in vitro* autoradiographic study, *J. Neurosci.*, 11, 563, 1991.

68. **Hollister, L. E. and Gillespie, H. K.,** Delta-8- and delta-9-tetrahydrocannabinol comparison in man by oral and intravenous administration, *Clin. Pharmacol. Ther.*, 14, 353, 1973.
69. **Ferraro, D. P.,** Acute effects of marijuana on human memory and cognition, in *Marijuana Research Findings: 1980*, Petersen, R. C., Ed., NIDA Research Monograph Series, Bethesda, MD, 1980, 98.
70. **Miller, L. L.,** Marijuana: acute effects on human memory, in *The Cannabinoids: Chemical, Pharmacologic, and Therapeutic Aspects*, Agurell, S., Dewey, W. L., and Willette, R. E., Eds., Academic Press, New York, 1984, 21.
71. **Campbell, K. A., Foster, T. C., Hampson, R. E., and Deadwyler, S. A.,** Δ^9-tetrahydrocannabinol differentially affects sensory-evoked potentials in the rat dentate gyrus, *J. Pharmacol. Exp. Ther.*, 239, 936, 1986.
72. **Campbell, K. A., Foster, T. C., Hampson, R. E., and Deadwyler, S. A.,** Effects of Δ^9-tetrahydrocannabinol on sensory-evoked discharges of granule cells in the dentate gyrus of behaving rats, *J. Pharmacol. Exp. Ther.*, 239, 941, 1986.
73. **Braude, M. C. and Ludford, J. P., Eds.,** *Marijuana Effects on the Endocrine and Reproductive Systems*, NIDA Research Monograph, 1984, 44.
74. **Schmeling, W. T. and Hosko, M. J.,** Hypothermic effects of intraventricular and intravenous administration of cannabinoids in intact and brainstem transected cats, *Neuropharmacology*, 19, 567, 1980.
75. **Fitton, A. G. and Pertwee, R. G.,** Changes in body temperature and oxygen consumption rate of conscious mice produced by intrahypothalamic and intracerebroventricular injections of Δ^9-tetrahydrocannabinol, *Br. J. Pharmacol.*, 75, 409, 1982.
76. **Pertwee, R. G.,** Effects of cannabinoids on thermoregulation: a brief review, in *Marihuana '84*, Harvey, D. J., Ed., IRL Press, Oxford, 1985, 263.
77. **Jones, R. T.,** Cardiovascular effects of cannabinoids, in *Marihuana '84*, Harvey, D. J., Ed., IRL Press, Oxford, 1985, 325.
78. **Hosko, M. J., Schmeling, W. T., and Hardman, H. F.,** Delta-9-tetrahydrocannabinol: site of action for autonomic effects, in *The Cannabinoids: Chemical, Pharmacologic and Therapeutic Aspects*, Agurell, S., Dewey, W. L., and Willette, R. E., Eds., Academic Press, New York, 1984, 635.
79. **Vollmer, R. R., Cavero, I., Ertel, R. J., Solomon, T. A., and Buckley, J. P.,** Role of the central autonomic nervous system in the hypotension and bradycardia induced by (–)-Δ^9-*trans*-tetrahydrocannabinol, *J. Pharm. Pharmacol.*, 26, 186, 1974.
80. **Borison, H. L., Borison, R., and McCarthy, L. E.,** Phylogenic and neurologic aspects of the vomiting process, *J. Clin. Pharmacol.*, 21, 23S, 1981.
81. **Vincent, B. J., McQuiston, D. J., Einhorn, L. H., Nagy, C. M., and Brames, M. J.,** Review of cannabinoids and their antiemetic effectiveness, *Drugs*, 25, 52, 1983.
82. **Segal, M.,** Cannabinoids and analgesia, in *Cannabinoids as Therapeutic Agents*, Mechoulam, R., Ed., CRC Press, Boca Raton, FL, 1986, 105.
83. **Yaksh, T. L.,** The antinociceptive effects of intrathecally administered levonantradol and desacetyllevonantradol in the rat, *J. Clin. Pharmacol.*, 21, 334S, 1981.
84. **McIlhenny, H. M., Mast, R. W., Johnson, M. R., and Milne, G. M.,** Nantradol hydrochloride: pharmacokinetics and behavioral effects after acute and chronic treatment, *J. Pharmacol. Exp. Ther.*, 219, 363, 1981.
85. **Merritt, J. C.,** Outpatient cannabinoid therapy for heterogenous glaucomas: guidelines for institution and maintenance of therapy, in *Marihuana '84*, Harvey, D. J., Ed., IRL Press, Oxford, 1985, 681.

86. **Herkenham, M., Lynn, A. B., DeCosta, B. R., and Richfield, E. K.,** Neuronal localization of cannabinoid receptors in the basal ganglia of the rat, *Brain Res.*, 1991, 547, 267, 1991.
87. **Gough, A. L. and Olley, J. E.,** Catalepsy induced by intrastriatal injections of Δ^9-THC and 11-OH-Δ^9-THC in the rat, *Neuropharmacology*, 17, 137, 1978.
88. **Montgomery, S. P., Moss, D. E., and Manderscheid, P. Z.,** Tetrahydrocannabinol and levonantradol effects on extrapyramidal motor behaviors: neuroanatomical location and hypothesis of mechanism, in *Marihuana '84*, Harvey, D. J., Ed., IRL Press, Oxford, 1985, 295.
89. **Conti, L. H. and Musty, R. E.,** The effects of delta-9-tetrahydrocannabinol injections to the nucleus accumbens on the locomotor activity of rats, in *The Cannabinoids: Chemical, Pharmacologic, and Therapeutic Aspects*, Agurell, S., Dewey, W. L., and Willette, R. E., Eds., Academic Press, New York, 1984, 649.
90. **Howlett, A. C., Bidaut-Russell, M., Devane, W. A., Melvin, L. S., Johnson, M. R., and Herkenham, M.,** The cannabinoid receptor: biochemical, anatomical and behavioral characterization, *TINS*, 13, 420, 1990.
91. **Bidaut-Russell, M. and Howlett, A. C.,** Cannabinoid receptor-regulated cAMP accumulation in the rat striatum, *J. Neurochem.*, 57, 1769, 1991.
92. **Hillard, C. J. and Bloom, A. S.,** Δ^9-Tetrahydrocannabinol-induced changes in β-adrenergic receptor binding in mouse cerebral cortex, *Brain Res.*, 235, 370, 1982.
93. **Hillard, C. J. and Bloom, A. S.,** Further studies of the interaction of delta-9-tetrahydrocannabinol with the beta-adrenergic receptor, in *The Cannabinoids: Chemical, Pharmacological and Therapeutics Aspects*, Agurell, S., Dewey, W. L., and Willette, R. E., Eds., Academic Press, Orlando, FL, 1984, 591.
94. **Bloom, A. S. and Hillard, C. J.,** Cannabinoids, neurotransmitter receptors, and brain membranes, in *Marihuana '84*, Harvey, D. J., Ed., IRL Press, Oxford, 1985, 217.
95. **Bloom, A. S.,** Effects of cannabinoids on neurotransmitter receptors in the brain, in *The Cannabinoids: Chemical, Pharmacological and Therapeutic Aspects,* Agurell, S., Dewey, W. L., and Willette, R. E., Eds., Academic Press, Orlando, FL, 1984, 585.
96. **Vaysse, P. J.-J., Gardner, E. L., and Zukin, R. S.,** Modulation of rat brain opioid receptors by cannabinoids, *J. Pharmacol. Exp. Ther.*, 241, 534, 1987.
97. **Tulunay, F. C., Ayhan, I. H., Portoghese, P. S., and Takemori, A. E.,** Antagonism by chlornaltrexamine of some effects of Δ^9-tetrahydrocannabinol in rats, *Eur. J. Pharmacol.*, 70, 219, 1981.
98. **Tulunay, F. C., Ayhan, I. H., and Sparber, S. B.,** The effects of morphine and Δ^9-tetrahydrocannabinol on motor activity in rats, *Psychopharmacology*, 78, 358, 1982.
99. **Kaymakcalan, S., Ayhan, I. H., and Tulunay, F. C.,** Naloxone-induced or postwithdrawal abstinence signs in Δ^9-tetrahydrocannabinol-tolerant rats, *Psychopharmacology*, 55, 243, 1977.
100. **Frederickson, R. C. A., Hewes, C. R., and Aiken, J. W.,** Correlation between the *in vivo* and an *in vitro* expression of opiate withdrawal precipitated by naloxone: their antagonism by 1-(–)-Δ^9-tetrahydrocannabinol, *J. Pharmacol. Exp. Ther.*, 199, 375, 1976.
101. **Hine, B., Friedman, E., Torrelio, M., and Gershon, S.,** Morphine-dependent rats: blockade of precipitated abstinence by tetrahydrocannabinol, *Science*, 187, 443, 1975.
102. **Bhargava, H. N.,** Inhibition of naloxone-induced withdrawal in morphine dependent mice by 1-trans-Δ^9-tetrahydrocannabinol, *Eur. J. Pharmacol.*, 36, 259, 1976.

103. **Milne, G. M., McIlhenny, H. M., Johnson, M. R., Koe, B. K., and Weissman, A.,** One way cross tolerance from morphine to nantradol — a potent, non-opioid analgetic, in *Endogenous and Exogenous Opiate Agonists and Antagonists*, Leong Way, E., Ed., Pergamon Press, New York, 1980, 367.
104. **Moss, D. E., Manderscheid, P. Z., Kobayashi, H., and Montgomery, S. P.,** Evidence for the nicotinic cholinergic hypothesis of cannabinoid action within the central nervous system: extrapyramidal motor behaviors, in *Marijuana '87*, Chesher, G., Consroe, P., and Musty, R., Eds., Australian Government Publication Service, Canberra, 1987, 359.
105. **Horne, W. C.,** Desacetyl-1-nantradol: a selective prostaglandin antagonist, *Prostaglandins Leukotrienes Med.*, 15, 129, 1984.
106. **Okey, A. B. and Bondy, G. P.,** Is delta-9-tetrahydrocannabinol estrogenic?, *Science*, 195, 904, 1977.
107. **Rawitch, A. B., Schultz, G. S., Ebner, K. E., and Vardaris, R. M.,** Competition of Δ^9-tetrahydrocannabinol with estrogen in rat uterine estrogen receptor binding, *Science*, 197, 1189, 1977.
108. **Sauer, M. A., Rifka, S. M., Hawks, R. L., Cutler, G. B., and Loriaux, D. L.,** Marijuana: interaction with the estrogen receptor, *J. Pharmacol. Exp. Ther.*, 224, 404, 1983.
109. **Eldridge, J. C. and Landfield, P. W.,** Cannabinoid interactions with glucocorticoid receptors in rat hippocampus, *Brain Res.*, 534, 135, 1990.
110. **Haubrich, D. R., Ward, S. J., Baizman, E., Bell, M. R., Bradford, J., Ferrari, R., Miller, M., Perrone, M., Pierson, A. K., Saelens, J. K., and Luttinger, D.,** Pharmacology of pravadoline: a new analgesic agent, *J. Pharmacol. Exp. Ther.*, 255, 511, 1990.
111. **Ward, S. J., Mastriani, D., Casiano, F., and Arnold, R.,** Pravadoline: profile in isolated tissue preparations, *J. Pharmacol. Exp. Ther.*, 255, 1230, 1990.
112. **Ward, S. J., Baizman, E., Bell, M., Childers, S., D'Ambra, T., Eissenstat, M., Estep, K., Haycock, D., Howlett, A., Luttinger, D., and Miller, M.,** *Aminoalkylindoles (AAIs)—A New Route to the Cannabinoid Receptor?*, NIDA Research Monograph, 105, 425, 1991.
113. **Nye, J. S., Voglmaier, S., Martenson, R. E., and Snyder, S. H.,** Myelin basic protein is an endogenous inhibitor of the high-affinity cannabinoid binding site in brain, *J. Neurochem.*, 50, 1170, 1988.
114. **Evans, D. M., Johnson, M. R., and Howlett, A. C.,** Ca^{2+} -dependent release from rat brain of cannabinoid receptor binding activity, *J. Neurochem.*, 58, 780, 1992.
115. **Engelman, D. M., Goldman, A., and Steitz, T. A.,** Identifying nonpolar transbilayer helices in amino acid sequences of membrane proteins, *Annu. Rev. Biophys. Biophys. Chem.*, 15, 321, 1982.
116. **Dixon, R. A. F., Kobilka, B. K., Strader, D. J., Benovic, J. L., Dohlman, H. G., Frielle, T., Bolanowski, M. A., Bennett, C. D., Rands, E., Diehl, R. E., Mumford, R. A., Slater, E. E., Sigal, I. S., Caron, M. G., Lefkowitz, R. J., and Strader, C. D.,** Cloning of the gene and cDNA for mammalian β-adrenergic receptor and homology with rhodopsin, *Nature (London)*, 321, 75, 1986.
117. **Weiss, E. R., Hadcock, J. R., Johnson, G. L., and Malbon, C. C.,** Antipeptide antibodies directed against cytoplasmic rhodopsin sequences recognize the β-adrenergic receptor, *J. Biol. Chem.*, 262, 4319, 1987.
118. **Matsuda, L. A., Lolait, S. J., Brownstein, M. J., Young, A. C., and Bonner, T. I.,** Structure of a cannabinoid receptor and functional expression of the cloned cDNA, *Nature (London)*, 346, 561, 1990.

119. **Gerard, C., Mollereau, C., Vassart, G., and Parmentier, M.,** Nucleotide sequence of a human cannabinoid receptor cDNA, *Nucleic Acids Res.*, 18, 7142.
120. **Boege, F., Ward, M., Jurss, R., Hekman, M., and Helmreich, E. J. M.,** Role of glycosylation for β_2-adrenoceptor function in A431 cells, *J. Biol. Chem.*, 263, 9040, 1988.
121. **O'Dowd, B. F., Hnatowich, M., Caron, M. G., Lefkowitz, R. J., and Bouvier, M.,** Palmitoylation of the human β_2-adrenergic receptor, *J. Biol. Chem.*, 264, 7564, 1989.
122. **Campbell, P. T., Hnatowich, M., O'Dowd, B. F., Caron, M. G., and Lefkowitz, R. J.,** Mutation of the human β_2-adrenergic receptor that impairs coupling to G_s interfere with receptor down-regulation but not sequestration, *Mol. Pharmacol.*, 39, 192, 1991.
123. **Kornfeld, R. and Kornfeld, S.,** Assembly of asparagine-linked oligosaccharides, *Annu. Rev. Biochem.*, 54, 631, 1985.
124. **Elbein, A. D.,** Inhibitors of the biosynthesis and processing of N-linked oligosaccharide chains, *Annu. Rev. Biochem.*, 56, 497, 1987.
125. **Liles, W. C. and Nathanson, N. M.,** Regulation of neuronal muscarinic acetylcholine receptor number by protein glycosylation, *J. Neurochem.*, 46, 89, 1986.
126. **George, S. T., Ruoho, A. E., and Malbon, C. C.,** *N*-glycosylation in expression and function of β-adrenergic receptors, *J. Biol. Chem.*, 261, 16559, 1986.
127. **Benovic, J. L., Staniszewski, C., Cerione, R. A., Codina, J., Lefkowitz, R. J., and Caron, M. G.,** The mammalian beta-adrenergic receptor: structural and functional characterization of the carbohydrate moiety, *J. Rec. Res.*, 7, 257, 1987.
128. **Cervantes-Olivier, P., Delavier-Klutchko, C., Durieu-Trautman, O., Kavieri, S. V., Desmandril, M., and Strosberg, A. D.,** The β_2-adrenergic receptors of human epidermoid carcinoma cells bear two different types of oligosaccharides which influence expression on the cell surface, *Biochem. J.*, 250, 133, 1988.
129. **Howlett, A. C., Champion-Dorow, T. M., McMahon, L. L., and Westlake, T. M.,** The cannabinoid receptor: biochemical and cellular properties in neuroblastoma cells, *Pharmacol. Biochem. Behavior*, 40, 565, 1991.
130. **Benovic, J. L., Bouvier, M., Caron, M. G., and Lefkowitz, R. J.,** Regulation of adenylyl cyclase-coupled β-adrenergic receptors, *Annu. Rev. Cell Biol.*, 4, 405, 1988.
131. **Bouvier, M., Hausdorff, W. P., DeBlasi, A., O'Dowd, B. F., Kobilka, B. K., Caron, M. G., and Lefkowitz, R. J.,** Removal of phosphorylation sites from the β_2-adrenergic receptor delays onset of agonist-promoted desensitization, *Nature (London)*, 333, 370, 1988.
132. **Hausdorff, W. P., Bouvier, M., O'Dowd, B. F., Irons, G. P., Caron, M. G., and Lefkowitz, R. J.,** Phosphorylation sites on two domains of the β_2-adrenergic receptor are involved in distinct pathways of receptor desensitization, *J. Biol. Chem.*, 264, 12657, 1989.
133. **Clark, R. B., Kunkel, M. W., Friedman, J., Goka, T. J., and Johnson, J. A.,** Activation of cAMP-dependent protein kinase is required for heterologous desensitization of adenylyl cyclase in S49 wild-type lymphoma cells, *Proc. Natl. Acad. Sci. USA*, 85, 1442, 1988.
134. **Lohse, M. J., Benovic, J. L., Caron, M. G., and Lefkowitz, R. J.,** Multiple pathways of rapid β_2-adrenergic receptor desensitization, *J. Biol. Chem.*, 265, 3202, 1990.
135. **Dill, J. A. and Howlett, A. C.,** Regulation of adenylate cyclase by chronic exposure to cannabimimetic drugs, *J. Pharmacol. Exp. Ther.*, 244, 1157, 1988.
136. **Benovic, J. L., Strasser, R. H., Caron, M. G., and Lefkowitz, R. J.,** β-adrenergic receptor kinase: identification of a novel protein kinase that phosphorylates the agonist occupied form of the receptor, *Proc. Natl. Acad. Sci. U.S.A.*, 83, 2797, 1986.

137. **O'Brien, P. J. and Zatz, M.,** Acylation of bovine rhodopsin by [^{3}H]palmitic acid, *J. Biol. Chem.*, 259, 5054, 1984.
138. **Ovchinnikov, Y. A., Abdulaev, N. G., and Bogachuk, A. S.,** Two adjacent cysteine residues in the C-terminal cytoplasmic fragment of bovine rhodopsin are palmitylated, *FEBS Lett.*, 230, 1, 1988.
139. **Karnik, S. S., Sakmar, T. P., Chen, H.-B., and Khorana, H. G.,** Cysteine residues 110 and 187 are essential for the formation of correct structure in bovine rhodopsin, *Proc. Natl. Acad. Sci. U.S.A.*, 85, 8459, 1988.
140. **Houston, D. B. and Howlett, A. C.,** High affinity ligand binding by the detergent-solubilized cannabinoid receptor, *Proc. Soc. Neurosci.*, 17, 605, 1991.
141. **Dewey, W. L.,** Cannabinoid pharmacology, *Pharmacol. Rev.*, 38, 151, 1986.
142. **Lemberger, L.,** Potential therapeutic usefulness of marijuana, *Annu. Rev. Pharmacol. Toxicol.*, 20, 151, 1980.
143. **Mechoulam, R.,** The pharmacohistory of cannabis sativa, in *Cannabinoids as Therapeutic Agents*, Mechoulam, R., Ed., CRC Press, Boca Raton, FL, 1986, 1.
144. **Hollister, L. E.,** Health aspects of cannabis, *Pharmacol. Rev.*, 38, 1, 1986.
145. **Martin, B. R.,** Cellular effects of cannabinoids, *Pharmacol. Rev.*, 38, 45, 1986.
146. **Pertwee, R. G.,** The central neuropharmacology of psychotropic cannabinoids, *Pharmacol. Ther.*, 36, 189, 1988.

3 Eicosanoids as Mediators of Cannabinoid Action

INTRODUCTION

Recent reports on the existence of putative cannabinoid receptors have significantly advanced our understanding of signal transduction by these drugs at the cellular level.[1-4] The question of the possible involvement of second messengers in the actions of cannabinoids, therefore, gains considerable importance. Prime candidates for this role are the various members of the eicosanoid family as well as the classical second messenger, cyclic AMP. The purpose of this review is to discuss the evidence supporting the hypothesis that the metabolites of arachidonic acid are important mediators in many of the actions of both psychoactive and nonpsychoactive cannabinoids.

The early studies suggesting a role for prostaglandins in the actions of cannabinoids were prompted by a consideration of some of their pharmacological effects.[5] Antiinflammatory, analgesic, and lowering of body temperature effects were reminiscent of other drugs such as the nonsteroidal antiinflammatory drugs (NSAIDs) which, at that time, had just been demonstrated to act by inhibiting prostaglandin synthesis. Moreover, it was already known by 1970 that arachidonic acid and its metabolites were abundant constituents of the CNS and therefore might be involved in regulating mood and perception, which are part of the spectrum of activities affected by cannabinoids. Much of the earlier work supporting prostaglandin involvement in cannabinoid action has been mentioned in other reviews;[6-8] this report, then, represents an update and expansion of the earlier discussions.

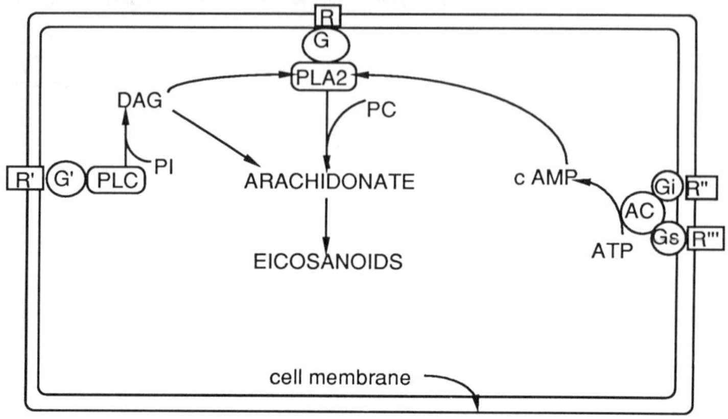

FIGURE 1. Cellular mechanisms believed to be involved in the release of arachidonic acid for eicosanoid synthesis. PLA_2, phospholipase A_2; PLC, phospholipase C; DAG, diacylglyceride; PI, phosphatidylinositol; PC, phosphatidylcholine; AC, adenylate cyclase; G, GTP-binding protein; R, receptor.

EICOSANOIDS IN THE CNS

Biosynthesis

The general pathways for the production of eicosanoids in the CNS appear to be the same as those at peripheral sites. These have been reviewed in great detail by a number of authors;[9] therefore, the presentation of this subject will be brief and is only meant to serve as a background for the subsequent discussion on the cannabinoids.

The polyunsaturated fatty acid precursors for the eicosanoids, such as arachidonic acid, are found in cells almost entirely in esterified form. These generally occur at position 2 of phospholipids such as phosphatidyl choline, phosphatidyl ethanolamine, or phosphatidyl inositol. Before biosynthesis leading to eicosanoids can take place, hydrolysis must be induced to release the free fatty acid. Phospholipase A_2 (PLA_2) or phospholipase C (PLC) mediates this process and seems to be the rate-limiting step for the entire cascade leading to the biologically active eicosanoids (Figure 1). Free arachidonate may also arise from diacylglycerides by the action of the appropriate lipase.

A number of stimuli can promote the release of arachidonic acid for eicosanoid synthesis. These each have specific cellular binding sites that are thought to be coupled to GTP-binding proteins that become activated when the receptor-ligand complex is formed. The activated G protein then stimulates PLA_2 (or PLC) mediated hydrolysis by a process that is not understood at this time. Intracellular calcium ion is required for full enzymic activity. A phospholipase activating protein (PLAP) specific for PLA_2 has recently been identified; however, its role is unclear at the moment. There is also evidence that cAMP and diacylglyceride levels may modulate PLA_2 activity.

The released free arachidonic acid can be utilized through several pathways leading to either prostaglandins or leukotrienes. The extent of conversion by a particular pathway in a specific cell type is probably determined by the level of expression of the relevant enzyme(s). Prostaglandins (PGs) are produced by the action of cyclooxygenase (CO) which leads to the unstable endoperoxides, PGG_2, and PGH_2 which are the common intermediates for all of the subsequently formed PGs. Specific enzymes then lead to the various PGs listed in Table 1. While CO does not seem to be subject to acute physiological control, it can be inhibited by a number of therapeutically useful drugs such as aspirin and other NSAIDs.

There are three known lipoxygenases that will oxidize arachidonate at either the 5, 12, or 15 positions. The most important is the 5-lipoxygenase system which initiates the formation of several potent mediators of biological activity. Among these are the leukotrienes (LTs) as well as 5-HETE (hydroxyeicosatetraenoic acid). A 5-lipoxygenase activating protein (FLAP) has recently been isolated and it has been suggested that it is involved in regulating cellular 5-lipoxygenase activity. Less is known about the 12-lipoxygenase pathway which is abundant in neural tissue.

Biological Effects

This aspect of the eicosanoids presents an exceedingly complex and often confusing picture.[9] Biphasic dose-response relationships are common and have resulted in seemingly conflicting reports in the literature. In addition, the same eicosanoid may produce different effects when studied in different species or even tissue sites within a species. The identification and characterization of receptors for each of the eicosanoids is still in its early stages so that the diversity of responses cannot be easily explained at the molecular level. Table 1 contains a partial listing of the more commonly reported biological activities for these substances.

The effects of the eicosanoids in the CNS may be the least-well understood of all. This is surprising since, as early as 1963, Ramwell and Shaw had reported evidence for the involvement of PGs in brain physiology.[10] On

TABLE 1
Major Eicosanoids and Some of Their Biological Activities

Eicosanoid	Activities
	Cyclooxygenase products
PGD_2 (primarily in rodents)	↑ cAMP, catalepsy, antinociception, sedation
PGE_2	↓↑ cAMP, sedation, ↓ locomotor activity, hyperthermia, smooth muscle relaxant, ↓ gastric acid, vasodilation
$PGF_{2\alpha}$	Uterine luteolysis, hyperthermia, smooth muscle contractant
PGI_2 (prostacyclin)	Vasodilation, ↑ cAMP, ↓ platelet aggregation, hyperthermia
TxA_2 (thromboxane)	Vasoconstrictor, ↑ platelet aggregation
	5, 12, or 15 Lipoxygenase products
5(S)-HETE	Chemotaxis, ↑ hexose uptake, ↑ Ca^{2+} flux, ↑ degranulation
LTB_4	Chemotaxis, hyperalgesia, vascular permeability, bronchoconstriction, ↑ hexose transport, ↑ Ca^{2+} flux, leukocyte adhesion
LTC_4, LTD_4, LTE_4	Smooth muscle contractant, affects vascular tone and permeability
12(*R*)-HETE	↑ Chemokinesis, ↑ leukocyte aggregation, ↓ Na^+, K^+ ATPase
12(*S*)-HETE	Chemotaxis, ↓ neuroblastoma growth
15(*S*)-HETE	Immunosuppressive, ↑ neovascularization, modulates 5- and 12-lipoxygenases

the other hand, the experimental difficulties encountered in working with brain tissue may account for the slow progress to date when compared, for example, to our understanding of the role of eicosanoids in the vascular system. The high lipid content in brain tissue, as well as extensive postmortem increases in brain eicosanoids, cause significant problems in making accurate measurements of these mediators that are present at nanomolar levels.

PGs are believed to be involved in temperature regulation, brain circulation, and neuromodulation. Injection of PGE_2 into the anterior hypothalamus raises body temperature, and elevated levels of PGE_2 are seen during pyrogen-induced fever. Consistent with this are studies that show that inhibition of CO reduces PG levels and fever. Prostaglandins E produce a dose-dependent inhibition of chemically-induced convulsion in mice and cats. The sedative and cataleptic effects of the E series have been well documented. As might be expected, the vasodilating prostaglandins can have effects on cerebral blood flow.

Neurotransmitter functions for eicosanoids have not been well established; however, they seem to interact extensively with many of the known transmitter systems. For example, $PGF_{2\alpha}$ inhibits noradrenaline release in cortex, as well in the peripheral nervous system. The neuropeptide, substance P, has its release from cultured neurons facilitated by both PGE_2 and PGD_2. There is indirect evidence that the release of vasoactive intestinal peptide (VIP) is mediated by a lipoxygenase product. Other examples of eicosanoid-neurotransmitter interactions can be found in the literature.

The release of the anterior pituitary hormones (TSH, ACTH, GH, FSH, LH, and prolactin) seems to be regulated to some degree by eicosanoids. The extreme potency of their actions is exemplified by a report that LTC_4 at 10^{-11} to 10^{-14} M causes secretion of LH from the anterior pituitary.

The above-mentioned effects are not a comprehensive list and are only meant to give a feeling for the types of activities in which eicosanoids may be mediators. It is anticipated that new activities for eicosanoids, especially lipoxygenase products, in the brain will be discovered in the future. Hopefully, this will aid in understanding the mechanisms of action of CNS-active drugs such as the cannabinoids.

HUMAN STUDIES

A great deal of information on cannabinoid experiments in humans has been published;[11] however, very little of this body of data increases our understanding of the mechanism of action of these drugs in man. Only two papers have appeared that are directed toward the question of eicosanoid mediation of cannabinoid effects in humans.

Perez-Reyes et al.[12] have recently reported that indomethacin pretreatment abolished the effect of THC on time estimation and productivity. There was also a reduction of the subjective "high" and heart rate acceleration; however, there was no change in the THC-induced effects on word recall. Blood levels of PGE_2 were measured during the experiment and are in good agreement with the indomethacin effects. Within 5 min after the beginning of marihuana smoking, PGE_2 levels rose 250% and remained elevated for at least 60 min. Pretreatment with indomethacin completely eliminated this increase in PGE_2. Conceivably, indomethacin could inhibit the THC responses by a mechanism other than inhibition of eicosanoid synthesis, such as altering the pharmacokinetics of THC. Measurements of THC blood levels showed that this was probably not the case, thus eliminating this alternative explanation. While the results seem clear, the small number of subjects (n = 4) and the low dose of indomethacin make it essential that a more elaborate study be carried out to confirm and extend these findings. Nevertheless, when viewed together with analogous experiments in mice (see below), it seems likely that eicosanoids may play an important role in mediating THC responses in humans.

Apparently contradictory results were reported by Laviolette and Belanger in their study on the mechanism of marijuana-induced bronchodilation.[13] They found that pretreatment of healthy subjects with either aspirin or mefenamic acid (Ponstan) had no significant effect on bronchodilation. The choice of inhibitors may be a major factor in reducing THC effects in humans. For example, in the above experiments by Perez-Reyes et al.,[12] it was observed that mefenamic acid was ineffective, whereas indomethacin produced complete inhibition (M. Perez-Reyes, personal communication). Various CO inhibitors probably have very different tissue distribution profiles, or other specificities, that could also explain the lack of inhibition observed by Laviolette and Belanger.[13] It would be interesting to see whether THC-induced bronchodilation would be inhibited using the conditions described for the inhibition of the CNS effects.

IN VIVO ANIMAL STUDIES

Effects on Brain Levels of Prostaglandins

Tissue levels of eicosanoids are normally very low and are only increased following stimulation by hormones, drugs, or other agents. Bhattacharya reported dose related rises of PGE_2 and $PGF_{2\alpha}$ in rat brain as measured by radioimmunoassay (RIA).[14] A temporal relationship was also demonstrated with peak increases of PGE_2 (215%) and $PGF_{2\alpha}$ (172%) at 2 h following the intraperitoneal injection of THC (2 mg/kg). Similar dose-response and time course data were reported for the THC-induced potentiation of hexabarbitone sleeping time, prompting the conclusion that this effect may involve eicosanoid mediation. It is noteworthy that such accurate data were obtained by RIA of brain tissue considering the methodological difficulties mentioned above.

Using a radiolabeled precursor approach, Hunter et al.[15] found that the oral administration of THC (50 mg/kg) to mice resulted in a 29% increase in brain radioactivity associated with PGs. There was a concomitant 17% decrease in phospholipids, suggesting the occurrence of THC activation of PLA_2 in mouse brain *in vivo*.

In an attempt to develop a more precise and specific assay method for measuring brain PGE_2, Hunter et al.[16] utilized the endogenous brain receptor for this eicosanoid. Rats or mice were given THC orally and, after 20 min, the animals were killed and a receptor-rich membrane fraction prepared. The extent of binding of $[^3H]$-PGE_2 was then measured and compared with binding by tissue for vehicle-treated animals. A significant decrease in binding of labeled ligand was seen following drug treatment; this was interpreted as being due to a rise in endogenous PGE_2 levels, leading to partial occupancy of the receptor sites. As expected, indomethacin treatment prior to THC prevented this decrease in binding, i.e., the rise in brain PGE_2 levels. More controversial was the finding that in rats injected i.v. with PGE_2 antiserum, the binding effect was abolished. One interpretation of this result is that much of the receptor-bound PGE_2 may have come from peripheral sites.

Prostaglandin Mediation of THC-Induced Catalepsy

Early reports had suggested that elevated levels of specific PGs could produce sedation or catalepsy in rodents.[9] In 1979, Fairbairn and Pickens postulated that THC-induced catalepsy was due to elevated PGE_2.[17] They reported data showing that several CO inhibitors effectively blocked the cannabinoid response and that this could be overcome by the subsequent injection of PGE_2, but not PGE_1, or $PGF_{2\alpha}$. Not discussed in the paper was their observation that injection of PGE_2 (3.3 μg/kg) alone failed to elicit a cataleptic response.

The above discrepancy was resolved in a subsequent report by Burstein et al.,[18] who showed a dose-related response for PGE_2-induced catalepsy in mice that paralleled that for THC, but was almost two orders of magnitude more potent. Again, the CO inhibitor indomethacin was effective in reducing THC-induced catalepsy, as was the PLA_2 inhibitor, mepacrine. It was also reported that while tolerance to THC was quickly reached, PGE_2 injections continued to produce catalepsy after three daily exposures. This suggests that cannabinoid tolerance in this model is at the level of PLA_2 stimulation or arachidonate metabolism and not at any subsequent stage.

The involvement of eicosanoids in THC-induced catalepsy was demonstrated by yet another experimental approach.[19] Mice actively immunized against PGE_2 and $PGF_{2\alpha}$ were challenged with THC (20 mg/kg p.o.) and the degree of catalepsy measured. In both cases, the response was reduced to about half of the control value. When challenged with PGE_2 (0.4 mg/kg i.v.), only the mice immunized against PGE_2 showed a significantly decreased response. Plasma levels of PGE_2 and PGI_2 in nonimmunized mice were monitored following the administration of THC. As anticipated, there was a substantial rise in both eicosanoids which could be blocked by prior treatment with indomethacin. A firm conclusion from these data is that PGs mediate THC-induced catalepsy. A second possible conclusion is that a large proportion of the PG may arise from peripheral sites. This is based on the assumption that circulating immunoglobulins do not enter the CNS or indirectly reduce CNS-generated levels of PGs.

A CO inhibitor of particular relevance to THC was also tested as an antagonist to its cataleptic effects.[20] The principal metabolite of THC, THC-7-oic acid, can inhibit the *in vitro* conversion of arachidonic acid to prostaglandins with a potency similar to that of naproxen or ibuprofen. The prior administration of THC-7-oic acid to mice effectively reduced THC-induced catalepsy when compared to controls. This lends further support to the other reports using inhibitors; however, there is added significance in this case. The possibility is raised that THC may generate an antagonist to at least some of its actions in the course of its metabolism. This might help explain the relatively low toxicity of cannabis.

Analgesia and Related Activities

The pain relieving properties of marijuana have been known since antiquity. Much effort has been expended in finding a cannabinoid that retains its analgesic properties without displaying undesirable CNS effects in the hope of discovering a potent but nonaddictive therapeutic agent. Unfortunately, little is known about the mechanism of cannabinoid analgesia other than the fact that it seems to differ from that of the opiates.

About 10 years ago, Milne and Johnson[21] proposed a mechanism for cannabinoid analgesia based on structural similarities between PGE_2 and levonantradol, a synthetic cannabinoid. They suggested that the cannabinoid could interfere with PGE_2 action on adenylate cyclase which, according to Collier and Roy's hypothesis,[22] mediates pain perception. Some support for this theory has come from *in vitro* studies showing that cannabinoid effects on platelet PG receptor binding and action on adenylate cyclase can be demonstrated (see below). Milne and Johnson also pointed out that levonantradol inhibits PG-induced diarrhea, in animals in agreement with their proposed mechanism.

A somewhat different hypothesis on the mechanism of cannabinoid-induced analgesia has recently emerged from studies on the THC-7-oic acids.[23] The CO inhibitory properties of these acids suggested that they may exhibit analgesic and antiinflammatory properties such as those shown by the NSAIDs. Using the mouse hot plate assay at 55°C, it was shown that the acids were comparable in activity to THC and naproxen. A time course experiment was done to address the question; is bioconversion required for THC-mediated antinociception? At early time points, it was found that THC actually decreased hot plate latency time, indicating a hyperalgesic response and only after 30 min did antinociception occur. In contrast, the acid produced increases in latency time throughout. This, as well as other data in the report, suggested that the acid metabolite(s) is responsible for cannabinoid analgesia by virtue of its ability to inhibit eicosanoid production.

In a subsequent report,[24] it was shown that the above effects had a strong dependence on copper levels as well as on the hot plate temperature and vehicle composition. Mice given copper-free drinking water for only 3 d showed a significant decrease in response to THC-7-oic acid, as did those treated with naproxen under similar conditions. *In vitro,* it was shown that $CuSO_4$ increased cannabinoid inhibition of PGE_2 synthesis in human lung fibroblasts, further supporting a role for eicosanoids in cannabinoid-induced analgesia.

The potent lipid mediator of inflammation, platelet activating factor (PAF) is believed to act, in part, by stimulating eicosanoid synthesis. Thus, it was reasoned by Burstein et al.[25] that some of the actions of PAF might be antagonized by THC-7-oic acid. Both PAF-induced paw edema and mortality were reduced in mice pretreated with the acid at doses of 20 to 40 mg/kg p.o. This was probably not due to PAF receptor antagonism (G. Carter, personal communication), but rather to the acid's property as a dual inhibitor of both CO and 5-lipoxygenase.

Effects on the Pituitary-Gonadal Axis

The effects of cannabinoids on both male and female reproductive functions have been the subject of extensive investigation and have been discussed elsewhere.[26] A role for eicosanoids in mediating these effects has been suggested and will be mentioned here for the sake of completeness in this review. Other authors have discussed the possible mechanisms for cannabinoid-induced changes in reproductive function and the reader is referred to them.[27-29]

Ayalon et al.[27] reported that the rise in ovarian PGE normally seen in rats on the evening of proestrus is inhibited by i.p. Δ^1-THC (10 mg/kg). This inhibitory response could be reversed by the subsequent administration of ovine LH. No effect of THC could be observed on tissue levels of hypothalamic PGE. It is possible that the route of injection precluded significant concentrations of THC from reaching the brain.

Dalterio et al.[28,29] have observed a complex response to THC, CBD, and CBN in male mice. PGF production *ex vivo* from the pituitary was reduced by exposure to cannabinoids in male mice housed with receptive females. No effect was seen when the mice were housed in all male groups. Testicular PGE was increased by cannabinoids in the all male groups with no change occurring in groups with females. This suggests a possible role for eicosanoids in the cannabinoid effects on male reproductive function; however, much more data will be needed to better define the nature of the involvement.

IN VITRO STUDIES

Intact Cells as Models of Cannabinoid CNS Activities

Two different methodological approaches have been used to study cannabinoid effects on eicosanoid production in whole cells *in vitro*. The more direct one utilizes RIA to measure the production of PGs from endogenous arachidonic acid. The other approach involves radiolabeling of the various lipid pools in cells by the addition of ^{14}C-labeled arachidonic acid to the culture media. There follows a spontaneous incorporation of labeled arachidonate into phospholipids, triglycerides, and cholesteryl esters, the relative distribution being somewhat dependent on conditions. Stimulation of the cells by cannabinoids then results in the turnover of arachidonate into various metabolites. By adding albumin to the culture medium, the released arachidonic acid can be trapped by binding to the protein, thereby providing a way of measuring lipase activity only. There is some correlation between the stimulation of endogenous PG synthesis and the release of arachidonic acid by cannabinoids; however, the differences between these processes must be kept in mind when interpreting data derived by either approach.

Early studies with epithelial cells from mouse mammary gland and the HeLa cell line[30] suggested a complex dose-response relationship for THC. At concentrations below 1.6 μ*M*, a modest inhibition of endogenous PGE and PGF production was observed for the epithelial cells; at 1.6 μ*M* there was a two- to threefold stimulation. CBN, which has less CNS activity than THC, was inhibitory at all doses. By contrast, THC was only stimulatory in HeLa cells.

Among peripheral tissues, the lung is a likely site for cannabinoid action and is known to be an important tissue for eicosanoid production. The WI-38 cell line was derived from human lung fibroblasts and several studies utilizing these cells for characterizing cannabinoid-induced effects on eicosanoid synthesis have been reported. THC was stimulatory at all doses tested and the effect was inhibited by aspirin, a CO inhibitor, and mepacrine, a phospholipase inhibitor.[31] A modest degree of stereospecificity was observed in which the natural (3*R*, 4*R*) isomer gave about a fivefold greater response than the "inactive" synthetic (3*S*, 4*S*) isomer.[32] Higher doses of cannabinoid and greater cell densities reduced the stereospecificity, suggesting that other nonreceptor-mediated mechanisms may become significant under those conditions.

The WI-38 cells also showed a ready disposition to develop tolerance upon repeated exposure to THC.[33] This is a common characteristic of *in vivo* responses to THC, suggesting the possibility of similar mechanisms at the molecular level. Structure-activity relationships in these cells were less clear cut.[34] When confined to THC and its metabolites, there was a good correlation between eicosanoid stimulation and the "high" in humans. However, when THC was compared with botanically related cannabinoids (CBD, CBN, CBG, etc.), no correlation could be seen. In fact, there are many other substances that stimulate eicosanoid production that are not psychotropic, raising the question of how far one can extend structure-activity comparisons between an *in vitro* model and a complex *in vivo* response.

Two cells lines of neural origin were also tested for cannabinoid-induced stimulation of arachidonate metabolism.[35] N2a neuroblastoma and C-6 glioma cells responded with a structure-activity profile similar to the WI-38 fibroblast (Table 2). The potent antiinflammatory agent, indomethacin, was highly effective in reducing the THC effect on N2a cells. Finally, N2a cells quickly develop a tolerance to THC, as was the case for the WI-38 lung cells.[33]

The antagonist activity of THC-7-oic acid (see above) was first reported in WI-38 cells.[36] It was reasoned that the lack of psychotropic activity, as

TABLE 2
Comparison of Responses to Cannabinoids in Neuroblastoma, Glioma, and Lung Fibroblast Cell Lines

	Cell type			
Cannabinoid	**N2a**	**C-6**	**WI-38**	**"High" in humans[a]**
Δ^1-THC	1.00	1.00	1.00	1.0
7-OXO-THC	1.79	0.71	0.73	—
7-OH-THC	1.32	1.14	0.45	0.88
6α-OH-THC	<0.63	1.10	0.09	<9.9
6β-OH-THC	0.63	<0.62	0.18	2.6
THC-7-oic acid	<0.58	<0.48	0.09	<16

Note: Values shown are relative concentrations of PGE_2 in culture media as measured by RIA. Cannabinoid concentrations were 3.2 μM for WI-38 (human lung fibroblast) cells and 32 μM for N2a (neuroblastoma) and C-6 (glioma) cells.

[a] Data taken from Reference 54. Values are the relative doses needed to obtain a "high".

well as its low potency in stimulating eicosanoid production *in vitro* (Table 2), made the acid a good candidate as a THC antagonist, especially considering the similarities in molecular structure. Cells preexposed to THC-7-oic acid were inhibited in a concentration-dependent manner when subsequently challenged with THC. Since the acid also appears to inhibit CO and 5-lipoxygenase activities, it is not clear whether the THC antagonism is functional or by competition for THC receptor binding sites.

Cannabinoid-induced stimulation of lipoxygenase pathways also seems to be a possible action of THC based on a study reported using mouse peritoneal cells.[37] These cells consist primarily of macrophages where arachidonate metabolism by 5-lipoxygenase is a major route of metabolism. The macrophages respond in the same dose range as the WI-38 cells and show a similar structure-activity relationship. On this basis, it could be predicted that leukotriene production *in vivo* may also be elevated following THC.

Studies with Isolated Tissues and Organs

Peripheral Sites

Perfused rabbit kidney and perfused guinea pig lung released PG-like substances when exposed to THC.[38] The effects could be inhibited by aspirin, suggesting CO mediation; however, the presence of eicosanoids was based solely on a bioassay procedure, thus making the conclusions somewhat uncertain. Uterine vein levels of PGE were elevated in a rat preparation following s.c. injection of THC.[39] The eicosanoids were measured 12 h following THC, which is an unusually long interval for a PG response.

Incubation of mouse testes with both THC and CBN reduced the levels of PGE and PGF compared with controls.[28] A single dose of each cannabinoid was used, raising the possibility that other doses may be stimulatory. Eicosanoid production by glucose-stimulated rat pancreatic islets was increased by THC exposure.[40] The radiolabeled arachidonate approach was used and both PGE_2 and 12-HETE synthesis were found to be elevated.

Central Sites

Reichman et al.[41] incubated slices from various regions of guinea pig brain with 0.8 μM THC, and PGE and PGF secretion was measured by RIA. Stimulation was observed in striatum, while inhibition was seen in cortical slices; in cerebellum and hippocampus, no significant differences from control values were found. These authors reported some degree of correlation with psychoactivity when several other cannabinoids were tested in the cortical slice preparation for inhibition of PG synthesis. They also showed that 0.8 μM THC could reduce acetylcholine-stimulated release of PGE and PGF. While the slice preparation is very different from the cell culture model described above, there is considerable agreement between the results obtained from both systems, the complexity of the response being a feature of both.

In a second report,[42] the above authors examined arachidonate release following cannabinoid exposure in cortical slices that had been labeled with ^{14}C-arachidonic acid. A simple dose-related response was seen with all of the cannabinoids tested over the range of 2 to 64 μM. While the structure-activity relationship was good, no stereospecificity was seen when enantiomeric THC was tested. There is no obvious explanation for this except that it may be related to the heterogeneity of cell types in the cortical slices.

A possible role for PGs in the THC-induced inhibition of LHRH release from the median eminence was suggested in a report by Rettori et al.[43] They showed that 10 to 100 nM THC was highly effective in reducing PG synthesis in hypothalamic fragments with both endogenously and exogenously added arachidonate. It was also suggested that the stimulation of somatostatin release by THC may be due to its inhibition of PG production.

Platelet Effects

Eicosanoids have a major role in platelet function and much research has been done in this area, making it of interest to examine for cannabinoid action as well. White and Tansik treated human platelets that had been equilibrated with radiolabeled arachidonic acid with THC and CBD.[44] Both cannabinoids were effective in promoting eicosanoid synthesis, which consisted primarily of lipoxygenase products.

Platelets contain binding sites for PGD_2 and PGE_1/I_2. It has been reported by Horne that the synthetic cannabinoid,[45] desacetylnantrodol, selectively reduces PGD_2 binding to its platelet receptor. It had been suggested earlier by Milne and Johnson that certain cannabinoids had many structural similarities with specific PGs.[21] This may be the basis for the platelet binding effect; however, the data are not conclusive for competitive binding.

In a brief communication, Formukong et al.[46] gave data showing that high concentrations of cannabinoids did inhibit platelet aggregation. Only one dose of each drug was used and these were levels that could never be attained *in vivo,* making these findings of limited significance. It would be of interest to see whether lower concentrations of cannabinoids had any effect on platelet function.

Synaptosomes and Other Cell Membrane Preparations

In a study on the release of PGs by rat brain synaptosomes, Raffel et al.[47] reported inhibitory activity for THC ($ID_{50} = 750\ \mu M$). It seems inconceivable that CNS levels of THC could ever reach such high values, so that these findings seem unrelated to the *in vivo* actions of marijuana. Unfortunately, they only measured $PGF_{2\alpha}$ production; it is possible that the synthesis of other eicosanoids would have been more sensitive to cannabinoid influence.

A $10{,}000 \times g$ supernate prepared from rat Leydig cells was used to study cannabinoid action on PLA_2 activity.[48] It was found that a greater than twofold stimulation of arachidonate release could be seen with 3.2 μM THC and was more than fourfold at 16 μM. In a more detailed study, using isolated mouse brain synaptosomes, the above findings were confirmed and extended by Hunter et al.[49] A structure-activity relationship was reported that showed similarity to those seen with the neuroblastoma and glioma cell culture models discussed above (Table 2). THC stimulation of PLA_2 activities in CNS myelin and mitochondria were also reported in this paper. It will be of some interest to examine all of these brain fractions for the presence of cannabinoid binding sites.

Prostaglandin Adenylate Cyclase Interactions

Hillard and Bloom observed that cannabinoids produced significant stimulation of adenylate cyclase activity in homogenates from mouse brain cortex.[50] The effect was biphasic, with a maximal response around 30 μM, and was strongly dependent on added GTP. Of particular interest to this review was their finding that indomethacin, aspirin, and mepacrine, all inhibitors of eicosanoid synthesis, were able to completely abolish the THC effect. This strongly suggests that the cannabinoid stimulation of cyclase activity is mediated by PGs. This is supported by the fact that certain PGs have potent agonist action on adenylate cyclase from a variety of sources including nervous tissues.

An apparently contradictory finding was reported by Howlett who used the neuroblastoma cell line N18TG2 to study cannabinoid effects on cAMP metabolism.[51] She observed that both basal and hormone stimulated adenylate cyclase activity was decreased by THC. Previous work suggested the existence of a prostacyclin receptor on these cells which was coupled to adenylate cyclase. In the present study, THC was able to inhibit prostacylin-induced cyclase activity in a dose-related manner. It was also shown that THC had no effect on phosphodiesterase activity.

A subsequent report,[52] in which N18TG2 membranes were used, further characterized the nature of the cannabinoid effect. The inhibition of adenylate cyclase was noncompetitive when either PGE_1 or PGI_2 were the agonists, as was the case for stimulation by the peptides secretin or vasoactive intestinal peptide (VIP). Howlett has suggested that the discrepancies between her findings and those reported by Hillard and Bloom may reside in the multiplicity of cell and receptor types in their cerebral cortex preparation. This suggestion is supported by a recent publication from Howlett's laboratory in which indomethacin did reduce cannabinoid inhibition of adenylate cyclase in cerebellar slices, but not in slices from rat brain striatum or hippocampus.[53]

SUMMARY AND CONCLUSIONS

For the most part, the reports discussed in this review support the possibility that eicosanoids have an important role in mediating at least some of the actions of the cannabinoids. The question remains: at what point in the sequence of molecular events following exposure to a cannabinoid does eicosanoid mediation occur? For example, cannabinoid-induced elevation of eicosanoid levels could occur directly as a result of the formation of a drug-receptor complex possibly involving G protein transduction of the signal to a phospholipase. On the other hand, it is not unreasonable to suggest that, in some systems, the drug-receptor interaction may alter cAMP levels which, in turn, could modulate PLA_2 activity. The possibility that high doses of cannabinoids act through nonreceptor type mechanisms needs to be considered as well. Cannabinoids are highly lipophylic and, at sufficient concentrations, are known to alter membrane characteristics. Such an effect could also impact on eicosanoid synthetic activity.

The existence of cannabinoid receptors has, at long last, received experimental support.[1-4] The best studied system is one that is coupled to adenylate cyclase through an inhibitory G protein.[1,3] Unfortunately, there are not sufficient data currently available to determine whether this pathway produces any change in phospholipase activity. A very recent report suggests that there may be several high-affinity binding sites for THC,[4] raising the possibility that more than one mechanism of action may be available at the molecular level.

The elevation of tissue eicosanoid levels can result in a wide variety of biological effects which include some of the responses produced by cannabinoids.[6,9,26] On the other hand, several of the cannabinoid activities seem to be better explained by an inhibition of eicosanoid production. The reports describing the antinociceptive and antiedema actions of THC-7-oic acid illustrate how a small structural modification can change cannabinoid action from agonist to antagonist with regard to eicosanoid synthesis.[23-25] This finding provides a novel approach for the development of therapeutically useful cannabinoids. While a mechanism can never be proven in the absolute sense, there is much evidence in favor of a process involving eicosanoid mediation of cannabinoid action. At the very least, it does allow a basis for understanding some of the phenomena associated with the effects produced by this very interesting group of substances.

ACKNOWLEDGMENTS

The author wishes to thank the National Institute on Drug Abuse for support in the form of a Research Scientist Award (DA 00043-15).

REFERENCES

1. **Devane, W. A., Dysarz, F. A., Johnson, M. R., Melvin, L. S., and Howlett, A. C.,** Determination and characterization of a cannabinoid receptor in rat brain, *Mol. Pharmacol.,* 34, 605, 1988.
2. **Herkenham, M., Lynn, A. B., Little, M. D., Johnson, M. R., Melvin, L. S., DeCosta, B. R., and Rice, K. C.,** Cannabinoid receptor localization in brain, *Proc. Natl. Acad. Sci. U.S.A.,* 87, 1932, 1990.
3. **Matsuda, L. A., Lolait, S. J., Brownstein, M. J., Young, A. C., and Bonner, T. I.,** Structure of a cannabinoid receptor and functional expression of the cloned DNA, *Nature (London),* 346, 561, 1990.
4. **Burstein, S. H., Audette, C. A., Charalambous, A., Doyle, S. A., Guo, Y., Hunter, S. A., and Makriyannis, A.,** Detection of cannabinoid receptors by photoaffinity labelling, *Biochem. Biophys. Res. Commun.,* 176, 492, 1991.
5. **Burstein, S. and Raz, A.,** Inhibition of prostaglandin E_2 biosynthesis by delta-1-tetrahydrocannabinol, *Prostaglandins,* 2, 369, 1972.
6. **Pertwee, R. G.,** The central neuropharmacology of psychotropic cannabinoids, *Pharmacol. Ther.,* 36, 189, 1988.
7. **Burstein, S.,** Inhibitory and stimulatory effects of cannabinoids on eicosanoid synthesis, *NIDA Res. Monogr. Ser.,* 79, 158, 1987.
8. **Martin, B. R.,** Cellular effects of cannabinoids, *Pharmacological Rev.,* 38, 45, 1986.
9. **Schaad, N. C., Magistretti, P. J., and Schorderet, M.,** Prostaglandins and their role in cell-cell interactions in the central nervous system, *Neurochem. Int.,* 18, 303, 1991.
10. **Ramwell, P. W. and Shaw, J. E.,** The spontaneous and evoked release of non-cholinergic substances from cerebral cortex of cats, *Life Sci.,* 2, 419, 1963.
11. **Waller, C. W., Urbanek, B. S., and Wall, G. M.,** *Marihuana, An Annotated Bibliography,* The Research Institute of Pharmaceutical Sciences, University of Mississippi, 1988.
12. **Perez-Reyes, M., Burstein, S. H., White, W. R., McDonald, S. A., and Hicks, R. E.,** Antagonism of marihuana effects by indomethacin, *Life Sci.,* 48, 507, 1991.
13. **Laviolette, M. and Belanger, J.,** Role of prostaglandins in marihuana-induced bronchodilation, *Respiration,* 49, 10, 1986.
14. **Bhattacharya, S. K.,** Delta-9-tetrahydrocannabinol increases brain prostaglandins in the rat, *Psychopharmacology,* 90, 499, 1986.
15. **Hunter, S. A., Latham, V., Renzulli, L., and Burstein, S.,** *In vivo* effects of Δ^1-tetrahydrocannabinol on rat and mouse prostaglandin tissue levels, in *Marihuana: An International Research Report,* Monograph Series 7, Chesher, G., Consroe, P., Musty, R., Eds., Australian Government Publishing Service, Canberra, 1988, 399.
16. **Hunter, S. A., Audette, C. A., and Burstein, S.,** Elevation of brain prostaglandin E_2 levels in rodents by Δ^1-THC, *Prost. Leuk. Essential Fatty Acids,* 43, 185, 1991.
17. **Fairbairn, J. W. and Pickens, J. T.,** The oral activity of Δ^1-THC and its dependence on prostaglandin E_2, *Br. J. Pharmacol.,* 67, 374, 1979.
18. **Burstein, S., Hunter, S. A., Latham, V., and Renzulli, L.,** Prostaglandin E_2 mediation of THC-induced catalepsy in the mouse, in *Marijuana: An International Research Report,* Chester, G., Consroe, P., and Musty, R., Eds., Australian Government Publishing Service, Canberra, 1988, 371.

19. **Burstein, S. H., Hull, K., Hunter, S. A., and Shilstone, J.,** Immunization against prostaglandins reduces Δ^1-THC-induced catalepsy in mice, *Mol. Pharmacol.,* 35, 6, 1989.
20. **Burstein, S., Hunter, S. A., Latham, V., and Renzulli, L.,** A major metabolite of Δ^1-THC reduces its cataleptic effect in mice, *Experientia,* 43, 402, 1987.
21. **Milne, G. M. and Johnson, M. R.,** Levonantrodol: a role for central prostanoid mechanisms, *J. Clin. Pharmacol.,* 21, 367, 1981.
22. **Collier, H. O. J. and Roy, A. C.,** Morphine-like drugs inhibit the stimulation by E prostaglandins of cyclic AMP formation by rat brain homogenate, *Nature (London),* 248, 24, 1981.
23. **Burstein, S. H., Hull, K., Hunter, S., and Latham, V.,** Cannabinoids and pain responses: a possible role for prostaglandins, *FASEB J.,* 2, 3022, 1988.
24. **Doyle, S. A., Burstein, S. H., Dewey, W. L., and Welch, S. P.,** Further studies on the antinociceptive effects of Δ^6-THC-7-oic acid, *Agents Actions,* 31, 157, 1990.
25. **Burstein, S. H., Audette, C. A., Doyle, S. A., Hull, K., Hunter, S. A., and Latham, V.,** Antagonism to the actions of PAF by a nonpsychoactive cannabinoid, *J. Pharmacol. Exp. Ther.,* 251, 531, 1989.
26. **Dewey, W. L.,** Cannabinoid pharmacology, *Pharmacol. Rev.,* 38, 151, 1986.
27. **Ayalon, D., Nir, I., Cordova, T., Bauminer, S., Pudar, M., Naor, Z., Kashi, R., Zor, W., Harell, A., and Lindner, H. R.,** Acute effect of Δ^1-THC on the hypothalamic pituitary-ovarian axis in the rat, *Neuroendocrinology,* 23, 31, 1977.
28. **Dalterio, S., Bartke, A., Roberson, C., Watson, D., and Burstein, S.,** Direct and pituitary-mediated effects of Δ^9-THC and cannabinol on the testis, *Pharmacol. Biochem. Behavior,* 8, 673, 1978.
29. **Dalterio, S., Bartke, A., Harper, M. J. K., Huffman, R., and Sweeney, C.,** Effects of cannabinoids and female exposure on the pituitary-testicular axis in mice: possible involvement of prostaglandins, *Biol. Reprod.,* 24, 315, 1981.
30. **Burstein, S. and Hunter, S. A.,** Prostaglandins and cannabis. VI. Release of arachidonic acid from HeLa cells by Δ^1-tetrahydrocannabinol and other cannabinoids, *Biochem. Pharmacol.,* 27, 1275, 1978.
31. **Burstein, S., Hunter, S. A., Sedor, C., and Shulman, S.,** Prostaglandins and cannabis. IX. Stimulation of prostaglandin E_2 synthesis in human lung fibroblasts by Δ^1-THC, *Biochem. Pharmacol.,* 31, 2361, 1982.
32. **Burstein, S., Hunter, S. A., Mechoulam, R., Melchior, D. L., and Renzulli, L.,** Prostaglandins and cannabis. XV. Comparison of enentiomeric cannabinoids in stimulating prostaglandin synthesis in fibroblasts, *Life Sci.,* 39, 1823, 1986.
33. **Burstein, S., Hunter, S. A., and Renzulli, L.,** Prostaglandins and cannabis. XIV. Tolerance to the stimulatory actions of cannabinoids on arachidonate metabolism, *J. Pharmacol. Exp. Ther.,* 235, 87, 1985.
34. **Burstein, S., Hunter, S. A., and Ozman, K.,** Prostaglandins and cannabis. XII. The effect of cannabinoid structure on the synthesis of prostaglandins by human lung fibroblasts, *Mol. Pharmacol.,* 23, 121, 1983.
35. **Burstein, S., Hunter, S. A., Ozman, K., and Renzulli, L.,** *In vitro* models of cannabinoid-induced psychoactivity, in *Marihuana '84: Proc. Oxford Conf. on Cannabis,* Harvey, D. J., Ed., IRL Press, Oxford, 1985, 399.
36. **Burstein, S., Hunter, S. A., Latham, V., and Renzulli, L.,** Prostaglandin and cannabis. XVI. Antagonism of Δ^1-THC action by its metabolites, *Biochem. Pharmacol.,* 35, 2553, 1986.

37. **Burstein, S., Hunter, S. A., Ozman, K., and Renzulli, L. A.,** Prostaglandins and cannabis. XIII. Cannabinoid induced elevation of lipoxygenase products in mouse peritoneal macrophages, *Biochem. Pharmacol.*, 33, 2653, 1984.
38. **Kaymakcalan, S., Ercan, Z. S., and Turker, R. K.,** Evidence of release of prostaglandin-like material form rabbit kidney and guinea-pig lung by (–)-trans-Δ^9-THC, *J. Pharm. Pharmacol.*, 27, 564, 1975.
39. **Jordan, V. C. and Castracane, V. D.,** The effect of reported PG synthetase inhibitors on estradiol-stimulated uterine prostaglandin biosynthesis *in vivo* in the ovariectomized rat, *Prostaglandins*, 12, 1073, 1976.
40. **Laychock, S. G., Hoffman, J. M., Meisol, E., and Bilgin, S.,** Pancreatic islet arachidonic acid turnover and metabolism and insulin release in response to Δ^9-THC, *Biochem. Pharmacol.*, 35, 2003, 1986.
41. **Reichman, M., Nen, W., and Hokin, L. E.,** Effects of Δ^9-THC on prostaglandin formation in brain, *Mol. Pharmacol.*, 32, 686, 1987.
42. **Reichman, M., Nen, W., and Hokin, L. E.,** Δ^9-THC increases arachidonic acid levels in guinea pig cerebral cortex slices, *Mol. Pharmacol.*, 34, 823, 1988.
43. **Rettori, V., Aguila, M. C., Gimeno, M. F., Franchi, A. M., and McCann, S. M.,** *In vitro* effect of Δ^9-THC to stimulate somatostatin release and block that of LH-RH by suppression of the release of prostaglandin E_2, *Proc. Natl. Acad. Sci. U.S.A.*, 87, 10063, 1990.
44. **White, H. L. and Tansik, R. L.,** Effects of Δ^9-THC and CBD on phospholipase and other enzymes regulating arachidonate metabolism, *Prostaglandins Med.*, 4, 409, 1980.
45. **Horne, W. C.,** Desacetyl-1-nantrodol: a selective prostaglandin antagonist, *Prost. Leuk. Med.*, 15, 129, 1984.
46. **Formukong, E. A., Lee, C. R., Roberts, M. F., Evan, A. T., and Evans, F. J.,** Cannabinoid, the active constituents of *Cannabis sativa* L. inhibit both human and rabbit platelet aggregation, *Br. J. Pharmacol.*, 92 (Suppl.), 601P, 1987.
47. **Raffel, G., Clarenback, P., Peskar, B. A., and Hertting, G.,** Synthesis and release of prostaglandins by rat brain synaptosomal fractions, *J. Neurochem.*, 26, 493, 1976.
48. **Burstein, S. and Hunter, S. A.,** Prostaglandins and cannabis. VIII. Elevation of phospholipase A_2 activity by cannabinoids in whole cells and subcellular preparation, *J. Clin. Pharmacol.*, 21, 240S, 1981.
49. **Hunter, S. A., Burstein, S., and Renzulli, L.,** Effects of cannabinoids on the activities of mouse brain lipases, *Neurochem. Res.*, 11, 1273, 1986.
50. **Hillard, C. J. and Bloom, A. S.,** Possible role in the effects of the cannabinoids on adenylate cyclase activity, *Eur. J. Pharmacol.*, 91, 21, 1983.
51. **Howlett, A. C. and Fleming, R. M.,** Cannabinoid inhibition of adenylate cyclase. Pharmacology of the response in neuroblastoma cell membranes, *Mol. Pharmacol.*, 26, 532, 1984.
52. **Howlett, A. C.,** Inhibition of neuroblastoma adenylate cyclase by cannabinoid and nantrodol compounds, *Life Sci.*, 35, 1803, 1984.
53. **Bidant-Russell, M., Devane, W. A., and Howlett, A. C.,** Cannabinoid receptors and modulation of cyclic AMP accumulation in the rat brain, *J. Neurochem.*, 55, 21, 1990.
54. **Perez-Reyes, M.,** Pharmacodynamics of certain drugs of abuse, in *Pharmacokinetics and Pharmacodynamics of Psychoactive Drugs*, Barnett, G. and Chiang, N. C., Eds., Biomedical Publishers, Foster City, 1985, 287.

4 Cannabinoid-Glucocorticoid Interactions in the Hippocampal Region of the Brain

INTRODUCTION

The psychoactive cannabinoid substances produced by smoked marijuana leaves are lipophilic molecules of very small size. Due to this hydrophobic nature, it was thought by many until recently that cannabinoids would be poorly capable of engaging in the covalent or hydrogen bonding activities used by most active hormones, neurotransmitters, and drugs; this belief hampered the search for a cannabinoid "receptor" system. Indeed, although a series of synthetic cannabimimetic compounds has now been devised that binds specifically and strongly to neuronal membranes, plant-derived cannabinoids are observed to possess a somewhat lower affinity for even this receptor.[1,2] Thus, it remains that natural cannabinoids must exert their myriad of often subtle effects in a mammalian environment that did not evolve a specific means to respond to these substances, although very active searches are presently well underway to identify "endogenous cannabinoids".

Due to the similarities in structure between cannabinoids and steroid hormones (Figure 1), it has occurred to many that at least a portion of the drug's mechanisms of action might include interaction with one or more existing steroid hormone receptor systems (reviewed in References 3 and 4). One author even described how cannabinoids could be classified as "false

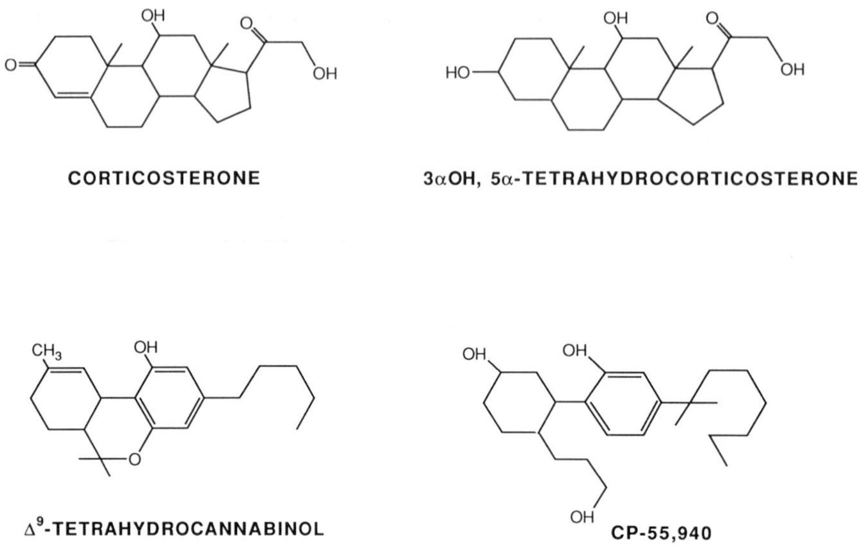

FIGURE 1. Steroid and cannabinoid structures.

hormones".[5] A major supportive reason for belief in cannabinoid-steroid interaction has been the continual appearance of research literature describing the effects of administered cannabinoids on endocrine function. Thus, although a membrane binding system that responds to synthetic cannabinoids has now been established, and this receptor system may well be linked to a variety of unconnected neuroendocrine and behavioral functions, there remain many experimental observations, particularly those involving steroid target effects, that continue to support the notion of cannabinoid action in multiple expression systems.

One important area of study of cannabinoid-endocrine interaction has concerned the effect on reproductive activity, a subject that has been explored in greater depth elsewhere in this volume. To summarize here: Δ^9-tetrahydrocannabinol (THC), the principal psychoactive compound of marijuana, has been reported to inhibit gonadotropin secretion in male[6-9] and female[10-13] rats and mice,[14] and to disrupt ovulation in rats[15] and rabbits.[16] Alterations of normal development and sexual maturation have also been reported in THC-treated neonatal[9,13,17] and immature[18] rodents. Although the vast majority of the aforementioned studies concentrated on neuroendocrine sites of action, it has also been reported that THC can exert direct gonadal actions that result in reduced spermatogenesis[19,20] and steroidogenesis.[14,21-24] Furthermore, smoked marijuana has been reported to suppress pituitary luteinizing hormone (LH) secretion in male[25] and female[26] human volunteers.

Additional studies of potential THC-endocrine interactions in rats have shown a stimulation of β-endorphin in plasma[27] and hypothalamus,[9,13,27] suppression of prolactin[12] and growth hormone[28-31] release, and suppression of thyroxine and pituitary thyroid stimulating hormone (TSH) release.[32,33] The primary site of action for cannabinoids, in these and in the reproductive studies, has been generally presumed to reside in the neuroendocrine brain. That is, neurochemical effects of the drug likely include additional actions on brain regions that eventually control endocrine secretory aspects but, in the absence of candidates for specific receptors, few detailed mechanisms of action have been proposed.

With regard to the pituitary-adrenal system, a similar context and accumulation of experiments and results have developed. A number of observations were reported over many years, but little ability existed until recently to explain or to explore specific mechanisms by which the numerous effects could occur. The neuroendocrine functions controlling adrenocortical secretion are very sensitive to changes in the surrounding environment. Both emotional and physical stressors stimulate glucocorticoid release and, in contrast, marijuana smoking is considered by its users to be a rather soothing and an anxiety-reducing experience. In the search for possible connections between cannabinoids and the neuroendocrine stress response, our initial hypothesis was that cannabinoids have an ability to interact with the glucocorticoid receptor system in the CNS. The study has been recently expanded to include the reverse hypothesis: that steroid hormones and their metabolites may be capable of influencing cannabinoid binding to its receptor.

CANNABINOID INTERACTION WITH GLUCOCORTICOID BINDING IN THE HIPPOCAMPUS

A number of reports over the past 20 years have amply demonstrated that THC administration to rodents stimulates adrenal corticosterone (CORT) secretion.[25,31,34-43] Although the acute response diminishes in magnitude with repeated exposure,[39,43] significant stimulation is still evident for up to 9 months of chronic cannabinoid treatment.[42] Both the maintenance of response and its accommodation over time have been reported by our laboratories in recently published studies;[42,43] this strongly suggests that the stimulation involves a specific mechanism of action. In other studies, animals subjected to a mild footshock stressor secreted greater amounts of CORT if pretreated with THC,[38] which implies that the neuronal pathways responding to cannabinoids and to stress are not the same, but are allied in some fashion. The enhancement of CORT secretion can be attributed to THC-induced release of pituitary ACTH.[34] Prior administration of barbiturate drugs has been shown to reduce the stimulation of ACTH or CORT by THC,[34,36] which underscores the role of CNS mechanisms in the cannabinoid action. In fact, it was reported that THC *suppressed* steroid release from cultured adrenal tumor cells;[44] thus, it appears that any increase of glucocorticoid secretion *in vivo* must be strongly driven by trophic factors of central origin. It also deserves mention that changes in pituitary-adrenal activity of humans administered THC have not been frequently observed or substantiated with confidence.[25,45,46]

Initial attempts to explain the ability of cannabinoids to stimulate ACTH and CORT secretion concentrated on the possible interaction of THC with glucocorticoid receptor systems that mediate steroid feedback control of ACTH. It was reported, for example, that selective uptake of [^{3}H]-CORT into limbic structures of rats was reduced by prior THC treatment.[47] However, another study found increased uptake of radiolabeled CORT after THC treatment, particularly in hippocampus.[37] The hippocampus is replete with glucocorticoid receptors and, although the complete picture of how the hippocampus participates in the homeostatic control of ACTH and CORT remains to be established, it seems clear that normal pituitary-adrenal function, particularly during episodes of stress exposure, requires an intact hippocampus.[48-50]

Several studies in other laboratories examined the ability of cannabinoids to interact directly with steroid receptor binding, most notably with estrogen receptors. THC was observed to be uterotrophic in ovariectomized rats,[51] and an ability of THC to compete with estrogen receptor binding *in vitro* was reported.[52] However, further examination of these findings called the results into question,[53,54] and this served to diminish the strength of the overall hypothesis. A single study[55] describing inhibition of androgen receptor binding by THC was reported, but no further research on that issue appears to have followed. With the exception of the aforementioned studies regarding *in vivo* uptake of radiolabeled steroid in THC-treated rats, little detailed examination of steroid receptor systems in brain has taken place.

Our studies of cannabinoid interaction with glucocorticoid receptors began to focus on the hippocampal formation of the brain for two reasons: (1) the hippocampus contains a very dense population of glucocorticoid receptors,[56,57] of two subtypes,[58-60] and this brain region is very labile to the destructive effects of glucocorticoid-mediated aging and stress;[61-64] and (2) earlier studies conducted in the authors' laboratories had found that long-term THC administration to rats had resulted in aging-like degenerative changes of hippocampal morphometric structure.[42] These degenerative changes (e.g., pyramidal cell loss and astrocyte reactivity) appeared to be similar to responses we and others had previously observed to be glucocorticoid dependent (i.e., retarded by adrenalectomy[65,66] and accelerated by chronic stress[64,67] or treatment with glucocorticoids[62]). It thus became a reasonable hypothesis that the reported effects of THC on hippocampal structure and function would occur through glucocorticoid receptors, and this led to additional studies in our laboratory with the objective of determining whether cannabinoids could have any direct effect on the steroid receptor system.

In a preliminary study, in which aliquots of cytosol prepared from adrenalectomized rat hippocampi were incubated with 2.5 n*M* [^{3}H]-dexamethasone (DEX) plus or minus 250 n*M* unlabeled THC, cannabidiol (CBD), a nonpsychoactive cannabinoid, or unlabeled DEX, approximately 31% of the specific DEX binding was displaced by THC. A more modest 11% of binding was displaced by CBD. When hippocampal cytosols were incubated with [^{3}H]-DEX plus or minus 100-fold excess THC or CBD, layered on sucrose gradients, and centrifuged at 325,000 *g*, approximately 28% of the 8 S peak was reduced by excess THC, and 9% reduced by excess CBD. These results suggested that cannabinoids were able to compete against glucocorticoid binding to hippocampal receptors but that, at 100-fold molar excess, the competition was neither strong nor complete.

It is now recognized that at least two types of corticosteroid-binding receptors exist in hippocampus.[58-61,68] Both receptor types bind to DEX, as well as to naturally occurring adrenal steroids, but with differing affinities.[58,60] However, certain synthetic steroids bind specifically to one or the other receptor type; thus, we conducted additional competitive binding studies with RU-28362, a glucocorticoid agonist that associates only with

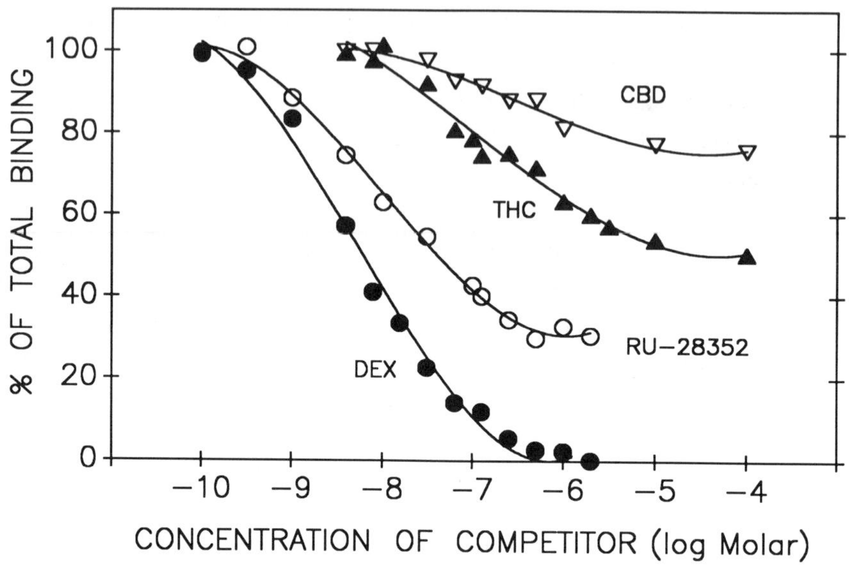

FIGURE 2. Dose-response curves of cannabinoid inhibition of binding by [^{3}H]-dexamethasone in hippocampal cytosol. Cytosol aliquots from adrenalectomized rats were incubated with 5 n*M* radiolabeled DEX plus doses of competitor that are indicated on the abscissa scale. Plotted points represent an average of duplicate incubations at each dose. The maximum binding of control tubes, containing no competitor, was set as 100% binding. Nonspecific binding, which remained after maximum displacement by nonlabeled DEX, was set as 0% specific binding; nonspecific counts were subtracted from all other readings before calculation of percent binding. (From Eldridge, J. C. and Landfield, P. W., Cannabinoid interactions with glucocorticoid receptors in rat hippocampus, *Brain Res.*, 534, 135, 1990. With permission.)

the type II receptor.[58-61,68-70] When aliquots of hippocampal cytosol were incubated with 5 n*M* radiolabeled DEX plus or minus a variety of doses of competitors (Figure 2), THC displaced a maximum of only 50% of total net sites. The CBD maximum displacement was 22%. Sites displaced by excess RU-28362 were 69% of the total capacity (type I binding was the remaining 31% of total). It was thus apparent that THC did not even displace all of the available type II capacity. Further study, conducted with RU-28362 together with THC or CBD, revealed that the addition of cannabinoid to RU-28362 did not displace further binding than did RU-28632 alone (data presented and discussed in Reference 71).

The initial interpretation of these results was that cannabinoids either completely suppress the association of glucocorticoids with an unrecognized portion of the type II receptor population in hippocampus, or that they

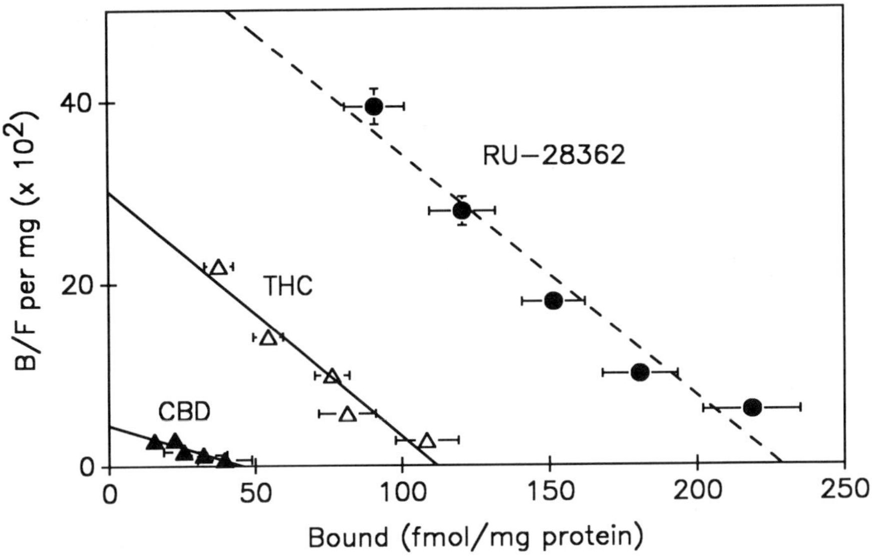

FIGURE 3. Scatchard plot of cannabinoid competition with [^{3}H]-DEX binding to hippocampal glucocorticoid receptors in cytosol. Aliquots of cytosols prepared from individual adrenalectomized rats were incubated with 1.25 to 20 n*M* radiolabeled DEX alone, or plus 100-fold molar excess of unlabeled RU-28362, Δ^9-THC, or cannabidiol (CBD). Points represented means ± S.E. of four animals assayed separately, triplicate tubes per dose (additional methodologic details in References 68 and 71). RU plot indicates total type II glucocorticoid receptor displacement (K_d = 3.41 n*M*, B_{max} = 224.4 fmol/mg). The slope of the THC plot was parallel to the RU plot (K_d = 3.29 n*M*) but THC displaced fewer sites (B_{max} = 108.7 fmol/mg). Unlabeled CBD produced a slight displacement of bound tracer (estimated B_{max} = 46 fmol/mg), but precision of the plot was poor (K_d = 17 n*M*, r = 0.622; all other plots showed r >0.95).

incompletely interfere with ligand binding to the entire type II population. It has been reported that the glucocorticoid receptor molecule contains secondary binding sites that are capable of exerting allosteric or noncompetitive action on the primary glucocorticoid binding domain.[72-74] Additional evidence for this property occurring for THC interaction was therefore sought by Scatchard analyses of the competition.

As shown in Figure 3, Scatchard plots of THC displacement of [^{3}H]-DEX binding were parallel to those of RU-28362 displacement, but with a lower B_{max}. This result is more consistent with a property of noncompetitive or allosteric interaction. As with all previous studies, CBD displacement was very small in comparison with that of THC and, in this case, produced an imprecise Scatchard plot. It is of course possible that THC fully occupies an undefined subgroup of the type II glucocorticoid receptor population that also retains an identical affinity with the subgroup not occupied by THC but which binds RU-28362. However, in the absence of other evidence that type II receptors can be separated further into subgroups, our tentative conclusion is that cannabinoids exert a suppressive effect on glucocorticoid binding to the type II hippocampal receptor by interaction at a distant binding site that becomes communicated with the steroid binding domain of the receptor. Although cannabinoid structure is similar to that of glucocorticoids, it is not extremely close; one might presume that a truly competitive antagonist would possess a structure more closely related to that of glucocorticoids. Furthermore, progesterone, which is similar but not identical to the classic glucocorticoid structure, has been observed to exert noncompetitive interaction with the glucocorticoid receptor.[72,73] Other known allosteric antagonists of glucocorticoid binding, such as ketoconazole,[73] are not at all similar to the glucocorticoid structure.

THC displacement of radiolabeled glucocorticoid binding in hippocampus is one observation that suggests that the cannabinoid is competitive at the glucocorticoid receptor. However, we had not been able to conduct the reverse experiment, i.e., direct measurement of cannabinoid binding in cytosol, until recently because radiolabeled THC is not commercially available. The National Institute of Drug Abuse was able to furnish a supply of $[^3H]$-labeled Δ^8-THC suitable for ligand studies. In comparison to Δ^9-THC, the Δ^8-isomer is psychoactive but less so.[3,4]

Cytosols were prepared from hippocampi dissected from adrenalectomized rats and incubated with either radiolabeled CORT or Δ^8-THC, plus or minus 100-fold molar excess of unlabeled Δ^9-THC, CP-55940, or CORT, and the receptor-bound fractions were counted. As shown in Figure 4 (lower panel), both THC and CP displaced significant amounts of radiolabeled THC binding, about 15% each. Excess CORT displaced 27% of radiolabeled THC binding, and this amount was significantly greater than that displaced by either THC or CP ($p < 0.05$). In the incubations containing $[^3H]$-CORT (upper panel), excess unlabeled THC displaced about 15% of total sites, but CP did not displace radiolabeled CORT. Perhaps THC interacts with a site on the glucocorticoid receptor that does not recognize CP-55940. On the other hand, cytosol proteins that bound radiolabeled THC (lower panel) seemed equally suppressed by unlabeled THC or CP. We are not yet sure that the two radioactive ligands bind to the same protein, nor do we know why so much "nonspecific" binding remained in the presence of 100-fold

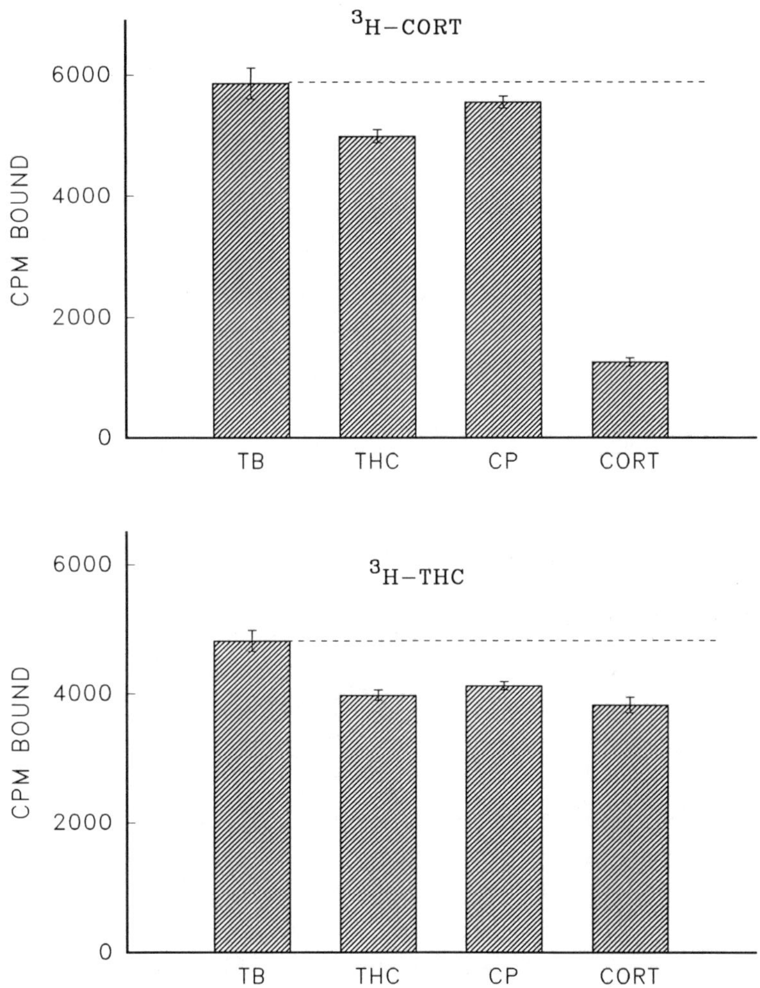

FIGURE 4. Comparison of binding of radiolabeled corticosterone (CORT) and radiolabeled Δ^8-tetrahydrocannabinol (THC) in hippocampal cytosol from adrenalectomized rats. Pooled cytosols were prepared and incubated as described in References 68 and 71 with 3 μ*M* unlabeled Δ^9-THC, CP-55940 (CP), or CORT for 30 min at 4°C, and then 3 n*M* tracer for 60 min at 4°C. Each bar represents the mean ± S.E. of 7 assay tubes. Cytosol protein = 2.2 mg/ml. TB, total binding.

molar excess of unlabeled ligand. One might attribute poor THC competition against CORT to its relatively lower affinity, but the lower panel in Figure 4 shows THC competition against radioactive THC binding. Perhaps only a small portion of the protein molecules that attracted radiolabeled

THC contained saturable receptor sites that 100-fold molar excess could displace. Nevertheless, the study provides evidence that THC can bind to one or more cytosolic proteins with some degree of specificity and saturability, at nanomolar concentrations, and that glucocorticoids can interact with this binding.

An effect of cannabinoids to inhibit or restrict glucocorticoid binding to its receptor in brain structures could be responsible for the observation reported on several occasions, by ourselves and others, that THC administration to rodents stimulated release of pituitary ACTH and adrenal corticosterone.[25,31,34-43] ACTH secretion is under a tonic negative feedback inhibition by glucocorticoids acting through sites in the pituitary, hypothalamus, and the hippocampus.[48-50] Stress-induced release of ACTH and corticosterone can be inhibited by prior administration of glucocorticoids[50] and, for that matter, THC-induced corticosterone release can also be reduced by pretreatment with DEX.[31] Therefore, it seems plausible that administered cannabinoids could "disinhibit" the negative feedback of endogenous levels of corticosterone, leading to a rise of ACTH secretion.

The observations that cannabinoids are capable of interaction with the hippocampal glucocorticoid receptor system are not completely restricted to inhibitory or antagonistic effects. We have also determined that treatment of adrenalectomized rats with THC for 14 d induced an apparent down-regulation of hippocampal glucocorticoid receptor content (results are presented and discussed more completely in Reference 71). This loss of binding, restricted to the type II receptor population, was caused by an apparent reduction of number of available sites, with no effect seen on affinity of receptor for ligand. The concentration of THC in the cytosols was not sufficiently high to interfere with the ligand binding assay, leading to a conclusion that the hippocampal target tissue of treated rats indeed contained fewer receptor molecules. Treatment of adrenalectomized rats with CORT has been shown to cause a down-regulation, or reduced content, of glucocorticoid receptors in the hippocampus;[62,63] thus, the finding that THC treatment produced a similar result suggests that the cannabinoid was capable of acting as a glucocorticoid *agonist* in this instance. It remains to be determined if receptor synthesis has been diminished in THC-treated animals, or if receptor function of an unchanged population has been altered in a fundamental way.

However, the possibility that cannabinoids can be both agonist and antagonist to the glucocorticoid receptor system in brain raises some interesting questions regarding the ability of THC to mediate, directly, the structural pathology observed earlier by us in hippocampi of THC-treated intact rats.[42] For example, are cannabinoids sufficiently agonistic, at chronic doses in the milligram per day range, to imitate glucocorticoids and induce (advance) the same neuropathologic damage associated with aging and stress exposure? Does chronic THC exposure, by inhibiting negative feedback control of ACTH, result in chronically elevated glucocorticoid secretion that becomes toxic to hippocampal neurons? Perhaps elements of both actions occur simultaneously.

Finally, the observation that chronic THC administration can reduce glucocorticoid binding in the hippocampus could be a component of the psychoactive actions of cannabinoids. Glucocorticoid binding in hippocampus is importantly involved in behavioral plasticity and a range of other neurologic functions,[49] and it seems plausible that cannabinoids might be able to alter the effects of adrenal steroids on behavior. This could result in a dampening of the brain's responsiveness to stressful stimuli. In turn, it seems conceivable that such a dampening could be perceived as "reinforcing", or it could contribute to the pronounced tolerance that develops for many pharmacological actions of cannabinoids.[3] Nevertheless, considerable additional research remains to be done in order to fully evaluate the possible role of the hippocampal glucocorticoid receptor system as a mediator of the various pharmacological actions of THC.

STEROID INTERACTION WITH CANNABINOID BINDING IN THE HIPPOCAMPUS

The search for specific cannabinoid binding sites in the brain has been neither simple nor direct. In addition to exploration of the possibility that THC could associate with intracellular sites such as steroid receptors, much attention has also been devoted to localizing cell membrane binding sites because of the relatively rapid onset of psychoactive responses to smoked marijuana (cf. reviews in References 2 through 4). Some years ago, Harris and co-workers described a displacable, high-affinity binding of [^{3}H]-Δ^8-THC in hepatoma cells;[75] however, this finding proved difficult to follow. Among other technical problems, naturally occurring cannabinoids are quite resinous, hydrophobic, and difficult to work with in aqueous buffered solutions. In 1985, another group, using radiolabeled trimethyl-ammonium-

Δ^8-THC (TMA), a less hydrophobic quaternary cannabinoid analog, was able to demonstrate a membrane binding site of modest affinity (K_d = 89 n*M*).[76] However, both psychoactive and nonpsychoactive cannabinoids displaced TMA binding, and this finding created uncertainty about the validity of the low specificity binding site.

It was not until the appearance of some synthetic compounds originally devised by the pharmaceutical industry as nonnarcotic analgesics that a specific membrane receptor site for cannabinoids could be localized (reviewed in Reference 2). Using [^{3}H]-CP-55940 (CP) (Figure 1), very high-affinity binding sites have now been demonstrated in rat brain.[1,2] Autoradiographic studies with [^{3}H]-CP have localized dense binding in the hippocampus, molecular layer of cerebellum, and olfactory bulb and basal ganglia regions.[77,78] This receptor, which has been cloned and transfected for expression in cultured neuroblastoma cells,[79,80] binds natural cannabinoids in a structure-activity relationship that matches biological cannabinoid effects, although natural cannabinoids display lower affinities for binding than do the most potent synthetic CP compounds.[1,2,81] The receptor complex appears to include a guanine nucleotide regulatory site, G_i, which inhibits adenylate cyclase activity in target cells.[81-83] As reported for ligand binding, enantioselective activity for cyclase inhibition has also been observed for both naturally occurring cannabinoids and for CP compounds.[81] It is now postulated that many of the psychoactive effects of smoked marijuana can be traced to the activation of this receptor system in various brain locations.[2,78]

Our laboratory has begun to study the CP binding system in hippocampus with a view to investigating relationships between steroids and cannabinoids at this site. We have determined that membrane preparations of hippocampus dissected from rat brain contain a very large number of high-affinity binding sites for [^{3}H]-CP (Figure 5), which is in agreement with published autoradiography studies.[77,78] The quantity and affinity of the sites in whole hippocampi of rats is approximately the same as those found in whole cerebellum preparations (Figure 5).

While developing CP binding assays in tissue dissected from individual animals, our laboratory observed a substantial variation in number, but not affinity, of receptor sites between animal subjects. Analysis of tissues obtained from adrenalectomized rats resulted in enhanced precision of the binding assay, which led to a postulation that stress or adrenocortical steroids could be modulators of the CP receptor. An experiment was conducted in which adrenalectomized rats were injected with corticosterone,

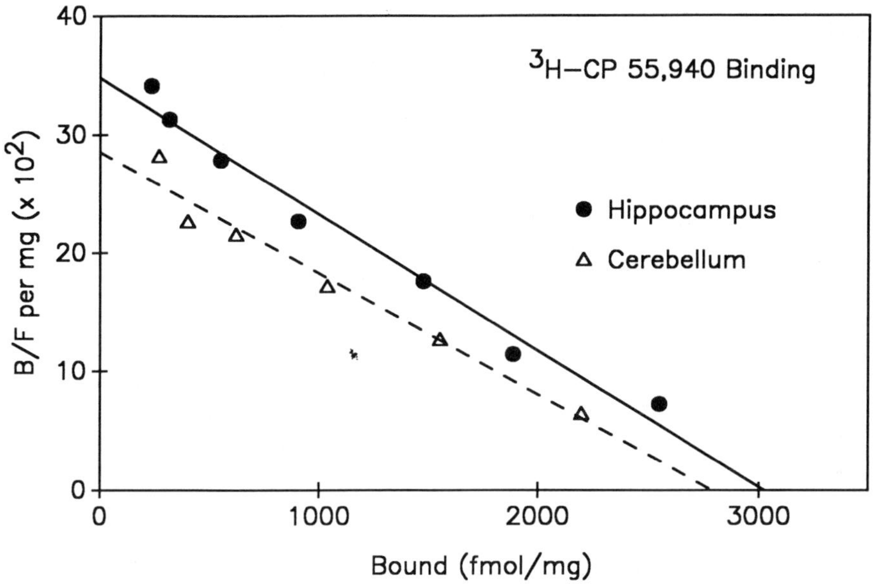

FIGURE 5. Scatchard plots of [^{3}H]-CP-55940 (CP) binding in membrane preparations from rat hippocampus or cerebellum. Each point represents the average difference of triplicate incubations of total binding and nonspecific binding (500-fold molar excess unlabeled CP) from two adrenalectomized rats. Methods for membrane preparation and binding assay adopted from Reference 1. Protein content = 50 μg/tube. Plot data for hippocampus: K_d = 87 p*M*, B_{max} = 3022 fmol/mg, r = 0.967. Plot data for cerebellum: K_d = 98 p*M*, B_{max} = 2791 fmol/mg, r = 0.979.

5 mg/kg, s.c. in oil, for 2 d. Analysis of CP binding in hippocampi revealed an apparent increase in binding after CORT treatment (Table 1). There was no change in the binding capacity of cerebellar tissue dissected from the same animals. This result suggested that the presence of glucocorticoids was able to enhance cannabinoid binding in hippocampus, but not cerebellum. Further studies of a kinetic nature will be necessary in order to determine if this observation was the result of altered binding affinity for the ligand or of a change in the number of available binding sites.

It has been recently reported that corticosterone, as well as other steroid hormones and a number of neurotransmitter and neuromodulator substances, failed to displace or augment binding of radiolabeled CP in whole rat brain membrane preparations.[83] Attempts by ourselves to determine whether *in vitro* incubation of hippocampal membranes with CORT could alter the amount of CP binding yielded inconsistent results. However, Howlett et al.

TABLE 1
Net Binding (dpm/μg protein) of [^{3}H]-CP-55940 to Membranes Prepared from the Hippocampi or Cerebellum of Adrenalectomized Rats Injected for 2 d with Corticosterone (5 mg/kg/d s.c. in oil) or with Vehicle Alone

	CORT-treated	Vehicle
Hippocampus	597.9 ± 19.6	500.8 ± 13.7[a]
Cerebellum	457.6 ± 28.3	453.8 ± 11.0

Note: Membrane preparation and binding assay conducted as described in Reference 1. Data are means ± S.E. of three animals assayed in duplicate in the same run. Hippocampus and cerebellum tissues were dissected and assayed from the same animals in each experiment. Protein content = 50 μg/tube; tracer = 2.5 n*M*/tube (300,000 dpm at 90 Ci/mmol); NSB at 500-fold molar excess unlabeled CP = 200 to 250 dpm/μg protein.

[a] Significantly different from CORT-treated, $p < 0.05$.

have frequently observed that cannabinoid binding is negatively influenced by GTP interaction at the G_i regulatory site.[82,83] We therefore decided to test whether the increased CP binding we had observed following *in vivo* glucocorticoid exposure (Table 1) could have been due to an effect of steroids to protect or stabilize the cannabinoid receptor complex in the face of guanine nucleotide inhibition.

In several experiments, aliquots of membranes prepared from hippocampi of adrenalectomized rats were dispensed into two sets of incubation vials. A steroid hormone, at a concentration range of 10^{-9} to 10^{-5} *M*, was added to both sets of tubes followed by introduction of guanylyl-5′-imidophosphate (GPP), a nonhydrolyzable analog of GTP, to one set of assay tubes. The laboratory of Childers has shown that GTP inhibits enkephalin binding to opiate receptors, in a fashion similar to that for CP binding to cannabinoid receptors, and that various substances are able to counteract GTP inhibition of opiate binding.[84,85] In our studies, following incubation with [^{3}H]-CP, the radioactivity remaining bound to membranes with or without GPP was determined. As shown in Figure 6 (upper panel), preincubation with corticosterone did not appreciably alter the number of available CP binding sites. However, in the tubes in which binding was inhibited by addition of GPP, the presence of CORT was able to restore total binding toward control levels.

Owing to our earlier finding of enhanced CP binding following *in vivo* exposure to CORT, and because a substantial interest has been generated

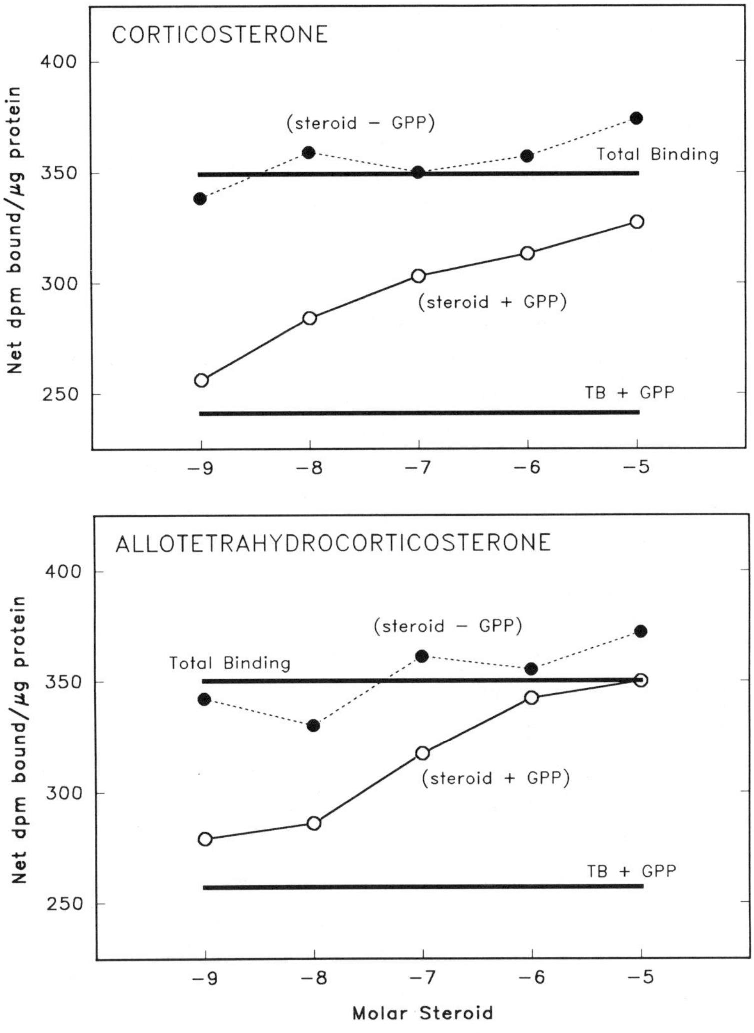

FIGURE 6. Effect of steroid incubation with hippocampal membranes on binding of [^{3}H]-CP-55940. Methods for membrane preparation and binding assay adopted from Devane et al.,[1] with additional methods for guanylyl-5′-imidophosphate (GPP) incubation taken from Childers.[84] Each point represents the average of two tubes. Four assay tubes (50 μg protein per tube) were incubated with each steroid, at each concentration, for 60 min at 25°C. GPP (8.33 m*M*) was then added to 2 of each set of 4 tubes for 10 min, still at 25°C, followed by radiolabeled CP (3 n*M*), into all tubes. After 30 min of additional incubation at 30°C, tubes were centrifuged and the pellets were counted. Total binding (TB) = membranes plus tracer, no steroid or GPP; TB + GPP = membranes with GPP and tracer, with no steroid added.

lately in a family of membrane-active steroids that are metabolites of gonadal and adrenal steroids,[86] it was decided to repeat the *in vitro* test model to examine the restoration of GPP-suppressed binding by the concomitant presence of GPP in the presence of several of these "neurosteroids". Indeed, many steroids have been shown to enhance [^{3}H]-muscimol binding to rat cortical membrane $GABA_A$ receptors (which are, however, not associated with G proteins). Results with steroid-modulated, GPP-inhibited CP binding were not positive, for the most part. When preincubated with alphaxalone, androsterone, tetrahydrodeoxycorticosterone (THDOC), 5α-dihydrotestosterone (DHT), 3α,5α-tetrahydroprogesterone, 3α,5β-tetrahydroprogesterone, or 3α,5α,20-hexahydroprogesterone, the hippocampal membranes failed either to increase [^{3}H]-CP binding alone, or to demonstrate a recovery of CP binding in the presence of GPP. However, one steroid metabolite did appear capable of influencing CP binding in hippocampus: 3α-hydroxy-5α-tetrahydrocorticosterone (allotetrahydrocorticosterone, ATHB) (Figure 1). The addition of ATHB without GPP enhanced total binding of CP in a slight but dose-related manner (Figure 6), and preincubation with ATHB resulted in a restoration of GPP-inhibited CP binding to the original levels. Most of the neurosteroids that have been demonstrated to possess modulatory action at the $GABA_A$ receptor are metabolites of progesterone rather than glucocorticoids. However, ATHB is a 5α-reduced metabolite of corticosterone, with hydroxyl groups at carbons 21 and 11 that typify glucocorticoids.

It is interesting to note that two of the more prominent pharmacological actions of cannabinoids, namely an anxiolytic effect and an anesthetic effect,[2,3] are both being reported as properties of many pregnane-derived neurosteroids.[87-91] As stated above, the brain receptor type usually identified as the mediator for these neurosteroids is the $GABA_A$ receptor, which also responds to barbiturates and which functions by controlling the chloride ionophore in target neurons. In studies of the neuroendocrine control of GnRH, the hypothalamic-releasing hormone that stimulates release of pituitary FSH and LH, it has been demonstrated that administered progesterone stimulates hypothalamic GnRH release.[92] A progesterone metabolite, 3α,5α-tetrahydroprogesterone (THP_4), an anesthetic more potent than barbiturates, also stimulates GnRH release through interaction with the $GABA_A$ receptor.[91] This progesterone metabolite is an identical reduction product, from progesterone, as is ATHB from corticosterone. THP_4 does not bind to cytosolic progesterone receptors;[92] ATHB does not bind to cytosolic glucocorticoid receptors.[89] However, it is also known that progesterone and opiates antagonize each other. Kalra and co-workers have demonstrated that morphine administration blocks the ability of progesterone to stimulate GnRH release,[93] and they have reported that progesterone administration diminishes opiate binding in the medial basal hypothalamus and the preoptic area where GnRH release is controlled.[94] Thus, if progesterone and its 5α,3α-hydroxy metabolite can interact with opiate receptors, which appear to contain a G protein subunit similar to the cannabinoid receptor, it seems reasonable to consider that a glucocorticoid metabolite might be likewise disposed to interact with hippocampal cannabinoid receptors.

It has recently been demonstrated that pregnenolone, the principal precursor to all steroid biosynthesis, is present in the brain at high concentrations that do not decrease following gonadectomy and adrenalectomy.[95] Furthermore, studies of primary cultures of rat glial cells have shown that the neurosteroids associated with membrane binding events are capable of being synthesized *in situ*,[96] so it appears that the brain may possess a paracrine steroid modulator system for rapid control of certain functions. It is intriguing to consider that glucocorticoids and/or their metabolites could be involved as modulators of an anxiolytic-analgetic system linked to a cannabinoid-binding receptor in the hippocampus.

SUMMARY AND CONCLUSIONS

For many years, the detailed mechanisms responsible for the psychoactive effects of cannabinoids have remained undefined. Progress on this issue was so slow that some were beginning to conclude that these extremely lipophilic compounds might not act through "typical receptor" systems, but were more likely interacting with cell membranes in a manner not dissimilar to that of ethanol. Nevertheless, the spectrum of observed behavioral and pharmacologic actions that occur following cannabinoid treatment of test animals, and after human subjects have inhaled marijuana smoke, supported the theoretical possibility that specific anatomical and biochemical loci must be central to THC actions in the brain. For example, a number of behavioral effects displayed enantioselectivity; indeed, the stereoisomers of Δ^9-THC itself vary widely in potency (reviewed in References 1 through 4).

This perplexing situation has caused many, including our laboratories, to explore the possibility that cannabinoid activity would be expressed, at least in part, through interaction with receptor systems that were already known and defined for endogenous ligands. Such a possibility is not at all unusual since many of our most important clinical drugs originate from plant-synthesized chemicals that "happen" to insert into a mammalian molecule with receptor-like transduction (e.g., morphine, cocaine, phenobarbital, atropine, and digoxin). While not all of these substances match with known endogenous ligands (as yet), it has always seemed likely that ample opportunity should exist for cannabinoids to find one or more preexisting locations with which to interact.

Our laboratories began to study cannabinoid interaction with the glucocorticoid receptor system of hippocampus because of very compelling results by Landfield et al.[42] showing that chronic THC administration induced aging-like degenerative changes in the rat brain that strongly resembled the effects of stress exposure and elevated corticosterone secretion. Our results indicate that cannabinoids can interact with the glucocorticoid receptor system of hippocampus. The interaction is probably one of weak antagonism, although an agonist activity has also been suggested. Enantioselective activity has likewise been observed because cannabidiol, a naturally-occurring cannabinoid of very weak pharmacologic activity[3,4] is also much weaker than THC in its glucocorticoid receptor interactive effects. If the type of morphologic degeneration observed in the hippocampus following stress, CORT treatment, or normal aging can be attributed to glucocorticoid actions toxic to those neurons, then we would conclude that the similar response we observed following chronic THC administration should also be due to cannabinoid-glucocorticoid receptor interaction. Either the cannabinoid exerts agonist-like properties and imitates the neurotoxic actions of glucocorticoids or else THC inhibits negative feedback control over pituitary ACTH secretion that results in a greatly elevated ACTH and glucocorticoid output, with the stimulated steroids being toxic. Additional experiments are presently underway to investigate specific cellular markers of glucocorticoid action in order to determine whether THC can specifically stimulate (or inhibit) biochemical events in the hippocampus.

Regarding the membrane receptor system that binds synthetic cannabinoid ligands with a very high affinity, the discovery is yet too new to associate the numerous pharmacologic actions of THC with this receptor.

The system appears to contain a guanine nucleotide regulatory element that modulates cannabinoid binding, and we have observed that certain steroids, particularly corticosterone and its reduced metabolite, 3α, 5α-tetrahydrocorticosterone, are capable of modulating GTP-induced reduction of CP binding. The purpose for this apparent stabilization or protection of the cannabinoid receptor complex remains unclear. One possibility is that stress conditions might, among other things, "up-regulate" the cannabinoid receptor so as to make an anxiolytic ligand more effective. A state of lowered stress sensation may be considered as a concomitant low-anxiety state, so an agent that serves to sedate or relax tension would have less of a target. It is likely that the state of stress and anxiety must involve an *induction* of systems that are aroused or in a heightened state of activity, which would then give anxiolytic substances a better target for action. Whether cannabinoids and glucocorticoids act in such concert in the hippocampal formation remains to be firmly established. Evidence suggests that this is possible; but at present, we know far too little about specific actions of either glucocorticoids or of cannabinoids and their potential to mediate the constellation of behavioral, pharmacological, and pathological events in the brain.

ACKNOWLEDGMENTS

These studies have been supported in part by grants DA-03637, DA-06218, and P50 DA-06634 from the National Institute on Drug Abuse. The authors also wish to express appreciation to Centre de Recherches Roussel-Uclaf for providing RU-28362.

REFERENCES

1. **Devane, W. A., Dysarz, F. A., III, Johnson, M. R., Melvin, L. S., and Howlett, A. C.,** Determination and characterization of a cannabinoid receptor in rat brain, *Mol. Pharmacol.,* 34, 605, 1988.
2. **Howlett, A. C., Bidaut-Russell, M., Devane, W. A., Melvin, L. S., Johnson, M. R., and Herkenham, M.,** The cannabinoid receptor: biochemical, anatomical and behavioral characterization, *Trends Neurosci.,* 13, 420, 1990.
3. **Dewey, W. L.,** Cannabinoid pharmacology, *Pharmacol. Rev.,* 38, 151, 1986.
4. **Martin, B. R.,** Cellular effects of cannabinoids, *Pharmacol. Rev.,* 38, 45, 1986.

5. **Szara, S.,** Δ^9-Tetrahydrocannabinol: potential precursor of "false hormones"?, in *Psychopharmacology, Sexual Disorders and Drug Abuse,* Ban T. A., Boissier, J. R., Gessa, G. J., Heimann, H., Hollister, L., Lehman, H. E., Steinberg, H., Sulser, F., Sundwall, A., and Vinar, O., Eds., North-Holland, Amsterdam, 1973, 707.
6. **Wenger, T., Rettori, V., Snyder, G. D., Dalterio, S., and McCann, S. M.,** Effects of Δ^9-tetrahydrocannabinol on the hypothalamic-pituitary control of luteinizing hormone and follicle-stimulating hormone secretion in adult male rats, *Neuroendocrinology,* 46, 488, 1987.
7. **Steger, R. W., Murphy, L. L., Bartke, A., and Smith, M. S.,** Effects of psychoactive and nonpsychoactive cannabinoids on the hypothalamic-pituitary axis of the adult male rat, *Pharmacol. Biochem. Behav.,* 37, 299, 1990.
8. **Murphy, L. L., Steger, R. W., Smith, M. S., and Bartke, A.,** Effects of Δ^9-tetrahydrocannabinol, cannabinol and cannabidiol alone and in combinations, on luteinizing hormone and prolactin release and on hypothalamic neurotransmitters in the male rat, *Neuroendocrinology,* 52, 316, 1990.
9. **Kumar, A. M., Haney, M., Becker, T., Thompson, M. L., Kream, R. M., and Miczek, K.,** Effects of early exposure to Δ^9-tetrahydrocannabinol on the levels of opioid peptides, gonadotropin releasing hormone and substance P in the adult male rat brain, *Brain Res.,* 525, 78, 1990.
10. **Tyrey, L.,** Δ^9-Tetrahydrocannabinol: a potent inhibitor of episodic LH secretion, *J. Pharmacol. Exp. Ther.,* 213, 306, 1980.
11. **Steger, R. W., Silverman, A. Y., Siler-Khodr, T. M., and Asch, R. H.,** The effect of Δ^9-tetrahydrocannabinol on the positive and negative feedback control of luteinizing hormone release, *Life Sci.,* 27, 1911, 1980.
12. **Steger, R. W., De Paolo, L., Asch, R. H., and Silverman, A. Y.,** Interactions of Δ^9-tetrahydrocannabinol with hypothalamic neurotransmitters controlling luteinizing hormone and prolactin release, *Neuroendocrinology,* 37, 361, 1983.
13. **Kumar, A. M., Solomon, J., Patel, V., Kream, R. M., Drieze, J. M., and Millard, W. J.,** Early exposure to Δ^9-tetrahydrocannabinol influences neuroendocrine and reproductive functions in female rats, *Neuroendocrinology,* 44, 260, 1986.
14. **Dalterio, S. L., Bernard, S. A., and Esquivel, C. R.,** Acute Δ^9-tetrahydrocannabinol exposure alters calcium ATPase activity in neuroendocrine and gonadal tissue in mice, *Eur. J. Pharmacol.,* 137, 91, 1987.
15. **Murphy, L. L. and Tyrey, L.,** Induction of luteinizing hormone release by electrochemical stimulation of the medial preoptic area in Δ^9-tetrahydrocannabinol-blocked proestrous rats, *Neuroendocrinology,* 43, 471, 1986.
16. **Asch, R. H., Fernandez, E. O., Smith, C. G., and Pauerstein, C. J.,** Pre-coital single doses of Δ^9-tetrahydrocannabinol block ovulation in the rabbit, *Fertil. Steril.,* 31, 331, 1979.
17. **Walters, D. E. and Carr, L. A.,** Perinatal exposure to cannabinoids alters neurochemical development in rat brain, *Pharmacol. Biochem. Behav.,* 29, 213, 1988.
18. **Field, E. and Tyrey, L.,** Delayed sexual maturation during prepubertal cannabinoid treatment: importance of the timing of treatment, *J. Pharmacol. Exp. Ther.,* 254, 171, 1990.
19. **Patra, P. B. and Wadsworth, R. M.,** Effect of the synthetic cannabinoid nabilone on spermatogenesis in mice, *Experientia,* 46, 852, 1990.
20. **Heindel, J. J. and Keith, W. B.,** Specific inhibition of FSH-stimulated cAMP accumulation by delta-9-tetrahydrocannabinol in cultures of rat Sertoli cells, *Toxicol. Appl. Pharmacol.,* 101, 124, 1989.

21. **Dalterio, S., Mayfield, D., Bartke, A., and Morgan, W.,** Effects of psychoactive and non-psychoactive cannabinoids on neuroendocrine and testicular responsiveness in mice, *Life Sci.,* 36, 1299, 1985.
22. **Dalterio, S., Steger, R., Peluso, J., and De Paolo, L.,** Acute Δ^9-tetrahydrocannabinol exposure: effects on hypothalamic-pituitary-testicular activity in mice, *Pharmacol. Biochem. Behav.,* 26, 533, 1987.
23. **Schurr, A.,** Marijuana: much ado about THC, *Comp. Biochem. Physiol.,* 80, 1, 1985.
24. **Adashi, E. Y., Jones, P. B., and Hsueh, A. J.,** Direct antigonadal activity of cannabinoids: suppression of rat granulosa cell functions, *Am. J. Physiol.,* 244, E177, 1985.
25. **Cone, E. J., Johnson, R. E., Moore, J. D., and Roache, J. D.,** Acute effects of smoking marijuana on hormones: subjective effects and performance in male human subjects, *Pharmacol. Biochem. Behav.,* 24, 1749, 1986.
26. **Mendelson, J. H., Mello, N. K., Ellingboe, J., Skupny, A. S., Lex, B. W., and Griffin, M.,** Marijuana smoking suppresses luteinizing hormone in women, *J. Pharmacol. Exp. Ther.,* 237, 862, 1986.
27. **Wiegant, V. M., Sweep, C. G., and Nir, I.,** Effect of acute administration of Δ^1-tetrahydrocannabinol on β-endorphin levels in plasma and brain tissue of the rat, *Experientia,* 43, 413, 1987.
28. **Tyrey, L.,** Reversal of the Δ^9-tetrahydrocannabinol inhibitory effect on prolactin secretion by rostral deafferentation of the medial basal hypothalamus, *Neuroendocrinology,* 44, 204, 1986.
29. **Rettori, V., Wenger, T., Snyder, G., Dalterio, S., and McCann, S. M.,** Hypothalamic action of Δ^9-tetrahydrocannabinol to inhibit the release of prolactin and growth hormone in the rat, *Neuroendocrinology,* 47, 498, 1988.
30. **Puder, M., Nir, I., Siegel, R. A., Weidenfield, J., and Avalon, D.,** The effect of Δ^1-tetrahydrocannabinol on LH release in castrated and hypothalamic deafferentated male rats, *Exp. Brain Res.,* 59, 213, 1985.
31. **Kokka, N. and Garcia, J. F.,** Effects of Δ^9-tetrahydrocannabinol on growth hormone and ACTH secretion in rats, *Life Sci.,* 15, 329, 1974.
32. **Nazar, B., Kairys, D. J., Fowler, R., and Harclerode, J.,** Effects of Δ^9-tetrahydrocannabinol on serum thyroxine concentration in the rat, *J. Pharm. Pharmacol.,* 29, 778, 1977.
33. **Hilliard, C. S., Farber, N. E., Hagen, T. C., and Bloom, A. S.,** The effects of Δ^9-tetrahydrocannabinol on serum thyroid-stimulating hormone levels in the rat, *Pharmacol. Biochem. Behav.,* 20, 547, 1984.
34. **Dewey, W. L., Peng, T. C., and Harris, L. S.,** The effect of l-*trans*-Δ^9-tetrahydrocannabinol on the hypothalamo-hypophyseal adrenal axis of rats, *Eur. J. Pharmacol.,* 12, 382, 1970.
35. **Kubena, R. K., Pernatch, J. L., and Barry, H.,** Corticosterone elevation mediated centrally by Δ^1-tetrahydrocannabinol in rats, *Eur. J. Pharmacol.,* 14, 89, 1971.
36. **Barry, H., Kubena, R. K., and Pernatch, J. L.,** Pituitary-adrenal activation and related responses to Δ^1-tetrahydrocannabinol, *Progr. Brain Res.,* 39, 323, 1973.
37. **Johnson, K. M., Dewey, W. L., Ritter, K. S., and Beckner, J. S.,** Cannabinoid effects on plasma corticosterone and uptake of ^{3}H-corticosterone by mouse brain, *Eur. J. Pharmacol.,* 47, 303, 1978.
38. **Jacobs, J. A., Dellarco, A. J., Manfredi, R. A., and Harclerode, J.,** The effects of Δ^9-tetrahydrocannabinol cannabidiol and shock on plasma corticosterone concentrations in rats, *J. Pharm. Pharmacol.,* 31, 341, 1979.

39. **Miczek, K. A. and Dixit, B. N.,** Behavioral and biochemical effects of chronic Δ^9-tetrahydrocannabinol in rats, *Psychopharmacology,* 67, 195, 1980.
40. **Puder, M., Weidenfeld, J., Chowers, I., Nir, I., Conforti, N., and Siegel, R. A.,** Corticotrophin and corticosterone secretion following Δ^9-tetrahydrocannabinol in intact and hypothalamic deafferentated male rats, *Exp. Brain Res.,* 46, 85, 1982.
41. **Zuardi, A. W., Teixeira, N. A., and Karniol, I. C.,** Pharmacologic interaction of the effects of Δ^9-tetrahydrocannabinol and cannabidiol on serum corticosterone levels in rats, *Arch. Int. Pharmacodyn. Ther.,* 269, 12, 1984.
42. **Landfield, P. W., Cadwallader, L. B., and Vinsant, S.,** Quantitative changes in hippocampal structure following long-term exposure to Δ^9-tetrahydrocannabinol: possible mediation by glucocorticoid systems, *Brain Res.,* 443, 47, 1988.
43. **Eldridge, J. C., Murphy, L. L., and Landfield, P. W.,** Cannabinoids and the hippocampal glucocorticoid receptor: recent findings and possible significance, *Steroids,* 56, 226, 1991.
44. **Warner, W., Harris, L. S., and Carchman, R. A.,** Inhibition of corticosteroidogenesis by Δ^9-tetrahydrocannabinol, *Endocrinology,* 101, 1815, 1977.
45. **Benowitz, N. L., Jones, R. T., and Lerner, C. B.,** Depression of cortisol and growth hormone response to insulin-induced hypoglycemia after prolonged oral Δ^9-tetrahydrocannabinol administration in man, *J. Clin. Endocrinol. Metab.,* 42, 938, 1976.
46. **Perez-Reyes, M.,** Clinical study of frequent marijuana use: adrenocortical reserve, metabolism of a contraceptive agent and development of tolerance, *Ann. N.Y. Acad. Sci.,* 282, 173, 1976.
47. **Drew, W. G. and Slagel, D. E.,** Δ^9-Tetrahydrocannabinol: selective impairment of corticosterone uptake by limbic structures of the rat, *Neuropharmacology,* 12, 909, 1973.
48. **Conforti, N. and Feldman, S.,** Effects of dorsal fornix section and hippocampectomy on adrenocortical responses to sensory stimulation in the rat, *Neuroendocrinology,* 22, 1, 1976.
49. **De Kloet, E. R.,** Brain corticosteroid receptor balance and homeostatic control, *Frontiers Neuroendocrinol.,* 12, 95, 1991.
50. **Jacobson, L. and Sapolsky, R.,** The role of the hippocampus in feedback regulation of the hypothalamic-pituitary-adrenocortical axis, *Endocrine Rev.,* 12, 118, 1991.
51. **Solomon, J., Cocchia, M. A., Gray, R., Shattuck, D., and Vossmer, A.,** Uterotropic effect of delta-9-tetrahydrocannabinol in ovariectomized rats, *Science,* 192, 559, 1976.
52. **Rawitch, A. B., Schultz, G. S., Ebner, K. E., and Vardaris, R. M.,** Competition of delta-9-tetrahydrocannabinol with estrogen in rat uterine estrogen receptor binding, *Science,* 197, 1189, 1977.
53. **Okey, A. B. and Bondy, G. P.,** Δ^9-Tetrahydrocannabinol and 17β-estradiol bind to different macromolecules in estrogen target tissues, *Science,* 200, 312, 1978.
54. **Smith, R. G., Besch, N. F., Besch, P. K., and Smith, C. G.,** Inhibition of gonadotropin by Δ^9-tetrahydrocannabinol: mediation by steroid receptors?, *Science,* 204, 325, 1979.
55. **Purohit, V., Ahluwahlia, B. S., and Vigersky, R. A.,** Marijuana inhibits dihydrotestosterone binding to the androgen receptor, *Endocrinology,* 107, 848, 1980.
56. **McEwen, B. S., Biegon, A., Davis, P. G., Krey, L. C., Luine, V. N., McGinnis, M. Y., Paden, C. M., Parsons, B., and Rainbow, T. C.,** Steroid hormones: humoral signals which alter brain cell properties and functions, *Rec. Progr. Horm. Res.,* 38, 41, 1982.
57. **Stumpf, W. E. and Sar, M.,** Anatomical distribution of corticosterone-concentrating neurons in rat brain, in *Anatomical Neuroendocrinology,* Stumpf, W. E. and Grant, L. D., Eds., Karger, Basel, 1975, 254.

58. **De Kloet, E. R. and Reul, J. M. H. M.,** Feedback action and tonic influence of corticosteroids on brain function: a concept arising from the heterogeneity of brain receptor systems, *Psychoneuroendocrinology,* 12, 83, 1987.
59. **Moguilewsky, M. and Raynaud, J. P.,** Evidence for a specific mineralocorticoid receptor in rat brain, *J. Steroid Biochem.,* 12, 309, 1980.
60. **Reul, J. M. H. M. and De Kloet, E. R.,** Two receptor systems for corticosterone in rat brain: microdistribution and differential occupation, *Endocrinology,* 117, 2505, 1985.
61. **Eldridge, J. C., Kute, T. E., Brodish, A., and Landfield, P. W.,** Apparent age-related resistance of type II hippocampal corticosteroid receptors to down-regulation during chronic escape training, *J. Neurosci.,* 9, 3237, 1989.
62. **Sapolsky, R. M. and McEwen, B. S.,** Down-regulation of neural corticosterone receptors by corticosterone and dexamethasone, *Brain Res.,* 339, 161, 1985.
63. **Tornello, S., Orti., F., De Nicola, A. F., Rainbow, T. C., and McEwen, B. S.,** Regulation of glucocorticoid receptors in rat brain by corticosterone treatment of adrenalectomized rats, *Neuroendocrinology,* 35, 411, 1982.
64. **Kerr, D. S., Campbell, L. W., Applegate, M. D., Brodish, A., and Landfield, P. W.,** Chronic stress-induced acceleration of electrophysiologic and morphometric biomarkers of hippocampal aging, *J. Neurosci.,* 11, 1316, 1991.
65. **Landfield, P. W.,** Modulation of brain aging correlates by long-term alterations of adrenal steroids and neurally-active peptides, in *Progress in Brain Research,* Vol. 72, De Kloet, E. R., Wiegant, V. M., and De Wied, D., Eds., Elsevier, Amsterdam, 1987, 279.
66. **Landfield, P. W., Baskin, R. K., and Pitler, T. A.,** Brain aging correlates: retardation by hormonal-pharmacological treatments, *Science,* 214, 581, 1981.
67. **Landfield, P. W., Sundberg, D. K., Smith, M. S., Eldridge, J. C., and Morris, M.** Mammalian brain aging: theoretical implications of changes in brain and endocrine systems during mid- and late-life, *Peptides,* 1 (Suppl. 1), 185, 1980.
68. **Eldridge, J. C., Fleenor, D. G., Kerr, W. S., and Landfield, P. W.,** Impaired up-regulation of type II corticosteroid receptors in hippocampus of aged rats, *Brain Res.,* 478, 248, 1989.
69. **Hermann, T., Schramm, K., and Ghraf, R.,** Photoaffinity labeling with [^{3}H]-RU-28362: a powerful tool for the study of rat brain glucocorticoid receptors, *J. Steroid Biochem.,* 26, 417, 1987.
70. **Raynaud, J. P., Ojasoo, T., Jouquey, A., Moguilewsky, M., and Teutsch, G.,** Probes for steroid receptors, in *Endocrinology,* Labrie, F. and Proulx, L., Eds., Elsevier, New York, 1984, 533.
71. **Eldridge, J. C. and Landfield, P. W.,** Cannabinoid interactions with glucocorticoid receptors in rat hippocampus, *Brain Res.,* 534, 135, 1990.
72. **Suthers, M. B., Pressley, L. A., and Funder, J. W.,** Glucocorticoid receptors: evidence for a second, non-glucocorticoid binding site, *Endocrinology,* 99, 260, 1976.
73. **Svec, F.,** Differences in the interaction of RU-486 and ketoconazole with the second binding site of the glucocorticoid receptor, *Endocrinology,* 123, 1902, 1988.
74. **Srivastava, D. and Thompson, E. B.,** Two glucocorticoid binding sites on the human glucocorticoid receptor, *Endocrinology,* 127, 1770, 1990.
75. **Harris, L. S., Carchman, R. A., and Martin, B. R.,** Evidence for the existence of specific cannabinoid binding sites, *Life Sci.,* 22, 1131, 1978.
76. **Nye, J. S., Seltzman, H. H., Pitt, C. G., and Snyder, S. H.,** High-affinity cannabinoid binding sites in brain membranes labeled with [^{3}H]-5′-trimethylammonium-delta-8-tetrahydrocannabinol, *J. Pharmacol. Exp. Ther.,* 234, 784, 1985.

77. **Herkenham, M., Lynn, A. B., Little, M. D., Johnson, M. R., Melvin, L. S., De Costa, B. R., and Rice, K. C.,** Cannabinoid receptor localization in brain, *Proc. Natl. Acad. Sci. U.S.A.*, 87, 1932, 1990.
78. **Herkenham, M., Lynn, A. B., Johnson, M. R., Melvin, L. S., De Costa, B. R., and Rice, K.C.,** Characterization and localization of cannabinoid receptors in rat brain: a quantitative *in vitro* autoradiographic study, *J. Neurosci.*, 11, 563, 1991.
79. **Gérard, C., Mollereau, C., Vassart, G., and Parmentier, M.,** Nucleotide sequence of a human cannabinoid receptor cDNA, *Nucleic Acids Res.*, 18, 7142, 1990.
80. **Matsuda, L. A., Lolait, S. J., Brownstein, M. J., Young, A. C., and Bonner, T. I.,** Structure of a cannabinoid receptor and functional expression of the cloned cDNA, *Nature (London)*, 346, 561, 1990.
81. **Howlett, A. C., Johnson, M. R., Melvin, L. S., and Milne, G. M.,** Nonclassical cannabinoid analgetics inhibit adenylate cyclase: development of a cannabinoid receptor model, *Mol. Pharmacol.*, 33, 297, 1988.
82. **Howlett, A. C., Qualy, J. M., and Khachatrain, L. L.,** Involvement of G_i in the inhibition of adenylate cyclase by cannabimimetic drugs, *Mol. Pharmacol.*, 29, 307, 1986.
83. **Bidaut-Russell, M., Devane, W. A., and Howlett, A. C.,** Cannabinoid receptors and modulation of cyclic AMP accumulation in the rat brain, *J. Neurochem.*, 55, 21, 1990.
84. **Childers, S. R.,** Interaction of opiate receptor binding sites and guanine nucleotide regulatory sites: selective protection from *N*-methyl-maleimide, *J. Pharmacol. Exp. Ther.*, 230, 684, 1984.
85. **Lambert, S. M. and Childers, S. R.,** Modification of guanine nucleotide-regulatory components in brain membranes. I. Changes in guanosine-5′-triphosphate regulation of opioid receptor binding sites, *J. Neurosci.*, 4, 2755, 1984.
86. **Baulieu, E. E. and Robel, P.,** Neurosteroids: a new brain function?, *J. Steroid Biochem. Mol. Biol.*, 37, 395, 1990.
87. **Turner, D. M., Ransom, R. W., Yang, J. S.-J., and Olson, R. W.,** Steroid anesthetics and naturally occurring analogs modulate δ-amino-butyric acid receptor complex at a site distinct from barbiturates, *J. Pharmacol. Exp. Ther.*, 248, 960, 1989.
88. **Gee, K. W., Bolger, M. B., Brinton, R. E., Coirini, H., and McEwen, B. S.,** Steroid modulation of the chloride ionophore in rat brain: structure-activity requirements, regional dependence and mechanism of action, *J. Pharmacol. Exp. Ther.*, 246, 803, 1988.
89. **Sutanto, W., Handelmann, G., De Bree, F., and De Kloet, E. R.,** Multifaceted interaction of corticosteroids with the intracellular receptors and with membrane $GABA_A$ receptor complex in the rat brain, *J. Neuroendocrinol.*, 1, 243, 1989.
90. **Crawley, J. N., Glowa, J. R., Majewska, M. D., and Paul, S. M.,** Anxiolytic activity of an endogenous adrenal steroid, *Brain Res.*, 398, 382, 1986.
91. **Majewska, M., Harrison, N., Schwartz, R., Baker, J., and Paul, S. M.,** Steroid hormone metabolites are barbiturate-like modulators of the GABA receptor, *Science*, 232, 1004, 1986.
92. **Brann, D.W. and Mahesh, V. B.,** Regulation of gonadotropin secretion by steroid hormones, *Frontiers Neuroendocrinol.*, 12, 165, 1991.
93. **Kalra, S. P. and Kalra, P. S.,** Opioid-adrenergic-steroid connection in regulation of luteinizing hormone secretion in the rat, *Neuroendocrinology*, 38, 418, 1984.
94. **Jacobson, W. and Kalra, S. P.,** Decreases in mediobasal hypothalamic and preoptic opioid ($[^3H]$-naloxone) binding are associated with the progesterone-induced luteinizing hormone surge, *Endocrinology*, 124, 199, 1989.

95. **Robel, P., Bourreau, E., Corpechot, C., Dang, D., Halberg, F., Clarke, C., Haug, M., Schlegel, M., Synguelakis, M., Vowch, C., and Baulieu, E. E.,** Neurosteroids: 3β-hydroxy-5 derivatives in rat and monkey brain, *J. Steroid Biochem.*, 27, 649, 1987.
96. **Jung-Testas, I., Hu, Z., Baulieu, E. E., and Robel, P.,** Neurosteroids: biosynthesis of pregnenolone and progesterone in primary cultures of rat glial cells, *Endocrinology*, 125, 2083, 1989.

5 Maternal Cannabinoid Exposure and Brain Development: Changes in the Ontogeny of Dopaminergic Neurons

PERINATAL DRUG EXPOSURE: GENERAL ASPECTS

Cannabinoids and other natural or synthetic psychoactive drugs are among the most important agents able to modify neural development in the unborn child when they are used by the mother during gestation and lactation periods.[1] A number of studies have established that these substances transferred from the mother to the developing fetus and neonate may lead to alterations in neural development during both prenatal and postnatal periods. This can produce long-term alterations in adult neurobiological processes controlled by different brain areas.[2-9] During development, psychotrophic drugs may interfere with the rigidly ordered temporal sequence of events that occur in the CNS, events which are known to be particularly vulnerable to the action of epigenetic factors.[10]

It is likely that one of the mechanisms involved in the neuroteratological effects of neurotrophic drugs is a disturbance of neurotransmitters[11] that may have important trophic and modifying influences on the development of the brain during critical periods.[12-15] The vulnerability of neurotransmitters to drug action has regional variations and also depends on the length of exposure. Mirmiran and Swaab[11] have suggested that the last week of prenatal and the first 3 weeks of post-natal life in the rat is the period of increased vulnerability of neurotransmitters to drug actions, as a result of which: (1) the development of brain neurotransmitter target areas is affected and (2) the activity of the neurotransmitter system itself and its receptor sensitivity become permanently hampered. In general, psychoactive drugs cause long-term effects when they are administered early, whereas they only produce transient effects after late exposures. This is because psychoactive drugs can produce indelible modifications, probably at the genomic level, in neurotransmitters and/or other endogenous factors involved in neural development. The phenotypic expression of a neurotransmitter, as well as neuronal migration and synaptogenesis, seem to depend on the presence of several endogenous factors in the microenvironment of neurons during certain critical periods of development.[16,17] For instance, steroid hormones secreted during pre- and perinatal development play "organizational" and "activational" roles in the maturation of various neuronal systems in areas related to neuroendocrine control[18-20] and also in areas involved in unrelated processes.[21] The presence of exogenous substances during these critical periods may affect normal neural development by acting directly and/or through modifications in endogenous factors, leading to behavioral abnormalities.[11,22] McEwen[23] has recently reviewed the developmental effects of a set of neurotropic substances, including nicotine, barbiturates, alcohol, opiates, and cannabinoids, which are able to modify brain development, probably by mimicking or modifying the effects of natural hormones.

Results of studies that document modifications in neural development after perinatal exposure to several types of neurotropic drugs are summarized in Table 1. We have concentrated on the effects on dopaminergic neurons because changes that appear early in development seem to play a possible "trophic" and "plastic" role in the differentiation of other neuronal systems[24] and their target cells.[25] In particular, it has been proposed that this neurotransmitter acting through D_1 dopaminergic receptors and, in part, D_2 receptors, might affect the expression of specific neuronal mRNAs in several target areas.[26,27]

In light of these proposed neurotropic and morphogenic actions of fetal dopamine (DA), it is possible that early alterations, mainly deficiencies, in this neurotransmitter could contribute to the abnormal pre- and postnatal development of one or more neurotransmitter systems. This fact, together with the observation that dopaminergic neurons are targets for the action of steroid hormones[28-30] and one of the neurochemical substrates for addictive behavior[31] in adulthood, supports their use as a valid model for the study of

TABLE 1
Effects of Perinatal Exposure to Neurotropic Drugs on Development of Brain Dopaminergic Systems

Drugs	Effect	Ref.
Alcohol	Adverse effects on dopaminergic activity: regional decreases in DA concentrations, DA uptake and D_1 and D_2 receptor density	5, 6, 35, 36
	Modifications in several DA-related behaviors	37–40
Amphetamine	Increased striatal DA synthesis and turnover	61–63
	Increased motor activity	
Caffeine	Decreased DA content in nucleus coeruleus	50
	Dose-, sex- and age-dependent effects on DA-related behaviors: open-field locomotor and rearing activities and emergence latency	46–47
Barbiturates	Hyperactivity, decreased male sex behavior and other behavioral effects related to changes in dopaminergic neurotransmission	53
Cocaine	Evidence for altered development of dopaminergic neurons and related behaviors	8, 54–60
Diazepam	Decreased DA turnover in nucleus accumbens and ventral tegmental area, accompanied by changes in motor activity, exploratory behavior, and learning ability	3
Methadone	Slowing monoaminergic development and behavioral abnormalities	43–45
Nicotine	Increased DA and related metabolite contents in fetal forebrain	9, 41
	Suppression of dopaminergic increases during cortical maturation	7
	DA-related behavioral modifications	42

the epigenetic effects of steroid hormones and psychotropic drugs during development.

In general, most of the effects that originate as a consequence of pre- and perinatal exposure to neurotrophic drugs conform to a common rule. They cause marked modifications, usually decreases, in the development of dopaminergic neurons, although sometimes it is very difficult to discriminate whether these effects originate as a consequence of direct effects of the drug on brain dopaminergic neurons or whether they result from a drug-induced anorexia and, consequently, nutritional deficits. Another important aspect derived from these studies is the fact that male offspring are more sensitive than females to the effects of perinatal exposure to these drugs. This sexually dimorphic response to perinatal cannabinoid exposure will be addressed later in this chapter.

Maternal alcoholism profoundly affects the brain development of offspring, producing mental retardation, severe hyperactivity, and behavioral abnormalities.[32] Prenatal ethanol exposure in rodents adversely affects the development of various neurotransmitter activities.[33,34] DA activity is decreased in certain brain areas, as reflected by regional decreases in DA concentrations,[5] DA uptake,[35] and D_1 and D_2 dopaminergic receptor density.[6,35,36] This DA deficiency already appears during late fetal brain development[35] and produces modifications in several DA-related behaviors.[37-40]

Acute or prolonged exposure to nicotine during development increases the content of DA and related metabolites in the fetal forebrain.[41] The drug-induced pattern of dopaminergic alterations changes during ontogeny until adulthood[41] and is sex dependent.[9] These effects correlate with behavioral modifications.[42] On the contrary, the increases in dopaminergic activity that normally occurs during cortical maturation are suppressed by nicotine exposure.[7] However, the possible involvement of central nicotinic receptors in these dopaminergic effects has not been elucidated.

A number of studies has examined the effects of perinatal methadone exposure at doses sufficient to produce maternal and fetal dependence. Animals prenatally exposed to methadone exhibit behavioral abnormalities[43] related to a slowing of neurotransmitter development, in particular monoaminergic[43,44] and cholinergic[45] neurons.

Prenatal exposure to low doses of caffeine produces sex- and age-dependent effects on DA-related behaviors. In fact, caffeine increases open-field locomotor and rearing activities and decreases emergence latency.[46] These changes were more typical of male rats, particularly when older. In contrast, other studies have demonstrated the opposite effects; namely, a decreased locomotor activity and an increased emotional reactivity after prenatal exposure to high doses of caffeine.[47] Although the mechanisms underlying the behavioral effects of caffeine are not yet known, a possible involvement of adenosine receptors has been suggested.[48] Chronic pre- and postnatal caffeine exposure increases the number of these receptors in the brain.[49] The possibility of the appearance of caffeine-induced modifications in the development of monoaminergic mechanisms seems more unlikely, although a decreased DA content in the locus coeruleus following caffeine treatment has been described,[50] in addition to other noradrenergic[51] and serotoninergic[52] modifications.

Chronic exposure to barbiturates during development produces hyperactivity, decreased masculine sexual behavior, REM sleep deprivation, impaired learning, and other behavioral effects that have been related to modifications in dopaminergic neurotransmission (see Reference 53 for a review).

Cocaine has been convincingly related to alterations in dopaminergic neurotransmission in adult rodents. Evidence for altered development of dopaminergic neurons[8,54-57] and different DA-related behaviors[8,58-60] after its administration during prenatal life has also been presented, although the effects were sometimes very slight.[8,54] Moreover, glucose metabolism in several brain areas decreases in adult males prenatally exposed to cocaine.[56] This was interpreted by these authors as lasting changes in the functional activity of neuronal systems related to neuroendocrine function, motor control, drug reinforcement, learning, and memory. However, postnatal cocaine exposure has no effect on glucose metabolism in adult males and causes slight increases in glucose metabolism in adult females.[2]

Another example of behavioral and neurotransmitter disturbances resulting from exposure to centrally acting drugs has been observed in rats exposed perinatally to amphetamine. This drug increases the synthesis and turnover of DA and norepinephrine, accompanied by increases in motor activity.[61-63]

Perinatal exposure to diazepam, a benzodiazepine whose consumption is now quite extensive, also affects DA-related behaviors, such as motor activity, exploratory behavior, and learning ability. These alterations have been associated, in addition to their well-known GABAergic effects, with a decrease in DA turnover in the nucleus accumbens and the ventral tegmental area.[3]

CANNABINOID EFFECTS ON DOPAMINERGIC NEURONS

Cannabis sativa preparations (hashish, marijuana) are among the most widely used psychoactive drugs in the world.[64] Their abuse produces multiple effects on a variety of neurobiological processes, including the neuroendocrine control of pituitary function, extrapyramidal activity, and limbic system-related behaviors. Cannabinoid consumption has been reported to produce changes in anterior pituitary hormone release,[65-67] modify motor control,[68,69] facilitate brain stimulation reward,[70] and exacerbate or precipitate psychotic disorders.[71]

A large number of experimental studies has related these neurological and psychiatric effects of cannabinoids to changes in the activity of neurotransmitter systems involved in the control of those processes in specific brain areas, particularly DA.[72-74] In general, cannabinoids produce a set of effects on brain neurotransmitter systems that differ as a function of the brain area, as well as the cannabinoid administered and its dose. Thus, modifications in various biochemical indices reflecting brain catecholaminergic,[75-82] serotoninergic,[78,83-85] cholinergic,[86] and GABAergic[87] activities have been found in experimental animals after acute or chronic administration of marijuana or Δ^9-tetrahydrocannabinol (THC), its main psychoactive constituent.[64,88]

Cannabinoids and Neuroendocrine Control

Administration of THC produces marked alterations in anterior pituitary hormone release. Thus, prolactin (PRL), growth hormone, and gonadotropin release decreases[65-67,89-94] and that of adrenocorticotrophic hormone increases.[95] These pituitary effects appear to result from THC-induced alterations in the activity of tuberoinfundibular dopaminergic neurons and other hypothalamic neurotransmitters, also involved in the control of anterior pituitary release, such as noradrenaline and serotonin.[67,89,90]

In the case of PRL, acute administration of THC reduces peripheral PRL levels in both normal and hyperprolactinemic animals.[65,96] This reduction can be prevented by surgical[90] or chemical[97] interference in serotoninergic hypothalamic influences, suggesting involvement of this neurotransmitter in the THC-induced decrease of PRL release. Other studies have described a parallelism between THC-induced decreases in PRL secretion and increases in the activity of tuberoinfundibular dopaminergic neurons, thus supporting a possible mediation of THC effects on PRL by this neurotransmitter.[82,97,98]

The effects of THC on gonadotropin release are inhibitory,[99-102] including decreases in basal plasma levels of luteinizing hormone (LH) and follicle-stimulating hormone (FSH),[91] suppression of the preovulatory peak of LH,[100] and a reduction in sex behavior-related LH release.[101] These effects of THC on pituitary function have been reported to be mediated through hypothalamic influences. Thus, pituitary responsiveness to luteinizing hormone-releasing hormone (LHRH) does not appear to be affected by exposure to the drug; whereas, THC appears to affect hypothalamic mechanisms underlying LHRH release,[103] namely opioid[74] and noradrenergic[89] influences. However, Murphy and co-workers[104] have recently suggested the existence of a possible direct effect of THC on the pituitary, suggesting presence of binding sites for THC in this gland. The significance of this possibility remains to be determined.

These changes at the hypothalamic-pituitary level lead to marked modifications in gonadal function (see References 105 and 106 for review): namely, (1) alterations in ovarian cycles in females, leading to anovulation; and (2) regression of Leydig cells, leading to decreases in testicular weight and testosterone production, a decrease in the size of other reproductive organs in males, an inhibition of male sexual behavior, reflected in increased latency of mounting and ejaculation, and a decreased competition for estrous females. However, direct effects of THC on reproductive organs in both males and females have also been demonstrated.[106,107]

Cannabinoids and Motor Activity

Several clinical studies have demonstrated that marijuana consumption produces alterations in the motor activity of healthy humans by affecting the extrapyramidal system. This effect has been subjected to clinical research for its possible therapeutic application in several motor disorders, such as Parkinson's disease or Huntington's chorea, in which cannabinoids might produce antidystonic effects. Nigrostriatal dopaminergic neurons are the apparent site of these extrapyramidal effects of cannabinoids. It has been reported that THC stimulates the activity of these neurons[72,108,109] by inhibiting neurotransmitter uptake[85,109] and stimulating its release.[110] Moreover, it has been shown that THC and cannabidiol, a relatively nonpsychoactive cannabinoid, are able to modify the binding of spiroperidol to D_2 receptors in striatal membranes in a dose-dependent manner[72] by decreasing receptor density and increasing their affinity. In addition to this effect on the nigrostriatal neurons, THC is also able to modify striatal GABAergic activity, producing a dystonic effect similar to that produced by diazepam.[111] These neurochemical alterations are accompanied by behavioral modifications, as indicated by catalepsy in mice,[112] altered motor behavior in a rat model of reserpine-induced Parkinson's disease,[111] and convulsions in rabbits.[113]

Cannabinoids and Motivational Behavior

Chronic abuse of cannabinoids may increase the incidence of various psychiatric syndromes.[71] Clinical observations point to the existence of an antagonism between cannabinoid consumption and neuroleptic treatment in schizophrenic patients that may lead to an exacerbation of the symptoms of this disease.[114] This seems to indicate that the psychiatric effects of cannabinoids could be related to alterations in the activity of dopaminergic neurons in limbic areas. In this regard, administration of THC to rats is followed by an increase in the content of homovanillic acid in the prefrontal cortex and the olfactory tubercle.[115] Gardner and co-workers[116] also found that THC enhances presynaptic DA release from limbic neurons in freely moving rats and that this effect is blocked by naloxone. In addition, other studies have disclosed THC-induced modifications in noradrenergic receptors, related to a decrease in social behavior.[117] These neurochemical effects were also accompanied by behavioral modifications in animals following THC treatment, as indicated by a facilitation of brain stimulation reward,[70] changes in active or passive avoidance behavior,[117] and increased aggressive behavior.[74,118]

PERINATAL CANNABINOID EXPOSURE: EFFECTS ON DEVELOPMENT

Most of the above-mentioned effects of cannabinoids on the brain have been studied during the adult life. Less information is available on the possibility of alterations in the development of the nervous system caused by maternal consumption of cannabinoid derivatives during gestation and lactation.[119,120] As mentioned above, the presence of drugs during the early stages of development may affect the differentiation and maturation of the brain.[11] Cannabinoids can be transferred from the mother to the offspring through placental blood during gestation[121] and through milk during lactation.[122,123] In both cases, they can reach the fetal and neonatal brain in substantial amounts because of the immaturity of the blood-brain barrier at these stages of development. Administration of radioactive THC to pregnant or lactating animals is followed by the appearance of labeled compounds in the brain of fetuses[121] and suckling newborns,[122] respectively.

TABLE 2
Developmental Effects of Maternal Exposure to Cannabinoids During Preconceptional, Early Prenatal, and Perinatal Periods

Exposure	Effect	Ref.
Preconceptional	Males: gametotoxicity, oligospermia, atrophy of testis and accessory sex organs, and chromosomal abnormalities	100, 105, 124–126
	Females: adverse effects on early embryonic development by greater incidence of abnormalities in fertilized ovocytes	100, 127
Early prenatal	Teratogenic effects on embryonic and fetal organogenesis	100, 105, 133, 134
	High incidence of embryonic resorptions, particularly female embryolethality;	105, 120
	reduced survival of fetuses;	105, 130, 131
	intrauterine growth retardation	132
Perinatal	Delay in pup development	100
	Small weight at birth, related to changes in body composition	100, 105
	Impairment of stress response, diminished learning ability and other behavioral disturbances in adulthood	4, 158
	Adverse effects on gonadal function and reproductive behavior	
	Males: decreased gonadotropin release, testicular weight and testosterone production	136, 144
	Females: delayed onset of puberty; decreased gonadotropin release; high frequency of irregular ovarian cycles; low number of mature ovocytes per cycle	137–141

Dalterio reviewed the effects of cannabinoids on development[4] and showed that pre- and perinatal exposure to THC, either alone or in combination with other cannabinoids, produces a variety of long-term abnormalities in brain, endocrine, immune, and hepatic functions.[4,119] Cannabinoids may be embryocidal; they may affect gestational length and labor, or induce delays in maturation or other functions. In general, prenatal exposure mainly affects embryonic and fetal organogenesis, whereas postnatal exposure results in damage to the growing offspring.[119,120] A brief review of the effects of cannabinoids on development, including the effects of exposure to cannabinoids during preconceptional, prenatal, and postnatal stages, is summarized in Table 2.

Preconceptional exposure to cannabinoids damages gametogenesis. In the male, this is manifested by decreases in the number of sperm cells and in their motility, as well as by the dose-related increase in the incidence of abnormally nonovoid-shaped sperm cells.[105,124,125] These effects on gametogenesis also extend to possible cannabinoid-induced chromosomic aberrations because THC exposure in male mice is associated with a series of developmental alterations in their offspring.[126] These effects seem to originate through both direct effects of THC on the testis, probably due to a direct suppressant action of cannabinoids on the germinal epithelium,[105] and indirect actions mediated via modifications in hypothalamic-anterior pituitary-testicular axis. In the female, the gametotoxic effect due to preconceptional exposure to THC is manifested in a greater incidence of abnormalities in fertilized ovocytes, producing adverse effects on the early stages of embryonic development.[127]

Prenatal exposure to cannabinoids mainly affects embryonic and fetal organogenesis because THC may cross the placental blood-barrier and be stored in the fetus where its metabolism is slower.[119] The cannabinoid effects are usually dose related and include:

1. Increases in embryonic resorption and fetal mortality,[105,128] probably associated with alterations in the maternal/fetal circulation. We have found that maternal exposure to a mixture of cannabinoids slightly increases the average duration of pregnancy, an effect reported also by Wenger and co-workers,[129] and correlating with a decrease in the cumulative maternal weight gain during pregnancy (Table 3). We have related this decreased maternal weight gain to a decrease in litter size (Table 3), which may be indicative of a decreased number of successful implantations of fertilized ovocytes or greater fetal mortality. Similar observations were reported by other authors,[130,131] although in contrast to their findings of a higher prenatal mortality in female offspring,[120] the sex ratio was not affected in our studies (Table 3).
2. Intrauterine growth retardation, which mainly occurs when dams are exposed during the last third of pregnancy in the rat.[132]
3. The presence of a set of developmental abnormalities (for a review, see References 105, 119, and 133). Teratogenic effects have frequently been noted in animals prenatally exposed to cannabinoids, although high doses are required to elicit their response. Oral administration of THC at doses up to 50 mg/kg weight was not terato-

TABLE 3
Effects of Maternal Exposure to Crude Hashish Extract on Several Parameters During Gestation and Lactation Periods

Parameters	Vehicle	Hashish
Gestation length (d)	22	22.6 ± 0.2[a]
Mother weight gain (g) during pregnancy[b]	117.5 ± 7.1	87.4 ± 5.6[a]
Mother weight gain/litter size (g) during pregnancy	8.2 ± 0.4	9.4 ± 1.1
Mother food intake (g)		
During pregnancy[c]	20.4 ± 0.6	20.5 ± 0.7
During lactation[c]	37.6 ± 6.7	40.8 ± 7.8
Mother water intake (ml)		
During pregnancy[c]	43.1 ± 0.8	48.0 ± 1.2
During lactation[c]	60.4 ± 6.5	70.8 ± 6.9
Litter size	14.4 ± 0.6	10.8 ± 1.3[a]
Females/males ratio	1.2 ± 0.3	1.0 ± 0.2
Postnatal mortality (%)	0.9 ± 5.6	27.4 ± 7.3[a]

Note: Values are means ± SEM of 6 to 10 determinations per group.

[a] Statistical differences were obtained by Student *t*-test, $p < 0.05$.
[b] This parameter was obtained as the difference between the weight on the day before delivery and the weight on the day of mating.
[c] Average value per day.

From Rodríguez de Fonseca, F., Hernández, M. L., de Miguel, R., Fernández-Ruiz, J. J., and Ramos, J. A., *Pharmacol. Biochem. Behav.*, 41, 1992. With permission.

genic.[105] The type and frequency of anomalies vary as a function of several variables, such as species or strain used, route, and timing of administration, as well as dosage, type of cannabis preparation and, in the hamster, the season in which the studies are conducted. In the rat, cannabinoid exposure is only teratogenic if administered i.p. and only if exposure occurs before day 7 of gestation.[134] Malformations in rodents include encephalocele, exencephaly, hydronephrosis, omphalocele, body edema, and abnormal skeletal development (cleft palate, fused ribs, accessory ribs, syndactyly, amelia, phocomelia, spina bifida, and myelocele).[105]

Cannabinoid exposure around birth and during postnatal development causes important long-term modifications, such as delays in development, small birth weight related to specific changes in body composition, alterations in pituitary-gonadal function and reproductive behavior, and diminished learning ability, among others.[119,130,135,136] The alteration of reproductive function is one of the most characteristic effects resulting from perinatal cannabinoid exposure.[137,138] For instance, females exposed to THC during the first days after birth display abnormal postpubertal reproductive performance, such as a delay of 2 d in vaginal opening, a high frequency of irregular ovarian cycles, a decreased number of matured ovocytes per cycle, and low plasma gonadotropin levels.[139-141] These long-term effects on reproductive function have been related to alterations in the normal functionality of several diencephalic neurotransmitters, including met-enkephalin, β-endorphin, and substance P, which through modifications in the release of LHRH from the medial basal hypothalamic, modulate gonadotropin release.[142,143] A similar antigonadotrophic effect was found by Dalterio and co-workers[136,144] in male mice pre- and postnatally exposed to cannabinoids.

It has been suggested that cannabinoid effects after either pre- or postnatal exposures, in particular the delay in development, could be due to several causes. First, a nutritional deficit in the developing offspring, due to either a decrease in water and food intake by the mother,[4,105] to cannabinoid-induced decreases in milk production, and/or the inhibition of suckling-induced milk ejection.[145] THC decreases PRL levels in lactating females and interferes with PRL release in response to the suckling stimulus. Second, a decrease in the brain synthesis of DNA, RNA, and/or proteins, starting at fetal ages, may occur.[127,146] However, as mentioned above, the most plausible reason would be the marked interference in the molecular microenvironment of the developing newborn.[4,11,23]

In this regard, we have not seen any significant differences in food and water intake between mothers exposed to cannabinoids and controls during gestation and lactation, as shown in Table 3. As mentioned above, we hypothesize that the cannabinoid-induced loss in maternal weight gain during pregnancy could be explained on the basis of a greater prenatal mortality rather than a decreased food intake. Abel[105] has shown that dams treated with marijuana gained less weight than their pair-fed controls. However, maternal undernutrition cannot be dismissed as a possible contributor to the effects of *in utero* cannabinoid exposure.

Postnatal mortality also increases after perinatal cannabinoid exposure (Table 3), especially if dams are exposed to cannabinoids during the third week of gestation.[105] In this regard, the possible existence of "sensitive periods" during gestation for various gestational effects of cannabinoid exposure has been suggested and could explain some discrepancies among the different studies. Thus, it has been proposed that day 8 of gestation is the most sensitive time for *in utero* exposure in rats as far as resorptions are concerned.[105,147] Similar sensitive periods have been proposed in mice.[148,149]

On the other hand, it is important to note that the increase in prenatal mortality could be due to a cannabinoid or noncannabinoid compound other than the psychoactive substances because its appearance depends on the hashish or marijuana extract used. We were unable to find any significant differences in fetal mortality and maternal weight gain in recent studies using a hashish extract with a different origin and different proportions of THC, cannabinol, and cannabidiol than that used in the initial studies (Fernández-Ruiz et al., unpublished observations). Differences in proportions of various cannabinoids, which may have antagonistic or synergistic actions, and/or the action of other noncannabinoid ingredients that include certain alkaloids, terpenes, phenols, and flavonoids present in the hashish extract should be also considered[150] because they may affect some biological actions of marijuana.[105,151] This view can be extended to other controversial responses such as food and water intake. Several examples can be given in support of this hypothesis. Thus, Abel[105] has demonstrated that the frequency of developmental anomalies varies with the geographical location in which the marijuana grew. Perinatal cannabichromene exposure failed to decrease adult male sex behavior in mice, but antagonized the THC-induced decrease when both cannabinoids were simultaneously administered.[152] In another example, Walters and Carr[130] found decreases in maternal weight gain during pregnancy and offspring weight at birth, together with an increased prenatal and postnatal mortality after marijuana extract exposure, but were unable to find any significant differences after administration of different cannabinoids separately.[153]

Cannabinoids and Neural Development

The brain is one of the most important targets of perinatal cannabinoid action. Perinatal cannabinoid exposure markedly affects neural development by producing modifications that will be manifested immediately during the developmental process itself, or by inducing long-term alterations in various brain-controlled processes in adulthood. Among other effects, perinatal exposure to cannabinoids produces a marked delay in the onset of puberty and alterations in adult male copulatory behavior, open-field activity, behavioral development, learning ability and stress response.[4,139,141,154-158]

As mentioned above, one of the possible mechanisms involved in the neuroteratology of cannabinoids could be the neurotransmitter disturbance due to the important "trophic" and "plastic" role played by them during critical periods.[11-13] Recent studies have shown that perinatal exposure to cannabinoids results in alterations in the normal development of several brain neurotransmitter systems.[130,153,159-161] Thus, administration of cannabinol or cannabidiol to female mice on day 18 of pregnancy[144] or within 12 h of parturition[136] decreased catecholamine contents in the brains of their offspring in adulthood. Walters and Carr have described that perinatal exposure of rats to a crude marijuana extract[130] or to either THC or cannabidiol[153] results in several changes in the development of catecholaminergic mechanisms in the striatum and cerebral cortex of the offspring. In fact, perinatal marijuana extract exposure reduced the number of D_2 dopaminergic receptors and the activity of tyrosine hydroxylase (TH) in the striatum of peripubertal rats, but did not affect α_1-adrenergic receptors in the cerebral cortex. However, some of the effects observed with marijuana extracts did not resemble those found with the pure compounds.[153] Similarly divergent effects of THC and other cannabinoids were found by Dalterio and co-workers in mice.[159]

As discussed above, one explanation for the foregoing may be that cannabinoids interact with or antagonize one another when administered together in a cannabis extract such that they would produce a greater or lesser effect than when administered individually. Evidence for these interactions has been reported previously. For example, cannabidiol was found to potentiate the hypothermic and analgesic effects[162] and anticonvulsant activity[163] of THC, as well as to inhibit several actions of this psychoactive constituent.[74]

Early postnatal exposure to cannabinoids also produces a long-lasting effect on the functionality of the endogenous opioid system in male rats.[143] Thus, postnatal exposure to low doses of THC increases the contents of β-endorphin and methionine-enkephalin in several brain areas. These changes correlated with greater sensitivity of perinatally cannabinoid-exposed rats to pain in adulthood, although the pain perception was decreased during early post-pubertal ages in these animals. Other brain neuropeptides, including tachykinins, were also affected by perinatal cannabinoid exposure.[164]

We have performed several studies to investigate the effects of maternal cannabis consumption on the maturation of brain dopaminergic neurons, that is, nigrostriatal, mesolimbic, and tuberoinfundibular systems, in developing offspring during both prenatal and early postnatal periods.[160,161] To this end, pregnant rats were fed daily with hashish extracts from day 5 of gestation until weaning. This period of treatment was chosen in accordance with the previously cited study of Mirmiran and Swaab.[11] Studies were carried out on different days before and after birth, analyzing several pre- and postsynaptic dopaminergic biochemical indices, in both male and female offspring with the purpose of examining the possible influence of gonadal status on cannabinoid effects. The results are summarized in Tables 4 to 8. Prior to discussing the results of these experiments, we will briefly summarize some pertinent information about the ontogeny of these neuronal systems.

Ontogeny of Dopaminergic Neurons

The development and maturation of dopaminergic neurons have been studied in depth.[165-171] In the rat, dopaminergic neurons differentiate between days 12 and 14 of embryonic development and the three major dopaminergic systems can be identified on day 15. Their different projections reach maturation during the first month after birth, with differences between areas.[167,172] As mentioned above, the early appearance of presynaptic activity and postsynaptic sensitivity for DA supports the hypothesis of a role for this neurotransmitter in the regulation of maturation of other neurotransmitter systems[24,26,27] and target cells.[25]

The normal development of dopaminergic neurons is not linear because it includes phases of maximum growth,[167,172] as reflected by increases in neurotransmitter synthesis and release and in the number of receptors.[171] This is a consequence of the transient appearance of certain groups of dopaminergic neurons that are specifically involved in development and which later disappear.[173] Thus, the transient expression of TH immunoreactivity in the rat neocortex during postnatal development[173] and D_2 dopaminergic receptors in the ventral spinal medulla[171] and the somatosensorial cortex[174] during prenatal ages has been described. In relation to this latter observation, it has been proposed that these receptors, together with serotoninergic receptors, might play a trophic role in the organization of the cortex.[174] We have also observed this particular pattern of development of dopaminergic neurons in our studies;[161] in this sense, the evolution of postsynaptic sensitivity and presynaptic activity for DA was mostly inverse, with high numbers of dopaminergic receptors on days 15 and 40 after birth, correlating with low presynaptic activity, and a lower number of binding sites and higher presynaptic activity on days 20 and 30. This particular type of development, especially in the number of dopaminergic receptors, is consistent with previous studies by other authors[166-168] and supports the importance of the interaction between presynaptic activity and postsynaptic sensitivity in the regulation of the direction and extension of axonal development.[12,168,175]

Cannabinoid Effects on Developing Dopaminergic Neurons

Our studies on the effects of perinatal cannabinoids on various developing dopaminergic neurons provided undeniable evidence that cannabinoids produce marked alterations in the development of several biochemical indices of the activity of this neurotransmitter measured at perinatal ages (see Tables 4, 5, 7, and 8). The most important observation was that these effects differed according to gender and the specific pathway studied, with more marked and consistent alterations appearing in males. Some of these alterations disappeared after weaning-induced drug withdrawal, whereas others were maintained, and "silent" effects, which did not appear during drug exposure, were manifested after termination of cannabinoid exposure. In addition, we have recently observed that the neurochemical modifications induced by perinatal cannabinoids were accompanied by changes in the functional expression of these neurons at the behavioral and neuroendocrine levels (Tables 6 and 8), probably producing long-term alterations in adulthood.

These cannabinoid-induced modifications in the maturation of dopaminergic neurons already appear at early perinatal ages.[176] In fact, at these ages, the sexual dimorphism in the effects of cannabinoids on dopaminergic activity can already be observed in the prosencephalic area, which contains an increasing population of dopaminergic terminals from the different dopaminergic pathways that become differentiated in the adult rat (Table 4).

Effects of Cannabinoids on Maturation of the Nigrostriatal System

The first axonal terminals containing TH activity reach the caudate-putamen area on embryonic days 14 and 15; they originate from the substantia nigra.[177] The contacts of these neurons with the ventrolateral neostriatum are developed during days 14 to 21 of fetal life.[178] Dopamine D_2 receptors are detected on day 14 and are functional (i.e., coupled to their transduction mechanisms) on day 15.[171] Their number increases from prenatal day 18 to postnatal day 30 and their affinity for DA is also increasing. Dopamine D_1 receptors are present at birth in the striatum[179] and their numbers increase

TABLE 4
Neurochemical Effects of Perinatal Cannabinoid Exposure on the Ontogeny of Maturing Prosencephalic Dopaminergic Neuronal Systems at Early Stages of Development

Days	Males	Females
–1[a]	Decreased DA and DOPAC content	No changes
5	Low TH activity and DA content	No changes
10	Increased TH activity	Increased DA content

[a] Taking the day of birth as zero.

From Rodríguez de Fonseca, F., Hernández, M. L., de Miguel, R., Fernández-Ruiz, J. J., and Ramos, J. A., *Pharmacol. Biochem. Behav.*, 41, 1992. With permission.

TABLE 5
Neurochemical Effects of Perinatal Cannabinoid Exposure on the Ontogeny of Nigrostriatal Dopaminergic Neurons

Days	Males	Females
15	High number of D_1 and D_2 receptors	No changes
20	High number of D_2 receptors Decreased TH activity	Low number of D_1 receptors High presynaptic activity
30	High number of D_1 receptors Decreased TH activity	No changes
40	High number of D_1 and D_2 receptors Decreased TH activity	Low number of D_1 receptors

Note: Days 15 and 20 were within the period of drug exposure, whereas days 30 and 40 were after drug withdrawal.

From Rodríguez de Fonseca, F., Cebeira, M., Fernández-Ruiz, J. J., Navarro, M., and Ramos, J. A., *Neuroscience,* 43, 713, 1991. With permission.

until postnatal day 40,[166] with a marked peak on day 15 coinciding with greater spontaneous motor activity.[180]

Prolonged exposure to cannabinoids from early development until weaning markedly affects the ontogeny of this neuronal system, with changes that are evident at immature and early postpubertal ages, particularly in male offspring.[160,161] The major effects are summarized in Table 5. Among others, perinatal exposure to cannabinoids adversely affects the pattern of postnatal development of TH activity in males (Table 5). A similar decrease in the activity of this enzyme in immature rats was also found by Walter and Carr,[130] although these authors did not differentiate between sexes. This decrease may reflect possible cannabinoid-induced changes in either of the following processes in male offspring: (1) a delay in the onset of differentiation of dopaminergic neurons, or (2) a decrease in the ability of dopaminergic neurons to synthetize DA. Although at present we have no data to critically test the first possibility, some arguments can be made in its support. First, it is interesting to note that the TH enzyme is present in the growing nigral axons before their contact with striatal neurons, and the enzyme may play an important role in neurochemical processes of neuronal recognition and synaptogenesis. It is possible that early cannabinoid exposure might affect a critical process in nigral neuronal development such as expression of the TH gene, resulting in slowing neuronal recognition and, therefore, a delay in the maturation of these neurons. This is supported by the fact that perinatal cannabinoid exposure was followed by a decrease in TH activity and DA content at early perinatal ages, although this observation was made in an area containing other dopaminergic neuronal systems as well (Table 4).

Perinatal cannabinoid administration also affects postsynaptic sensitivity in the striatum of immature and early postpubertal males. In particular, cannabinoids increased the number of both D_1 and D_2 dopaminergic binding sites (Table 5). It has not been demonstrated whether this inverse relationship between cannabinoid-induced effects on presynaptic and postsynaptic parameters reflects an up-regulation of postsynaptic receptors as a consequence of a cannabinoid-induced presynaptic hypoactivity, or whether the decrease in TH activity is due to a cannabinoid-induced activation of presynaptic receptors, which are of the D_2 type in the striatum[181] and which are known to be involved in the control of neurotransmitter synthesis and release.[182] However, it is probable that this represents a compensatory mechanism, leading to attenuated behavioral expression of these neurochemical effects. In this regard, we have recently examined the effects of perinatal cannabinoid exposure on development of spontaneous motor activity, which is the most important behavioral expression of nigrostriatal neurons.[183] The study was conducted in rats and motor activity was measured in an actimeter (M. Navarro et al., unpublished results). The results are summarized in Table 6. It is interesting that, although spontaneous motor activity was not significantly altered in male offspring during the developmental period observed, it did show a nonsignificant trend to decrease. This tendency correlates with the decreased presynaptic activity

TABLE 6
Behavioral Effects of Perinatal Cannabinoid Exposure on the Ontogeny of Nigrostriatal Dopaminergic Neurons: Spontaneous Motor Activity and Stereotypic Behavior

Days	Males	Females
15	Low spontaneous motor activity[a]	No change
20	Low spontaneous motor activity[a]	No change
30	Low spontaneous motor activity[a]	No change
40	Low spontaneous motor activity	Low spontaneous motor activity[a]
70	No change in spontanous motor activity	No change in spontaneous motor activity
	High frequency of self-grooming and shaking	High frequency of self-grooming[b]

Note: Days 15 and 20 were within the period of drug exposure, whereas days 30, 40, and 70 were after drug withdrawal.

[a] Nonsignificant trend.
[b] Less marked than males.

From M. Navarro et al., unpublished results. With permission.

indicated by low TH activity (Table 5), although the effect was presumably compensated for by the hypersensitivity of dopaminergic receptors.

This is concordant with the observations of Vardaris et al.[156] who did not find any differences in open-field activity when perinatally THC-exposed animals were tested at 18 d of age. However, it contrasts with the studies of other authors who observed lower activities at 9[184] and 18[185] d of age in rats prenatally exposed to cannabinoids. Despite this, it should be noted that the open-field test reflects indications of both motor activity and emotionality. An analysis of the ontogeny of motor activity in rats prenatally exposed to cannabis, using other kinds of motor performance tests, was also done in other studies, but no conclusive results were obtained (for review, see Reference 105).

Another important aspect to note is the fact that these neurochemical and behavioral changes were maintained in spite of the drug withdrawal caused by weaning on day 24. This supports the hypothesis that early cannabinoid exposure produces a marked modification in the functionality of nigrostriatal neurons which, later on, does not depend on the presence of the drug. Previous studies have indicated that THC and its metabolites disappear from the whole rat with a half-life of approximately 17 h.[186] It is therefore unlikely that significant amounts of cannabinoids would have been present in the brain of offspring at 30 and 40 d after birth. Our results contrast with the observations of Walter and Carr,[130] who observed the disappearance of marijuana-induced neurochemical effects after weaning, although these authors did not analyze males and females separately.

At this time, it is not possible to assert that these marked modifications in the maturation of nigrostriatal neurons at immature ages are completely irreversible and lead to a permanent change in the adult functionality of these neurons. Since dopaminergic systems undergo a significant amount of maturation during the postnatal period, a "normalization" due to compensatory mechanisms during the course of development is not an unreasonable assumption. We have no data concerning the persistence of these neurochemical effects in the adult male rat because animals over 40 d of age were not included in our studies.[160,161] However, we did observe that the tendency for decreased spontaneous motor activity in male offspring perinatally exposed to hashish disappears in adulthood. This agrees with the results obtained from open-field studies in rats that were prenatally exposed to cannabis extracts and tested at 75 d of age.[187] However, preliminary data obtained with other tests support possible motor alterations. Indeed, perinatally hashish extract-exposed males show an increase in the frequency of stereotypic behaviors, such as water-induced self-grooming and shaking (Table 6; Navarro et al., unpublished observations).

The effects of perinatal exposure to cannabinoids in female offspring were transient and less marked than those in males, suggesting the existence of a profound sexual dimorphism in the sensitivity of growing nigrostriatal neurons to cannabinoid exposure. Changes in both presynaptic activity and the number of postsynaptic receptors in males were observed on most of the days studied, whereas cannabinoid effects in females appeared on individual days and were not normally maintained after the end of the period of drug exposure, except for a decrease in the number of D_1 dopaminergic receptors, which appeared on day 40 after birth (Table 5).[161] This effect corresponds to a significant decrease in spontaneous motor activity of perinatally cannabinoid-exposed rats at early postpubertal ages and to an increased frequency of stereotypic behaviors at adult age (Table 6; Navarro et al., unpublished observations).

TABLE 7
Neurochemical Effects of Perinatal Cannabinoid Exposure on the Ontogeny of Mesolimbic Dopaminergic Neurons

Days	Males	Females
15	No changes	Decreased number of D_1 receptors
20	No changes	Increased presynaptic activity
30	Increased TH activity	No changes
40	Increased presynaptic activity	No changes

Note: Days 15 and 20 were within the period of drug exposure, whereas days 30 and 40 were after drug withdrawal.

From Rodríguez de Fonseca, F., Cebeira, M., Fernández-Ruiz, J. J., Navarro, M., and Ramos, J. A., *Neuroscience*, 43, 713, 1991. With permission.

Effects of Cannabinoids on Maturation of the Mesolimbic System

Developing axonal fibers of the mesolimbic-mesocortical system of the rat reach their first targets, coursing through the striatum, on embryonic days 15 and 16. During the following 3 d (i.e., by day 19) all their targets in the cortex—including limbic areas—have dopaminergic innervation.[168,188,189] However, the adult pattern of this neuronal activity is not reached until the second month after birth.[168] This last observation can be related to the fact that cannabinoid-induced effects on the maturation of mesolimbic neurons in male offspring perinatally exposed to hashish extracts[161] became evident only after the drug withdrawal caused by weaning on day 24. These effects consisted of an increased presynaptic activity, but with no effect on postsynaptic sensitivity at peripubertal ages (Table 7). However, the neurochemical substrate at adult age in males perinatally exposed to cannabinoids has not been studied until now, although the fact that these animals exhibit several behavioral disturbances strongly supports the possible persistence of peripubertal changes until adulthood. In particular, we have observed an inhibition of motivational behavior, reflected by increases in emergence latency against stressor stimuli together with marked changes in behavioral patterns executed in a sociosexual approach test (i.e., a delay expressed early in the adaptability of the animals to new environments M. Navarro et al., unpublished results).

The observation of a reduced habituation to stressor stimuli (i.e., light or new environments) in marijuana-exposed offspring has also been observed in studies with primates.[190,191] In rodents, there is a decrease in rearing in immature animals exposed postnatally to cannabinoids.[192] The inhibition of motivational behavior induced by perinatal cannabinoid exposure is also consistent with the above-quoted studies concerning decreases in open-field activity at 9[184] and 18[185] d of age in rats prenatally exposed to cannabinoids because the open-field test reflects indications of both motor activity and emotionality.

The existence of sexually dimorphic sensitivity of developing dopaminergic neurons to cannabinoid exposure also extends to mesolimbic neurons. Thus, whereas cannabinoid effects on the maturation of the mesolimbic system in male offspring became evident only after drug withdrawal, the effects in female offspring disappeared after the end of treatment (Table 7). Moreover, perinatally cannabinoid-exposed females did not exhibit any significant differences at adult age in the motivational behavior expressed in both the emergence test and social-sexual approach test (M. Navarro et al., unpublished results).

Effects of Cannabinoids on Maturation of the Tuberoinfundibular System

The hypothalamus has a complex catecholaminergic innervation.[193] This includes: (1) adrenergic and noradrenergic fibers of extrahypothalamic origin (mainly the ventral-tegmental area of brain stem) which lag behind in development, and (2) dopaminergic neurons which appear early in development, arising from the arcuate nucleus and reaching mainly the median eminence, both located inside the medial basal hypothalamus. This, besides the fact that various hypothalamic dopaminergic cell groups appear to play a purely transient role in developmental process,[194] makes it difficult to relate the immunohistochemical or biochemical detection of TH to any particular group of neurons.

Tyrosine hydroxylase, presumably in dopaminergic neurons, appears on embryonic day 13 and TH-positive neurons migrate to their final locations, mainly the hypothalamic vessels in the median eminence, on day 18.[195,196] This phenomenon is preceded by the appearance of D_2 receptors in anterior pituitary lactotroph cells, the main target cells for tuberoinfundibular neurons, on embryonic day 16,[171] although it has also been proposed that hypothalamic dopaminergic neurons play a role in inducing the develop-

TABLE 8
Neurochemical Effects of Perinatal Cannabinoid Exposure on the Ontogeny of Tuberoinfundibular Dopaminergic Neurons

Days	Males	Females
15	No changes	No changes
20	No changes	No changes
30	Increased TH activity	Increased presynaptic activity
40	Increased presynaptic activity	Decreased presynaptic activity
	High anterior pituitary DA content	Low anterior pituitary DA content
	Decreased PRL release	Increased PRL release[a]

Note: Days 15 and 20 were within the period of drug exposure, whereas days 30 and 40 were after drug withdrawal.

[a] Nonsignificant trend.

From Rodriguez de Fonseca, F., Cebeira, M., Fernandez-Ruiz, J. J., Navarro, M., and Ramos, J. A., *Neuroscience,* 43, 713, 1991. With permission.

ment of PRL cells in the anterior pituitary.[25] It has been proposed that this early appearance of TH-containing neurons may contribute to the differentiation of other hypothalamic neurons, especially peptidergic neurons, which are also targets for tuberoinfundibular neurons.[197]

On the other hand, the development of tuberoinfundibular dopaminergic neurons, as well as that of other hypothalamic systems, depends on fetal and neonatal production of sex steroids which exert a marked effect on the prepubertal development of these neurons and produce the well-documented sexual dimorphism in adult hypothalamic function,[198-200] which is also related to the differences in male and female reproductive functions. In fact, this sexual dimorphism can appear transiently in relation to maturational processes. For instance, the dopaminergic control of LH release is sexually dimorphic during prepubertal development,[201] as are other hypothalamic influences of gonadotropin release,[202-204] although it disappears in adulthood.

Hypothalamic sexual dimorphism is also evident in the sensitivity of developing tuberoinfundibular neurons[161] and other hypothalamic monoaminergic neurons[136] to perinatal cannabinoid exposure in rodents. Dalterio et al.[136] reported that maternal exposure to various cannabinoids on the day of parturition results in long-term alterations in hypothalamic and extrahypothalamic activities of DA, norepinephrine, and serotonin in mice, leading to alterations in neuroendocrine function. These authors found that these alterations differed as a function of sex. In our studies, cannabinoid-induced effects on development of tuberoinfundibular activity became evident only after drug withdrawal, also being clearly different depending on the sex of the animals (Table 8).

In the male offspring, we have found that maternal exposure to cannabinoids produces an increase in the activity of tuberoinfundibular neurons at early postpubertal ages, as reflected by increases in TH activity and DOPAC production. This is correlated with a rise in the anterior pituitary content of DA, as well as with an expected decrease in PRL release (Table 8).[161] These findings are concordant with the report of increased DA and decreased NE turnovers in the hypothalamus of adult male mice perinatally exposed to cannabidiol and cannabinol or THC, respectively.[136] The cannabinoid-induced decrease in PRL secretion observed in our studies agrees with previous observations from several laboratories, including our own group,[82] in adult rodents and supports the view that the effects of cannabinoids on PRL secretion may be mediated through an increase in the dopaminergic hypothalamic input.

In contrast, the responses of developing tuberoinfundibular dopaminergic neurons of female offspring to perinatal cannabinoid exposure were markedly different. In particular, the nerve activity of these neurons decreased at early postpubertal age and was accompanied by an expected decrease in the anterior pituitary content of DA, although the apparent rise in PRL release was not significant (Table 8).[161]

Sexual Dimorphism in Perinatal Cannabinoid Effects

As was emphasized heretofore, one of the most interesting findings from the studies of perinatal cannabinoid exposure was the fact that the cannabinoid-induced changes in the development of brain dopaminergic neurons are different as a function of sex.[136,160,161] These sex-dependent differences appear early, even before the complete differentiation and maturation of dopaminergic projections into their target areas (Table 4).[176] In general, the effects observed in males are more marked and constant than those in females, which suggests that either females are more protected from the pernicious effects of cannabinoid exposure or males are more responsive to cannabinoids. In both cases, it can be suspected that the hormonal milieu plays an important role in protecting or sensitizing these neurons to cannabinoid exposure during development.

The observation of sexual differences in the sensitivity of developing hypothalamic and extrahypothalamic dopaminergic neurons to perinatal cannabinoids is not surprising because it has become increasingly apparent that numerous sexual dimorphisms exist in the organization of the mammalian CNS. The classical sexual dimorphism in the activity of the hypothalamic nuclei-preoptic area,[205-207] which is responsible for the sex differences in the neuroendocrine control of pituitary-gonadal function and sex behavior,[18-21,208] has also been extended to extrahypothalamic neurons.[21,209-212] There is also clinical evidence for the existence of sexual differences in the adult functionality of nigrostriatal and mesolimbic dopaminergic pathways, particularly in relation to pathologies of these neuronal systems (for a review, see Reference 212). For instance, (1) the incidence of chorea and other extrapyramidal disorders is greater in females, in particular during gestation and oral contraceptive use; (2) the appearance of tardive dyskinesia, the false Parkinsonism induced by neuroleptic therapy, is increased by estrogens; and (3) the incidence of several affective disorders depends on circumstances modifying the gonadal status, i.e., gender, gestation, menstrual cycle, oral contraceptive use, or menopause.

Sexually Dimorphic Sensitivity to Perinatal Neurotropic Drugs

Recently, the existence of sex differences in the mammalian brain has been extended to the sensitivity of neuroendocrine and non-neuroendocrine regions during developmental periods to exposure to various neurotropic drugs.[213,214] A common finding in these studies is the fact that male offspring are normally more sensitive to neurotropic drugs such as cannabinoids. For instance, male rats perinatally exposed to ethanol exhibited more marked behavioral modifications in the open-field test,[37,38] spontaneous motor activity,[39] and maturation of D_1 antagonist-induced cataleptic response;[40] whereas, perinatal exposure to nicotine abolishes several sexually dimorphic behaviors, accompanied by an increased DA turnover in the male striatum.[9,215] Similar sex-dependent differences have been shown after perinatal exposure to other drugs,[41,45] such as the differences in cocaine-induced effects on glucose metabolism.[2,56]

The demonstration of a similarly sexually dimorphic sensitivity of developing dopaminergic neurons to neurotropic drugs during perinatal ages allows two conclusions. First, the appearance of sex differences in the CNS during development[216] can be markedly altered, apparently through the ability of these drugs to interfere with fundamental processes of sexual differentiation in the brain during critical periods, inducing incomplete masculinization or having a demasculinizing effect.[214] Second, these sexually dimorphic effects could well be produced via a similar mechanism. The existence of such a mechanism remains to be demonstrated, but this conclusion is supported by the observations that alcohol or nicotine-induced loss of male saccharin preference[216,217] can also be produced by other kinds of prenatal manipulations, including maternal stress.[218,219] Both situations, alcohol or nicotine exposures and prenatal stress, produce a deficit in the fetal production of testosterone, which plays a basic role in the sexual differentiation of the brain,[19,220] and thereby reduces the masculinizing effect of this androgen.[217,221,222] In addition, it has been shown that cocaine exposure at birth is able to inhibit sex steroid incorporation into the hypothalamus during a critical period for sexual differentiation of the brain and it produces the demasculinization of various sexually dimorphic behaviors.[214]

Therefore, environmental or pharmacological manipulations that interfere with the prenatal surge of testosterone on days 18 and 19 of gestation or with the postnatal surge beginning at parturition result in an incomplete masculinization or a feminization of sexually dimorphic behavior patterns in males in adulthood.[218,221] McEwen[23] has recently reviewed the effects of a set of neurotropic drugs, including nicotine, neuroleptics, barbiturates, amphetamine, cannabinoids, opiates, and alcohol, which are able to influence gonadal function when administered during development. In particular, nicotine, THC, morphine, and alcohol are all reported to reduce serum testosterone levels in developing male fetuses and therefore can reduce the masculinization and defeminization of affected individuals.[223]

With regard to perinatal cannabinoid exposure, the reason for its sex-related effect is unclear at the moment. The possibility of cannabinoid-induced changes in fetal and neonatal gonadal function, such as those reported by Dalterio and Bartke after prenatal exposure of male mice to THC,[224] might explain the sexual dimorphism observed. However, other possible explanations can also be proposed; their importance is suggested by the effects observed in male dopaminergic ontogeny after perinatal exposure to cannabinoids[161] that do not appear to result in female-like dopaminergic neurons.

One of the possible interpretations of sexual dimorphism in response to drugs can be found in the studies of Richmond and Sachs who reported that male pups receive preferential care from the dams in the first 24 h postpartum.[225] This could result in increased nursing behavior, with males taking in more cannabinoids than their female littermates and thereby exhibiting a greater cannabinoid effect. A similar explanation has been used by others[45] to explain sex-dependent differences induced by prenatal exposure to other drugs. This possibility is supported by the fact that cannabinoid exposure reliably disrupts maternal behavior in rodents.[105] However, contrary to this explanation, it has also been shown that cannabinoid metabolism is faster in males than females,[226] leading to a lower degree of cannabinoid exposure.

A second explanation surmises that developing dopaminergic neurons of males would be, in general, more sensitive to epigenetic factors. This hypothesis is supported by two observations. First, it would appear that any exogenous effects on monoaminergic activity, particularly DA, during sensitive periods of development can compromise masculine development. This view is based on the findings of several studies showing that perinatal interference in the development of dopaminergic neurons with specific drugs for this neurotransmitter can demasculinizate several sexually dimorphic behaviors in male rats[227] and can produce sex-dependent changes in DA-related behaviors.[228] Thus, the administration of haloperidol, apomorphine, or α-methyl-*p*-tyrosine to pregnant or lactating rats impaired the sex behavior of their male offspring without affecting fetal testosterone production.[227]

The second observation is based on recent suggestions that the critical period for sexual differentiation of the brain, which has been described as the "time of greatest hormonal vulnerability", does not depend exclusively on fetal and neonatal androgen production. According to the classical concept of the sexual differentiation of the brain, the expression of sex differences in brain structure and function is dictated by the gonadal steroid environment during critical periods of pre- and postnatal development. Development of the male brain is brought about by the epigenetic actions of androgens, which are secreted by the testes after differentiation of the Leydig cells. The organizational effect of sex hormones appears to depend upon the aromatization of testosterone to estradiol within the brain. In the rat, the period during which sexually dimorphic regions of the brain are irreversibly programmed is thought to correspond to the perinatal period.[205] However, the studies of Reisert and co-workers have recently introduced new considerations into this scheme. These authors have reported the existence of subtle sex differences in the number and distribution of both mesencephalic and diencephalic TH-immunoreactive neurons at early ages.[194,229] These differences appear prior to the classical periods of sexual differentiation in the rat brain[230,231] and seem to be initiated independently of the action of androgens, although they are modified later by sex steroids.[231] Moreover, there is controversy concerning the parallelism between the increase in fetal testosterone production, which is an index of masculinizing ability, and the activation of the aromatase complex, which can be used as an index of steroid brain susceptibility.[232]

Based on these observations and on the fact that cannabinoids have a molecular structure that probably allows them to interact with steroid mechanisms, we hypothesize the fact that perinatal cannabinoid effects are more marked in males could be due to the possible existence of a "period of maximal susceptibility" in the developing male brain prior to the elevation of fetal testosterone production rather than to the adverse effects of cannabinoids on the production of this androgen.

Finally, one of the most plausible explanations could be the existence of a sexually dimorphic distribution of recently described cannabinoid binding sites in the brain.[233] However, there is no information either about their presence at early ages of development or their sex-dependent distribution. The possibility and significance of the involvement of these receptors in perinatal cannabinoid effects and their sexually dimorphic response, as well as the molecular mechanisms and the physiological implications of this phenomenon, will be the subject of a future and more detailed analysis.

Cannabinoid-Sexual Steroid Interactions

As mentioned above, the reason(s) for the sexual dimorphism observed in the responsiveness to perinatal cannabinoid exposure has not been clarified. Some of the hypotheses proposed can be supported by experimental observations dealing with the possibility of interrelated effects of THC and sexual steroids. In this regard, some studies have suggested that THC has an affinity for cytoplasmic steroid receptors, mainly for estrogens,[234] although this possibility is controversial.[235] In addition, recent studies by our group suggest the possibility of an interaction between estrogens and THC in the control of several physiological processes.[236,237]

The possibility of THC and steroid interactions at the receptor level is controversial. Cannabinoids have a molecular structure that might allow them to bind or interact with steroid receptors. This has led several groups of researchers to suggest that THC can bind to cytoplasmic sexual steroid receptors,[234] producing both estrogenic[234,238] and antiestrogenic[239] effects and also antiandrogenic effects.[240] However, the facts that THC only slightly decreases the binding of estradiol to its receptors *in vitro*[234,241] and that it binds to a different fraction than estrogens after gradient centrifugation[242] have shed considerable doubt on this matter.[235]

Another possibility is that estrogens and THC might interact at the post-receptor level. This is supported by the recent demonstration of several interactions between estrogens and THC in the control of various physiological processes.[234,236,237] Thus, Murphy and co-workers[236] have found that THC antagonizes the estrogen-induced increase in PRL release at the pituitary level in immature female rats. This antagonism was apparently not mediated through changes either in the dopaminergic hypothalamic influence or at the level of anterior pituitary D_2 receptors, suggesting a direct pituitary effect at the post-receptor level. Moreover, the possibility of estrogen and THC interactions at the level of the adrenal medulla has been suggested by recent findings (F. Rodríguez de Fonseca, L. L. Murphy, R. W. Steger, and A. Bartke, unpublished results). Administration of THC to immature females abolishes the estradiol-induced increase in adrenomedullary TH activity and increases its catecholamine contents, which are not increased by estradiol or THC alone. Dalterio et al.[136] have also proposed the existence of interrelationships among gonadal steroids and cannabinoids regarding their effects on adult neuroendocrine control of reproductive function in perinatally exposed mice. These authors found that cannabinoid effects on the neuroendocrine axis became more apparent after the removal of steroid negative feedback. Finally, we have recently found that blockade of estrogenic receptors with tamoxifen leads to the disappearance of several effects of THC treatment and the appearance of new modifications in dopaminergic activity in several brain areas (Table 9).[237] Although this latter study was performed in adult rats, it is reasonable to assume that a similar kind of effect could occur during development.

Molecular Mechanism(s) of Perinatal Cannabinoid Effects: Possible Involvement of Cannabinoid Receptors

The mechanism through which maternal cannabinoid consumption might interfere with the development of dopaminergic systems is not presently known. However, studies of the effects of other neurotropic drugs (including ethanol[35]) suggest several possibilities. For example, cannabinoids, especially THC, could be toxic to development of dopaminergic nerve cells, although relatively large doses of these compounds, much greater than those used in our studies, must be administered for toxicity to be observed.[105] Other possibilities include cannabinoid-induced effects on one or more

TABLE 9
Neurochemical Parameters Indicating Presynaptic Activity and Postsynaptic Sensitivity of Striatal, Limbic, and Hypothalamic Dopaminergic Terminals after Δ^9-Tetrahydrocannabinol (THC) Administration to Intact and Tamoxifen (TMX)-Pretreated Rats

Parameter	THC	THC to TMX-pretreated rats
Striatum		
TH activity	+	=
DOPAC/DA ratio	=	=
D_1 and D_2 density	=	=
Limbic		
TH activity	=	=
DOPAC/DA ratio	=	+
D_1 density	–	=
Hypothalamic		
TH activity	=	=
DOPAC/DA ratio	=	=
D_2 density[a]	–	=
PRL levels[b]	–	=

Note: All the animals were ovariectomized and implanted with estradiol silastic capsules. +: increase; –: decrease; =: no change.

[a] Measured in anterior pituitaries.
[b] Values in plasma.

From Bonnin, A., Fernández-Ruiz, J. J., Martin, M., Rodríguez de Fonseca, F., de Miguel, R., and Ramos, J. A., *Mol. Cell. Neurosci.*, in preparation, 1992.

embryonic neurotrophic factors, second messengers, hormones, or other cellular components required for the processes of nerve growth, neuronal migration, axonal guidance, and synaptogenesis.

Within this context, it is interesting to note the recent description of cannabinoid binding sites in the brain[233,243-245] that may act as possible molecular mediators of neurobiological and neuroendocrine effects of THC. This could be particularly relevant to the effects of cannabinoids on dopaminergic neurons because cannabinoid receptors are present at a higher concentration in brain areas closely related to dopaminergic neurotransmission, including the substantia nigra, the caudate-putamen, and the pallidum.[246]

One of these receptors has been described by Nye and co-workers[244] and it appears to be involved in the nonpsychoactive effects of cannabinoids.

The other receptor, described by Howlett and co-workers,[233] was characterized from preliminary studies on the possible inhibitory action of cannabinoids on adenylate cyclase activity in neuroblastoma cells[247,248] and the functional expression of its cloned DNA has been characterized.[245] This receptor appears to be negatively coupled to adenylate cyclase and it produces a decrease in cAMP production that can be blocked by pertussis toxin, thus suggesting involvement of a G_i protein.[249]

To our knowledge, no published data are available concerning the presence of these receptors during early development. However, the fact that prenatal cannabinoid exposure produces important effects on the ontogeny of brain dopaminergic neurons[161] suggests that cannabinoid receptors could be operative during fetal development. In this regard, recent studies in progress in our laboratory reveal the presence of specific and saturable binding sites, using [^{3}H]CP-55940, in the prosencephalic area of both male and female newborns (F. Rodríguez de Fonseca et al., unpublished results).

On the other hand, as suggested above, one of the most plausible explanations for the differences observed in the perinatal cannabinoid effects between males and females[161] could be that these effects are mediated via a physiological mechanism that could have heterogeneous and sexually dimorphic distribution in the brain and be related to gonadal status. Thus, the possibility exists that sex-dependent differences in the distribution, density, or functionality of cannabinoid receptors in the developing brain could explain the sex-dependent differences in the effects of perinatal cannabinoid exposure. Such a notion is attractive but remains to be experimentally tested.

SUMMARY

This chapter addresses the effects produced by maternal consumption of cannabinoids during gestation and lactation periods on neural development, with emphasis on the maturation of brain dopaminergic neurons in the rat. We have emphasized the differences in the effects of cannabinoids on three major dopaminergic pathways: namely, the nigrostriatal, mesolimbic, and tuberoinfundibular systems, and to the possible modifications in the functional expression of these neurons at behavioral or neuroendocrine levels. In addition, we have analyzed the possible existence of a sexual dimorphism in these ontogenic effects of cannabinoids, as well as the possible molecular mechanism underlying such effects. In general, the results discussed support the view that exposure to drugs during critical periods of development produces marked modifications in the functional expression of these neurons in adulthood. We have also briefly related these effects to those observed in other studies using perinatal exposure to other neurotropic and psychoactive drugs in an attempt to establish a possible parallelism with the effects of cannabinoids, specifically in the existence of a sex-dependent response.

ACKNOWLEDGMENTS

This work has been supported by two grants from DGICYT SM89-0003 and from the Rich Foundation. The authors are indebted to the Spanish Administration (Servicio de Restricción de Estupefacientes y Psicótropos; Dirección General de Farmacia y Productos Sanitarios, Ministerio de Sanidad y Consumo, Madrid, Spain) for the supply of hashish.

REFERENCES

1. **Zuckerman, B., Frank, D. A., Hingson, R., Amaro, H., Levenson, S. M., Kayne, H., Parker, S., Vinci, R., Aboagye, K., Fried, L. E., Cabral, H., Timperi, R., and Bauchner, H.,** Effects of maternal marijuana and cocaine use on fetal growth, *N. Engl. J. Med.*, 320, 762, 1989.
2. **Dow-Edwards, D. L., Freed, L. A., and Milhorat, T. H.,** Stimulation of brain metabolism by perinatal cocaine exposure, *Dev. Brain Res.*, 42, 137, 1988.
3. **Gruen, R. J., Deutch, A. Y., and Roth, R. H.,** Perinatal diazepam exposure: alterations in exploratory behavior and mesolimbic dopamine turnover, *Pharmacol. Biochem. Behav.*, 36, 169, 1990.
4. **Dalterio, S. L.,** Cannabinoid exposure: effects on development, *Neurobehav. Toxicol. Teratol.*, 8, 345, 1986.
5. **Cooper, J. D. and Rudeen, P. K.,** Alterations in regional catecholamine content and turnover in the male rat brain in response to *in utero* ethanol exposure, *Alcohol Clin. Exp. Res.*, 12, 282, 1988.
6. **Druse, M. J., Tajuddin, N. S., and Kuo, A. P.,** D-1 and D-2 receptors in ethanol-exposed offspring, *Alcohol*, 5, 455, 1988.
7. **Navarro, H. A., Seidler, F. J., Whitmore, W. L., and Slotkin, T. A.,** Prenatal exposure to nicotine via maternal infusions: effects on development of catecholamine systems, *J. Pharmacol. Exp. Ther.*, 244, 940, 1988.
8. **Spear, L. P., Kirstein, C. L., and Frambes, N. A.,** Cocaine effects on the developing central motor system: behavioral, psychopharmacological and neurochemical studies, *Ann. N.Y. Acad. Sci.*, 562, 290, 1989.

9. **Lichtensteiger, W., Ribary, U., Schlumpf, M., Odermatt, B., and Widmer, H. R.,** Prenatal adverse effects of nicotine on the developing brain, in *Progress in Brain Research,* Vol. 73, Boer, G. J., Feenstra, M. G. P., Mirmiran, M., Swaab, D. F., and van Haaren, F., Eds., Elsevier, Amsterdam, 1988, 137.
10. **Arenander, A. T. and de Vellis, J.,** Development of the nervous system, in *Basic Neurochemistry,* Siegel, G., Agranoff, B., Albers, R. W., and Molinoff, P., Eds., Raven Press, New York, 1989, 479.
11. **Mirmiran, M. and Swaab, D. F.,** Influence of drugs on brain neurotransmitter and behavioral states during development, *Dev. Pharmacol. Ther.,* 10, 377, 1987.
12. **Kalsbeek, A., Buijs, R. M., Hofman, M. A., Matthijssen, M. A. H., Pool, C. W., and Uylings, H. B. M.,** Effects of neonatal thermal lesioning of the mesocortical dopaminergic projection on the development of the rat prefrontal cortex, *Dev. Brain Res.,* 32, 123, 1987.
13. **Kasamatsu, T. and Pettigrew, J. D.,** Preservation of binocularity after monocular deprivation in the striate cortex of kittens treated with 6-hydroxydopamine, *J. Comp. Neurol.,* 185, 139, 1979.
14. **Mirmiran, M., Feenstra, M. G. P., Dijcks, F. A., Bos, N. P. A., and van Haaren, F.,** Functional deprivation of noradrenaline neurotransmission: effects of clonidine on brain development, in *Progress in Brain Research,* Vol. 73, Boer, G. J., Feenstra, M. G. P., Mirmiran, M., Swaab, D. F., and van Haaren, F., Eds., Elsevier, Amsterdam, 1988, 159.
15. **Kostrzewa, R. M. and Iqbal Saleh, M.,** Impaired ontogeny of striatal dopamine D-1 and D-2 binding sites after postnatal treatment of rats with SCH-23390 and spiroperidol, *Dev. Brain Res.,* 45, 95, 1989.
16. **Le Douarin, N. M., Teillet, M. A., Ziller, C., and Smith, J.,** Adrenergic differentiation of cells of the cholinergic ciliary and Remak ganglia in avian embryo after *in vivo* transplantation. *Proc. Natl. Acad. Sci. U.S.A.,* 75, 106, 1978.
17. **Fishman, M. C. and Nelson, P. G.,** Activity and competition-dependent synapse repression in culture, *Adv. Exp. Med. Biol.,* 181, 247, 1984.
18. **Toran-Allerand, C. D.,** On the genesis of sexual differentiation of the central nervous system: morphogenetic consequences of steroidal exposure and possible role of alpha-fetoprotein, in *Progress in Brain Research,* Vol. 61, De Vries, G. J., De Bruin, J. P. C., Uylings, H. B. M., and Corner, M. A., Eds., Elsevier, Amsterdam, 1984, 63.
19. **Arnold, P. and Gorski, R. A.,** Gonadal steroid induction of strutural sex differences in the central nervous system, *Annu. Rev. Neurosci.,* 7, 413, 1984.
20. **Williams, C. L.,** A reevaluation of the concept of separable periods of organizational and activational actions of estrogens in development of brain and behavior, *Ann. N.Y. Acad. Sci.,* 474, 282, 1986.
21. **Diamond, M. C.,** Sex differences in the rat forebrain, *Brain Res. Rev.,* 12, 235, 1987.
22. **Mirmiran, M., Brenner, E., Van der Gugten, J., and Swaab, D. F.,** Neurochemical and electrophysiological disturbances mediate developmental behavioral alterations produced by medicines, *Neurobehav. Toxicol. Teratol.,* 7, 677, 1985.
23. **McEwen, B. S.,** Steroid hormones and brain development: some guidelines for understanding actions of pseudohormones and other toxic agents, *Environ. Health Persp.,* 74, 177, 1987.
24. **Lankford, K. L., DeMello, F. G., and Klein, W. L.,** D1-type dopamine receptors inhibit growth cone motility in cultured retina neurons: evidence that neurotransmitters act as morphogenic growth regulators in the developing central nervous system, *PNAS (U.S.A.),* 85, 2839, 1988.

25. **Hagino, N. and Inoue, K.,** Hypothalamic dopamine may play a role in inducing prolactin in pituitary cells, *Int. J. Dev. Neurosci.*, 4, 119, 1986.
26. **Young, W. S., Bonner, T. I., and Braun, M. R.,** Mesencephalic dopamine neurons regulate the expression of neuropeptide mRNAs in the rat forebrain, *PNAS (U.S.A.)*, 83, 9827, 1986.
27. **De Vitry, F., Hillion, J., Catelon, J., Thibault, J., Benoliel, J. J., and Hamon, M.,** Dopamine increases the expression of tyrosine hydroxylase and aromatic aminoacid decarboxylase in primary cultures of fetal neurons, *Dev. Brain Res.*, 59, 123, 1991.
28. **Fernández-Ruiz, J. J., Amor, J. C., and Ramos, J. A.,** Time-dependent effects of estradiol and progesterone on the number of striatal dopaminergic D2-receptors, *Brain Res.*, 476, 388, 1989.
29. **Fernández-Ruiz, J. J., de Miguel, R., Hernández, M. L., and Ramos, J. A.,** Time-course of the effects of ovarian steroids on the activity of limbic and striatal dopaminergic neurons in female rat brain, *Pharmacol. Biochem. Behav.*, 36, 603, 1990.
30. **Dluzen, D. E. and Ramírez, V. D.,** Bimodal effect of progesterone on *in vitro* dopamine function of the rat corpus striatum, *Neuroendocrinology*, 39, 149, 1984.
31. **Herz, A. and Shippenberg, S. T.,** Neurochemical aspects of addiction: opioids and other drug of abuse, in *Molecular and Cellular Aspects of the Drug Addictions*, Goldstein, A., Ed., Springer-Verlag, New York, 1989, 111.
32. **Streissguth, A. P.,** Alcohol and pregnancy: an overview and an update, *Subst. Alcohol Actions Misuse*, 4, 149, 1983.
33. **Rathbun, W. and Druse, M.,** Dopamine, serotonin and acid metabolites in brain regions from the developing offspring of ethanol-treated rats, *J. Neurochem.*, 44, 57, 1985.
34. **Serbus, D. C., Stull, R. E., and Light, K. E.,** Neonatal ethanol exposure to rat pups: resultant alterations of cortical muscarinic and cerebellar H1-histaminergic receptor binding dynamics, *Neurotoxicology*, 7, 257, 1986.
35. **Druse, M. J., Tajuddin, N., Kuo, A., and Connerty, M.,** Effects of in utero ethanol exposure on the developing dopaminergic system in rats, *J. Neurosci. Res.*, 27, 233, 1990.
36. **Lucchi, L., Covelli, V., Petkov, V. V., Spano, P. F., and Trabucchi, M.,** Effects of ethanol, given during pregnancy, on the offspring dopaminergic system, *Pharmacol. Biochem. Behav.*, 19, 566, 1983.
37. **Chen, J. J. and Smith, E. R.,** Effects of perinatal alcohol on sexual differentiation and open field behavior in rats, *Horm. Behav.*, 13, 219, 1979.
38. **Blanchard, B. A., Hannighan, J. H., and Riley, E. P.,** Amphetamine-induced activity after fetal alcohol exposure and undernutrition in rats, *Neurotoxicol. Teratol.*, 9, 113, 1987.
39. **Zimmerberg, B., Riley, E. P., and Glick, S. D.,** Differential effects of prenatal exposure to alcohol on activity and circling behavior in rats, *Pharmacol. Biochem. Behav.*, 25, 1021, 1986.
40. **Hannighan, J. H.,** The ontogeny of SCH 23390-induced catalepsy in male and female rats exposed to ethanol *in utero*, *Alcohol*, 7, 11, 1990.
41. **Ribary, V. and Lichtensteiger, W.,** Effects of acute and chronic prenatal nicotine treatment on central catecholamine systems of male and female rat fetuses and offspring, *J. Pharmacol. Exp. Ther.*, 248, 786, 1989.
42. **Peters, D. A. V. and Tang, S.,** Sex-dependent biological changes following prenatal nicotine exposure in the rat, *Pharmacol. Biochem. Behav.*, 17, 1077, 1982.

43. **Hutchings, D. E. and Fitter, W. P.,** Neurobehavioral effects in human and animal offspring following prenatal exposure to methadone, in *Handbook of Behavioral Teratology*, Riley, E. P. and Vorhees, C. V., Eds., Plenum Press, New York, 1986, 141.
44. **Slotkin, T. A., Whitmore, W. L., Salvaggio, M., and Seidler, F. J.,** Perinatal methadone addiction affects brain synaptic development of biogenic amine systems in the rats, *Life Sci.,* 24, 1223, 1979.
45. **Guo, H., Enters, E. K., McDowell, K. P., and Robinson, S. E.,** The effect of prenatal exposure to metadone on neurotransmitters in neonatal rats, *Dev. Brain Res.,* 57, 296, 1990.
46. **Hughes, R. N. and Beveridge, I. J.,** Sex- and age-dependent effects of prenatal exposure to caffeine on open-field behavior, emergency latency and adrenal weights in rats, *Life Sci.,* 47, 2075, 1990.
47. **Hughes, R. N. and Beveridge, I. J.,** Effects of prenatal exposure to chronic caffeine on locomotor and emotional behavior, *Psychobiology,* 15, 179, 1987.
48. **Lombardelli, G., Balduini, W., Feduzi, A., Peruzzi, G., and Cattabeni, F.,** Long-lasting tolerance to stimulatory effects of perinatal caffeine treatment, *Psychopharmacology,* 84, 285, 1984.
49. **Marangos, P. J., Boulenger, J. P., and Patel, J.,** Effects of chronic caffeine on brain adenosine receptors: regional and ontogenic studies, *Life Sci.,* 34, 899, 1984.
50. **Enslen, M., Milon, H., and Wurzner, H. P.,** Brain catecholamines and sleep states in offspring of caffeine-treated rats, *Experientia,* 36, 1105, 1980.
51. **Berkowitz, B. A., Tarver, J., and Spector, S.,** Release of norepinephrine in the central nervous system by theophylline and caffeine, *Eur. J. Pharmacol.,* 10, 64, 1970.
52. **Berkowitz, B. A. and Spector, S.,** The effect of caffeine and theophylline on the disposition of brain serotonin in the rat, *Eur. J. Pharmacol.,* 16, 322, 1971.
53. **Swaab, D. F. and Mirmiran, M.,** Possible mechanisms underlying the teratogenic effects of medicines on the developing brain, in *Neurobehavioral Teratology*, Yanai, J., Ed., Elsevier, Amsterdam, 1984, 55.
54. **Fung, Y. K., Reed, J. A., and Lau, Y. S.,** Prenatal cocaine exposure fails to modify neurobehavioral responses and the striatal dopaminergic system in newborn rats, *Gen. Pharmacol.,* 20, 689, 1989.
55. **Segal, D. H., Freed, L. A., Hughes, H. E., Milhorat, T. H., and Dow-Edwards, D. L.,** Alteration in ^{3}H-SCH23390 binding in rat brain following perinatal exposure to cocaine, *Soc. Neurosci.,* Abstr. 15, 519.19, 1989.
56. **Dow-Edwards, D. L., Freed, L. A., and Fico, T. A.,** Structural and functional effects of prenatal cocaine exposure in adult rat brain, *Dev. Brain Res.,* 57, 263, 1990.
57. **Ali, S. F., Scalzo, F. M., Trambes, N. A., and Spear, L. P.,** Prenatal exposure of cocaine in rat produces significant alterations in striatal dopamine (D2) receptor binding in offspring, *Teratology*, 41, 616, 1990.
58. **Hutchings, D., Fico, T. A., and Dow-Edwards, D. L.,** Prenatal cocaine: maternal toxicity, fetal effects and locomotor activity in rat offspring, *Neurotoxicol. Teratol.,* 11, 65, 1989.
59. **Foss, J. A. and Riley, E. P.,** Behavioral evaluation of animals exposed prenatally to cocaine, *Teratology,* 37, 517, 1988.
60. **Smith, R. F., Mattran, K. M., Kurkjian, M. F., and Kurtz, S. L.,** Alterations in offspring behavior by chronic prenatal cocaine dosing, *Neurotoxicol. Teratol.,* 11, 35, 1989.
61. **Middaugh, L. D.,** Prenatal amphetamine effects on behavior: possible mediation by brain monoamines, *Ann. N.Y. Acad. Sci.,* 562P, 308, 1989.

62. **Nasello, A. G. and Ramírez, O. A.,** Brain catecholamine metabolism on offspring of amphetamine treated rats, *Pharmacol. Biochem. Behav.,* 9, 17, 1978.
63. **Ramírez, O. A., Carrier, H. F., and Nasello, A. G.,** Prenatal amphetamine exposure: ovulation, sexual behavior and hypothalamic monoamine content in rats, *Pharmacol. Biochem. Behav.,* 11, 605, 1979.
64. **Mendelson, J. H.,** Marihuana, in *Psychopharmacology: The Third Generation of Progress,* Meltzer, H. Y., Ed., Raven Press, New York, 1987, 1565.
65. **Hughes, C. L., Everett, J. W., and Tyrey, L.,** Delta-9-tetrahydrocannabinol suppression of prolactin secretion in the rat: lack of direct pituitary effect, *Endocrinology,* 109, 876, 1981.
66. **Tyrey, L.,** Delta-9-tetrahydrocannabinol suppression of episodic luteinizing hormone secretion in the ovariectomized rat, *Endocrinology,* 102, 1808, 1978.
67. **Rettori, V., Wenger, T., Snyder, G., Dalterio, S., and McCann, S. M.,** Hypothalamic action of delta-9-tetrahydrocannabinol to inhibit the release of prolactin and growth hormone in the rat, *Neuroendocrinology,* 47, 498, 1988.
68. **Consroe, P., Sandyk, R., and Snider, R. S.,** Open label evaluation of cannabidiol in dystonic movement disorders, *Int. J. Neurosci.,* 30, 277, 1986.
69. **Conti, L. H. and Musty, R. E.,** The effects of delta-9-tetrahydrocannabinol injections to the nucleus accumbens on the locomotor activity of rats, in *The Cannabinoids: Chemical, Pharmacologic and Therapeutic Aspects,* Agurell, S., Dewey, W. L., and Willette, R. E., Eds., Academic Press., Orlando, FL, 1984, 649.
70. **Gardner, E. L., Paredes, W., Smith, D., Donner, A., Milling, C., Cohen, D., and Morrisson, D.,** Facilitation of brain stimulation reward by delta-9-tetrahydrocannabinol, *Psychopharmacology,* 96, 142, 1988.
71. **Andreasson, S., Allebeck, P., Engström, A., and Rydberg, V.,** Cannabis and schizophrenia: a longitudinal study of Swedish conscripts, *Lancet,* ii, 1483, 1987.
72. **Bloom, A. S.,** Effects of cannabinoids on neurotransmitter receptors in the brain, in *The Cannabinoids: Chemical, Pharmacologic and Therapeutic Aspects,* Agurell, S., Dewey, W. L., and Willette, R. E., Eds., Academic Press, Orlando, FL, 1984, 575.
73. **Taylor, D. A., Sitaram, B. R., and Elliot-Baker, S.,** Effect of delta-9-tetrahydrocannabinol on the release of dopamine in the corpus striatum of the rat, in *Marihuana '87, Proc. Melbourne Symp. Cannabis,* 1987, 405.
74. **Dewey, W. L.,** Cannabinoid pharmacology, *Pharmacol Rev.,* 38, 151, 1986.
75. **Aulakh, C. S., Bhattacharyya, A. K., Hossain, M. A., and Pradham, S. N.,** Behavioral and neurochemical effects of repeated administration of delta-9-tetrahydrocannabinol in rats, *Neuropharmacology,* 19, 97, 1980.
76. **Bloom, A. S.,** Effect of delta-9-tetrahydrocannabinol on the synthesis of dopamine and norepinephrine in mouse brain synaptosomes, *J. Pharmacol. Exp. Ther.,* 221, 97, 1982.
77. **Mazurkiewiez-Kwilecki, I. M. and Fidezewski, M.,** The effects of chronic treatment with delta-9-tetrahydrocannabinol on catecholamine synthesis in the rat, *Psychopharmacology,* 33, 71, 1973.
78. **Ho, B. T., Taylor, A., Englert, L. F., and McIsaac, W. M.,** Neurochemical effects of L-delta-9-tetrahydrocannabinol in rats following repeated inhalation, *Brain Res.,* 31, 233, 1971.
79. **Hershkowitz, M. and Szechtman, H.,** Pretreatment with delta-1-tetrahydrocannabinol and psychoactive drugs: effects on uptake of biogenic amines and on behaviour, *Eur. J. Pharmacol.,* 59, 267, 1979.
80. **Bloom, A. S., Johnson, K. M., and Dewey, W. L.,** The effects of cannabinoids on body temperature and brain catecholamine synthesis, *Res. Commun. Chem. Pathol. Pharmacol.,* 20, 51, 1978.

81. **Hillard, C. J. and Bloom, A. S.,** Further studies of the interaction of delta-9-tetrahydrocannabinol with the beta-adrenergic receptor, in *The Cannabinoids: Chemical, Pharmacologic and Therapeutic Aspects*, Agurell, S., Dewey, W. L., and Willette, R. E., Eds., Academic Press, Orlando, FL, 1984, 591.
82. **Rodríguez de Fonseca, F., Fernández-Ruiz, J. J., Murphy, L. L., Cebeira, M., Steger, R. W., Bartke, A., and Ramos, J. A.,** Acute effects of delta-9-tetrahydrocannabinol on dopaminergic activity in several rat brain areas, *Pharmacol. Biochem. Behav.,* 41, 1982.
83. **Sofia, R. D., Dixit, B. N., and Barry, H., III,** The effects of delta-1-tetrahydrocannabinol on serotonin metabolism in the rat brain, *Life Sci.,* 10, 425, 1971.
84. **Ho, B. T., Taylor, D., Fritchie, G. E., Englert, L. F., and McIsaac, W. M.,** Neuropharmacological study of delta-9- and delta-8-L-tetrahydrocannabinols in monkeys and mice, *Brain Res.,* 38, 163, 1972.
85. **Banerjee, S. P., Snyder, S. H., and Mechoulam, R.** Cannabinoids: influence on neurotransmitter uptake in rat brain synaptosomes, *J. Pharmacol. Exp. Ther.,* 194, 74, 1975.
86. **Tripathi, H. L., Vocci, F. J., Brase, D. A., and Dewey, W. L.,** Effects of cannabinoids on levels of acetylcholine and choline and on turnover rate of acetylcholine in various regions of mouse brain, *Alcohol Drug Res.,* 7, 525, 1987.
87. **Shan-Ching, S. and Jakubovic, A.,** Interaction of a water derivative of delta-9-tetrahydrocannabinol with [^{3}H]-diazepam and [^{3}H]-flunitrazepam binding to rat brain membranes, *Progr. Neuro-Psychopharmacol. Biol. Psychiatr.,* 11, 335, 1987.
88. **Mechoulam, R.,** Marihuana chemistry, *Science,* 168, 1159, 1970.
89. **Steger, R. W., De Paolo, L., Asch, R. H., and Silverman, A. Y.,** Interaction of delta-9-tetrahydrocannabinol (THC) with hypothalamic neurotransmitters controlling luteinizing hormone and prolactin release, *Neuroendocrinology,* 37, 361, 1983.
90. **Tyrey, L.,** Reversal of the delta-9-tetrahydrocannabinol inhibitory effect on prolactin secretion by rostral deafferentation of the medial basal hypothalamus, *Neuroendocrinology,* 44, 204, 1986.
91. **Wenger, T., Croix, D., West, M., and Tram, G.,** Acute effect of intracerebroventricularly injected delta-9-tetrahydrocannabinol on serum LH and FSH levels in adult male rats, *Neuroendocrinol. Lett.,* 6, 327, 1984.
92. **Murphy, L. L., Steger, R. W., Smith, M. S., and Bartke, A.,** Effects of delta-9-tetrahydrocannabinol, canabidiol and cannabidiol, alone and in combinations, on luteinizing hormone and prolactin release and on hypothalamic neurotransmitters in the male rat, *Neuroendocrinology,* 52, 316, 1990.
93. **Dalterio, S. L., Michael, S. D., MacMillan, B. T., and Bartke, A.,** Differential effects of cannabinoid exposure and stress on plasma prolactin, growth hormone and corticosterone levels in male mice, *Life Sci.,* 28, 761, 1981.
94. **Steger, R. W., Silverman, A. Y., Siler-Khodr, T. M., and Asch, R. H.,** The effect of delta-9-tetrahydrocannabinol on the positive and negative feedback control of luteinizing hormone release, *Life Sci.,* 27, 1911, 1980.
95. **Kubena, R. K., Perhack, J. C., and Barry, H.,** Corticosterone elevation mediated centrally by delta-1-tetrahydrocannabinol in rats, *Eur. J. Pharmacol.,* 14, 89, 1971.
96. **Kramer, J. and Ben-David, M.,** Supression of prolactin secretion by acute administration of delta-9-tetrahydrocannabinol in rats, *Proc. Soc. Exp. Biol. Med.,* 147, 482, 1974.
97. **Kramer, J. and Ben-David, M.,** Prolactin supression by (–)-delta-9-tetrahydrocannabinol: involvement of serotoninergic and dopaminergic pathways, *Endocrinology,* 103, 452, 1978.

98. **Navarro, M., Fernández-Ruiz, J. J., de Miguel, R., Hernández, M. L., Cebeira, M., and Ramos, J. A.,** An acute dose of Δ^9-tetrahydrocannabinol affects behavioral and neurochemical indices of mesolimbic dopaminergic activity, *Neuroscience,* submitted.
99. **Ayalon, D., Nir, I., Cordova, T., Bauminger, S., Puder, M., Naor, Z., Kashi, R., Zor, U., Harrell, A., and Lindner, H. R.,** Acute effect of delta-1-tetrahydrocannabinol on the hypothalamo-pituitary-ovarian axis in the rat, *Neuroendocrinology,* 23, 31, 1977.
100. **Nir, I., Ayalon, D., Tsafriri, A., Cordova, T., and Lindner, H. R.,** Suppression of the cyclic surge of luteinizing hormone secretion of ovulation in the rat by delta-1-tetrahydrocannabinol, *Nature (London),* 243, 470, 1973.
101. **Gher, J., Murphy, L. L., Steger, R. W., and Bartke, A.,** Delta-9-tetrahydrocannabinol inhibition of the neuroendocrine response of adult male rats to sexually receptive female conspecifics, *Soc. Neurosci. Abstr.,* 16 (Part 2), 1101, 1990.
102. **Murphy, L. L. and Tyrey, L.,** Induction of luteinizing hormone release by electrochemical stimulation of the medial preoptic area in delta-9-tetrahydrocannabinol blocked proestrous rats, *Neuroendocrinology,* 43, 471,1986.
103. **Chakravarty, I., Shah, P. G., Sheth, A. R., and Ghosh, J. J.,** Mode of action of delta-9-tetrahydrocannabinol on hypothalamo-pituitary function in adult female rats, *Reprod. Fertil.,* 57, 113, 1979.
104. **Murphy, L. L., Newton, S. C., Dhali, J., and Chavez, D.,** Evidence for a direct anterior pituitary site of delta-9-tetrahydrocannabinol action, *Pharmacol. Biochem. Behav.,* 40, 603, 1991.
105. **Abel, E. L.,** *Marihuana, Tobacco, Alcohol and Reproduction,* CRC Press, Boca Raton, FL, 1983.
106. **Murphy, L. L., Steger, R. W., and Bartke, A.,** Psychoactive and non-psychoactive cannabinoids and their effects on reproductive neuroendocrine parameters, in *Biochemistry and Physiology of Substance Abuse*, Vol. 2, Watson, R. R., Ed., CRC Press, Boca Raton, FL, 1990, 73.
107. **Dalterio, S., Bartke, A., Roberson, C., Watson, D., and Burstein, S.,** Direct and pituitary-mediated effects of delta-9-tetrahydrocannabinol and cannabinol on the testis, *Pharmacol. Biochem. Behav.,* 8, 673, 1978.
108. **Poddar, M. K. and Dewey, W. L.,** Effects of cannabinoids on catecholamine uptake and release in hypothalamic and striatal synaptosomes, *J. Pharmacol. Exp. Ther.,* 214, 63, 1980.
109. **Sakurai-Yamashita, Y., Kataoka, Y., Fujiwara, M., Mine, K., and Ueki, S.,** Delta-9-tetrahydrocannabinol facilitates striatal dopaminergic transmission, *Pharmacol. Biochem. Behav.,* 33, 397, 1989.
110. **Ton, J. M., Gerhardt, G. A., Friedeman, M., Etgen, A. M., Rose, G. M., Sharless, N. S., and Gardner, E. L.,** The effect of delta-9-tetrahydrocannabinol on potassium-evoked release of dopamine in the rat caudate nucleus: an *in vivo* electrochemical and *in vivo* microdialysis study, *Brain Res.,* 451, 59, 1988.
111. **Moss, D. E., Montgomery, S. P., Salo, A. A., and Steger, R. W.,** THC effects on extrapyramidal motor behaviors in an animal model of Parkinson's disease, in *The Cannabinoids: Chemical, Pharmacologic and Therapeutic Aspects*, Agurell, S., Dewey, W. L., and Willette, R. E., Eds., Academic Press, Orlando, FL, 1984, 815.
112. **Pertwee, R. G.,** The ring test: a quantitative method for assessing the cataleptic effect of cannabis in mice, *Br. J. Pharmacol.,* 46, 753, 1972.
113. **Consroe, P. and Fish, B.,** Rabbit behavioral model of marihuana psychoactivity in humans, *Med. Hypot.,* 7, 1079, 1981.
114. **Knudsen, P. and Vilmar, T.,** Cannabis and neuroleptic agents in schizophrenia, *Acta Psychiatr. Scand.,* 69, 162, 1984.

115. **Bowers, M. B. and Hoffman, F. J.,** Regional homovanillic acid following delta-9-tetrahydrocannabinol and cocaine, *Brain Res.,* 366, 405, 1986.
116. **Cheng, J., Paredes, W., Li, J., Smith, D., Lowinson, J., and Gardner, E. L.,** Δ^9-tetrahydrocannabinol produces naloxone-blockable enhancement of presynaptic basal dopamine efflux in nucleus accumbens of concious, freely moving rats as measured by intracerebral microdialysis, *Psychopharmacology,* 102, 156, 1990.
117. **Razdan, R. K.,** Structure-activity relationships in cannabinoids, *Pharmacol. Rev.,* 38, 75, 1986.
118. **Sassenrath, E. N.,** Assesment of acute versus long-term effects of delta-9-tetrahydrocannabinol on social behavior and adaptability in group-living macaques, in *Ethopharmacology: Primate Models of Neuropsychiatric Disorders*, Alan R. Liss, New York, 1983, 277.
119. **Nahas, G. G. and Frick, H. C.,** Developmental effects of cannabis, *Neurotoxicology,* 7, 381, 1986.
120. **Hutchings, D. E., Morgan, B., Brake, S. C., Shi, T., and Lasalle, E.,** Delta-9-tetrahydrocannabinol during pregnancy in the rat. I. Differential effects on maternal nutrition, embryotoxicity and growth in the offspring, *Neurotoxicol. Teratol.,* 9, 39, 1987.
121. **Martin, B. R., Dewey, W. L., Harris, L. S., and Beckner, J. S.,** ^{3}H-delta-9-tetrahydrocannabinol distribution in pregnant dogs and their fetuses, *Res. Commun. Chem. Pathol. Pharmacol.,* 17, 457, 1977.
122. **Jakubovic, A., Hattori, T., and McGeer, P. L.,** Radioactivity in suckled rats after giving ^{14}C-tetrahydrocannabinol to the mother, *Eur. J. Pharmacol.,* 22, 221, 1973.
123. **Jakubovic, A., Hattori, T., and McGeer, P. L.,** Excretion of THC and its metabolites in ewe's milk, *Toxicol. Appl. Pharmacol.,* 28, 38, 1974.
124. **Hembree, W. C., Zeidenberg, P., and Nahas, G. G.,** Marihuana effects on human gonadal function, in *Marihuana: Chemistry, Biochemistry and Cellular Effects*, Nahas, G. G., Ed., Springer-Verlag, New York, 1976, 521.
125. **Issidoriedes, M. R.,** Observations in chronic hashish users: nuclear observations in blood and sperm and abnormal acrosomes in spermatozoa, in *Marihuana: Biological Effects*, Nahas, G. G. and Paton, W. D., Eds., Pergamon Press, Oxford, 1980, 377.
126. **Dalterio, S. L., Badr, F., Bartke, A., and Mayfield, D.,** Cannabinoids in mice: effects on fertility and spermatogenesis, *Science,* 216, 315, 1982.
127. **Nogawa, T., Shinohaza, O., Henrich, R. T., and Morishima, A.,** The effects of chronic administration of delta-9-tetrahydrocannabinol on the early embryonic development of mice, in *The Cannabinoids: Chemical, Pharmacologic and Therapeutic Aspects*, Agurell, S., Dewey, W. L., and Willette, R. E., Eds., Academic Press, Orlando, FL, 1984, 649.
128. **Abel, E. L. and Tan, S. E.,** Effects of tetrahydrocannabinol, chlordiazepoxide (librium) and their combination on pregnancy and offspring in rats, *Reprod. Toxicol.,* 1, 37, 1987.
129. **Wenger, T., Aranyi, Z., Croix, D., Bereczky, M., and Tramn, G.,**The effect of marihuana (delta-9-tetrahydrocannabinol) on pregnancy in rat, *Neuroendocrinol. Lett.,* 10, 39, 1988.
130. **Walters, D. E. and Carr, L. A.,** Changes in brain catecholamine mechanisms following perinatal exposure to marihuana, *Pharmacol. Biochem. Behav.,* 25, 763, 1986.
131. **Abel, E. L., Dintcheff, B. A., and Day, N.,** Effects of marihuana on pregnant rats and their offspring, *Pychopharmacology,* 71, 71, 1980.
132. **Abel, E. L., Bush, R., Dintcheff, B. A., and Ernst, C. A. S.,** Critical periods for marihuana-induced intrauterine growth retardation in the rat, *Neurobehav. Toxicol. Teratol.,* 3, 351, 1981.

133. **Qazi, Q. H., Mariano, E., Milman, D. H., Bellez, E., and Crombleholme, W.,** Abnormalities in offspring associated with prenatal marihuana exposure, *Dev. Pharmacol. Ther.,* 8, 141, 1985.
134. **Haley, S. L., Wright, P. L., Plank, J. B., Keplinger, M. L., Braude, M. C., and Calandra, J. C.,** The effect of natural and synthetic delta-9-tetrahydrocannabinol on fetal development, *Toxicol. Appl. Pharmacol.,* 25, 450, 1975.
135. **Dalterio, S. L. and Bartke, A.,** Perinatal exposure to cannabinoids alters male reproductive function in mice, *Science,* 205, 1420, 1979.
136. **Dalterio, S., Steger, R. W., Mayfield, D., and Bartke, A.,** Early cannabinoid exposure influences neuroendocrine and reproductive functions in male mice. II. Postnatal effects, *Pharmacol. Biochem. Behav.,* 20, 115, 1984.
137. **Albert, M. and Solomon, J.,** Effect of neonatal administration of delta-9-tetrahydrocannabinol and/or estradiol benzoate (EB) on reproductive development and function of the male rat, in *The Cannabinoids: Chemical, Pharmacologic and Therapeutic Aspects,* Agurell, S., Dewey, W. L., and Willette, R. E., Eds., Academic Press, Orlando, FL, 1984, 339.
138. **Dalterio, S. L.,** Perinatal or adult exposure to cannabinoids alters male reproductive functions in mice, *Pharmacol. Biochem. Behav.,* 12, 143, 1980.
139. **Wenger, T., Croix, D., and Tramm, G.,** The effect of chronic prepuberal administration of marihuana (delta-9-tetrahydrocannabinol) on the onset of puberty and the postpuberal reproductive functions in female rats, *Biol. Reprod.,* 39, 540, 1988.
140. **Field, E. and Tyrey, L.,** Delayed sexual maturation in the female rat during chronic exposure to delta-9-tetrahydrocannabinol, *Life Sci.,* 35, 1725, 1984.
141. **Field, E. and Tyrey, L.,** Delayed sexual maturation during prepubertal cannabinoid treatment: importance of the timing of treatment, *J. Pharmacol. Exp. Ther.,* 254, 171, 1990.
142. **Kumar, A. M., Solomon, J., Patel, V., Kream, R. M., Drieze, J. M., and Millard, W. J.,** Early exposure to delta-9-tetrahydrocannabinol influences neuroendocrine and reproductive functions in female rats, *Neuroendocrinology,* 44, 260, 1986.
143. **Kumar, A. M., Haney, M., Becker, T., Thompson, M. L., Kream, R. M., and Miczek, K.,** Effect of early exposure to delta-9-tetrahydrocannabinol on the levels of opioid peptides, gonadotropin-releasing hormone and substance P in the adult male rat brain, *Brain Res.,* 525, 78, 1990.
144. **Dalterio, S., Steger, R. W., Mayfield, D., and Bartke, A.,** Early cannabinoid exposure influences neuroendocrine and reproductive functions in male mice. I. Prenatal effects, *Pharmacol. Biochem. Behav.,* 20, 107, 1984.
145. **Tyrey, L. and Murphy, L. L.,** Inhibition of suckling induced milk ejections in the lactating rat by delta-9-tetrahydrocannabinol, *Endocrinology,* 123, 469, 1988.
146. **Morgan, B., Brake, S. C., Hutchings, D. E., Miller, N., and Gamagaris, Z.,** Delta-9-tetrahydrocannabinol during pregnancy in the rat: effects on development of RNA, DNA and protein in offspring brain, *Pharmacol. Biochem. Behav.,* 31, 365, 1988.
147. **Fleischman, R. W., Naqui, E. H., and Rosenkrantz, H.,** The embryotoxic effects of cannabinoids in rats and mice, *J. Environ. Pathol. Toxicol.,* 4, 471, 1980.
148. **Joneja, M. G.,** A study of teratological effects of intravenous, subcutaneous, and intragastric administration of delta-9-tetrahydrocannabinol in mice, *Toxicol. Appl. Pharmacol.,* 36, 151, 1976.
149. **Mantilla-Plata, B., Clewe, C. L., and Harbison, R. D.,** Delta-9-tetrahydrocannabinol-induced changes in prenatal growth and development of mice, *Toxicol. Appl. Pharmacol.,* 33, 333, 1975.

150. **Turner, C. E.,** Marihuana research and problems: an overview, *Pharm. Int.,* 1, 93, 1980.
151. **Martin, B. R., Compton, D. R., Little, P. J., Martin, T. J., and Beardsley, P. M.,** Pharmacological evaluation of agonistic and antagonistic activity of cannabinoids, *NIDA Res. Monogr. Ser.,* 79P, 108, 1987.
152. **Hatoum, N. S., Davis, W. M., Elsohly, M. A., and Turner, C. E.,** Perinatal exposure to cannabichromene and delta-9-tetrahydrocannabinol: separate and combined effects on viability of pups and on male reproductive system at maturity, *Toxicol. Lett.,* 8, 141, 1981.
153. **Walters, D.E. and Carr, L. A.,** Perinatal exposure to cannabinoids alters neurochemical development in rat brain, *Pharmacol. Biochem. Behav.,* 29, 213, 1988.
154. **Abel, E. L.,** Prenatal exposure to cannabis: a critical review of effects on growth, development and behavior, *Behav. Neural Biol.,* 29, 137, 1980.
155. **Borgen, L. A., Davis, W. N., and Pace, H. B.,** Effects of prenatal delta-9-tetrahydrocannabinol on the development of rat offspring, *Pharmacol. Biochem. Behav.,* 1, 203, 1973.
156. **Vardaris, R. M., Weisz, D. J., Fazel, A., and Rawitch, A. B.,** Chronic administration of delta-9-tetrahydrocannabinol to pregnant rats: studies of pup behavior and placental transfer, *Pharmacol. Biochem. Behav.,* 4, 249, 1976.
157. **Dalterio, S., Thomford, P. J., Michael, S. D., De Angelo, L., and Mayfield, D.,** Perinatal cannabinoid exposure: effects on hepatic cytochrome P-450 and plasma protein levels in male mice, *Teratology,* 33, 195, 1986.
158. **Mokler, D. A., Robinson, S. E., Johnson, J. H., Hong, J. S., and Rosecrans, J. A.,** Neonatal administration of delta-9-tetrahydrocannabinol alters the neurochemical response to stress in the adult Fischer-344 rat, *Neurotoxicol. Teratol.,* 9, 321, 1987.
159. **Dalterio, S., Steger, R. W., and Bartke, A.,** Maternal or paternal exposure to cannabinoids affects central neurotransmitter levels and reproductive function in male offspring, in *The Cannabinoids: Chemical, Pharmacologic and Therapeutic Aspects,* Agurell, S., Dewey, W. L., and Willette, R. E., Eds., Academic Press, Orlando, FL, 1984, 649.
160. **Rodríguez de Fonseca, F., Cebeira, M., Hernández, M. L., Ramos, J. A., and Fernández-Ruiz, J. J.,** Changes in brain dopaminergic indices induced by perinatal exposure to cannabinoids in rats, *Dev. Brain Res.,* 51, 237, 1990.
161. **Rodríguez de Fonseca, F., Cebeira, M., Fernández-Ruiz, J. J., Navarro, M., and Ramos, J. A.,** Effects of pre- and perinatal exposure to hashish extracts on the ontogeny of brain dopaminergic neurons, *Neuroscience,* 43, 713, 1991.
162. **Karnil, I. G. and Carlini, E. A.,** Pharmacological interaction between cannabidiol and delta-9-tetrahydrocannabinol, *Psychopharmacologia,* 33, 53, 1973.
163. **Turkanis, S. A., Cely, W., Olsen, D. M., and Karler, R.,** Anticonvulsant properties of cannabidiol, *Res. Commun. Chem. Pathol. Pharmacol.,* 8, 231, 1974.
164. **Wenger, T., Croix, D., Tramm, G., and Leonardelli, J.,** Prenatally administered delta-9-tetrahydrocannabinol temporary inhibits the developing hypothalamic-pituitary system in rats, *Marihuana '90. Proceedings of the International Conference on Cannabis and Cannabinoids,* Crete, Greece, Abstr. 8-4, 1990.
165. **Coyle, J. T. and Henry, D.,** Catecholamines in fetal and newborn rat brain, *J. Neurochem.,* 21, 61, 1973.
166. **Gelbard, H. A., Teicher, M. H., Faedda, G., and Baldessarini, R. J.,** Postnatal development of dopamine D1 and D2 receptors sites in rat striatum, *Brain Res.,* 49, 123, 1989.

167. **Murrin, L. C.,** *In vivo* studies of dopamine receptor ontogeny, *Life Sci.,* 31, 971, 1982.
168. **Noisin, E. L. and Thomas, W. E.,** Ontogeny of dopaminergic function in the rat midbrain tegmentum, corpus striatum and frontal cortex, *Dev. Brain Res.,* 41, 241, 1988.
169. **Parés-Herbuté, N., Tapia-Arancibia, L., and Artier, H.,** Ontogeny of the metencephalic, mesencephalic and diencephalic content of catecholamines as measured by high performance liquid chromatography with electrochemical detection, *Int. J. Dev. Neurosci.,* 7, 73, 1989.
170. **Coyle, J. T. and Axelrod, J.,** Tyrosine hydroxylase in the rat brain: development characteristics, *J. Neurochem.,* 19, 1117, 1972.
171. **Sales, N., Martres, M. P., Bouthenet, M. L., and Schwartz, J. C.,** Ontogeny of dopaminergic D-2 receptors in the rat nervous system: characterization and detailed autoradiographic mapping with ^{125}I-iodosulpiride, *Neuroscience,* 28, 673, 1989.
172. **Giorgi, O., De Montis, G., Porceddu, M. L., Mele, S., Calderini, G., Toffano, G., and Biggio, G.,** Developmental and age-related changes in D1-dopamine receptors and dopamine content in the rat striatum, *Dev. Brain Res.,* 35, 283, 1987.
173. **Berger, B., Verney, C., Gaspar, P., and Febvret, A.,** Transient expression of tyrosine hydroxylase immunoreactivity in some neurons of the rat neocortex during postnatal development, *Dev. Brain Res.,* 23, 141, 1987.
174. **Killackey, H. P. and Bedford, G. R.,** The formation of afferent patterns in the somatosensory cortex of the neonatal rat, *J. Comp. Neurol.,* 183, 285, 1979.
175. **Whitaker-Azmitia, P. M. and Azmitia, E. C.,** Autoregulation of fetal neuronal development: role of high affinity serotonin receptors, *Neurosci. Lett.,* 67, 302, 1986.
176. **Rodríguez de Fonseca, F., Hernández, M. L., de Miguel, R., Fernández-Ruiz, J. J., and Ramos, J. A.,** Early changes in the development of dopaminergic neurotransmission after maternal exposure to cannabinoids, *Pharmacol. Biochem. Behav.,* 41, 1992.
177. **Specht, L. A., Pickel, W. M., Jooh, T. H., and Reis, D. J.,** Light microscopic immunocytochemical localization of tyrosine hydroxylase in prenatal rat brain. I. Early ontogeny, *J. Comp. Neurol.,* 199, 233, 1981.
178. **Specht, L. A., Pickel, V. M., Jooh, T. H., and Reis, D. J.,** Light microscopic immunocytochemical localization of tyrosine hydroxylase in prenatal rat brain. II. Late ontogeny, *J. Comp. Neurol.,* 199, 255, 1981.
179. **Murrin, L. C. and Zeng, W.,** Dopamine D-1 receptor development in the rat striatum: early localization in striosomes, *Brain Res.,* 480, 170, 1989.
180. **Randall, P. K. and Campbell, B. A.,** Ontogeny of behavioral arousal in rats: effect of maternal and sibling presence, *J. Comp. Physiol. Psychol.,* 90, 435, 1975.
181. **Nilsson, J. L. G. and Carlsson, A.,** Dopamine receptor agonists with apparent selectivity for autorreceptors: a new principle for antipsychotic action?, *TIPS,* 3, 322, 1982.
182. **Nowycky, M. C. and Roth, R. H.,** Dopaminergic neurons: role of presynaptic receptors in the regulation of transmitter biosynthesis, *Progr. Neuropsychopharmacol.,* 2, 139, 1978.
183. **Bloom, F. E., Schulman, J. A. and Koob, G. F.,** Catecholamines and behavior, in *Handbook of Experimental Pharmacology*, Vol. 90/II, Trendelenburg, W. and Weiner, N., Eds., Springer-Verlag, New York, 1988, 27.
184. **Fried, P. A.,** Short and long-term effects of prenatal cannabis inhalation upon rat offspring, *Psychopharmacologia,* 50, 285, 1976.
185. **Kawash, G. F., Yeung, D. L., and Berg, S. D.,** Effects of administration of cannabis resin during pregnancy on emotionality and learning in rat offspring, *Percept. Motor Skills,* 50, 359, 1980.

186. **Klausner, H. A. and Dingell, J. V.,** The metabolism and excretion of delta-9-tetrahydrocannabinol in the rat, *Life Sci.,* 10, 49, 1971.
187. **Abel, E. L.,** Behavioral teratology of marihuana extract in rats, *Neurobehav. Toxicol.,* 1, 285, 1979.
188. **Kalsbeek, A., Voorn, P., Buijs, R. M., Pool, C. W., and Vylings, H. B. M.,** Development of the dopaminergic innervation in the prefrontal cortex of the rat, *J. Comp. Neurol.,* 269, 58, 1988.
189. **Verney, C., Berger, B., Adrien, J., Vigny, A., and Gay, M.,** Development of dopaminergic innervation of the rat cerebral. A light-microscopic immunocytochemical study, using anti-tyrosine hydroxylase antibodies, *Dev. Brain Res.,* 5, 41, 1982.
190. **Fried, P. A.,** Marihuana use by pregnant women: neurobehavioral effects in neonates, *Drug. Alcohol Depend.,* 6, 415, 1980.
191. **Golub, M. S., Sassenrath, E. N., and Chapman, L. F.,** Regulation of visual attention in offspring of female monkeys treated chronically with delta-9-tetrahydrocannabinol, *Dev. Psychobiol.,* 14, 507, 1981.
192. **Abel, E. L., Day, N., Dintcheff, B. A., and Ernst, C. A. S.,** Inhibition of postnatal maternal performance in rats treated with marihuana extract during pregnancy, *Bull. Psychol. Soc.,* 4, 353, 1979.
193. **Palkovits, M. and Brownstein, M. J.,** Catecholamines in the central nervous system, in *Handbook of Experimental Pharmacology,* Vol 90/II, Trendelenburg, U. and Weiner, N., Eds., Springer-Verlag, New York, 1988, 1.
194. **Reisert, I., Schuster, R., Zienecker, R., and Pilgrim, Ch.,** Prenatal development of mesencephalic and diencephalic dopaminergic systems in the male and female rat, *Dev. Brain Res.,* 53, 222, 1990.
195. **Ugrumov, M. V., Taxi, J., Tixier-Vidal, A., Thibault, J., and Mitskevich, M. S.,** Ontogenesis of tyrosine hydroxylase-immunopositive structures in the rat hypothalamus: an atlas of neuronal cell bodies, *Neuroscience* 29, 35, 1989.
196. **Daikoku, S., Kawaro, H., Okamura, Y., Tokuzen, M., and Nagatsu, I.,** Ontogenesis of immunoreactive tyrosine hydroxylase-containing neurons in rat hypothalamus, *Dev. Brain Res.,* 28, 85, 1986.
197. **Wray, S. and Hoffman, G.,** Catecholamine innervation of luteinizing hormone-releasing hormone neurons: a developmental study, *Brain Res.,* 399, 327, 1986.
198. **Demarest, K. T., McKay, D. W., Riegle, G. D., and Moore, K. E.,** Sexual differences in tuberoinfundibular dopamine nerve activity induced by neonatal androgen exposure, *Neuroendocrinology,* 32, 108, 1981.
199. **Lookingland, K. J. and Barraclough, C. A.,** Changes in plasma hormone profiles and in hypothalamic catecholamine turnover rates in neonatally androgenized rats during the transition phase from ciclicity to persistent estrus (delayed anovulatory syndrome), *Biol. Reprod.,* 27, 282, 1982.
200. **Simerly, R. B.,** Hormonal control of the development and regulation of tyrosine hydroxylase expression within a sexually dimorphic population of dopaminergic cells in the hypothalamus, *Mol. Brain Res.,* 6, 297, 1989.
201. **Lacau de Mengido, I., Becú-Villalobos, D., and Libertun, C.,** Sexual differences in the dopaminergic control of luteinizing hormone secretion in the developing rat, *Dev. Brain Res.,* 35, 91, 1987.
202. **Becú-Villalobos, D. and Libertun, C.,** Ontogenesis of [^{3}H]-serotonin binding sites in the hypothalamus of the female rat: relation to serotonin-induced luteinizing hormone release in moxestrol-pretreated rats, *Dev. Brain Res.,* 25, 111, 1986.
203. **Blank, M. S., Panerai, A. E., and Friesen, H. G.,** Opioid peptides modulate luteinizing hormone secretion during sexual maturation, *Science,* 203, 1129, 1979.

204. **Ojeda, S. R., Jameson, H. E., and McCann, S. M.,** Developmental changes in pituitary responsiveness to luteinizing hormone-releasing hormone (LHRH) in the female rat: ovarian-adrenal influence during the infantile period, *Endocrinology,* 100, 440, 1977.
205. **Weisz, J. and Ward, I. L.,** Plasma testosterone and progesterone litters of pregnant rats, their male and female fetuses and neonatal offspring, *Endocrinology,* 106, 306, 1980.
206. **Gorski, A. R., Harlan, R. E., Jacobson, C. D., Shryne, J. E., and Southam, A. M.,** Evidence for the existence of a sexually dimorphic nucleus in the preoptic area of the rat, *J. Comp. Neurol.,* 193, 529, 1980.
207. **Swaab, D. F. and Fliers, E.,** A sexually dimorphic nucleus in the human brain, *Science,* 228, 1112, 1985.
208. **Goy, R. W. and McEwen, B. S.,** *Sexual Differentiation of the Brain,* MIT Press, Cambridge, MA, 1989.
209. **Chen, Y. F. and Ramírez, V. D.,** Prolactin stimulates dopamine release from male but not from female rat striatal tissue superfused *in vitro, Endocrinology,* 111, 1740, 1982.
210. **Shepard, R. A., Nielsen, E. B., and Broadhurst, P. L.,** Sex and strain differences in benzodiazepine receptor binding in Roman rat strains, *Eur. J. Pharmacol.,* 77, 327, 1982.
211. **Maggi, A. and Zucchi, I.,** Sexual differentiation of mammalian frontal cortex, *Life Sci.,* 40, 1155, 1987.
212. **Maggi, A. and Pérez, J.,** Role of female gonadal hormones in the central nervous system: clinical and experimental aspects, *Life Sci.,* 37, 906, 1985.
213. **Sparber, S. B.,** Postnatal behavior effects of *in utero* exposure to drugs which modify catecholamines and/or serotonin, in *Drugs and the Developing Brain,* Vernadakis, A. and Weiner, N., Eds., Plenum Press, New York, 1974, 81.
214. **Raum, W. J., McGivern, R. F., Peterson, M. A., Shryne, J. G., and Gorski, R. A.,** Prenatal inhibition of hypothalamic sex steroid uptake by cocaine: effects on neurobehavioral sexual differentiation in male rats, *Dev. Brain Res.,* 53, 230, 1990.
215. **Fung, Y. K.,** Postnatal effects of maternal nicotine exposure on the striatal dopaminergic system in rats, *J. Pharm. Pharmacol.,* 41, 576, 1989.
216. **McGivern, R. F., Clancy, A. N., Hill, M. A., and Noble, E. P.,** Prenatal alcohol exposure alters adult expression of sexually dimorphic behavior in the rat, *Science,* 224, 896, 1984.
217. **Lichtensteiger, W. and Schlumpf, M.,** Prenatal nicotine affects fetal testosterone and sexual dimorphism of saccharin preference, *Pharmacol. Biochem. Behav.,* 23, 439, 1985.
218. **McGivern, R. F., Poland, R. E., Taylor, A. N., Branch, B. J., and Raum, W. J.,** Prenatal stress feminizes adult male saccharin preference and maze learning, *Monogr. Neural Sci.,* 12, 172, 1986.
219. **Ward, I. L.,** Prenatal stress feminizes and demasculinizes the behavior of males, *Science,* 175, 82, 1972.
220. **Beyer, C. and Feder, H. H.,** Sex steroids and afferent input: their roles in brain sexual differentiation, *Annu. Rev. Physiol.,* 49, 349, 1987.
221. **McGivern, R. F., Raum, W. J., Salido, E., and Redei, E.,** Lack of prenatal testosterone surge in fetal rats exposed to alcohol: alterations in testicular morphology and physiology, *Alcohol Clin. Exp. Res.,* 12, 243, 1988.
222. **Ward, I. L. and Weisz, J.,** Maternal stress alters plasma testosterone in fetal males, *Science,* 207, 328, 1980.

223. **Lichtensteiger, W. and Schlumpf, M.,** Modification of early neuroendocrine development by drugs and endogenous peptides, in *Progress in Neuroendocrinology,* Vol. 1, Parvez, H., Ed., VNU Science Press, Paris, 1985, 153.
224. **Dalterio, S. and Bartke, A.,** Fetal testosterone in mice: effect of gestational age and cannabinoid exposure, *J. Endocrinol.,* 91, 509, 1981.
225. **Richmond, G. and Sachs, B. D.,** Maternal discrimination of pup sex in rats, *Dev. Psychobiol.,* 17, 87, 1984.
226. **Burstein, S. and Kupfer, D.,** Hydroxylation of *trans*-delta-1-tetrahydrocannabinol by hepatic microsomal monooxygenase, *Ann. N.Y. Acad. Sci.,* 191, 61, 1971.
227. **Hull, E. N., Nishita, J. K., Bitran, D., and Dalterio, S.,** Perinatal dopamine-related drugs demasculinize rats, *Science,* 224, 1011, 1984.
228. **Felicio, L. F., Palermo-Neto, J., and Nasello, A. G.,** Perinatal bromopride treatment: effects on motor activity and stereotyped behavior of offspring, *Physiol. Behav.,* 45, 1081, 1989.
229. **Engele, J., Pilgrim, Ch., and Reisert, I.,** Nigrostriatal dopaminergic neurons in culture are sexually dimorphic, *Neuroscience,* 22, S235, 1987.
230. **Reisert, I., Engele, J., and Pilgrim, Ch.,** Early sexual differentiation of diencephalic dopaminergic neurons of the rat *in vitro, Cell. Tissue Res.,* 255, 411, 1989.
231. **Engele, J., Pilgrim, Ch., and Reisert, I.,** Sexual differentiation of mesencephalic neurons *in vitro*: effects of sex and gonadal hormones, *Int. J. Dev. Neurosci.,* 7, 603, 1989.
232. **Weisz, J., Brown, B. L., and Ward, I. L.,** Maternal stress decreases aromatase activity in brains of male and female rat fetuses, *Neuroendocrinology,* 35, 374, 1982.
233. **Howlett, A. C., Bidaut-Russell, M., Devane, W. A., Melvin, L. S., Johnson, M. R., and Herkenham, M.,** The cannabinoid receptor: biochemical, anatomical and behavioral characterization, *TINS,* 13, 420, 1990.
234. **Rawitch, A. B., Schultz, G. S., and Kurt, E. E.,** Competition of delta-9-tetrahydrocannabinol with estrogens in rat uterine estrogen receptor binding, *Science,* 197, 1189, 1977.
235. **Virgo, B. B.,** The estrogenicity of delta-9-tetrahydrocannabinol (THC): THC neither blocks nor induces ovum implantation, nor does it effect uterine growth, *Res. Commun. Chem. Pathol. Pharmacol.,* 25, 65, 1979.
236. **Murphy, L. L., Rodríguez de Fonseca, F., and Steger, R. W.,** Delta-9-tetrahydrocannabinol antagonism of the anterior pituitary response to estradiol in immature female rats, *Steroids,* 56, 97, 1991.
237. **Bonnin, A., Fernández-Ruiz, J. J., Martin, M., Rodríguez de Fonseca, F., de Miguel, R., and Ramos, J. A.,** Estrogenic modulation of Δ^9-tetrahydrocannabinol effects on nigrostriatal dopaminergic activity in the female rat brain, *Mol. Cell. Neurosci.,* in revision.
238. **Solomon, J., Cocchia, M. A., and DiMartino, R.,** The effects of delta-9-tetrahydrocannabinol on uterine and vaginal cytology, *Science,* 195, 875, 1977.
239. **Chakravarty, I., Sengupta, D., Bhattacharya, P., and Ghosh, J. J.,** Effect of treatment with cannabis extract on the water and glycogen content of the uterus in normal and estradiol-treated rats, *Toxicol. Appl. Pharmacol.,* 34, 513, 1975.
240. **Purohit, V. D., Ahluwahlia, B. S., and Vigersky, R. A.,** Marihuana inhibits dihydrotestosterone binding to the androgen receptor, *Endocrinology,* 107, 848, 1980.
241. **Okey, A. B. and Bondy, G. P.,** Is delta-9-tetrahydrocannabinol estrogenic?, *Science,* 195, 904, 1977.
242. **Okey, A. B. and Bondy, G. P.,** Delta-9-tetrahydrocannabinol and 17β-estradiol bind to different macromolecules in estrogen target tissues, *Science,* 200, 312, 1978.

243. **Devane, W. A., Dysarz, F. A., III, Johnson, M. R., Melvin, L. S., and Howlett, A. C.,** Determination and characterization of a cannabinoid receptor in rat brain, *Mol. Pharmacol.,* 34, 605, 1988.
244. **Nye, J. S., Seltzman, H. H., Pitt, C. G., and Snyder, S. H.,** High-affinity cannabinoid binding sites in brain membranes labelled with [^{3}H]-5′-trimethylammonium delta-8-tetrahydrocannabinol, *J. Pharmacol. Exp. Ther.,* 234, 784, 1985.
245. **Matsuda, L. A., Lolait, S. F., Brownstein, M. J., Young, A. C., and Bonner, T. I.,** Structure of a cannabinoid receptor and functional expression of the cloned cDNA, *Nature (London),* 346, 561, 1990.
246. **Herkenham, M., Lynn, A. B., Little, M. D., Johnson, M. R., Melvin, L. S., de Costa, B. R., and Rice, K. C.,** Cannabinoid receptor localization in brain, *Proc. Natl. Acad. Sci. U.S.A.,* 87, 1932, 1990.
247. **Howlett, A. C.,** Cannabinoid inhibition of adenylate cyclase. Biochemistry of the response in neuroblastoma cell membranes, *Mol. Pharmacol.,* 27, 429, 1985.
248. **Howlett, A. C., Johnson, M. R., Melvin, L. S., and Milne, G. M.,** Nonclassical cannabinoid analgetics inhibit adenylate cyclase: development of a cannabinoid receptor model, *Mol. Pharmacol.,* 33, 297, 1987.
249. **Howlett, A. C., Qualy, J. M., and Kachatrian, L. L.,** Involvement of Gi in the inhibition of adenylate cyclase by cannabimimetic drugs, *Mol. Pharmacol.,* 29, 307, 1986.

6 *In Vivo* Interactions between Psychotropic Cannabinoids and Other Drugs Involving Central and Peripheral Neurochemical Mediators

INTRODUCTION

There is good evidence from studies directed at determining the effects of cannabinoids on the concentration, turnover, or release of endogenous substances that the central effects of psychotropic cannabinoids are mediated by several putative neurotransmitters or neuromodulators (see review by Pertwee[1]). To date, experiments have focused on norepinephrine (NE), dopamine (DA), 5-hydroxytryptamine (5-HT), acetylcholine (ACh), γ-aminobutyric acid (GABA), histamine, opioid peptides, and prostaglandins (PGs). Although the results obtained suggest that all these substances have some role in the neuropharmacology of cannabinoids, they provide little information about the precise nature of this involvement.

One approach that has been used in an attempt to establish the neuropharmacological basis for at least some of the effects of psychotropic cannabinoids has been to search for synergistic or antagonistic interactions between cannabinoids and drugs that are known to activate certain types of receptors or to alter the biosynthesis, storage, release, fate, or actions of certain transmitters and modulators. This chapter describes how such studies have helped to identify some of the neurochemical systems that mediate particular effects of the cannabinoids and how they have helped to formulate hypotheses about the mechanisms by which cannabinoids interact with these systems. The chapter concentrates on cannabinoid-induced changes in motor function and on cannabinoid-induced hypothermia and antinociception partly because these are the effects that have been most investigated in drug interaction studies and partly because they form the basis of a battery of tests used extensively in structure-activity studies to distinguish cannabinoids from other classes of drugs.[2]

MOTOR FUNCTION

Psychotropic cannabinoids give rise to a mixture of excitant and depressant effects in animals, the former including hyperreactivity to tactile and auditory stimuli, particularly in rats and mice, and the latter hypokinesia and catalepsy. Cannabinoids also affect posture, impair coordination, and can produce various kinds of stereotyped behavior.

The motor response most studied in experiments involving interactions between cannabinoids and other drugs has been catalepsy. The most probable sites of action for the production of this response are the basal ganglia and, indeed, it has been shown that (1) cannabinoid binding sites are present in particularly high concentration in the corpus striatum and (mainly on striatal projection neurons) in the globus pallidus, substantia nigra, and entopeduncular nucleus,[3-5] and that (2) cannabinoids can induce catalepsy in rats when injected directly into some of these brain areas (corpus striatum, globus pallidus, and substantia nigra).[6-8]

The intensity of the cataleptic response to cannabinoids is normally measured using either the ring immobility test[9] or the bar test;[10] the first of these tests is more sensitive.[11] Although usually conducted with mice, it can also be performed with rats and is carried out by placing an animal across an elevated horizontal wire ring attached to the end of a vertical stalk and determining the proportion of time the animal remains immobile over a set period, usually 5 min. The bar test is performed by placing a rat or a mouse in an extended position with both front limbs over an elevated horizontal bar and both hind limbs on the bench. The degree of catalepsy is measured by noting the length of time that the animal remains in this abnormal position ("descent latency"). In both tests, hypokinetic animals showing no other sign of catalepsy sometimes achieve "positive scores" (see, for example, the section on "Drug Interactions Involving Dopamine or Norepinephrine"). Consequently, it is often best when referring to bar test or ring test data to avoid the term "catalepsy" and to use the terms "descent latency" or "ring immobility" instead.

Performance of Rats and Mice in the Bar Test or Ring Immobility Test

Drug Interactions Involving Dopamine or Norepinephrine

There is good evidence that the ability of cannabinoids to induce catalepsy or hypokinesia is enhanced by drugs that depress dopaminergic transmission and opposed by drugs that facilitate it (Table 1). More specifically, there are reports that the ability of cannabis and certain cannabinoids to induce immobility in the ring test or delay descent in the bar test can be (1) attenuated by drugs that activate DA receptors either directly (i.e., apomorphine) or indirectly by increasing DA release (i.e., amphetamine), and (2) augmented by DA antagonists and by drugs that inhibit DA synthesis (i.e., α-methyl-*p*-tyrosine) or prevent the vesicular storage of DA in nerve terminals (i.e., reserpine and 6-hydroxydopamine). Apart from apomorphine, all the drugs that have been used in these studies can also alter the activity of noradrenergic pathways. Indeed, one of them, reserpine, can depress transmission along NE, DA, and 5-HT-releasing pathways. It is important to note, therefore, that cannabinoid-induced catalepsy or hypokinesia does not seem to be enhanced by drugs that can antagonize NE or 5-HT without antagonizing DA, or by drugs that can inhibit the synthesis of NE from DA (i.e., diethyldithiocarbamate) or 5-HT from tryptophan (Tables 1 and 2).

The synergism between Δ^9-THC and reserpine noted by Moss et al.[13,14] (Table 1) was particularly intense in experiments in which the cannabinoid was given first. Indeed, it was reported that rats that had received Δ^9-THC before reserpine showed no response when their rigid bodies were tapped gently on a table, had no righting reflex, and adopted such a rigid curled posture that they could not be subjected to the bar test.[14] In contrast, it was found that rats given reserpine before Δ^9-THC retained their righting response.[13] The same animals showed postural rigidity and seemed unwilling or unable to initiate voluntary movements. Since the animals did not show "waxy flexibility", the response was described as "hypokinesia" rather than "catalepsy".[13]

TABLE 1
Drug-Induced Changes in the Ability of Cannabinoids to Increase Descent Latency in the Bar Test or Immobility in the Ring Test: Interactions Involving Norepinephrine or Dopamine

Species	Pretreatment	Treatment	Response to treatment or pretreatment	Effect of treatment or pretreatment on response	Ref.
Inhibitors of DA or NE synthesis					
Mouse	α-Methyl-*p*-tyrosine (i.p.)	Cannabis (i.p.)	Ring immobility	Enhancement	12
Mouse	Diethyldithiocarbamate (i.p.)	Cannabis (i.p.)	Ring immobility	None	12
Drugs that impair vesicular storage of catecholamines					
Mouse	Reserpine (i.p.)	Cannabis (i.p.)	Ring immobility	Enhancement	12
Rat	Reserpine (i.p.)	Δ^9-THC (p.o.)	Descent delay[a]	Enhancement[b]	13, 14
Rat	Reserpine (i.p.)	Levonantradol (p.o.)	Descent delay[a]	Enhancement[b]	7, 14
Rat	Reserpine (i.p.)	Levonantradol (i.c.)[c]	Descent delay[a]	Enhancement[b]	7
Rat	6-Hydroxydopamine (i.c.v.)	Δ^9-THC (i.p.)	Descent delay	Enhancement	15
Rat	6-Hydroxydopamine (i.c.v.)	Cannabis (i.p.)	Descent delay	Attenuation	16
Rat	6-Hydroxydopamine (i.c.v.)	Δ^9-THC (i.p.)	Descent delay	Attenuation	17
Catecholamine antagonists					
Mouse	Haloperidol (i.p.)	Cannabis (i.p.)	Ring immobility	Enhancement	12
Mouse	Propranolol (i.p.)	Cannabis (i.p.)	Ring immobility	None	12
	Phenoxybenzamine (i.p.)			None	
Mouse	*cis*(Z)-Flupentixol (s.c.)	Δ^9-THC (i.p.)	Descent delay	Enhancement	18
	trans(E)-Flupentixol (s.c.)			None	
Rat	Δ^9-THC (p.o.)	Haloperidol (i.p.)	Descent delay	Enhancement	14
		Chlorpromazine (i.p.)		None	

TABLE 1 (continued)
Drug-Induced Changes in the Ability of Cannabinoids to Increase Descent Latency in the Bar Test or Immobility in the Ring Test: Interactions Involving Norepinephrine or Dopamine

Species	Pretreatment	Treatment	Response to treatment or pretreatment	Effect of treatment or pretreatment on response	Ref.
Rat	Fluphenazine (i.p.)	Δ^9-THC (p.o.)	Descent delay[a]	Enhancement[b]	19
Rat	Δ^9-THC (i.v.)	Yohimbine (i.t.)[d]	Ring immobility[a]	None[b]	20
DA agonist					
Mouse	Apomorphine (i.p.)	Cannabis (i.p.)	Ring immobility	Attenuation	12
Precursor of DA and NE					
Mouse	L-Dopa (i.p.) and benserazide[e]	Cannabis (i.p.)	Ring immobility	Attenuation	12
Drug that facilitates neuronal release of DA and NE					
Mouse	Amphetamine (i.p.)	Cannabis (i.p.)	Ring immobility	Attenuation	12
Rat	Δ^9-THC (i.p.)	(+)-Amphetamine (i.p.)	Descent delay[a]	Attenuation[b]	11
Rat	Δ^9-THC (i.p.)	(+)-Amphetamine (i.c.)[f]	Descent delay[a]	Attenuation[b]	6
		(+)-Amphetamine (i.c.)[g]		Enhancement[b]	
Rat	Δ^9-THC (i.p.)	(±)-Amphetamine (i.p.)	Catalepsy[a] (rating scale)	Attenuation[b]	21

Note: i.c. denotes intracerebral administration via chronically implanted cannulae terminating in the substantia nigra,[c] dorsomedial nucleus of the thalamus,[c] striatum,[c,f] or globus pallidus.[g] i.t. indicates that injections were made intrathecally, and i.c.v. that they were made intracerebroventricularly.

[a] Response to pretreatment.
[b] Effect of treatment.
[c] See *Note*.
[d] The immobility response to Δ^9-THC was unaffected by doses of yohimbine known to attenuate Δ^9-THC-induced antinociception.
[e] An inhibitor of L-aromatic amino acid decarboxylase that does not readily cross the blood/brain barrier.

TABLE 2
Drug-Induced Changes in the Ability of Cannabis or Δ⁹-THC to Produce Immobility in the Ring Test: Interactions Involving 5-Hydroxytryptamine

Species	Pretreatment	Treatment(s)	Response to cannabis or Δ^9-THC	Effect of treatment or pretreatment on response to cannabis or Δ^9-THC	Ref.
Precursors of 5-HT					
Mouse	L-Tryptophan (i.p.) 5-Hydroxtryptophan (i.p.)	Cannabis (i.p.)	Ring immobility	Enhancement Enhancement	12
Drug that inhibits 5-HT synthesis					
Mouse	*p*-Chlorophenylalanine (i.p.)	Cannabis (i.p.)	Ring immobility	Attenuation	12
Rat	*p*-Chlorophenylalanine (i.p.)	Cannabis (i.p.) after restraint stress	Ring immobility	Attenuation	22
Drug that impairs vesicular storage of 5-HT (see also reserpine in Table 1)					
Mouse	Dihydroxytryptamine (i.p. to newborn mice)	Cannabis (i.p.)	Ring immobility	Attenuation	12
Rat	Dihydroxytryptamine (i.c.v.)	Cannabis (i.p.) after restraint stress	Ring immobility	Attenuation	22
5-HT antagonists					
Mouse	Methysergide (i.p.) Cyproheptadine (i.p.)	Cannabis (i.p.)	Ring immobility	Attenuation Attenuation	12
Rat	Methysergide (i.p.) Cyproheptadine (i.p.)	Cannabis (i.p.) after restraint stress	Ring immobility	Attenuation Attenuation	22
Rat	Δ^9-THC (i.v.)	Methysergide (i.t.)	Ring immobility	None	20

Note: i.c.v. indicates that injections were made intracerebroventricularly and i.t. that they were made intrathecally.

A few experiments, all performed with rats, have failed to detect any enhancement of cannabinoid-induced catalepsy or hypokinesia by drugs known to reduce transmission along DA-releasing pathways (Table 1). Moss et al.[14] were unable to detect any synergism between Δ^9-THC and chlorpromazine in animals subjected to the bar test. Ferri et al.[16] and Kataoka et al.[17] found that the ability of cannabis or Δ^9-THC to increase descent latency in the bar test was attenuated by intracerebroventricular pretreatment with 6-hydroxydopamine. The results obtained by Kataoka et al.[17] suggested that this attenuation was caused by loss from the brain of NE rather than DA. Thus, they found that the effect of 6-hydroxydopamine on the response to Δ^9-THC could be abolished by desipramine, a drug which prevents the uptake of 6-hydroxydopamine into noradrenergic nerve terminals without affecting the neurotoxin's uptake into dopaminergic neurons.

Drug Interactions Involving 5-Hydroxytryptamine

There are reports from experiments with rats and mice that the ability of cannabis to increase immobility in the ring test is enhanced by drugs that hasten the synthesis of 5-HT (i.e., its precursors, tryptophan, and 5-hydroxytryptophan) and attenuated by 5-HT antagonists and by drugs that inhibit 5-HT synthesis (i.e., *p*-chlorophenylalanine) or prevent the vesicular storage of 5-HT in nerve terminals (i.e., dihydroxytryptamine) (Table 2). Reserpine, which also impairs vesicular storage in 5-HT-releasing nerve terminals, has been shown to augment the effect of cannabis on ring immobility. However, this effect probably results from the drug's ability to prevent the vesicular storage of DA (see previous section).

Drug Interactions Involving Acetylcholine

Results obtained from experiments performed in several laboratories show that the ability of cannabinoids to induce catalepsy or hypokinesia can be enhanced by drugs that facilitate cholinergic transmission and opposed by drugs that depress it (Table 3). Thus, the ability of cannabis and Δ^9-THC to induce immobility in the ring test or to delay descent in the bar test can be (1) attenuated by ACh antagonists and (2) augmented by drugs that activate cholinoceptors either directly (i.e., tremorine and oxotremorine) or indirectly by increasing synaptic concentrations of ACh (i.e., acetylcholinesterase inhibitors). Consistent with these observations is evidence that somnolence and lethargy induced by Δ^9-THC in human subjects can be augmented by physostigmine.[24]

TABLE 3
Drug-Induced Changes in the Ability of Cannabis or Δ⁹-THC to Increase Descent Latency in the Bar Test or Immobility in the Ring Test: Interactions Involving Acetylcholine

Species	Pretreatment	Treatment(s)	Response to pretreatment or to treatment(s)	Effect of pretreatment or treatment(s) on response	Ref.
Cholinoceptor agonists					
Mouse	Tremorine (i.p.)	Cannabis (i.p.)	Ring immobility	Enhancement	12
Mouse	Oxotremorine (s.c.)	Δ^9-THC (i.p.)	Descent delay	Enhancement	23
Rat	Δ^9-THC (i.p.)	RS-86 (i.p.)	Descent delay	Enhancement	11
Rat	Δ^9-THC (i.p.)	RS-86 (i.c.)	Descent delay	Enhancement	6
Muscarinic antagonists					
Mouse	Atropine (i.p.)	Cannabis (i.p.)	Ring immobility	Attenuation	12
Mouse	Atropine (s.c.) Scopolamine (s.c.) Methylatropine (s.c.)	Oxotremorine (s.c.) and Δ^9-THC (i.p.)	Descent delay	Attenuation Attenuation None	23
Mouse	Atropine (s.c.) Scopolamine (s.c.) Methylatropine (s.c.)	Physostigmine (s.c.) before Δ^9-THC (i.p.)	Descent delay	Attenuation Attenuation None	23
Mouse	Atropine (s.c.) Scopolamine (s.c.) Methylatropine (s.c.)	Chlordiazepoxide (s.c.) before Δ^9-THC (i.p.)	Descent delay	Attenuation Attenuation None	23
Rat	Atropine (i.p.)	Cannabis (i.p.)	Descent delay	Attenuation	16
Rat	Atropine (i.p.) Scopolamine (i.p.) Quinuclidinyl benzilate (i.p.) Methylatropine (i.p.)	Δ^9-THC (i.p.)	Descent delay	Attenuation Attenuation Attenuation None	15
Rat	Reserpine (i.p.) before Δ^9-THC (p.o.)	Ethopropazine (i.p.) Scopolamine (i.p.)	Descent delay[a]	Attenuation[b] Slight attenuation[b]	13
Nicotinic antagonist					
Mouse	Mecamylamine (s.c.)	Chlordiazepoxide (s.c.) before Δ^9-THC (i.p.)	Descent delay	None	23

TABLE 3 (continued)
Drug-Induced Changes in the Ability of Cannabis or Δ⁹-THC to Increase Descent Latency in the Bar Test or Immobility in the Ring Test: Interactions Involving Acetylcholine

Species	Pretreatment	Treatment(s)	Response to pretreatment or to treatment(s)	Effect of pretreatment or treatment(s) on response	Ref.
Rat	Mecamylamine (s.c.)	Fluphenazine (i.p.)[c] before Δ^9-THC (p.o.)	Descent delay	Attenuation	19
Acetylcholinesterase inhibitors					
Mouse	Physostigmine (i.p.)	Cannabis (i.p.)	Ring immobility	Enhancement	12
Mouse	Physostigmine (s.c.)	Δ^9-THC (i.p.)	Descent delay	Enhancement	23
	Neostigmine (s.c.)			None	
Rat	TIDC (i.p.)[d]	Δ^9-THC (i.p.)	Descent delay	Enhancement	15
	DPDC (i.p.)[e]			Enhancement	
	Neostigmine (i.p.)			None	
Rat	Reserpine (i.p.) before Δ^9-THC (p.o.)	Physostigmine (i.p.)	Descent delay[a]	Enhancement[b]	13

Note: i.c. denotes intracerebral administration via chronically implanted cannulae terminating in the caudate/putamen or globus pallidus.

[a] Response to pretreatment.
[b] Effect of treatment.
[c] Fluphenazine was administered at –3 h, mecamylamine at –1 h, and Δ^9-THC at time zero.
[d] 5-(1,3,3-trimethyl) indolinyl-*N,N*-dimethyl carbamate.
[e] *m*-(dimethylamino)-ethoxyphenyl-*N,N*-dimethyl carbamate.

The finding that synergistic interactions between Δ^9-THC and cholinergic agonists or acetylcholinesterase inhibitors can be prevented by muscarinic antagonists (Table 3) supports the idea that this synergism depends on the activation of cholinoceptors and that these receptors are muscarinic in nature. Results from experiments with drugs that do not readily penetrate the blood/brain barrier (i.e., methylatropine and neostigmine) suggest that these receptors are located within the CNS (Table 3; see also "Conclusions and Summary" in this section).

There is some evidence for an involvement of cholinergic pathways in synergistic interactions that have been observed to occur between Δ^9-THC and certain noncholinergic drugs (i.e., chlordiazepoxide, reserpine, and fluphenazine). The Δ^9-THC-chlordiazepoxide interaction, which has been demonstrated in mice, seems to depend on the activation of central muscarinic receptors as it can be prevented by atropine and scopolamine, but not by methylatropine or by the nicotinic antagonist, mecamylamine (Table 3). The other interactions, studied in rats, appear to depend on the activation of muscarinic and nicotinic receptors (Table 3). Thus, the Δ^9-THC-reserpine interaction can be attenuated both by mecamylamine and by the muscarinic antagonist, ethopropazine. Although the effect of muscarinic antagonists on the Δ^9-THC-fluphenazine interaction has not been reported, it is known that this interaction can be attenuated by mecamylamine.[19] Consistent with an involvement of nicotinic receptors in these interactions with reserpine and fluphenazine is the finding that nicotine can also interact synergistically with these drugs in the bar test.[7,19]

Drug Interactions Involving γ-Aminobutyric Acid

There is evidence that the ability of cannabinoids to delay descent in the bar test can be greatly enhanced by drugs that facilitate transmission along GABAergic pathways (Table 4). For example, experiments with mice have shown that the effect of Δ^9-THC on bar test performance can be augmented by drugs expected to activate GABAergic receptors either directly (GABA agonists) or indirectly by inhibiting the metabolism or neuronal uptake of GABA so as to increase its concentration in the synapse. Benzodiazepines, which facilitate the interaction of GABA with $GABA_A$ receptors through allosteric mechanisms, have also been shown to interact synergistically with cannabinoids in rats and mice (Table 4).

There are several reasons for believing that both $GABA_A$ and $GABA_B$ receptors are involved in these interactions. These may be summarized as follows (see also Table 4):

1. Δ^9-THC undergoes synergistic interactions with both $GABA_A$ and $GABA_B$ agonists (i.e., muscimol and (–)-baclofen, respectively).
2. Δ^9-THC does not interact synergistically with (+)-baclofen which is much less potent as a GABA agonist than its (–)-isomer.

TABLE 4
Drug-Induced Changes in the Ability of Cannabinoids to Increase Descent Latency in the Bar Test: Interactions Involving γ-Aminobutyric Acid

Species	Pretreatment	Treatment	Response to pretreatment or treatment	Effect of treatment or pretreatment on response to cannabinoid	Ref.
$GABA_A$ and $GABA_B$ Agonists					
Mouse	Muscimol (s.c.)	Δ^9-THC (i.p.)	Descent delay	Enhancement	18
Mouse	(–)-Baclofen (i.p.)	Δ^9-THC (i.p.)	Descent delay	Enhancement	18
	(+)-Baclofen (i.p.)			None	
Drugs that facilitate the activation of $GABA_A$ receptors					
Mouse	Chlordiazepoxide (s.c.)	Δ^9-THC (i.p.)	Descent delay	Enhancement	18, 25
	Flurazepam (s.c.)			Enhancement	
Rat	Chlordiazepoxide (s.c.)	Δ^9-THC (i.v.)	Descent delay	Enhancement	8
		(–)-DMH (i.v.)[c]		Enhancement	
Rat	Chlordiazepoxide (s.c.)	Δ^9-THC (i.c.)	Descent delay	Enhancement	8
		(–)-DMH (i.c.)[c]		Enhancement	
Rat	Chlordiazepoxide (i.c.)	Δ^9-THC (i.v.)	Descent delay	Enhancement	8
Drug that inhibits the neuronal uptake of GABA					
Mouse	NO-328 (i.p.)	Δ^9-THC (i.p.)	Descent delay	Enhancement	18
Drug that inhibits GABA metabolism					
Mouse	Aminooxyacetic acid (s.c.)	Δ^9-THC (i.p.)	Descent delay	Enhancement	18
Antagonists of GABA, glycine, or benzodiazepines					
Mouse	Δ^9-THC (i.p.) after NO-328 (i.p.) or aminooxyacetic acid (s.c.)	(+)-Bicuculline (i.p.)	Descent delay[a]	Attenuation[b]	18
		Homotaurine (i.p.)		Attenuation[b]	
		Strychnine (i.p.)		None[b]	

TABLE 4 (continued)
Drug-Induced Changes in the Ability of Cannabinoids to Increase Descent Latency in the Bar Test: Interactions Involving γ-Aminobutyric Acid

Species	Pretreatment	Treatment	Response to pretreatment or treatment	Effect of treatment or pretreatment on response to cannabinoid	Ref.
Mouse	Δ^9-THC (i.p.) after muscimol (s.c.)	(+)-Bicuculline (i.p.)	Descent delay[a]	Attenuation[b]	18
		Homotaurine (i.p.)		None[b]	
		Strychnine (i.p.)		None[b]	
Mouse	Δ^9-THC (i.p.) after (–)-Baclofen (i.p.)	(+)-Bicuculline (i.p.)	Descent delay[a]	None[b]	18
		Homotaurine (i.p.)		Attenuation[b]	
Mouse	Δ^9-THC (i.p.) after flurazepam (s.c.)	(+)-Bicuculline (i.p.)	Descent delay[a]	Attenuation[b]	18
		Homotaurine (i.p.)		Attenuation[b]	
		Flumazenil (i.p.)		Attenuation[b]	
		Strychnine (i.p.)		None[b]	

Note: i.c. denotes intracerebral administration via chronically implanted cannulae terminating in the globus pallidus.

[a] Response to pretreatment.
[b] Effect of treatment.
[c] DMH, 11-hydroxy-Δ^8-dimethylheptyl-THC; the psychotropically inactive enantiomer, (+)-DMH, did not interact synergistically with chlordiazepoxide when given i.v. or i.c.

3. The Δ^9-THC-muscimol interaction can be attenuated by (+)-bicuculline (a $GABA_A$ antagonist) but not by homotaurine (a $GABA_B$ antagonist) or by strychnine (a glycine antagonist).
4. The Δ^9-THC-baclofen interaction can be attenuated by homotaurine but not by (+)-bicuculline or strychnine.
5. The interactions with inhibitors of GABA uptake or metabolism can be attenuated by (+)-bicuculline and homotaurine but not by strychnine.
6. The Δ^9-THC-benzodiazepine interaction can be attenuated by bicuculline and homotaurine, but not by strychnine.

The Δ^9-THC-benzodiazepine interaction can also be prevented by flumazenil, pointing to an involvement of benzodiazepine receptors,[18] and by antimuscarinic agents (Table 3).

Drug Interactions Involving Eicosanoids

There are reports from experiments with rats and mice that the ability of cannabis or Δ^9-THC to increase immobility in the ring test can be attenuated by drugs capable of preventing eicosanoid synthesis through the inhibition of phospholipases, lipoxygenases, or cyclooxygenases (Table 5). Other evidence pointing to an involvement of eicosanoids in the cataleptic response to cannabinoids is summarized below.

1. Cannabinoids are known to release the PG precursor, arachidonic acid, by stimulating phospholipase A_2.[1]
2. The ability of cannabis to increase immobility in the ring test can be enhanced by pretreatment with PGE_1 (Table 5).
3. The effect of Δ^9-THC on performance in the ring test can be enhanced by yeast given at a dose thought to release PGs.[32]
4. The ability of cyclooxygenase inhibitors to attenuate the effect of Δ^9-THC on immobility in the ring test can be counteracted by administering a subcataleptic dose of PGE_2.[28]
5. The ability of Δ^9-THC to increase immobility in the ring test can be attenuated by maintaining animals on a fat-free diet.[32]
6. It is possible to reverse the effect of fat-deprivation on Δ^9-THC-induced immobility by administering arachidonic acid or by supplying the animals with a normal diet.[32]

There is also evidence that Δ^9-THC can elevate PG levels in the blood and that the ability of Δ^9-THC to increase immobility in the ring test can be attenuated by anti-PGE_2 antibodies.[31,33] As it is unlikely that the PG antibodies entered the CNS, these observations have led Burstein et al.[26] to postulate that Δ^9-THC acts peripherally to stimulate the synthesis of PGE_2 and that this then passes into the brain to produce catalepsy. However, the nature of the link between PGs and cannabinoid-induced catalepsy is probably more complex than this since the nonpsychotropic cannabinoid, cannabidiol (CBD), which does not produce catalepsy, has been shown to be more potent than Δ^9-THC as a stimulator of phospholipase A_2 in several *in vitro* models.[34-37] Moreover, although Δ^9-THC may well stimulate the production of PGs in peripheral tissues, results from *in vitro* experiments suggest that it can also stimulate PG formation in the corpus striatum.[38]

TABLE 5
Drug-Induced Changes in the Ability of Cannabis or Δ^9-THC to Increase Immobility in the Ring Test: Interactions Involving Eicosanoids

Species	Pretreatment	Treatment	Response to cannabis	Effect of treatment or pretreatment on response to cannabis	Ref.
Phospholipase inhibitors					
Mouse	Mepacrine (i.p.)	Δ^9-THC (p.o.)	Ring immobility	Attenuation	26
Rat	Hydrocortisone (i.c.v.)	Δ^9-THC (i.p.)	Ring immobility	Attenuation	27
Lipoxygenase inhibitor					
Mouse	Gossypol (p.o.)	Δ^9-THC (p.o.)	Ring immobility	Attenuation	26
Cyclooxygenase inhibitors					
Mouse	Δ^9-THC (p.o.)	Aspirin (p.o.)	Ring immobility	Attenuation	28
		Indomethacin (p.o.)		Attenuation	
		Phenylbutazone (p.o.)		Attenuation	
		Diflunisal (p.o.)		Attenuation	
Mouse	Indomethacin (i.p. or p.o.)	Δ^9-THC (p.o.)	Ring immobility	Attenuation	26, 29
Mouse	Diclofenac (i.p.)	Cannabis (i.p.)	Ring immobility	Attenuation	30
Rat	Diclofenac (i.p.)	Cannabis (i.p.) after restraint stress	Ring immobility	Attenuation	22
	Mefenamic acid (i.p.)			Attenuation	
	Paracetamol (i.p.)			Attenuation	
Rat	Diclofenac (i.c.v.)	Δ^9-THC (i.p.)	Ring immobility	Attenuation	27
	Paracetamol (i.c.v.)			Attenuation	
Prostaglandins					
Mouse	PGE_1 (400 μg/kg i.p.)[a]	Cannabis (i.p.)	Ring immobility	Enhancement	30
	$PGF_{2\alpha}$ (2000 μg/kg i.p.)[a]			Attenuation	
Mouse	Δ^9-THC (p.o.) before aspirin, indomethacin, phenylbutazone, or diflunisal (p.o.)	PGE_2 (3.3 μg/kg i.p.)	Ring immobility	Enhancement[b]	28
		PGE_1 (3.3 μg/kg i.p.)		None	
		$PGF_{2\alpha}$ (3.3 μg/kg i.p.)		None	

TABLE 5 (continued)
Drug-Induced Changes in the Ability of Cannabis or Δ^9-THC to Increase Immobility in the Ring Test: Interactions Involving Eicosanoids

Species	Pretreatment	Treatment	Response to cannabis	Effect of treatment or pretreatment on response to cannabis	Ref.
Prostaglandin antibodies					
Mouse	PGE_2 antibodies	Δ^9-THC (p.o.)	Ring immobility	Attenuation	31

Note: i.c.v. indicates that injections were made intracerebroventricularly.

[a] The effects of these drugs in the absence of cannabis were not reported.

[b] PGE_2, at a dose not active in the ring test, reversed the attenuation of Δ^9-THC-induced immobility by cyclooxygenase inhibitors.

There is some evidence that the ability of PGs to interact with cannabinoids is structure dependent. Bhattacharya et al.[30] reported that whereas cannabis-induced ring immobility was enhanced by PGE_1, it was reduced by $PGF_{2\alpha}$. However, the PG doses used were rather high and their effects in the absence of cannabis were not given. In experiments in which much smaller doses of PGs were used (see Table 5), Fairbairn and Pickens[28] found that attenuation of the effect of Δ^9-THC on ring immobility by cyclooxygenase inhibitors could be counteracted by PGE_2 but not by PGE_1 or $PGF_{2\alpha}$.

Other "Catalepsy" Tests

Moss et al.[14] reported that Δ^9-THC interacted synergistically with reserpine to produce signs of marked catalepsy in a stump-tailed macaque. Grunfeld and Edery[21] found that the cataleptic response of rats to Δ^9-THC, measured using a rating scale, could be abolished by amphetamine. More recently, it was found that doses of Δ^9-THC and benzodiazepines (i.e., flurazepam or chlordiazepoxide) that showed synergism in the bar test (see section above on "Drug Interactions Involving γ-Aminobutyric Acid") also interacted synergistically to abolish the righting response in rats and mice.[1,8,25] The loss of the righting response observed in these experiments probably indicates the presence of extreme catalepsy rather than general anesthesia since the affected animals could still move about, retained their corneal reflex, showed no major reduction in skeletal muscle tone or grip strength, and exhibited a "waxy flexibility" of the hind legs.

Locomotor Activity

Grunfeld and Edery[21] reported that decreases in spontaneous motor activity induced in rhesus monkeys by Δ^8- or Δ^9-THC could be totally reversed by amphetamine. Similarly, experiments with methamphetamine have shown that it can reverse the depressant effect of Δ^9-THC on exploratory behavior in rabbits.[39,40] Pryor et al.[41] detected a mutual antagonism between the depressant effect of Δ^9-THC and the stimulant effect of cocaine on rat locomotor activity.

Jones et al.[42,43] found that in rabbits, grooming, locomotion, rearing, and exploration were increased by physostigmine and depressed by Δ^9-THC. When both drugs were given, they showed a mutual antagonism. Pryor et al.[41] reported that the ability of Δ^9-THC to reduce locomotor activity and to impair rotarod performance in rats could be markedly enhanced by nicotine. Interestingly, this drug did not affect the ability of Δ^9-THC to impair performance in a conditioned avoidance task. In an earlier study, Pryor et al.[44] found that Δ^9-THC and chlordiazepoxide exerted an additive depressant effect on rat locomotor activity. In other experiments with rats, Tulunay et al.[45] reported that the depressant effect of Δ^9-THC on spontaneous motor activity could be partially reversed by naloxone (5 mg/kg i.p.).

In experiments with mice, Pickens[46] showed that the depressant effects of cannabis, Δ^9-THC and CBD on spontaneous motor activity in a jiggle cage could be attenuated by aspirin. She also demonstrated that this attenuation could be reversed by PGE_2. Although PGE_2 also depressed motor activity when given by itself, this effect was much less than that produced in the presence of the other drugs.

Posture and Coordination

There are reports that the ability of cannabis or Δ^9-THC to produce ataxia in dogs or rabbits can be markedly enhanced by methamphetamine or amphetamine.[39,40,47] The experiments with rabbits showed that Δ^9-THC and methamphetamine could also interact synergistically in the production of rigid standing postures and uncoordinated stereotyped movements. However, Δ^9-THC-induced sprawling in rabbits was attenuated by methamphetamine, although it was enhanced by apomorphine. Ferri et al.[16] found that in rats, a pretreatment with 6-hydroxydopamine, known to decrease the effect of cannabis on descent latency in the bar test, did not lessen the ability of cannabis to produce atonic muscular prostration. In other studies with rats, it was found that prostration produced by Δ^9-THC was attenuated by amphetamine.[11,48]

Jones et al.[42,43] showed that the duration of Δ^9-THC-induced sprawling in rabbits was transiently reversed by physostigmine and by the muscarinic agonist, arecoline, but enhanced by nicotine. Physostigmine and arecoline also reversed the depressant effect of Δ^9-THC on standing duration.

Other Types of Motor Response

Lew and Richardson[48] reported that rats treated with Δ^9-THC tended to circle backwards and that this effect could be attenuated by amphetamine. Sakurai et al.[49] found that rats that had been given injections of 6-hydroxydopamine unilaterally into the substantia nigra showed ipsilateral postural asymmetry or ipsilateral circling when they were injected systemically with Δ^9-THC 3 weeks later. This behavior could be abolished by haloperidol. Hine et al.[50] noted that naloxone produced circling behavior in morphine-dependent rats pretreated 60 min earlier with Δ^8- or Δ^9-THC, and that this circling could be reduced by haloperidol but not by promethazine. Duncan and Dagirmanjian[51] observed contralateral circling behavior in rats that had been given an injection of carbachol directly into the left lateral septal nucleus shortly after an intraperitoneal injection of Δ^9-THC. Neither drug produced circling when given alone.

Ferri et al.[16] reported that stereotyped behavior (i.e., rhythmic head movements, gnawing, and sniffing) and hyperreactivity (i.e., hypersensitivity to touch) induced in rats by cannabis could be attenuated by both pimozide and 6-hydroxydopamine. The stereotypy (but not the hyperreactivity) could also be reduced by atropine.

In experiments with grouped mice, Sabelli et al.[52,53] found that pretreatment with pargyline or nialamide could markedly enhance the excitatory effects of Δ^9-THC, causing it to induce episodes of jumping, running, and vocalization ("popcorn effect"). The excitatory effects could be further enhanced by the injection of 2-phenylethylamine shortly after Δ^9-THC. Both *p*-chlorophenylalanine and α-methyl-*p*-tyrosine failed to prevent the "popcorn effect". It was also found that pargyline increased the excitatory effects of Δ^9-THC in rabbits. Carder and Deikel[54] reported that pretreatment with nialamide caused Δ^9-THC to increase the rate at which rats with electrodes implanted bilaterally in the lateral hypothalamic region would bar press for electrical self-stimulation. By itself, nialamide had little effect on self-stimulation, while Δ^9-THC had a depressant effect on the response when given alone.

Summary and Conclusions

Drug interaction experiments involving the production of catalepsy or the inhibition of locomotor activity by cannabinoids have yielded results that strongly suggest that these responses are mediated by ACh acting at muscarinic and nicotinic receptors, by GABA acting at $GABA_A$ and $GABA_B$ receptors, and by PGs of the E series. There is also evidence that NE and 5-HT are involved in the cataleptic response. In addition, there is evidence from a single study that points to an involvement of opioid peptides in the depressant effect of cannabinoids on locomotion. Negative results obtained in experiments in which methysergide or yohimbine were given intrathecally suggest that cannabinoid-induced catalepsy is not mediated by 5-HT receptors or α_2-adrenoceptors in the spinal cord (Tables 1 and 2).

The finding that cannabinoid-induced catalepsy and hypokinesia can be markedly potentiated by drugs that have in common the ability to suppress dopaminergic transmission has led Moss and colleagues[14] to propose that "the THC influence on extrapyramidal motor function is normally a minor subtle effect, probably on a nondopaminergic system, that is masked by the function of the major extrapyramidal dopamine system. In the absence of the regulatory effect of the main dopamine system, however, the normally occult effect(s) of THC become observable." In line with this hypothesis is evidence from drug interaction studies that cannabinoid-induced catalepsy can be attenuated by drugs that are known to augment central dopaminergic transmission (see above section on "Drug Interactions Involving Dopamine or Norepinephrine"), from behavioral studies that Δ^9-THC increases locomotor activity in rats when injected directly into the nucleus accumbens[55] and from microdialysis experiments that Δ^9-THC increases DA release in the corpus striatum and nucleus accumbens.[56,57] Such an effect on DA release is supported by the observation that Δ^9-THC induces ipsilateral circling in rats with unilateral nigral lesions (see the above section on "Other Types of Motor Response") since this behavioral response is typical of a drug that can elevate extracellular concentrations of DA at dopaminergic nerve terminals in the striatum. Interestingly, there appear to be no cannabinoid receptors on dopaminergic cell bodies or nerve terminals in the substantia nigra, corpus striatum, or nucleus accumbens, suggesting that Δ^9-THC does not increase DA release in the corpus striatum or nucleus accumbens through a direct effect on dopaminergic neurons.[5] Consistent with this conclusion is the finding that the ability of Δ^9-THC to elevate extracellular concentrations of DA is opposed by naloxone.[57]

Some of the depressant effects of cannabinoids on motor function appear neither to depend on the activation of muscarinic cholinoceptors nor to be opposed by drugs that augment dopaminergic transmission. Among these changes are ataxia in dogs and ataxia and sprawling behavior in rabbits (see above section on "Posture and Coordination"). There is also evidence from experiments with amphetamine that although Δ^9-THC-induced catalepsy in rats can be reversed by activating dopaminergic pathways in the corpus striatum, it can be potentiated by increasing dopaminergic transmission in the globus pallidus.[6]

The evidence that cannabinoids can induce catalepsy by interacting with sites located in the corpus striatum, globus pallidus, and substantia nigra has already been described. In the globus pallidus, the production of the response may well depend on the release of ACh and GABA and, in the striatum, on the release of ACh. Thus, (1) a marked degree of synergism in the production of catalepsy occurs between Δ^9-THC and the muscarinic agonist, RS-86, when the latter is injected into the corpus striatum or globus pallidus,[6] and (2) a dose of the benzodiazepine, chlordiazepoxide, known to enhance the cataleptic response to an intrapallidal injection of the GABA agonist, THIP, also potentiates the cataleptic response to an intrapallidal injection of Δ^9-THC.[8] Consistent with an involvement of striatal ACh in the production of catalepsy by cannabinoids is the finding that nicotine, injected intrastriatally, shares the ability of Δ^9-THC, injected intrastriatally, to interact synergistically with reserpine in the bar test.[7]

Although cannabinoid-induced catalepsy may be mediated by GABA-releasing neurons in the globus pallidus, it is unlikely to depend on the release of this transmitter in those other areas of the brain that form part of the extrapyramidal system and are thought to be particularly rich in cannabinoid receptors (substantia nigra pars reticulata, entopeduncular nucleus, and dorsolateral striatum).[3,4] Thus, whereas $GABA_A$ or $GABA_B$ agonists elicit catalepsy when injected into the globus pallidus, they are known to produce excitatory changes in motor function when injected into the entopeduncular nucleus or substantia nigra pars reticulata and no detectable motor effects when injected into the dorsolateral striatum.[58] Indeed, it could well be that while GABA-releasing neurons terminating in the globus pallidus contribute to the expression of cannabinoid-induced catalepsy, other GABA-releasing neurons in the brain mediate some of the concomitant excitatory effects of cannabinoids.

BODY TEMPERATURE

Although there are some reports that psychotropic cannabinoids are hyperthermic at low doses, most experiments have focused on the ability of higher doses of this group of drugs to induce hypothermia (see review by Pertwee[59]). Cannabinoids appear to elicit this response by decreasing the thermoregulatory set point, central as well as peripheral administration of Δ^9-THC having been found to elicit changes in heat gain and loss of the sort expected to promote hypothermia.[59] Thus, results from *in vivo* experiments with mice have demonstrated that hypothermic doses of Δ^9-THC, injected systemically or directly into the preoptic/anterior hypothalamic area, decreased heat production in the cold, as measured by a reduction in oxygen consumption, and augmented peripheral blood flow, as measured by a change in skin temperature.[59-62] It has also been found that mice pretreated with hypothermic doses of THC, either peripherally or centrally, show a greater preference for the cold than control animals.[63,64]

There is evidence that cannabinoids can elicit hypothermia by interacting with sites located in the preoptic/anterior hypothalamic area and in the caudal brain stem.[59] Interestingly, both these brain areas contain significantly fewer cannabinoid binding sites than many other parts of the brain.[3,4] Even so, it remains possible that the actions of cannabinoids in the hypothalamus and brain stem are cannabinoid receptor mediated. Thus, it could well be that cannabinoid receptors are concentrated at only a limited number of critical sites within these brain areas or that these parts of the brain contain another type(s) of cannabinoid receptor lacking high affinity for the radioligand that was used to determine the distribution of cannabinoid binding sites in the brain (^{3}H-CP 55,940). It is also worth noting that even in those areas of the CNS that are least populated with cannabinoid binding sites, the density of these sites is quite high compared to the density of binding sites for "classical" neurotransmitters (see review by Snyder[65]); the level of specific cannabinoid binding in most of the "low-density" areas being in the range of 0.5 to 2.0 pmol/mg protein.[4]

Drug Interactions Involving Dopamine or Norepinephrine

Mice

It has been shown that the hypothermic response to Δ^9-THC can be potentiated by drugs having in common the ability to block α- or β-

adrenoceptors (Table 6). A number of these antagonists (haloperidol, chlorpromazine, piflutixol, flupentixol, and phenoxybenzamine) can block several types of nonadrenergic receptors, including DA D_1 and D_2 receptors, 5-HT receptors, histamine H_1 receptors, and muscarinic cholinoceptors, and some of them can also inhibit the neuronal uptake of NE or act as local anesthetics. In view of this lack of selectivity, it is worth noting that the ability of these drugs to potentiate Δ^9-THC-induced hypothermia does seem to stem from their ability to block NE (at α-adrenoceptors) rather than from any of their other pharmacological properties. Thus, synergism has not been observed in mice between Δ^9-THC and drugs that inhibit NE uptake (desipramine), produce local anesthesia (lignocaine), or selectively antagonize DA D_1 receptors (SCH 23390), DA D_2 receptors (pimozide and sulpiride), 5-HT receptors (methysergide), histamine H_1 receptors (mepyramine), or central muscarinic cholinoceptors (atropine)[70,71,76] (see also Tables 6 to 8). Whenever negative results of this kind are obtained, the possibility always exists that the drugs used have been given at inappropriate doses. It should be noted, therefore, that the dose levels of pimozide and sulpiride used in these studies were sufficient to attenuate apomorphine-induced hypothermia and that SCH 23390 was used at a dose level well above the range known to block DA D_1 receptors in mice.[70,71] Similarly, methysergide was given at a dose that was known to decrease the frequency of head shaking induced by tryptophan in pargyline-pretreated mice,[59,71] desipramine at a dose known to accelerate the rate of recovery from reserpine-induced hypothermia,[71] and atropine at a dose known to prevent oxotremorine-induced hypothermia.[76] Although desipramine has been shown to have no effect on the hypothermic response to Δ^9-THC, there is a report that Δ^9-THC-induced hypothermia can be enhanced by clomipramine.[72] However, this has been attributed to the drug's ability to block the neuronal uptake of 5-HT rather than to its inhibitory effect on NE uptake (see section below on "Drug Interactions Involving 5-Hydroxytryptamine").

It has been found that chlorpromazine is no more potent (in milligram-per-kilogram terms) as an enhancer of Δ^9-THC-induced hypothermia when injected directly into the cerebral ventricles or preoptic/anterior hypothalamic area than when injected subcutaneously,[70 71] an observation which supports the notion that chlorpromazine and perhaps other α-adrenoceptor antagonists potentiate Δ^9-THC by acting at sites located outside the CNS. One possibility is that these drugs augment Δ^9-THC-induced hypothermia by antagonizing α-adrenoceptors located on cutaneous blood vessels, as this would be expected to decrease the capacity of animals to minimize periph-

TABLE 6
Drug-Induced Changes in the Ability of Cannabis or Δ9-THC to Produce Hyperthermia or Hypothermia: Interactions Involving Norepinephrine or Dopamine

Species	Pretreatment	Treatment	Response to cannabis or Δ9-THC	Effect of treatment or pretreatment on response to cannabis or Δ9-THC	Ref.
Inhibitors of DA or NE synthesis					
Rat	α-Methyl-*p*-tyrosine (i.p.)	Cannabis (p.o.)	Hypothermia	Attenuation	66
	Diethyldithiocarbamate (i.p.)			Attenuation	
Drug that impairs vesicular storage of catecholamines					
Rat	6-Hydroxydopamine (i.c.v.)	Cannabis (p.o.)	Hypothermia	Attenuation	66
Rat	6-Hydroxydopamine (i.c.v.)	Cannabis (i.p.)	Hyperthermia Hypothermia	Attenuation Attenuation	16
Rat	6-Hydroxydopamine (i.p.)	Δ9-THC (i.p.)	Hypothermia	Attenuation	67
Antagonists of α- or β-adrenoceptors					
Mouse	Propranolol (i.v.)	Δ9-THC (i.v.)	Hypothermia	Enhancement	68
	Practolol (i.v.)			Enhancement	
	Phentolamine (i.v.)			Enhancement	
Mouse	Propranolol (i.p.)	Δ9-THC (i.v.)	Hypothermia	Enhancement	69
Mouse	Phentolamine (s.c.)	Δ9-THC (i.p.)	Hypothermia	Enhancement	70, 71
	Prazosin (s.c.)			Enhancement	
	Indoramin (s.c.)			Enhancement	
Mouse	Phentolamine (PO/AH)	Δ9-THC (i.p.)	Hypothermia	None	70, 71
	Phentolamine (i.c.v.)			None	
Rat	Propranolol (i.p.)	Cannabis (p.o.)	Hypothermia	Attenuation	66
	Phentolamine (i.p.)			Attenuation	
Rat	Δ9-THC (i.v.)	Yohimbine (i.t.)[a]	Hypothermia	None	20

TABLE 6 (continued)
Drug-Induced Changes in the Ability of Cannabis or Δ^9-THC to Produce Hyperthermia or Hypothermia: Interactions Involving Norepinephrine or Dopamine

Species	Pretreatment	Treatment	Response to cannabis or Δ^9-THC	Effect of treatment or pretreatment on response to cannabis or Δ^9-THC	Ref.
Antagonists of DA D_1 or D_2 receptors					
Mouse	Pimozide (s.c.)	Δ^9-THC (i.p.)	Hypothermia	None	70,71
	Sulpiride (s.c.)			None	
	SCH 23390 (s.c.)			None	
Rat	Pimozide (i.p.)	Cannabis (i.p.)	Hyperthermia	None	16
			Hypothermia	Attenuation	
Less selective NE and DA antagonists					
Mouse	Chlorpromazine (s.c.)	Δ^9-THC (i.p.)	Hypothermia	Enhancement	70,71
	cis(Z)-Piflutixol (s.c.)			Enhancement	
	cis(Z)-Flupentixol (s.c.)			Enhancement	
Mouse	Chlorpromazine (PO/AH)	Δ^9-THC (i.p.)	Hypothermia	Enhancement	70,71
	Chlorpromazine (i.c.v.)			Enhancement	
Mouse	Haloperidol (i.v.)	Δ^9-THC (i.v.)	Hypothermia	Enhancement	68
Mouse	Haloperidol (i.p.)	Δ^9-THC (p.o.)	Hypothermia	Enhancement	72
	Phenoxybenzamine (i.p.)			Enhancement	
Mouse	Phenoxybenzamine (i.v.)	Δ^9-THC (i.v.)	Hypothermia	Enhancement	68
Rat	Phenoxybenzamine (i.p.)	Cannabis (p.o.)	Hypothermia	Attenuation	66
DA agonist					
Mouse	Δ^9-THC (i.p.)	Apomorphine (s.c.)	Hypothermia	None[b]	73

TABLE 6 (continued)
Drug-Induced Changes in the Ability of Cannabis or Δ^9-THC to Produce Hyperthermia or Hypothermia: Interactions Involving Norepinephrine or Dopamine

Species	Pretreatment	Treatment	Response to cannabis or Δ^9-THC	Effect of treatment or pretreatment on response to cannabis or Δ^9-THC	Ref.
Drugs that inhibit neuronal uptake of NE					
Mouse	Desipramine (s.c.)	Δ^9-THC (i.v.)	Hypothermia	None	71
Mouse	Clomipramine (i.p.)	Δ^9-THC (p.o.)	Hypothermia	Enhancement	72
Mouse	Clomipramine (i.p.)	Δ^9-THC (i.v.)	Hypothermia	None	59, 74
Rat	Clomipramine (i.p.)	Δ^9-THC (i.v.)	Hypothermia	Attenuation	75
Drug that inhibits the metabolism of NE and DA					
Rat	Nialamide (i.p.)	Cannabis (p.o.)	Hypothermia	Enhancement	66

Note: PO/AH indicates that injections were made directly into the preoptic/anterior hypothalamic area. i.c.v. indicates that injections were made intracerebroventricularly and i.t. that they were made intrathecally.

[a] The hypothermic response to D^9-THC was unaffected by doses of yohimbine known to attenuate Δ^9-THC-induced antinociception.
[b] See legend to Figure 1.

eral blood flow by vasoconstriction and, hence, to compensate for the cannabinoid's known ability to inhibit autonomic and behavioral heat gain mechanisms. Consistent with such a mechanism is evidence from experiments with mice that the ability of Δ^9-THC to increase heat loss from the tail is enhanced by the α-adrenoceptor antagonist phentolamine.[61]

There is no reason to suppose that β-adrenoceptor antagonists can also enhance Δ^9-THC-induced hypothermia by augmenting peripheral blood flow and, indeed, Haavik et al.[61] have reported that propranolol and practolol do not share the ability of phentolamine to increase the effect of Δ^9-THC on heat loss from the mouse tail. Even so, the site at which β-adrenoceptor antagonists act to potentiate the hypothermic effect of Δ^9-THC need not be located in the CNS. It is possible, for example that they interact with Δ^9-THC by reducing the ability of NE to activate β-adrenoceptors in peripheral tissues responsible for nonshivering thermogenesis. One likely site of action is brown adipose tissue since (1) Δ^9-THC is known to accumulate in brown fat, (2) the activation of β-adrenoceptors in brown fat can trigger the production of heat by this tissue, (3) β-adrenoceptor antagonists can decrease the thermogenic activity of brown fat, and (4) in the absence of a β-adrenoceptor antagonist, Δ^9-THC does not seem to affect the ability of brown adipose tissue to produce heat.[82,83]

Rats

Studies with antagonists of NE or DA have yielded results quite different from those obtained in experiments with mice (Table 6). Ferri et al.[16] found that although pimozide did not affect the ability of a subhypothermic dose of cannabis to produce hyperthermia, it did attenuate cannabis-induced hypothermia. Singh and Das[66] found that a normally hypothermic dose of cannabis did not decrease the body temperatures of rats that had already been rendered hypothermic by prior injection of an α- or β-adenoceptor antagonist (phentolamine, phenoxybenzamine, or propranolol). Consistent with these observations that, in rats, cannabinoid-induced hypothermia can be attenuated by drugs that exert an inhibitory action on catecholamine-releasing pathways, is the finding that this effect can also be attenuated by 6-hydroxydopamine administered either intracerebroventricularly or by peripheral injection (Table 6). When taken together, these data provide evidence for the notion that, in rats at least, NE- and/or DA-releasing neurons play an important part in the production of hypothermia by cannabinoids. The involvement of catecholamine-releasing neurons in cannabinoid-induced hypothermia receives further support from experiments showing that the production of hypothermia by cannabis can be potentiated by the monoamine oxidase inhibitor, nialamide (Table 6). The finding that cannabis-induced hypothermia can be prevented by 6-hydroxydopamine when injected directly into the ventricular system of the brain suggests that some of these neurons are located centrally. Similarly, the observation that the hypothermic response to Δ^9-THC is not altered by intrathecally administered yohimbine (Table 6) makes it unlikely that this response is mediated by NE acting at α_2-adrenoceptors in the spinal cord.

Drug Interactions Involving 5-Hydroxytryptamine

Mice

Davies and Graham[72] found that Δ^9-THC-induced hypothermia could be augmented by clomipramine, a drug expected to increase synaptic concen-

TABLE 7
Drug-Induced Changes in the Ability of Cannabinoids to Produce Hypothermia: Interactions Involving 5-Hydroxytryptamine

Species	Pretreatment	Treatment	Response to cannabinoid	Effect of treatment or pretreatment on response to cannabinoid	Ref.
Drug that inhibits 5-HT synthesis					
Mouse	*p*-Chlorophenylalanine (i.p.)	Δ^9-THC (p.o.)	Hypothermia	Attenuation	72
Mouse	*p*-Chlorophenylalanine (i.p.)	Δ^8-THC (i.v.)	Hypothermia	None[a]	77
		11-OH-Δ^8-THC (i.v.)	Hypothermia	None[a]	
Rat	*p*-Chlorophenylalanine (i.p.)	Cannabis (p.o.)	Hypothermia	Enhancement	66
Drug that impairs vesicular storage of 5-HT					
Mouse	Dihydroxytryptamine (i.c.v.)	Δ^8-THC (i.v.)	Hypothermia	None[a]	77
		11-OH-Δ^8-THC (i.v.)		Enhancement[a]	
Rat	Dihydroxytryptamine (i.c.v.)	Cannabis (p.o.)	Hypothermia	None	66
5-HT antagonists					
Mouse	Methysergide (i.p.)	Δ^9-THC (p.o.)	Hypothermia	Attenuation	72
Mouse	Methysergide (s.c.)	Δ^9-THC (i.p.)	Hypothermia	None	71
Mouse	Methysergide (s.c.)	Δ^9-THC (i.v.)	Hypothermia	None	59, 78
Mouse	Cinanserine (i.p.)	Δ^8-THC (i.v.)	Hypothermia	None	77
		11-OH-Δ^8-THC (i.v.)	Hypothermia	None	
Rat	Methysergide (i.p.)	Δ^9-THC (i.v.)	Hypothermia	Enhancement	79
Rat	Δ^9-THC (i.v.)	Methysergide (i.t.)	Hypothermia	Enhancement	20
Drug that inhibits neuronal uptake of 5-HT (see also clomipramine and desipramine in Table 6)					
Mouse	Fluoxetine (i.p.)	Δ^9-THC (i.v.)	Hypothermia	None	59, 74

Note: i.c.v. indicates that injections were made intracerebroventricularly and i.t. that they were made intrathecally.

[a] Effect of the pretreatment alone was not reported.

trations of 5-HT through inhibition of its neuronal uptake, and attenuated by drugs known to inhibit the synthesis or actions of 5-HT (Tables 6 and 7). These observations led to the conclusion that the hypothermic response of mice to Δ^9-THC is mediated by 5-HT systems. However, this conclusion is not supported by results obtained in other studies. Experiments by Gray et al.[71] and by Tavendale[78] (see also review by Pertwee[59]) showed that the hypothermic response of mice to Δ^9-THC was not modified by the 5-HT antagonist, methysergide, when this drug was given at a dose known to attenuate a behavioral response thought to be triggered by increased 5-HT release (tryptophan-induced head shaking in pargyline-pretreated mice). The hypothermic response was also unaffected by two inhibitors of 5-HT uptake, fluoxetine and clomipramine, administered at doses known to augment the head shaking response.[59,74] Watanabe et al.[77] reported that a hypothermic dose of the 5-HT antagonist, cinanserine, had no effect on hypothermia induced by Δ^8-THC or by its 11-hydroxy metabolite. They also found that in mice, pretreatment with drugs expected to deplete the brain of 5-HT had no inhibitory effect on hypothermia induced by either cannabinoid (Table 7). Indeed one of the drugs, dihydroxytryptamine, enhanced the hypothermic response to 11-hydroxy-Δ^8-THC.

Rats

Studies with a 5-HT antagonist or with drugs that inhibit the synthesis or neuronal uptake of 5-HT have produced results that are essentially opposite to those obtained in experiments with mice. Thus, there are reports that hypothermia induced by cannabis or Δ^9-THC can be potentiated by methysergide or *p*-chlorophenylalanine and attenuated by clomipramine (Tables 6 and 7). These findings support the notion that in rats, cannabinoids may induce hypothermia by suppressing the activity of 5-HT-releasing neurons. Against this idea is an observation that intracerebroventricular pretreatment with dihydroxytryptamine did not modify the ability of cannabis to decrease body temperature in rats (Table 7).

Drug Interactions Involving Acetylcholine

Mice

Three nicotinic cholinoceptor antagonists, hexamethonium, pentolinium, and mecamylamine, have each been shown to potentiate the hypothermic response to Δ^9-THC (Table 8). Two of these drugs, hexamethonium and pentolinium, are bisquaternary ammonium compounds and would therefore

TABLE 8
Drug-Induced Changes in the Ability of Cannabis or Δ^9-THC to Produce Hyperthermia or Hypothermia: Interactions Involving Acetylcholine

Species	Pretreatment	Treatment	Response to cannabis	Effect of treatment or pretreatment on response to cannabis	Ref.
Cholinoceptor agonists					
Mouse	Δ^9-THC (i.p.)	Oxotremorine (s.c.)	Hypothermia	Enhancement[a,b]	73
Rat	Δ^9-THC (p.o.)	Nicotine (i.p.)	Hypothermia	Enhancement	41
Muscarinic antagonists					
Mouse	Atropine (s.c.)	Δ^9-THC (i.p.)	Hypothermia	Attenuation	76
	Atropine (i c.v.)			Attenuation	
Mouse	Atropine (i.p.)	Δ^9-THC (i.v.)	Hypothermia	Attenuation	59, 78
	Methylatropine (i.p.)			Attenuation	
Mouse	Atropine (i.v.)	Δ^9-THC (i.v.)	Hypothermia	None	68
	Methylatropine (i.v.)			Enhancement	
Rat	Atropine (i.p.)	Cannabis (i.p.)	Hypothermia	None	16
Rat	Atropine (i.p.)	Δ^9-THC (i.v.)	Hypothermia	None	79
Nicotinic antagonist					
Mouse	Mecamylamine (i.v)	Δ^9-THC (i.v.)	Hypothermia	Enhancement	68
	Hexamethonium (i.v.)			Enhancement	
Mouse	Pentolinium (s.c.)	Δ^9-THC (i.p.)	Hypothermia	Enhancement	70
Rat	Hexamethonium (i.p.)	Δ^9-THC (i.p.)	Hypothermia	Attenuation	67
Acetylcholinesterase inhibitors[c]					
Rat	Physostigmine (i.p.)	Δ^9-THC (i.p.)	Hyperthermia	Attenuation	81
	Neostigmine (i.p.)			None	

Note: i.c.v. indicates that injections were made intracerebroventricularly.

[a] See Figure 1.

[b] This enhancement was prevented by atropine sulfate (2 mg/kg s.c.) injected 30 min before Δ^9-THC and oxotremorine.

[c] Coadministered with Δ^9-THC.

be expected to penetrate the blood/brain barrier rather slowly. Consequently, it seems likely that all three antagonists modified the response to Δ^9-THC by acting at sites located somewhere outside the CNS. One possibility is that they augmented the hypothermic response to Δ^9-THC by blocking transmission in sympathetic ganglia responsible for maintaining vasomotor tone in cutaneous blood vessels and/or for activating nonshivering thermogenesis. Indeed, it has been found that the ability of Δ^9-THC to increase heat loss from the tail of the mouse is enhanced by both hexamethonium and mecamylamine.[61]

Studies that have been carried out with muscarinic agents fall essentially into two groups (Table 8). One group of studies has yielded data suggesting that the hypothermic response to Δ^9-THC depends on the activation of muscarinic cholinoceptors. These studies have shown that Δ^9-THC-induced hypothermia can be attenuated by atropine and augmented by oxotremorine[59,73,76,78] (see also Table 8 and Figure 1). Methylatropine which crosses the blood/brain barrier only with difficulty was also found to attenuate Δ^9-THC-induced hypothermia,[59,78] implying that if muscarinic receptors do indeed mediate the hypothermic response to Δ^9-THC, they must be located either outside the CNS or in one of the areas of the brain where the blood/brain barrier is known to be relatively ineffective. In favor of the second of these ideas is the finding that atropine can attenuate Δ^9-THC hypothermia not only when injected peripherally but also when injected directly into the ventricular system of the brain.[59,76]

The second group of studies has generated data that suggest that ACh, acting at muscarinic cholinoceptors, is not involved in the production of hypothermia by Δ^9-THC. Thus, it was found that the hypothermic response to Δ^9-THC was unaffected by atropine and enhanced by methylatropine.[68] The methylatropine could well have been acting peripherally to reduce sympathetic tone since, unlike atropine, it can suppress not only muscarinic transmission but also nicotinic transmission in sympathetic ganglia. Evidence that methylatropine can enhance the ability of Δ^9-THC to increase heat loss from the tail of the mouse is consistent with this idea,[61] although it is important to note that such an effect on heat loss does not always lead to an enhancement of Δ^9-THC-induced hypothermia (see section below on "Drug Interactions Involving Opioids or Histamine"). Clearly, further experiments are still needed both to explain the conflicting results obtained in the experiments with muscarinic antagonists and to elucidate the role of central cholinergic pathways in the hypothermic response of mice to Δ^9-THC.

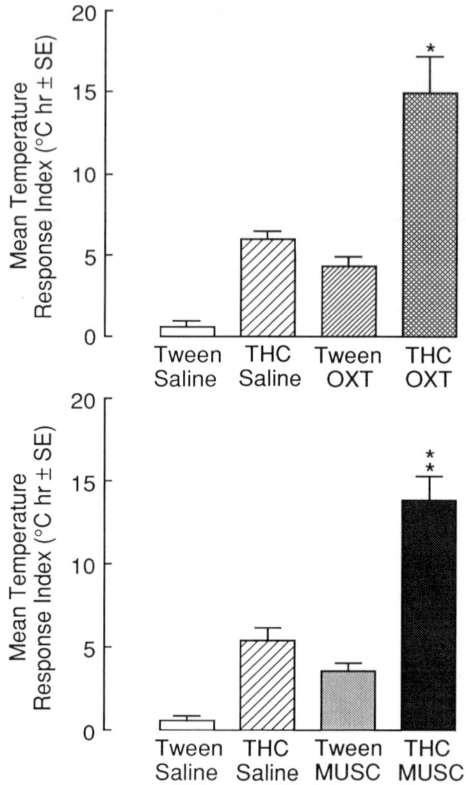

FIGURE 1. The effect of Δ^9-THC on the ability of oxotremorine (upper panel; n = 6) and muscimol (lower panel; n = 9) to produce hypothermia in mice.[73] Changes in body temperature were measured as described elsewhere[70] and have been expressed in terms of the temperature response index.[70] Oxotremorine (OXT; 0.1 mg/kg s.c.), muscimol (MUSC; 0.5 mg/kg s.c.), and saline were injected immediately after Δ^9-THC (THC; 10 mg/kg i.p.) or its vehicle, Tween 80 (Tween; 20 mg/kg i.p.), in saline. The asterisks indicate that the interactions between the pretreatment drugs and THC were more than additive (*p <0.05; **p <0.01; Mann-Whitney U-test applied to the procedure of multi-way analysis described by Bradley.[80] Similar experiments with submaximal hypothermic doses of baclofen (5 or 7 mg/kg s.c.) or apomorphine (2 mg/kg s.c.) showed that these drugs did not interact synergistically with Δ^9-THC (10 mg/kg i.p.) in the production of hypothermia.

Rats

In line with the results obtained with mice by Haavik and Hardman,[68] Ferri et al.[16] found that atropine had no effect either on hyperthermia produced by a subhypothermic dose of cannabis or on cannabis-induced hypothermia. Similarly, Fennessy et al.[79] reported that atropine had no effect on Δ^9-THC-induced hypothermia. Experiments with a nicotinic antagonist have yielded results that conflict with the corresponding mouse data, Schmeling and Hosko[67] having found that hexamethonium attenuates the hypothermic response of rats to Δ^9-THC. Consistent with this observation is a report that in rats, nicotine and Δ^9-THC interact synergistically in the production of hypothermia.[41]

Drug Interactions Involving γ-Aminobutyric Acid

Rats and Mice

Experiments with mice have shown that the hypothermic effect of Δ^9-THC is not potentiated by submaximal hypothermic doses of the $GABA_B$ agonist baclofen (see legend to Figure 1) or by the GABA uptake inhibitor, NO-328, given at a dose known to enhance the ability of Δ^9-THC to delay descent in the bar test.[84] On the other hand, a synergistic interaction has been detected between Δ^9-THC and the $GABA_A$ agonist, muscimol (Figure 1).

Other experiments have been carried out to investigate interactions between Δ^9-THC and benzodiazepines, drugs which can act through their own receptors to facilitate the interaction between GABA and $GABA_A$ receptors. Pryor et al.[44] showed that in rats, chlordiazepoxide interacts synergistically with Δ^9-THC in the production of hypothermia. Pertwee et al.[84] found that in mice, a dose of flurazepam (3 mg/kg s.c.) known to increase the ability of Δ^9-THC to delay descent in the bar test did not modify the hypothermic response to this cannabinoid. However, the response was found to be enhanced by higher doses of flurazepam.[84,85] Additional experiments[86] have now shown that this enhancement could not be prevented by the benzodiazepine receptor antagonist, flumazenil, given at a dose known to attenuate the ability of flurazepam to augment the effect of Δ^9-THC on descent latency in the bar test (10 mg/kg i.p. immediately before flurazepam and Δ^9-THC). This finding indicates that the potentiation of Δ^9-THC-induced hypothermia by benzodiazepines may not be mediated by benzodiazepine receptors, in which case it is unlikely to have arisen from a benzodiazepine-induced facilitation of GABAergic transmission.

Drug Interactions Involving Opioids or Histamine

Rats and Mice

Haavik et al.[69] reported that pretreatment of mice with a dose of nalorphine, known to antagonize the hypothermic effect of morphine, had no detectable effect on the hypothermic response to Δ^9-THC. However, other studies have shown that Δ^9-THC-induced hypothermia can be attenuated in mice by naloxone (1 mg/kg s.c.) and in rats by the intracerebroventricular administration of β-chlornaltrexamine.[87,88]

In experiments directed at investigating the role of histamine in the hypothermic response of mice to Δ^9-THC, Haavik and Hardman[68] found that the response could be attenuated by pretreatment with the histamine H_1 antagonists, diphenhydramine and chlorpheniramine. Paradoxically, the same antagonists were also found to augment Δ^9-THC-induced increases in peripheral heat loss.[61] Other experiments with mice employing the histamine H_1 antagonist, mepyramine, have shown that this drug does not prevent Δ^9-THC-induced hypothermia.[71] Consistent with this observation is a report by Fennessy et al.[79] that mepyramine has no effect on the hypothermic response of rats to Δ^9-THC.

Summary and Conclusions

NE, DA, 5-HT, ACh, GABA, histamine, and opioid peptides are all implicated as mediators of hypothermia in rats or mice.[89-95] However, there is evidence that the central release of some of these substances can also give rise to increases in deep body temperature; hence, it is not altogether unexpected that the data obtained from the drug interaction experiments described in this section are too inconsistent to allow firm conclusions to be drawn about the neuronal pathways that are involved in the production of hypothermia by cannabinoids. Some of these inconsistencies seem to arise from species differences. For example, whereas the rat data suggest that cannabinoid-induced hypothermia is mediated by noradrenergic and/or dopaminergic pathways and can be opposed by 5-HT, the mouse data suggest that the response is neither mediated by dopaminergic pathways nor opposed by 5-HT. Indeed, some of the data support the idea that cannabinoid-induced hypothermia in mice is mediated by 5-HT-releasing pathways. There is some evidence for an involvement of ACh, histamine, and opioid peptides in the production of hypothermia by cannabinoids, although again, the data are somewhat inconsistent. The response may also be mediated by GABA acting at $GABA_A$ receptors.

As discussed earlier, a significant proportion of the drug interactions described in this section may have been initiated at sites located outside the CNS. Since cannabinoids are thought to decrease body temperature by acting centrally, this serves to illustrate the point that changes in body temperature triggered at sites within the brain can be modified not only by drugs that are known to influence central transmission, but also by drugs that can alter transmission along peripheral pathways innervating tissues responsible for thermoregulatory heat production or heat loss.

ANTINOCICEPTION

Experiments with rats and mice have shown psychotropic cannabinoids to be active in three antinociceptive tests: the tail-flick test, the hot plate test, and the abdominal stretching test.[96] There is evidence that cannabinoids can produce antinociception by acting at sites located in the brain stem and spinal cord.[97-103] It is noteworthy that both of these areas have been reported to contain far fewer cannabinoid binding sites than many other parts of the CNS.[3,4] However, this does not rule out the possibility that the antinociceptive action of cannabinoids in the brain stem and spinal cord is mediated by cannabinoid receptors (see introduction to section on "Body Temperature").

Drug Interactions Involving Norepinephrine, Dopamine, 5-Hydroxytryptamine, Acetylcholine, or γ-Aminobutyric Acid

Mice

Jacob et al.[104] found that the antinociceptive effect of levonantradol in animals subjected to the hot plate test could be attenuated by intracerebroventricular pretreatment with 5,7-dihydroxytryptamine. However, they also reported the occurrence of synergism between levonantradol and the 5-HT antagonist, cyproheptadine. In the same study, it was found that levonantradol-induced antinociception was affected neither by phenoxybenzamine nor by atropine. Pertwee et al.[84] showed that the antinociceptive effect of Δ^9-THC, measured using the tail-flick response, was unaltered by the benzodiazepine, flurazepam, or the GABA uptake inhibitor, NO-328, when these were given at doses equal to or larger than ones known to enhance the ability of Δ^9-THC to delay descent in the bar test.

Rats

Milne and Johnson[105] found that a pretreatment with *p*-chlorophenylalanine known to attenuate morphine-induced antinociception had no effect on the antinociceptive response to nantradol. In contrast, Ghosh and Bhattacharya[106] obtained results supporting the notion that the antinociceptive effect of cannabis in rats is mediated by 5-HT. They found that cannabis-induced antinociception, measured using the tail-flick test, could be potentiated by 5-hydroxytryptophan, the immediate precursor of 5-HT, and imipramine, an inhibitor of 5-HT (and NE) uptake, and attenuated by drugs having in common the ability to reduce the activity of 5-HT-releasing pathways. In the same study, it was found that the antinociceptive effect of cannabis was not altered by drugs capable of decreasing the activity of noradrenergic or dopaminergic pathways through antagonism of NE or DA receptors, reduction of vesicular storage of NE and DA in central neurons, or inhibition of catecholamine synthesis. Additional experiments carried out in the same laboratory have yielded data supporting the hypothesis that cannabis-induced antinociception is mediated by 5-HT rather than by DA or NE.[107,108] More recently, however, Ferri et al.[109] obtained evidence that drug-induced depletion of catecholamines from central nerve terminals reduced the effect of cannabis on the tail flick response. It has also been shown that Δ^9-THC-induced antinociception can be attenuated by yohimbine when this α_2-adrenoceptor antagonist is injected directly into the lumbar region of the spinal cord, although not when it is injected into the upper thoracic region.[20] Intrathecally injected methysergide has been reported to have no effect on Δ^9-THC-induced antinociception.[20]

Drug Interactions Involving Opioids

Mice

Chesher et al.[110] found that a dose of nalorphine (5 mg/kg i.v.), known to reverse the antinociceptive effect of pethidine in the hot plate test, did not reverse the antinociceptive effect of cannabis. Essentially similar results have been obtained in more recent experiments, performed in the same laboratory with Δ^9-THC and naloxone[111] (see also Table 9). These experiments showed that the antinociceptive effect of Δ^9-THC, measured using the formic acid abdominal stretching test, was not modified by naloxone administered after the cannabinoid at a dose known to antagonize the effect of morphine. In contrast, there are reports that when given before Δ^9-THC, naloxone can decrease the cannabinoid's antinociceptive effect in both the

TABLE 9
Drug-Induced Changes in the Ability of Cannabinoids to Produce Antinociception: Interactions Involving Opioids

Species	Pretreatment	Treatment	Antinociceptive test	Effect of pretreatment or treatment on cannabinoid-induced antinociception	Ref.
Opioid μ receptor antagonists					
Mouse	Naloxone (1 mg/kg s.c.)	Δ⁹-THC (i.p.)	Tail flick	None	112
Mouse	Naloxone (5 mg/kg s.c.)	Δ⁹-THC (i.v.)	Tail flick	None	96
	Naloxone (10 or 20 mg/kg s.c.)			Attenuation	
Mouse	Naloxone (2 mg/kg s.c.)	β-HHC[a] (s.c.)	Tail flick	Attenuation	113
Rat	Naloxone) (1 mg/kg i.p.)	Cannabis (i.p.)	Tail flick	Attenuation	106
Rat	Naloxone (2 mg/kg s.c.)	Cannabis (p.o.)	Tail flick	None	109
	Naloxone (10 mg/kg s.c.)			Attenuation	
	Naltrexone (2 mg/kg s.c.)			None	
	Naltrexone (10 mg/kg s.c.)			Attenuation	
Mouse	Δ⁹-THC (i.t.)[b]	Naloxone (1 mg/kg s.c. or 1 μg i.t.)	Tail flick	None	99,114
	Δ⁸-THC (i.t.)			None	
	11-OH-Δ⁹-THC (i.t.)			None	
	Levonantradol (i.t.)			None	
	CP 55,940 (i.t.)			None	
	CP 56,667 (i.t.)[c]			None	
Mouse	Naloxone (1 mg/kg s.c.)	Δ⁹-THC (i.v.)	Hot plate	Attenuation	96
Mouse	Naloxone (1 mg/kg s.c.)	11-OH-Δ⁸-THC (s.c.)	Hot plate	Attenuation	115
Mouse	Levonantradol (s.c.)	Naloxone (0.1 mg/kg s.c.)	Hot plate	Attenuation	104
Rat	Naloxone (2 mg/kg i.p.)	Levonantradol (i.t.)	Hot plate	None	98
	Levonantradol (i.t.)	Naloxone (2 mg/kg i.p.)		None	

TABLE 9 (continued)
Drug-Induced Changes in the Ability of Cannabinoids to Produce Antinociception: Interactions Involving Opioids

Species	Pretreatment	Treatment	Antinociceptive test	Effect of pretreatment or treatment on cannabinoid-induced antinociception	Ref.
Rat	β-Chlornaltrexamine (2 μg i.c.v.)	Δ^9-THC (i.p.)	Hot plate	Attenuation	88
Mouse	Δ^9-THC (p.o.)	Naloxone (1 to 8 mg/kg i.v.)	Abdominal stretching	None	111
Mouse	Naloxone (20 mg/kg s.c.)	Δ^9-THC (i.v.)	Abdominal stretching	Attenuation	96
Opioid δ receptor antagonists					
Rat	ICI 154,129 (2 mg/kg s.c.)	Cannabis (p.o.)	Tail flick	None	109
	ICI 154,129 (10 mg/kg s.c.)			Attenuation	
Opioid κ receptor antagonists					
Rat	MR 1452 (2 mg/kg s.c.)	Cannabis (p.o.)	Tail flick	None	109
	MR 1452 (10 mg/kg s.c.)			Attenuation	
Opioid agonist					
Mouse	Δ^9-THC (i.t.)	Morphine (i.t.)	Tail flick	Enhancement[d,e]	99, 114
	Δ^8-THC (i.t.)	Morphine (i.t.)		Enhancement[d]	
	11-OH-Δ^9-THC (i.t.)	Morphine (i.t.)		Enhancement[d]	
	Levonantradol (i.t.)	Morphine (i.t.)		Enhancement[d]	
	Dextronantradol (i.t.)[f]	Morphine (i.t.)		None[d]	
	CP 55,940 (i.t.)	Morphine (i.t.)		None[d]	
	CP 56,667 (i.t.)[c]	Morphine (i.t.)		None[d]	
Rat	Cannabis (i.p.)	Morphine (2 mg/kg i.p.)	Tail flick	Enhancement	107
Rat	Δ^9-THC (s.c.)	Morphine (4.5 mg/kg s.c.)	Hot plate	Enhancement	116

TABLE 9 (continued)
Drug-Induced Changes in the Ability of Cannabinoids to Produce Antinociception: Interactions Involving Opioids

Note: i.t. indicates that injections were made intrathecally and i.c.v. that they were made intracerebroventricularly.

[a] 9-Nor-9b-hydroxyhexahydrocannabinol.
[b] Similar results were obtained using the hot plate test.
[c] CP 56,667 is the (+)-isomer of CP 55,940.
[d] Effect of subthreshold or submaximal antinociceptive dose of cannabinoid on morphine-induced antinociception.
[e] Attenuated or prevented by naloxone (0.1 to 1.0 mg/kg s.c.).
[f] Dextronantradol is the (+)-isomer of levonantradol.

tail-flick test and the *p*-phenylquinone abdominal stretching test (Table 9). Naloxone has also been shown to attenuate or partially reverse antinociception induced by certain other cannabinoids in either the hot plate test or the tail-flick test (Table 9). In some of these studies, naloxone was effective only at rather high doses (10 or 20 mg/kg s.c.), a finding consistent with the notion that cannabinoids do not interact directly with opioid receptors.[96]

Welch and Stevens[99,114] have shown that certain cannabinoids, administered intrathecally, can potentiate the antinociceptive effect of intrathecally injected morphine (Table 9). Interestingly, they detected no synergism between morphine and the two structurally atypical cannabinoids, CP 55,940 and its (+)-enantiomer, CP 56,667. They also found that CP 55,940, which has been reported to be significantly more potent than CP 56,667 as an antinociceptive agent when administered systemically,[117] was only marginally more potent than the (+)-isomer when injected intrathecally. While naloxone had no effect on the antinociceptive effect of cannabinoids when these were administered intrathecally in the absence of morphine, it did attenuate or block the ability of Δ^9-THC to potentiate the action of morphine (Table 9). In view of the hypothesis that the putative cannabinoid receptor interacts with G_i protein to inhibit adenylate cyclase,[4,118] it is noteworthy that Welch[99] has also reported that antinociception induced by intrathecal administration of Δ^9-THC can be attenuated by dibutyryl cyclic AMP.

Rats

There is evidence that in rats, as in mice, cannabinoids can interact synergistically with morphine in the production of antinociception and that the antinociceptive response to cannabinoids can be reduced by opioid antagonists (Table 9). Ferri et al.[109] found that the antinociceptive effect of cannabis could be decreased by naloxone and certain other opioid receptor antagonists only at the rather high dose of 10 mg/kg s.c. However, Ghosh and Bhattacharya[106] reported the attenuation of cannabis-induced antinociception by a much smaller dose of naloxone (Table 9). Other experiments, performed in the same laboratory, demonstrated that this dose of naloxone could also prevent the enhancement of cannabis-induced antinociception by restraint stress.[108] Not all experiments with rats have demonstrated cannabinoid-induced antinociception to be naloxone-sensitive, Yaksh[98] having shown that the antinociception response to intrathecal administration of levonantradol is unaltered by naloxone (Table 9).

Drug Interactions Involving Eicosanoids

Mice

Martin[96] reported that the antinociceptive effect of Δ^9-THC in the tail-flick test was not affected by indomethacin. Burstein et al.[119] found that Δ^9-THC exhibited a time-dependent biphasic effect on performance in the hot plate test, the eventual antinociceptive effect of the drug being preceded by a hypernociceptive effect similar to that which can be produced by PGE_2. This hypernociceptive response to Δ^9-THC was attenuated by indomethacin. It was also attenuated by Δ^9-THC-11-oic acid, a major metabolite of Δ^9-THC. Since Δ^9-THC-11-oic acid shares the ability of indomethacin to decrease the production of eicosanoids by inhibiting cyclooxygenase, the suggestion has been made that it is this metabolite that causes antinociception in Δ^9-THC-treated mice; the main effect of the parent compound being to produce the initial hypernociceptive response by raising tissue eicosanoid levels.[119]

Rats

Using the tail-flick test, Bhattacharya et al.[30] reported that the antinociceptive effect of cannabis could be potentiated by pretreatment (i.p.) with PGE_1 and attenuated by pretreatment (i.p.) with $PGF_{2\alpha}$ or with the cyclooxygenase inhibitor, diclofenac. Similarly, experiments using the tail-flick test or the hot plate test have shown that pretreatment (i.p.) with diclofenac, mefenamic acid, or paracetamol can decrease enhancement of cannabis-induced antinociception by restraint stress and that diclofenac or paracetamol, injected directly into the right lateral cerebral ventricle, can reduce the antinociceptive effect of Δ^9-THC.[27,108] In one of these studies, it was also found that the antinociceptive effect of Δ^9-THC could be reduced by intracerebroventricular pretreatment with hydrocortisone,[27] one of whose actions is to inhibit the activity of phospholipase A_2.

Summary and Conclusions

Central transmitters and modulators that are thought to serve as endogenous inhibitors of nociception include NE, 5-HT, ACh, GABA, opioid peptides and the PGs PGE_1 and PGD_2;[120-124] there is evidence, from the drug interaction studies described in this section, that some of these substances contribute to the production of antinociception by cannabinoids. More specifically, the data support an involvement of PGE_1, although not of ACh or GABA. Some experiments, but not others, also support an involvement of catecholamines, 5-HT and opioid peptides. The role of PGD_2 in cannabinoid-induced antinociception has yet to be investigated.

The part that PGs seem to play in the effects of cannabinoids on nociception is complex. This is not surprising inasmuch as cannabinoids are stimulators of phospholipase A_2 and inhibitors of cyclooxygenases (see review by Pertwee[1]) and as PGs enhance nociception at peripheral sites and oppose it centrally.[121,125] The ability of cannabinoids to stimulate PG release may contribute both to their hypernociceptive effect (released PGs acting at peripheral sites) and to their antinociceptive effect (released PGs acting at central sites). There is also evidence that the antinociceptive effect of cannabinoids depends on the ability of these drugs, or of certain of their metabolites, to inhibit PG production, presumably in peripheral tissues.

FINAL DISCUSSION AND SUMMARY

Interpretation of the Data

In interpreting the data from the drug interaction studies described in this review, it is important to consider whether any of the many interactions that have been observed stem from interference by one drug with the absorption, distribution, or fate of another. There are several reasons why this is especially important for data obtained in experiments that have been carried out either with cannabis or with individual cannabinoids. First, the nonpsychotropic cannabinoid, cannabidiol, which is usually present in cannabis preparations, is known to be a potent inhibitor of drug metabolism.[126-130] Second, there is one report in the literature suggesting that cannabis increases the permeability of the blood/brain barrier,[131] a finding of particular relevance to the interpretation of results from those experiments described in this chapter that rely on the ability of the blood/brain barrier to minimize the entry of certain drugs into the brain. For this and other reasons, it is clearly important that the constituents of cannabis responsible for this effect on blood/brain barrier permeability be identified. Third, there is evidence that the intensity of at least some of the effects of psychotropic cannabinoids, antinociception, impairment of operant behavior, and prolongation of barbital sleeping time, is altered when their metabolism is inhibited.[132-134] Finally, for the production of several effects including catalepsy, hypothermia, and antinociception, certain cannabinoid metabolites have been found to be more potent than their parent compounds[111,115,135-137] (see also review by Agurell et al.[138]). There is also evidence that cannabinoid metabolites can possess different pharmacological "profiles" from their parent compounds, Burstein et al.[139] having shown that Δ^9-THC is less potent than its nonpsychotropic metabolite, Δ^9-THC-11-oic acid, as an inhibitor of cyclooxygenase but more potent as a stimulator of phospholipase A_2.

Although there has been little attempt to determine the extent to which the interactions described in this chapter arose from changes in the absorption, distribution, metabolism, or excretion of the administered drugs, there is some indirect evidence which suggests that these interactions did not arise in this way but were caused instead by drug-induced modifications in the synthesis, storage, release, actions, and/or fate of central or peripheral neurochemical mediators. Thus, there are many instances where several drugs have produced an alteration in a cannabinoid response that can be explained most simply in terms of some change in a given neurochemical system, the drugs that alter the response sharing the ability to modify the activity of a particular neuronal system and yet differing markedly from each other in their modes of action and in their chemical and other pharmacological properties. There are also several instances where interactions have been detected between a cannabinoid and a second drug following the administration of either or both drugs directly into the CNS.

It seems likely, therefore, that the drug interactions described in this review resulted from drug-induced changes in neuronal transmission and that consequently they provide important clues about the identity of the neuronal pathways that mediate cannabinoid-induced changes in motor function, body temperature, and nociception. More specifically, the drug interaction data presented in this chapter provide support for the notions (1) that cannabinoid-induced catalepsy and hypokinesia are mediated by ACh, GABA, 5-HT, and PGs and opposed by DA, (2) that cannabinoid-induced hypothermia (in rats) is mediated by central NE- and/or DA-releasing pathways, and (3) that cannabinoid-induced antinociception depends on the release in the brain or spinal cord of NE, 5-HT, opioids, and PGs. Results from recent experiments with indomethacin suggest that some effects of cannabis in man are also PG mediated.[140]

Evidence that rats or mice rendered tolerant to the antinociceptive effect of morphine show tolerance to cannabinoids supports an involvement of opioids in cannabinoid-induced antinociception.[87,141-143] Results from other cross-tolerance experiments with morphine also support an involvement of opioids in the depressant effect of cannabinoids on locomotor activity.[45] Similarly, results obtained in cross-tolerance experiments with oxotremorine and Δ^9-THC[76] provide further evidence for the idea that cannabinoid-induced hypothermia (in mice) is mediated by cholinergic pathways.

Possible Actions Responsible for the Apparent Ability of Cannabinoids to Augment Transmission Along Certain Types of Neuronal Pathways

These fall essentially into two categories. In the first are actions that enhance the synthesis and/or release of neurochemical mediators or that inhibit the metabolic inactivation or tissue uptake of such substances after their release. These actions have in common the ability to augment transmission by increasing the extracellular concentrations of neurochemical mediators.

Actions in the second category are those that enhance the effects of a chemical mediator after its release through an effect either on the affinity of the mediator for its receptor or on the mediator's efficacy. Neither of the latter effects would need to involve occupancy of the chemical mediator's receptors by a cannabinoid as it is possible that cannabinoids can act through an allosteric mechanism, each mediator molecule acting at its normal site of action (the recognition site of its receptor) and each cannabinoid molecule at a different site (perhaps the recognition site of a cannabinoid receptor), the two sites of action being in close proximity to each other and functionally linked. A similar hypothesis has been put forward by Welch and Stevens[114] to explain their observation that certain cannabinoids potentiate morphine-induced antinociception. There is already a precedent for such a mechanism since it is generally accepted that benzodiazepines act on their own receptors to enhance the responsiveness of $GABA_A$ receptors for GABA.[144]

Evidence that Cannabinoids Can Augment Transmission by Increasing the Extracellular Concentrations of Neurochemical Mediators

The evidence may be summarized as follows:

1. Experiments, performed either *in vivo* or *in vitro* have yielded results indicating that psychotropic cannabinoids can accelerate the rate of formation of NE and DA from tyrosine, increase the synthesis of 5-HT by elevating brain levels of the 5-HT precursor, tryptophan, and enhance the production of eicosanoids from arachidonic acid.[1]
2. Cannabinoids can stimulate the release of DA in certain brain areas including the rat corpus striatum, nucleus accumbens, and medial prefrontal cortex.[1,56,57,145,146]
3. Δ^9-THC can augment GABA turnover in rat substantia nigra and septum (although not in rat corpus striatum or nucleus accumbens).[1]

There is also evidence from *in vitro* experiments with brain tissue[1] that cannabinoids can inhibit the neuronal uptake of NE, DA, 5-HT, and GABA, an effect which would be expected to intensify and prolong the actions of these transmitters after their release. However, the role of this inhibitory effect in the psychopharmacology of cannabinoids is not clear since cannabinoids can also increase the neuronal uptake of NE, DA, and 5-HT and since this ability correlates with psychotropic potency, whereas the ability to inhibit uptake does not.[1]

Although cannabinoids may well enhance transmission in systems that release NE, DA, 5-HT, GABA, or PGs through an effect on the synthesis, release, or fate of these chemical mediators, there is no evidence at present to suggest that cannabinoids alter other types of neuronal transmission through such mechanisms. Indeed, in the case of cholinergic transmission, while the data produced in drug interaction studies suggest that central cholinergic transmission can be augmented by cannabinoids, results from other kinds of experiments have shown that cannabinoids depress the synthesis and release of ACh in the brain and have no effect on its metabolic inactivation.[1]

Evidence that Cannabinoids Can Augment Transmission by Potentiating the Actions of Neurochemical Mediators

As summarized below, there is evidence that cannabinoids can potentiate certain actions of NE, ACh, GABA, opioids, and eicosanoids and that they can alter the properties of receptors for NE, DA, or ACh by producing changes either in the affinity of their recognition sites or in the functioning of their second messenger systems.

1. Cannabinoids have been reported to interact synergistically with the cholinergic agonists nicotine, tremorine, oxotremorine, and carbachol in the production of catalepsy, tremor, circling, salivation, lacrimation, hypothermia, and drinking[6,11,12,23,41,51,73,131] (see also sections on "Motor Function" and "Body Temperature"), and with GABA agonists, opioids, benzodiazepines, and PGs in the production of catalepsy, excitement (in cats), hypothermia, or antinociception[8,18,28,30,73,99,107,114,116] (see also sections on "Motor Function", "Body Temperature", and "Antinociception").
2. A single treatment with cannabis known to potentiate oxotremorine-induced tremor in mice has been reported to increase the affinity of binding sites in mouse striatal tissue for [^{3}H]-quinuclidinyl benzylate, a ligand with selectivity for muscarinic cholinoceptors.[131]
3. A repeated treatment with cannabis (*in vivo*) in a regimen known to produce tolerance to the effect of cannabis on oxotremorine-induced tremor in mice has also been found to abolish the ability of cannabis to enhance [^{3}H]-quinuclidinyl benzylate binding to mouse striatal tissue.[131]
4. SP-111, a water-soluble derivative of Δ^9-THC has been shown to enhance responses of rat cerebellar neurons to GABA and NE when it is coadministered iontophoretically with one or other of these transmitters.[147]
5. In some systems (but not others), cannabinoids have been shown to increase the affinity of certain ligands for dopamine receptors and to augment the ability of adrenergic agonists to stimulate the production of the second messenger, cyclic AMP.[1]

In addition, there are reports that cannabinoids can increase the binding of benzodiazepines to rat and mouse brain.[148,149] However, there is also evidence (1) that *in vitro* administration of cannabis to striatal tissue does not affect the binding of [^{3}H]-quinuclidinyl benzylate to muscarinic sites,[131] (2) that SP-111 does not alter the excitatory action of ACh on hippocampal neurons,[147] (3) that cannabinoids do not increase the activity of GABA- or opioid-releasing systems by enhancing the affinity of GABA or opioid receptors,[1,153] and (4) that cannabinoids do not augment the efficacy of ACh, opioids, or eicosanoids by increasing the ability of these mediators to stimulate cyclic AMP production.[1,150-153] Indeed, in some experiments, cannabinoids have been shown to inhibit the binding of noradrenergic, dopaminergic and opioid ligands to brain tissue and to decrease the ability of certain chemical mediators to stimulate cyclic AMP production.[1,150-152]

Final Comments

The drug interaction studies described in this chapter provide further support for the notion that the central effects of cannabinoids are mediated by several neurochemical mediators, including NE, DA, 5-HT, GABA, histamine, opioid peptides, and PGs (see "Introduction"). More importantly, they give some indication as to the identity of the chemical mediators responsible for producing four characteristic effects of psychotropic cannabinoids: catalepsy, hypokinesia, hypothermia, and antinociception. Although further studies are still required to establish the precise relationship between the cannabinoids and these mediators, it seems likely from the data already available that in some areas of the CNS at least, this relationship is a direct one. The cannabinoids may act either to alter the synaptic concentrations of these mediators through an effect on their synthesis, release, or fate, or to modulate mediator-receptor interactions. It is possible that such modulation is brought about allosterically via the putative cannabinoid receptor, a mechanism that might well confer some degree of selectivity since the cannabinoids would only modulate the actions of transmitters in brain areas in which there is a functional link between the two types of receptor and in which there is on-going mediator release. While this modulation could take place postsynaptically, it might also involve presynaptically located autoreceptors, when it could give rise to an alteration in the rate of transmitter release. In some cases, the involvement of a particular chemical mediator is no doubt somewhat less direct, reflecting for example its role as a transmitter at synapses located somewhere downstream of the cannabinoids' site(s) of action.

It is quite possible that psychotropic cannabinoids alter the activities of at least some neurochemical systems in the CNS both by changing the concentration of a chemical mediator in the synapse, through an effect on its synthesis, release, and/or fate, and by modulating the interaction of the same mediator with its receptors. This idea, that psychotropic cannabinoids have more than one action on a particular system, can be developed into a hypothesis that reconciles the observation that some nonpsychotropic cannabinoids are no less effective than psychotropic cannabinoids as stimulators of PG release (see review by Pertwee[1]) with the evidence that PGs mediate certain effects produced only by cannabinoids that possess psychotropic activity (see the section on "Drug Interactions Involving Eicosanoids"). Thus, it could well be that the production of such effects by psychotropic cannabinoids depends at least as much on their ability to potentiate the actions of PGs (a property so far only detected in cannabinoids with psychotropic activity) as on their ability to stimulate PG release.

One feature common to many of the interactions described in this chapter is that all the drugs under investigation were administered peripherally. As already discussed, it is necessary when interpreting data generated by such experiments to consider whether or not the observed interactions are metabolic/dispositional in nature (see section on "Interpretation of Data"). There is also a risk that experiments of this kind fail to detect important interactions or yield results that are difficult to interpret correctly because they involve drug-induced changes in the extracellular concentrations and/or actions of some neurochemical mediator that has a dual effect on a particular response, augmenting it when released at one site but opposing it when released at another (e.g., see section on "Motor Function"). One approach that minimizes these problems and that has already been successfully adopted by some investigators is to administer drugs directly into discrete areas of the nervous system either *in vivo* or *in vitro*.

The finding that the ability of cannabinoids to produce hypokinesia or catalepsy can be markedly enhanced by drugs that decrease dopaminergic transmission or augment cholinergic or GABAergic transmission could eventually be exploited clinically; for example, in the treatment of dystonias and hyperkinetic motor disorders (see reviews by Moss et al.[154] and Pertwee[1]). Cannabinoids can also interact synergistically with opioids (see section on "Drug Interactions with Opioids") and other analgesics,[1] and it is possible that this synergism might also be exploited to clinical advantage. More specifically, it is possible that if given together with certain conventional therapeutic agents, cannabinoids will reduce the dose levels of the conventional agents that are required to alleviate pain or to treat particular motor disorders. The concurrent administration of a cannabinoid with a second drug would only be advantageous if it led to an increase in efficacy, if the cannabinoid acted selectively, enhancing the ability of the other drug to produce its sought-after effects without also enhancing its ability to produce some or all of its adverse effects, and if, at the dose used, the cannabinoid itself did not give rise to a new set of significant adverse effects. It is important that further experiments be carried out with cannabinoids to explore their use as adjuncts in drug therapy.

REFERENCES

1. **Pertwee, R. G.,** The central neuropharmacology of psychotropic cannabinoids, in *Psychotropic Drugs of Abuse*, Balfour, D. J. K., Ed., Pergamon Press, New York., 1990, 355.
2. **Martin, B. R., Compton, D. R., Thomas, B. F., Prescott, W. R., Little, P. J., Razdan, R. K., Johnson, M. R., Melvin, L. S., Mechoulam, R., and Ward, S. J.,** Behavioral, biochemical, and molecular modeling evaluations of cannabinoid analogs, *Pharmacol. Biochem. Behav.,* 40, 471, 1991.
3. **Herkenham, M., Lynn, A. B., Little, M. D., Johnson, M. R., Melvin, L. S., de Costa, B. R., and Rice, K. C.,** Cannabinoid receptor localization in brain, *Proc. Natl. Acad. Sci. U.S.A.,* 87, 1932, 1990.
4. **Herkenham, M., Lynn, A. B., Johnson, M. R., Melvin, L. S., de Costa, B. R., and Rice, K. C.,** Characterization and localization of cannabinoid receptors in rat brain: a quantitative *in vitro* autoradiographic study, *J. Neurosci.,* 11, 563, 1991.
5. **Herkenham, M., Lynn, A. B., de Costa, B. R., and Richfield, E. K.,** Neuronal localization of cannabinoid receptors in the basal ganglia of the rat, *Brain Res.,* 547, 267, 1991.
6. **Gough, A. L. and Olley, J. E.,** Catalepsy induced by intrastriatal injections of delta-9-THC and 11-hydroxy-delta-9-THC in the rat, *Neuropharmacology,* 17, 137, 1978.
7. **Montgomery, S. P., Moss, D. E., and Manderscheid, P. Z.,** Tetrahydrocannabinol and levonantradol effects on extrapyramidal motor behaviors: neuroanatomical location and hypothesis of mechanism, in *Marihuana '84,* Harvey, D. J., Ed., IRL Press, Oxford, 1985, 295.
8. **Pertwee, R. G. and Wickens, A. P.,** Enhancement by chlordiazepoxide of catalepsy induced in rats by intravenous or intrapallidal injections of enantiomeric cannabinoids, *Neuropharmacology,* 30, 237, 1991.
9. **Pertwee, R. G.,** The ring test: a quantitative method for assessing the "cataleptic" effect of cannabis in mice, *Br. J. Pharmacol.,* 46, 753, 1972.
10. **Costall, B. and Olley, J. E.,** Cholinergic- and neuroleptic-induced catalepsy: modification by lesions in the caudate-putamen, *Neuropharmacology,* 10, 297, 1971.
11. **Gough, A. L. and Olley, J. E.,** Delta-9-tetrahydrocannabinol and the extrapyramidal system, *Psychopharmacology,* 54, 87, 1977.
12. **Ghosh, P., Bose, R., and Bhattacharya, S. K.,** Cannabis-induced catalepsy in mice—role of putative neurotransmitters, *Ind. J. Med. Res., 72,* 605, 1980.

13. **Moss, D. E., McMaster, S. B., and Rogers, J.,** Tetrahydrocannabinol potentiates reserpine-induced hypokinesia, *Pharmacol. Biochem. Behav.,* 15, 779, 1981.
14. **Moss, D. E., Montgomery, S. P., Salo, A. A., and Steger, R. W.,** Tetrahydrocannabinol effects on extrapyramidal motor behaviors in an animal model of Parkinson's disease, in *The Cannabinoids: Chemical, Pharmacologic and Therapeutic Aspects,* Agurell, S., Dewey, W. L., and Willette, R. E., Eds., Academic Press, Orlando, FL, 1984, 815.
15. **Gao, X.-K. and Song, B.,** Comparison between the catalepsies induced by delta-9-tetrahydrocannabinol and haloperidol in rats, *Acta Pharmacol. Sinica,* 8, 203, 1987.
16. **Ferri, S., Costa, G., Murari, G., Panico, A. M., Rapisarda, E., Speroni, E., and Arrigo-Reina, R.,** Investigations on behavioral effects of an extract of *Cannabis sativa* L in the rat, *Psychopharmacology,* 75, 144, 1981.
17. **Kataoka, Y., Ohta, H., Fujiwara, M., Oishi, R., and Ueki, S.,** Noradrenergic involvement in catalepsy induced by delta-9-tetrahydrocannabinol, *Neuropharmacology,* 26, 55, 1987.
18. **Pertwee, R. G., Greentree, S. G., and Swift, P. A.,** Drugs which stimulate or facilitate central GABAergic transmission interact synergistically with delta-9-tetrahydrocannabinol to produce marked catalepsy in mice, *Neuropharmacology,* 27, 1265, 1988.
19. **Moss, D. E., Manderscheid, P. Z., Kobayashi, H., and Montgomery, S. P.,** Evidence for the nicotinic cholinergic hypothesis of cannabinoid action within the central nervous system: extrapyramidal motor behaviors, in *Marijuana: An International Research Report, Monograph Series No. 7,* Chesher, G., Consroe, P., and Musty, R., Eds., Australian Government Publishing Service, Canberra, Australia, 1988, 359.
20. **Lichtman, A. H. and Martin, B. R.,** Cannabinoid-induced antinociception is mediated by a spinal α_2-noradrenergic mechanism, *Brain Res.,* 559, 309, 1991.
21. **Grunfeld, Y. and Edery, H.,** Psychopharmacological activity of the active constituents of hashish and some related cannabinoids, *Psychopharmacologia,* 14, 200, 1969.
22. **Bhattacharya, S. K. and Ghosh, P.,** Effect of restraint stress on cannabis-induced catalepsy in rats, *Ind. J. Physiol. Pharmacol.,* 26, 162, 1982.
23. **Pertwee, R. G. and Ross, T. M.,** Drugs which stimulate or facilitate central cholinergic transmission interact synergistically with delta-9-tetrahydrocannabinol to produce marked catalepsy in mice, *Neuropharmacology,* 30, 67, 1991.
24. **Freemon, F. R., Rosenblatt, J. E., and El-Yousef, M. K.,** Interaction of physostigmine and delta-9-tetrahydrocannabinol in man, *Clin. Pharmacol. Ther.,* 17, 121, 1975.
25. **Pertwee, R. G. and Greentree, S. G.,** Delta-9-tetrahydrocannabinol-induced catalepsy in mice is enhanced by pretreatment with flurazepam or chlordiazepoxide, *Neuropharmacology,* 27, 485, 1988.
26. **Burstein, S., Hunter, S. A., Latham, V., and Renzulli, L.,** Prostaglandin E_2 mediation of THC-induced catalepsy in the mouse, in *Marijuana: An International Research Report, Monograph Series No. 7,* Chesher, G., Consroe, P., and Musty, R., Eds., Australian Government Publishing Service, Canberra, Australia, 1988, 371.
27. **Bhattacharya, S. K., Sen, A. P., Mohan Rao, P. J. R., and Dasgupta, G.,** Role of prostaglandins in some neuropharmacological actions of delta-9-tetrahydrocannabinol (THC) in the rat, *Asia Pacific J. Pharmacol.,* 4, 179, 1989.
28. **Fairbairn, J. W. and Pickens, J. T.,** The oral activity of delta-1-tetrahydrocannabinol and its dependence on prostaglandin E_2, *Br. J. Pharmacol.,* 67, 379, 1979.
29. **Burstein, S., Hunter, S. A., Latham, V., and Renzulli, L.,** A major metabolite of delta-1-tetrahydrocannabinol reduces its cataleptic effect in mice, *Experientia,* 43, 402, 1987.

30. **Bhattacharya, S. K., Ghosh, P., and Sanyal, A. K.,** Effects of prostaglandins on some central actions of cannabis, *Ind. J. Med. Res.,* 71, 955, 1980.
31. **Burstein, S. H., Hull, K., Hunter, S. A., and Shilstone, J.,** Immunization against prostaglandins reduces delta-1-tetrahydrocannabinol-induced catalepsy in mice, *Mol. Pharmacol.,* 35, 6, 1989.
32. **Fairbairn, J. W. and Pickens, J. T.,** The effect of conditions influencing endogenous prostaglandins on the activity of delta-1-tetrahydrocannabinol in mice, *Br. J. Pharmacol.,* 69, 491, 1980.
33. **Hunter, S. A., Latham, V., Renzulli, L., and Burstein, S.,** *In vivo* effects of delta-1-tetrahydrocannabinol (delta-1-THC) on rat and mouse prostaglandin tissue levels, in *Marijuana: An International Research Report, Monograph Series No. 7,* Chesher, G., Consroe, P., and Musty, R., Eds., Australian Government Publishing Service, Canberra, Australia, 1988, 399.
34. **Burstein, S., Hunter, S. A., and Ozman, K.,** Prostaglandins and cannabis. XII. The effect of cannabinoid structure on the synthesis of prostaglandins by human lung fibroblasts, *Mol. Pharmacol.,* 23, 121, 1983.
35. **Burstein, S., Hunter, S. A., Ozman, K., and Renzulli, L.,** *In vitro* models of cannabinoid-induced psychoactivity, in *Marihuana '84,* Harvey, D. J., Ed., IRL Press, Oxford, 1985, 399.
36. **Evans, A. T., Formukong, E., and Evans, F. J.,** Activation of phospholipase A_2 by cannabinoids. Lack of correlation with CNS effects, *FEBS Lett.,* 211, 119, 1987.
37. **White, H. L. and Tansik, R. L.,** Effects of delta-9-tetrahydrocannabinol and cannabidiol on phospholipase and other enzymes regulating arachidonate metabolism, *Prostaglandins Med.,* 4, 409, 1980.
38. **Reichman, M., Nen, W., and Hokin, L. E.,** Effects of delta-9-tetrahydrocannabinol on prostaglandin formation in brain, *Mol. Pharmacol.,* 32, 686, 1987.
39. **Consroe, P. F., Jones, B. C., and Akins, F.,** Delta-9-tetrahydrocannabinol-methamphetamine interaction in the rabbit, *Neuropharmacology,* 14, 377, 1975.
40. **Consroe, P., Jones, B., and Laird, H.,** EEG and behavioral effects of delta-9-tetrahydrocannabinol in combination with stimulant drugs in rabbits, *Psychopharmacology,* 50, 47, 1976.
41. **Pryor, G. T., Larsen, F. F., Husain, S., and Braude, M. C.,** Interactions of delta-9-tetrahydrocannabinol with *d*-amphetamine, cocaine, and nicotine in rats, *Pharmacol. Biochem. Behav.,* 8, 295, 1978.
42. **Jones, B. C., Consroe, P. F., and Akins, F.,** Physostigmine-induced reversal of EEG and behavioral effects of delta-9-tetrahydrocannabinol, *Bull. Psychon. Soc.,* 6, 204, 1975.
43. **Jones, B. C., Consroe, P. F., and Laird, H. E.,** The interaction of delta-9-tetrahydrocannabinol with cholinomimetic drugs in an agonist-antagonist paradigm, *Eur. J. Pharmacol.,* 38, 253, 1976.
44. **Pryor, G. T., Larsen, F. F., Carr, J. D., and Braude, M. C.,** Interactions of delta-9-tetrahydrocannabinol with phenobarbital, ethanol and chlordiazepoxide, *Pharmacol. Biochem. Behav.,* 7, 331, 1977.
45. **Tulunay, F. C., Ayhan, I. H., and Sparber, S. B.,** The effects of morphine and delta-9-tetrahydrocannabinol on motor activity in rats, *Psychopharmacology,* 78, 358, 1982.
46. **Pickens, J. T.,** Sedative activity of cannabis in relation to its delta-1-trans-tetrahydrocannabinol and cannabidiol content, *Br. J. Pharmacol.,* 72, 649, 1981.

47. **Loewe, S.,** Studies on the pharmacology of marihuana, in *The Marihuana Problems in the City of New York*, The Mayor's Committee on Marihuana, Jaques Cattell Press, Lancaster, PA, 1944, 149.
48. **Lew, E. O. H. and Richardson, J. S.,** Neurochemical and behavioural correlates of the interaction between amphetamine and delta-9-tetrahydrocannabinol in the rat, *Drug Alcohol Dep.*, 8, 93, 1981.
49. **Sakurai, Y., Ohta, H., Shimazoe, T., Kataoka, Y., Fujiwara, M., and Ueki, S.,** Delta-9-tetrahydrocannabinol elicited ipsilateral circling behavior in rats with unilateral nigral lesions, *Life Sci.*, 37, 2181, 1985.
50. **Hine, B., Friedman, E., Torrelio, M., and Gershon, S.,** Tetrahydrocannabinol-attenuated abstinence and induced rotation in morphine-dependent rats: possible involvement of dopamine, *Neuropharmacology*, 14, 607, 1975.
51. **Duncan, G. E. and Dagirmanjian, R.,** Delta-9-tetrahydrocannabinol sensitization of the rat brain to direct cholinergic stimulation, *Psychopharmacology*, 60, 237, 1979.
52. **Sabelli, H. C., Pedemonte, W. A., Whalley, C., Mosnaim, A. D., and Vazquez, A. J.,** Further evidence for a role of 2-phenylethylamine in the mode of action of delta-9-tetrahydrocannabinol, *Life Sci.*, 14, 149, 1974.
53. **Sabelli, H. C., Vazquez, A. J., Mosnaim, A. D., and Madrid-Pedemonte, L.,** 2-Phenylethylamine as a possible mediator for delta-9-tetrahydrocannabinol-induced stimulation, *Nature (London)*, 248, 144, 1974.
54. **Carder, B. and Deikel, S. M.,** Similarities between delta-9-tetrahydrocannabinol (delta-9-THC) and reserpine-like drugs, *Behav. Biol.*, 17, 313, 1976.
55. **Conti, L. H. and Musty, R. E.,** The effects of delta-9-tetrahydrocannabinol injections to the nucleus accumbens on the locomotor activity of rats, in *The Cannabinoids: Chemical, Pharmacologic and Therapeutic Aspects*, Agurell, S., Dewey, W. L., and Willette, R. E., Eds., Academic Press, Orlando, FL, 1984, 649.
56. **Taylor, D. A., Sitaram, B. R., and Elliot-Baker, S.,** Effect of delta-9-tetrahydrocannabinol on the release of dopamine in the corpus striatum of the rat, in *Marijuana: An International Research Report, Monograph Series No. 7*, Chesher, G., Consroe, P., and Musty, R., Eds., Australian Government Publishing Service, Canberra, Australia, 1988, 405.
57. **Chen, J., Paredes, W., Li, J., Smith, D., Lowinson, J., and Gardner, E. L.,** Delta-9-tetrahydrocannabinol produces naloxone-blockable enhancement of presynaptic basal dopamine efflux in nucleus accumbens of conscious, freely-moving rats as measured by intracerebral microdialysis, *Psychopharmacology*, 102, 156, 1990.
58. **Scheel-Krüger, J.,** Dopamine-GABA interactions: evidence that GABA transmits, modulates and mediates dopaminergic functions in the basal ganglia and the limbic system, *Acta Neurol. Scand.* 73(Suppl. 107), 1, 1986.
59. **Pertwee, R. G.,** Effects of cannabinoids on thermoregulation: a brief review, in *Marihuana '84*, Harvey, D. J., Ed., IRL Press, Oxford, 1985, 263.
60. **Fitton, A. G. and Pertwee, R. G.,** Changes in body temperature and oxygen consumption rate of conscious mice produced by intrahypothalamic and intracerebroventricular injections of delta-9-tetrahydrocannabinol, *Br. J. Pharmacol.*, 75, 409, 1982.
61. **Haavik, C. O., Hardman, H. F., Collins, F. G., and Skibba, J.,** The effect of delta-9-tetrahydrocannabinol on oxygen consumption, body temperature, and heat loss, *Res. Commun. Subst. Abuse*, 1, 381, 1980.
62. **Pertwee, R. G. and Tavendale, R.,** Effects of delta-9-tetrahydrocannabinol on the rates of oxygen consumption of mice, *Br. J. Pharmacol.*, 60, 559, 1977.

63. **Pertwee, R. G. and Tavendale, R.,** Effects of delta-9-tetrahydrocannabinol, 2,4-dinitrophenol and pentolinium tartrate on behavioural thermoregulation in mice, *Br. J. Pharmacol.,* 66, 39, 1979.
64. **Pertwee, R. G., Moochhala, S., Nyemitei-Addo, I., and Todd, D. A.,** The effects of intrahypothalamic injections of delta-9-tetrahydrocannabinol, dopamine and prostaglandin E_2 on behavioural thermoregulation in mice, in *Environment, Drugs and Thermoregulation,* Lomax, P. and Schönbaum, E., Eds., Karger, Basel, 1983, 82.
65. **Snyder, S. H. and Bennett, J. P.,** Neurotransmitter receptors in the brain: biochemical identification, *Annu. Rev. Physiol.,* 38, 153, 1976.
66. **Singh, P. P. and Das, P. K.,** Role of catecholamines in the hypothermic activity of cannabis in albino rats, *Psychopharmacology,* 50, 199, 1976.
67. **Schmeling, W. T. and Hosko, M. J.,** Blockade of delta-9-THC induced hypothermia in rats, *Arch. Int. Pharmacodyn.,* 227, 302, 1977.
68. **Haavik, C. O. and Hardman, H. F.,** An investigation of the hypothermic action of delta-9-tetrahydrocannabinol by use of pharmacological blocking agents, *Res. Commun. Subst. Abuse,* 1, 353, 1980.
69. **Haavik, C. O., Collins, F. G., and Hardman, H. F.,** Studies on the mechanism of hypothermic action of tetrahydrocannabinols, in *Temperature Regulation and Drug Action,* Lomax, P., Schönbaum, E., and Jacob, J., Eds., Karger, Basel, 1975, 293.
70. **Pertwee, R. G., Hedley, D., McQueen, A. S. M., and Gentleman, S. M.,** The hypothermic response of mice to delta-9-tetrahydrocannabinol is enhanced by chlorpromazine, thioxanthenes, alpha-adrenoceptor antagonists and pentolinium but not by SCH 23390 or sulpiride, *Neuropharmacology,* 27, 149, 1988.
71. **Gray, G. A., Hedley, D., and Pertwee, R. G.,** Enhancement of the hypothermic response of mice to delta-9-tetrahydrocannabinol by subhypothermic doses of chlorpromazine and phentolamine, *Neuropharmacology,* 26, 229, 1987.
72. **Davies, J. A. and Graham, J. D. P.,** The mechanism of action of delta-9-tetrahydrocannabinol on body temperature in mice, *Psychopharmacology,* 69, 299, 1980.
73. **Pertwee, R. G. and Hunter-Addey, P.,** unpublished results, 1991.
74. **Fitton, A. G.,** The Effects of Delta-9-Tetrahydrocannabinol on Thermoregulation in the Mouse, Ph.D. thesis, University of Aberdeen, 1983.
75. **Fennessy, M. R. and Taylor, D. A.,** Antagonism of the effects on thermoregulation of delta-9-tetrahydrocannabinol by clomipramine in the rat, *Br. J. Pharmacol.,* 63, 267, 1978.
76. **Hedley, D., Pertwee, R. G., and Sutherland, J. F.,** Involvement of acetylcholine in the hypothermic response of mice to delta-9-tetrahydrocannabinol, *Br. J. Pharmacol.,* 83, 423, 1984.
77. **Watanabe, K., Yamamoto, I., Narimatsu, S., and Yoshimura, H.,** Is serotonin involved in hypothermic effect and its tolerance development of delta-8-tetrahydrocannabinol and 11-hydroxy-delta-8-tetrahydrocannabinol in the mouse?, *Res. Commun. Subst. Abuse,* 5, 89, 1984.
78. **Tavendale, R.,** The Effects of Delta-9-Tetrahydrocannabinol on Temperature Regulation in Mice, Ph.D. thesis, University of Aberdeen, 1979.
79. **Fennessy, M. R., Lewis, S. J., Taylor, D. A., and Verberne, A. J. M.,** Histaminergic and serotonergic interaction in the delta-9-THC induced hypothermia of the rat, *Br. J. Pharmacol.,* 74, 862, 1981.
80. **Bradley, J. V.,** *Distribution-Free Statistical Tests,* Prentice-Hall, Englewood Cliffs, NJ, 1968.

81. **Johansson, J. O., Järbe, T. U. C., and Henriksson, B. G.,** Physostigmine attenuation of delta-9-tetrahydrocannabinol induced hyperthermia in rats, *Experientia,* 30, 779, 1974.
82. **Pertwee, R. G., Nash, K., and Trayhurn, P.,** Evidence that the hypothermic response of mice to delta-9-tetrahydrocannabinol is not mediated by changes in thermogenesis in brown adipose tissue, *Can. J. Physiol. Pharmacol.,* 69, 767, 1991.
83. **Ryrfeldt, A., Ramsay, C. H., Nilsson, I. M., Widman, M., and Agurell, S.,** Whole-body autoradiography of delta-1-tetrahydrocannabinol and delta-1(6)-tetrahydrocannabinol in mouse, *Acta Pharm. Suecica,* 10, 13, 1973.
84. **Pertwee, R. G., Browne, S., Ross, T. M., and Stretton, D.,** An investigation of the involvement of GABA in certain pharmacological effects of delta-9-tetrahydrocannabinol, *Pharmacol. Biochem. Behav.,* 40, 581, 1991.
85. **Morrison, K. J. and Pertwee, R. G.,** A possible link between delta-9-tetrahydrocannabinol and GABA in mice, *Br. J. Pharmacol.,* 86, 684, 1985.
86. **Pertwee, R. G. and Ross, T. M.,** unpublished results, 1991.
87. **Bloom, A. S. and Dewey, W. L.,** A comparison of some pharmacological actions of morphine and delta-9-tetrahydrocannabinol in the mouse, *Psychopharmacology,* 57, 243, 1978.
88. **Tulunay, F. C., Ayhan, I. H., Portoghese, P. S., and Takemori, A. E.,** Antagonism by chlornaltrexamine of some effects of delta-9-tetrahydrocannabinol in rats, *Eur. J. Pharmacol.,* 70, 219, 1981.
89. **Bruinvels, J.,** Norepinephrine, in *Body Temperature. Regulation, Drug Effects, and Therapeutic Implications,* Lomax, P. and Schönbaum, E., Eds., Marcel Dekker, New York, 1979, 257.
90. **Cox, B.,** Dopamine, in *Body Temperature. Regulation, Drug Effects, and Therapeutic Implications,* Lomax, P. and Schönbaum, E., Eds., Marcel Dekker, New York, 1979, 231.
91. **Jacob, J. J. and Girault, J.-M. T.,** 5-Hydroxytryptamine, in *Body Temperature. Regulation, Drug Effects, and Therapeutic Implications,* Lomax, P. and Schönbaum, E., Eds., Marcel Dekker, New York, 1979, 183.
92. **Crawshaw, L. I.,** Acetylcholine, in *Body Temperature. Regulation, Drug Effects, and Therapeutic Implications,* Lomax, P. and Schönbaum, E., Eds., Marcel Dekker, New York, 1979, 305.
93. **Bligh, J.,** Amino acids as central synaptic transmitters or modulators in mammalian thermoregulation, *Fed. Proc.,* 40, 2746, 1981.
94. **Lomax, P. and Green, M. D.,** Histamine, in *Body Temperature. Regulation, Drug Effects, and Therapeutic Implications,* Lomax, P. and Schönbaum, E., Eds., Marcel Dekker, New York, 1979, 289.
95. **Adler, M. W., Geller, E. B., Rosow, C. E., and Cochin, J.,** The opioid system and temperature regulation, *Annu. Rev. Pharmacol. Toxicol.,* 28, 429, 1988.
96. **Martin, B. R.,** Characterization of the antinociceptive activity of intravenously administered delta-9-tetrahydrocannabinol in mice, in *Marihuana '84,* Harvey, D. J., Ed., IRL Press, Oxford, 1985, 685.
97. **Gilbert, P. E.,** A comparison of THC, nantradol, nabilone, and morphine in the chronic spinal dog, *J. Clin. Pharmacol.,* 21(Suppl.), 311, 1981.
98. **Yaksh, T. L.,** The antinociceptive effects of intrathecally administered levonantradol and desacetyllevonantradol in the rat, *J. Clin. Pharmacol.,* 21(Suppl.), 334, 1981.

99. **Welch, S. P.,** Interactions of morphine and cannabinoids in the production of antinociception at the spinal level, paper presented at 4th Meeting Int. Cannabinoid Study Group, Palm Beach, FL, 1991.

100. **Lichtman, A. H., and Martin, B. R.,** Pharmacological effects of CP- 55, 940 injected into periaqueductal gray (PAG) in rats, paper presented at 4th Meeting Int. Cannabinoid Study Group, Palm Beach, FL, 1991.

101. **Lichtman, A. H. and Martin, B. R.,** Spinal and supraspinal components of cannabinoid-induced antinociception, *J. Pharmacol. Exp. Ther.,* 258, 517, 1991.

102. **Smith, P. B., and Martin, B. R.,** 4th Meeting Int. Cannabinoid Study Group, Palm Beach, FL, 1991.

103. **Smith, P. B., Lichtman, A. H., and Martin, B. R.,** Comparison of lipophilicity and antinociceptive effect of intrathecally administered cannabinoids, paper presented at 53rd Annu. Scientific Meeting of the Committee on Problems of Drug Dependence, Palm Beach, FL, 1991.

104. **Jacob, J. J., Ramabadran, K., and Campos-Madeiros, M.,** A pharmacological analysis of levonantradol antinociception in mice, *J. Clin. Pharmacol.,* 21(Suppl.), 327, 1981.

105. **Milne, G. M. and Johnson, M. R.,** Levonantradol: a role for central prostanoid mechanisms?, *J. Clin. Pharmacol.,* 21(Suppl.), 367, 1981.

106. **Ghosh, P. and Bhattacharya, S. K.,** Role of putative neurotransmitters in cannabis-induced analgesia in rats, *Ind. J. Med. Res.,* 72, 449, 1980.

107. **Ghosh, P. and Bhattacharya, S. K.,** Cannabis-induced potentiation of morphine analgesia in rat—role of brain monoamines, *Ind. J. Med. Res.,* 70, 275, 1979.

108. **Bhattacharya, S. K. and Ghosh, P.,** Effect of restraint stress on cannabis-induced analgesia in rats, *Ind. J. Exp. Biol.,* 20, 788, 1982.

109. **Ferri, S., Cavicchini, E., Romualdi, P., Speroni, E., and Murari, G.,** Possible mediation of catecholaminergic pathways in the antinociceptive effect of an extract of *Cannabis sativa* L, *Psychopharmacology,* 89, 244, 1986.

110. **Chesher, G. B., Dahl, C. J., Everingham, M., Jackson, D. M., Marchant-Williams, H., and Starmer, G. A.,** The effect of cannabinoids on intestinal motility and their antinociceptive effect in mice, *Br. J. Pharmacol.,* 49, 588, 1973.

111. **Sanders, J., Jackson, D. M., and Starmer, G. A.,** Interactions among the cannabinoids in the antagonism of the abdominal constriction response in the mouse, *Psychopharmacology,* 61, 281, 1979.

112. **Bhargava, H. N. and Matwyshyn, G. A.,** Influence of thyrotropin releasing hormone and histidyl-proline diketopiperazine on spontaneous locomotor activity and analgesia induced by delta-9-tetrahydrocannabinol in the mouse, *Eur. J. Pharmacol.,* 68, 147, 1980.

113. **Bloom, A. S., Dewey, W. L., Harris, L. S., and Brosius, K. K.,** 9-nor-9 β-hydroxyhexahydrocannabinol, a cannabinoid with potent antinociceptive activity: comparisons with morphine, *J. Pharmacol. Exp. Ther.,* 200, 263, 1977.

114. **Welch, S. P. and Stevens, D. L.,** Antinociceptive activity of intrathecally administered cannabinoids alone, and in combination with morphine, in mice, *J. Pharmacol. Exp. Ther.,* submitted.

115. **Wilson, R. S. and May, E. L.,** Analgesic properties of the tetrahydrocannabinols, their metabolites and analogs, *J. Med. Chem.,* 18, 700, 1975.

116. **Kaymakcalan, S. and Deneau, G. A.,** Some pharmacologic properties of synthetic delta-9-tetrahydrocannabinol, *Acta Medica Turcica,* Suppl. 1, 5, 1972.

117. **Little, P. J., Compton, D. R., Johnson, M. R., Melvin, L. S., and Martin, B. R.,** Pharmacology and stereoselectivity of structurally novel cannabinoids in mice., *J. Pharmacol. Exp. Ther.,* 247, 1046, 1988.
118. **Howlett, A. C.,** Reverse pharmacology applied to the cannabinoid receptor, *Trends Pharmacol. Sci.,* 11, 395, 1990.
119. **Burstein, S. H., Hull, K., Hunter, S. A., and Latham, V.,** Cannabinoids and pain response: a possible role for prostaglandins, *FASEB J.,* 2, 3022, 1988.
120. **Besson, J.-M. and Chaouch, A.,** Peripheral and spinal mechanisms of nociception, *Physiol. Rev.,* 67, 67, 1987.
121. **Bhattacharya, S. K.,** The antinociceptive effect of intracerebroventricularly administered prostaglandin D_2 in the rat, *Psychopharmacology,* 89, 121, 1986.
122. **Green, P. G. and Kitchen, I.,** Antinociception opioids and the cholinergic system, *Prog. Neurobiol.,* 26, 119, 1986.
123. **Sanyal, A. K., Srivastava, D. N., and Bhattacharya, S. K.,** The antinociceptive effect of intracerebroventricularly administered prostaglandin E_1 in the rat, *Psychopharmacology,* 60, 159, 1979.
124. **Sawynok, J.,** GABAergic mechanisms of analgesia: an update, *Pharmacol. Biochem. Behav.,* 26, 463, 1987.
125. **Ferreira, S. H. and Vane, J. R.,** New aspects of the mode of action of nonsteroid anti-inflammatory drugs, *Annu. Rev. Pharmacol.,* 14, 57, 1974.
126. **Paton, W. D. M. and Pertwee, R. G.,** Effect of cannabis and certain of its constituents on pentobarbitone sleeping time and phenazone metabolism, *Br. J. Pharmacol.,* 44, 250, 1972.
127. **Burke, M. D., Kilgour, C., and Pertwee, R. G.,** Potent but selective inhibition of cytochrome P-450 by cannabidiol, *Br. J. Pharmacol.,* 83, 445P, 1984.
128. **Bornheim, L. M. and Correia, M. A.,** Effect of cannabidiol on cytochrome P-450 isozymes, *Biochem. Pharmacol.,* 38, 2789, 1989.
129. **Bornheim, L. M. and Correia, M. A.,** Selective inactivation of mouse liver cytochrome P-450IIIA by cannabidiol, *Mol. Pharmacol.,* 38, 319, 1990.
130. **Narimatsu, S., Watanabe, K., Matsunaga, T., Yamamoto, I., Imaoka, S., Funae, Y., and Yoshimura, H.,** Inhibition of hepatic microsomal cytochrome P450 by cannabidiol in adult male rats, *Chem. Pharm. Bull.,* 38, 1365, 1990.
131. **Agrawal, A. K., Kumar, P., Gulati, A., and Seth, P. K.,** Cannabis-induced neurotoxicity in mice: effect on cholinergic (muscarinic) receptors and blood brain barrier permeability, *Res. Commun. Subst. Abuse,* 10, 155, 1989.
132. **Sofia, R. D. and Barry, H.,** Depressant effect of delta-1-tetrahydrocannabinol enhanced by inhibition of its metabolism, *Eur. J. Pharmacol.,* 13, 134, 1970.
133. **Sofia, R. D. and Barry, H.,** The effects of SKF 525-A on the analgesic and barbiturate-potentiating activity of delta-9-tetrahydrocannabinol in mice and rats, *Pharmacology,* 27, 223, 1983.
134. **Peterson, D. W., Cohen, G. M., and Sparber, S. B.,** The delay of the behavioural effects of delta-9-tetrahydrocannabinol in rats by 2-diethylaminoethyl 2,2-diphenylvalerate HCl (SKF 525-A), *Life Sci.,* 10, 1381, 1971.
135. **Christensen, H. D., Freudenthal, R. I., Gidley, J. T., Rosenfeld, R., Boegli, G., Testino, L., Brine, D. R., Pitt, C. G., and Wall, M. E.,** Activity of delta-8 and delta-9-tetrahydrocannabinol and related compounds in the mouse, *Science,* 172, 165, 1971.
136. **Gill, E. W., Jones, G., and Lawrence, D. K.,** Contribution of the metabolite 7-hydroxy-delta-1-tetrahydrocannabinol towards the pharmacological activity of delta-1-tetrahydrocannabinol in mice, *Biochem. Pharmacol.,* 22, 175, 1973.

137. **Watanabe, K., Kijima, T., Narimatsu, S., Nishikami, J., Yamamoto, I., and Yoshimura, H.,** Comparison of pharmacological effects of tetrahydrocannabinols and their 11-hydroxy-metabolites in mice, *Chem. Pharm. Bull.,* 38, 2317, 1990.

138. **Agurell, S., Halldin, M., Lindgren, J.-E., Ohlsson, A., Widman, M., Gillespie, H., and Hollister, L.,** Pharmacokinetics and metabolism of delta-1-tetrahydrocannabinol and other cannabinoids with emphasis on man, *Pharmacol. Rev.,* 38, 21, 1986.

139. **Burstein, S., Hunter, S. A., Latham, V., and Renzulli, L.,** Prostaglandins and cannabis. XVI. Antagonism of delta-1-tetrahydrocannabinol action by its metabolites, *Biochem. Pharmacol.,* 35, 2553, 1986.

140. **Perez-Reyes, M., Burstein, S. H., White, W. R., McDonald, S. A., and Hicks, R. E.,** Antagonism of marihuana effects by indomethacin in humans, *Life Sci.,* 48, 507, 1991.

141. **Chesher, G. B.,** Cannabis and the opiate receptor—any link?, *Proc. Austr. Physiol. Pharmacol. Soc.,* 11, 24, 1980.

142. **Milne, G. M., McIlhenny, H. M., Johnson, M. R., Koe, B. K., and Weissman, A.,** One way cross tolerance from morphine to nantradol—a potent, non-opioid analgetic, in *Endogenous and Exogenous Opiate Agonists and Antagonists,* Way, E. L., Ed., Pergamon Press, New York, 1980, 367.

143. **Hine, B.,** Morphine and delta-9-tetrahydrocannabinol: two-way cross tolerance for antinociceptive and heart rate responses in the rat, *Psychopharmacology,* 87, 34, 1985.

144. **Haefely, W.,** Tranquilizers, in *Psychopharmacology 2. Part 1: Preclinical Psychopharmacology,* Grahame-Smith, D. G. and Cowen, P. J., Eds., Elsevier, Amsterdam, 1985, 92.

145. **Ng Cheong Ton, J. M., Gerhardt, G. A., Friedemann, M., Etgen, A. M., Rose, G. M., Sharpless, N. S., and Gardner, E. L.,** The effects of delta-9-tetrahydrocannabinol on potassium-evoked release of dopamine in the rat caudate nucleus: an *in vivo* electrochemical and *in vivo* microdialysis study, *Brain Res.,* 451, 59, 1988.

146. **Chen, J., Paredes, W., Lowinson, J. H., and Gardner, E. L.,** Delta-9-tetrahydrocannabinol enhances presynaptic dopamine efflux in medial prefrontal cortex, *Eur. J. Pharmacol.,* 190, 259, 1990.

147. **Segal, M.,** The effects of SP-111, a water soluble THC derivative, on neuronal activity in the rat brain, *Brain Res.,* 139, 263, 1978.

148. **Koe, B. K. and Weissman, A.,** Facilitation of benzodiazepine binding by levonantradol, *J. Clin. Pharmacol.,* 21(Suppl.), 397, 1981.

149. **Koe, B. K., Milne, G. M., Weissman, A., Johnson, M. R., and Melvin, L. S.,** Enhancement of brain [^{3}H]flunitrazepam binding and analgesic activity of synthetic cannabimimetics, *Eur. J. Pharmacol.,* 109, 201, 1985.

150. **Kelly, L. A. and Butcher, R. W.,** The effects of delta-1-tetrahydrocannabinol on cyclic AMP levels in W1-38 fibroblasts, *Biochim. Biophys. Acta,* 320, 540, 1973.

151. **Kelly, L. A. and Butcher, R. W.,** Effects of delta-1-tetrahydrocannabinol on cyclic AMP in cultured human diploid fibroblasts, *J. Cyclic Nucleotide Res.,* 5, 303, 1979.

152. **Howlett, A. C. and Fleming, R. M.,** Cannabinoid inhibition of adenylate cyclase. Pharmacology of the response in neuroblastoma cell membranes, *Mol. Pharmacol.,* 26, 532, 1984.

153. **Devane, W. A., Spain, J. W., Coscia, C. J., and Howlett, A. C.,** An assessment of the role of opioid receptors in the response to cannabimimetic drugs, *J. Neurochem.,* 46, 1929, 1986.

154. **Moss, D. E., Manderscheid, P. Z., Montgomery, S. P., Norman, A. B., and Sanberg, P. R.,** Nicotine and cannabinoids as adjuncts to neuroleptics in the treatment of Tourette syndrome and other motor disorders., *Life Sci.,* 44, 1521, 1989.

7 Behavioral, Neurochemical, and Neurohistological Effects of Chronic Marijuana Smoke Exposure in the Nonhuman Primate

INTRODUCTION

This study was designed to determine whether chronic marijuana (MJ) smoke exposure results in residual behavioral, neuropathological, or neurochemical alterations in rhesus monkeys. The monkey was selected as the animal model because it can perform several complex behavioral tasks and is physiologically similar to humans. Prior to the initiation of a year of chronic MJ smoke exposure, 62 periadolescent male rhesus monkeys were trained for 1 year to perform 5 operant behavioral tasks and then divided, according to their performance in these tasks, into 4 exposure groups (n = 15 to 16/group): (1) a high dose (HI) group, exposed 7 d/week to the smoke of one standard MJ cigarette (2.6% THC); (2) a low dose (LO) group, exposed on weekend days only to the smoke of a standard MJ cigarette and to sham conditions on other days; (3) an extracted MJ cigarette (EX) group, exposed 7 d/week to the smoke of one ethanol-extracted MJ cigarette; and (4) a sham (SH) group, exposed 7 d/week to sham smoke conditions. Daily exposures to smoke for 1 year were accomplished using masks that covered the subject's nose and mouth. Average body weights (initially 3.7 ± 0.5 kg, mean ± SD) and rates of weight gain (approximately 0.1 kg/month) were the same for all groups throughout the entire experiment. Plasma levels of Δ^9-tetrahydrocannabinol (THC) and 11-nor-9-carboxy-THC (11-NOR) were monitored by radioimmunoassay. During the first week of exposure, plasma THC and 11-NOR concentrations in the HI group were 59 ± 7 and 5.5 ± 1.5 ng/ml, respectively, 45 min after MJ smoke administration. Similar plasma levels of both THC and 11-NOR were observed after subsequent exposures throughout the remainder of the year. Whole blood carboxyhemoglobin (COHb) levels increased to a maximum mean of 13% 1 min after exposure to smoke from either MJ or EX cigarettes, while increases in COHb were undetectable in sham animals. Comparison of blood chemistry values collected from the monkeys before, during, and after exposure indicated no differences between treatment groups for most parameters and only minor alterations in a few (<30%). These data indicate that long-term inhalation of MJ smoke does not compromise the general health of the rhesus monkey. This long-term, daily MJ smoke exposure served as a chronic physiological stressor, however, as evidenced by sustained increases in urinary free cortisol excretion. Behavioral assessment of the monkeys suggested that during daily exposure to MJ smoke, a decrease in "motivation" to respond occurred. This chronic MJ smoke exposure in the monkey failed to result in any neurochemical alterations 7 months after exposure. Neurohistological and electronmicroscopic studies of the monkeys 7 months after the cessation of MJ smoke exposure failed to show any differences in hippocampal volume, neuronal size, number or length of CA3 pyramidal cell dendrites, synaptic width, or density. Together, these data indicate that while behavioral and neuroendocrine effects are observed during MJ smoke exposure in the monkey, no residual neuropathological or neurochemical effects of MJ exposure were observed 7 months after the year-long MJ smoke exposure regimen.

PLASMA CANNABINOID AND BLOOD CARBOXYHEMOGLOBIN CONCENTRATIONS AND CLINICAL CHEMISTRIES

Only a few animal studies have been conducted to examine the possibility that chronic exposure to MJ smoke, extracts of MJ, or to Δ^9-THC, the major psychoactive component of MJ, produces residual neurological effects. Harper et al.[1] and Heath et al.[2] reported that two rhesus monkeys exposed chronically to either MJ smoke or THC showed evidence of neurophysiological alterations and ultrastructural changes in several brain regions including the hippocampus. Stiglick and Kalant[3-6] and Stiglick et al.[7] reported altered behavior in rats 1 to 6 months after a 3-month oral dosing regimen of MJ extract or THC. They concluded that the behavioral alterations were reminiscent of behavioral changes seen after damage to the hippocampus.[4,7] Scallet and co-workers[8] used quantitative neuropathological techniques to examine the brains of rats 7 to 8 months after 3 months of chronic oral exposure to THC. The CA3 (Cornu Ammonis subarea 3 pyramidal cells; Lorente de No[9]) area of hippocampus was noted to be markedly altered by such THC treatment, as indicated by reduced cytoplasmic and nuclear volumes and decreased synaptic density. Another group of rats similarly treated, but analyzed 2 months after the last treatment, demonstrated reduced lengths of distal CA3 pyramidal cell dendrites as measured by quantitative Golgi analysis.[8] A study by Landfield et al.[10] showed that chronic exposure to THC (given s.c.) reduced the number of nucleoli per unit length of CA1 stratum pyramidal in the rat hippocampus.

In humans, the most common route of exposure to THC is via the inhalation of MJ smoke.[11] Few animal studies of long-term MJ smoke exposure have been completed because of the difficulty in obtaining efficient smoke delivery to the subject and the concern over carbon monoxide toxicity. Therefore, most chronic animal studies have been conducted using THC or MJ extracts.[3,4,6,7] Appropriate controls, including those which mimic the carbon monoxide exposure experienced during the smoking of MJ, are necessary.

Experimental Methods

Cigarettes

The MJ cigarettes contained approximately 2.6% Δ^9-THC, and the EX cigarettes (controls for carbon monoxide exposure) were essentially devoid of cannabinoids and other ethanol-extractable materials.

Smoke Exposure

MJ and EX cigarettes were placed in ADL/II smoke generating machines (Arthur D. Little, Inc., Cambridge, MA) and 35-ml puffs of smoke were delivered to the animals via nose-and-mouth-only face masks (developed in-house) at a rate of three puffs per minute. Cigarettes were burned to approximately a 10-mm butt length (generally 19 to 21 total puffs over a period of 6 to 8 min).

Carboxyhemoglobin (COHb), THC, and 11-Nor Measurements

Blood samples (approximately 0.5 ml) were obtained approximately monthly from all subjects throughout the course of the study. Samples were collected by venipuncture using heparinized syringes just prior to and immediately after exposure (MJ, EX, and SH). Carboxyhemoglobin (COHb) concentrations were determined using an IL282 Co-Oximeter (Instrumentation Laboratory, Lexington, MA).

Blood samples (approximately 1.5 ml) for THC and 11-NOR analyses were collected as above, 45 min after exposure, at weeks 1, 2, 5, 6, 10, 16, 18, 22, 26, 30, 34, 38, 42, 46, 50, and 52. THC and 11-NOR plasma concentrations were determined by radioimmunoassay (RIA) as previously described using kits supplied by NIDA.[12,13]

Dose Selection

Maximal THC plasma concentrations occurring in monkeys immediately after exposure to the smoke generated from one MJ cigarette were approximately 800 to 1000 ng/ml and decreased rapidly to about 60 to 80 ng/ml by 45 min after dosing. These peak THC plasma concentrations are considerably higher than those observed in humans (150 to 200 ng/ml) after smoking one relatively low-potency MJ cigarette, and thus would correspond to the human inhalation of smoke from about four to five low to medium potency MJ cigarettes.[14]

Subjects

Periadolescent monkeys, 2 to 3 years old and weighing 3 to 4 kg (3.7 ± 0.5, mean ± SD) at the start of the study, were selected since, at the beginning of chronic exposure, they would be roughly equivalent in age to teenage humans, members of a high MJ use group.[11]

Experimental Groups

Each of four dose groups consisted of 15 to 16 animals. The groups were: (1) high dose (HI), subjects received smoke from one MJ cigarette per day, 7 d/week; (2) low dose (LO), subjects received smoke from one MJ cigarette per day, 2 d/week on weekends only plus a sham exposure once per day Monday through Friday; (3) extracted marijuana (EX), subjects received smoke from one ethanol-extracted MJ cigarette per day, 7 d/week; and (4) sham (SH), subjects were placed in the exposure situation only (nose/mouth-only mask held in place and smoking machine operating but generating no smoke) once per day, 7 d/week.

Clinical Chemistries and Hematologies

The following analytes were measured: alkaline phosphatase, total protein, cholesterol, γ-glutamyl transferase, aspartate aminotransferase, and alanine aminotransferase. Plasma proteins were separated by electrophoresis using cellulose acetate and Ponceau S-stained plates scanned by a Helena Laboratories Cliniscan Densitometer (Helena Laboratories, Beaumont, TX).

Statistical Methods

Weight gain, THC, and COHb concentrations: analysis of variance was used to test for the effects of smoke conditions on weight gain and THC, 11-NOR, and COHb concentrations. Blood chemistries and hematologies: an analysis of variance (ANOVA) was performed for each variable included in the data set. A separate analysis was performed for animals prior to dosing (PREDOSE), during dosing (DOSE), and following discontinuation of dosing (POSTDOSE). PREDOSE values were divided by the median value for "SH" animals for the entire experiment. DOSE values were divided by mean PREDOSE values for the two bleeding periods prior to dosing, and POSTDOSE values were divided by mean DOSE values for the last two bleeding periods of dosing. These ratios-to-baselines were then analyzed. Bonferroni *t*-tests were used to compare mean values of each dose level in cases where the ANOVA found a significant dose effect (5% significance level). In a separate analysis, individual *t*-tests with a Bonferroni-type correction were applied to each dose condition at each sample time.

Results and Discussion

Weight Gain

Weight gain during the exposure period (approximately 0.1 kg/month) was the same for all dose groups (Figure 1) and was about normal for rhesus monkeys of this age range.[15]

Plasma THC and 11-NOR Concentrations

Plasma THC and 11-NOR concentrations in the HI group collected 45 min after smoke exposure averaged approximately 60 and 5.5 ng/ml, respectively (Figures 2 and 3).

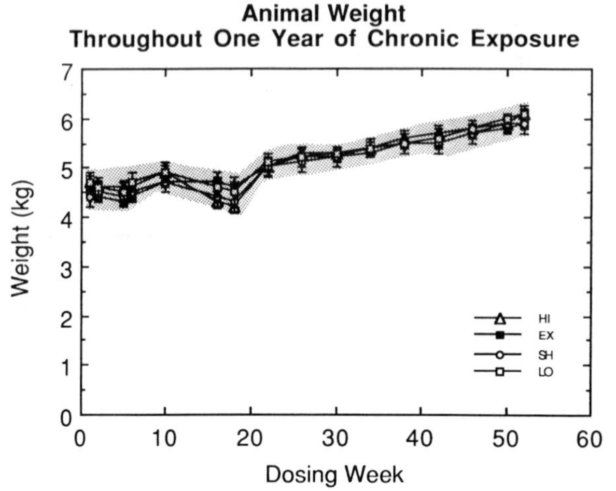

FIGURE 1. Mean body weights (kg) for monkeys in each dose group measured throughout the exposure period of 52 weeks.

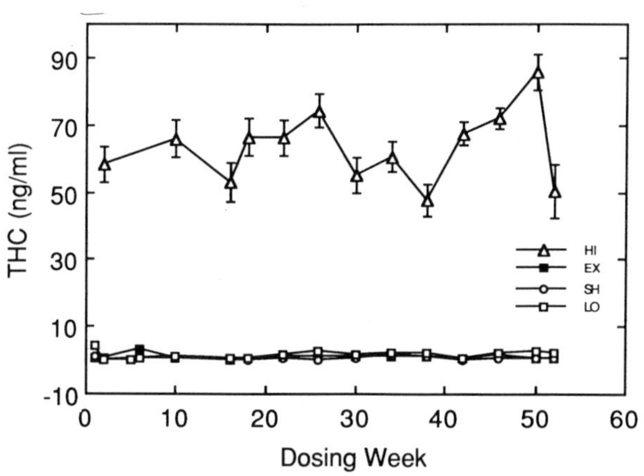

FIGURE 2. Plasma THC concentrations (ng/ml) 45 min after exposure (mean ± S.E. for each dose group). No THC would be expected in the LO group because the samples were drawn during the week while they were being sham exposed.

COHb Concentrations

Increases in % COHb in whole blood samples collected immediately after exposure averaged 13% and never exceeded 15% (Figure 4). The EX group had the same COHb values as the HI group.

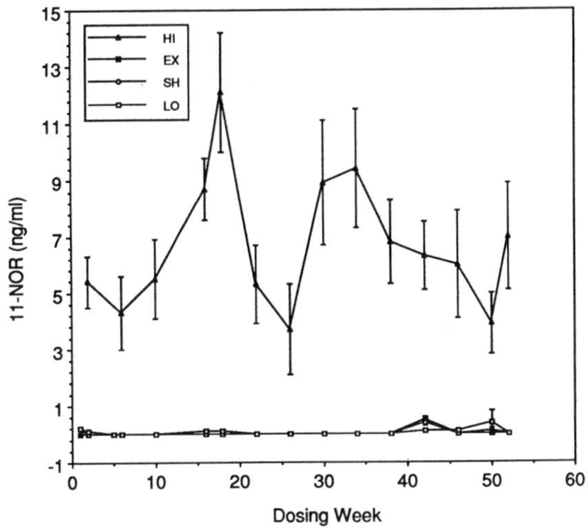

FIGURE 3. Plasma 11-NOR concentrations (ng/ml) 45 min after exposure (mean ± S.E. for each dose group). No 11-NOR would be expected in the LO group because the samples were drawn during the week while they were being sham exposed.

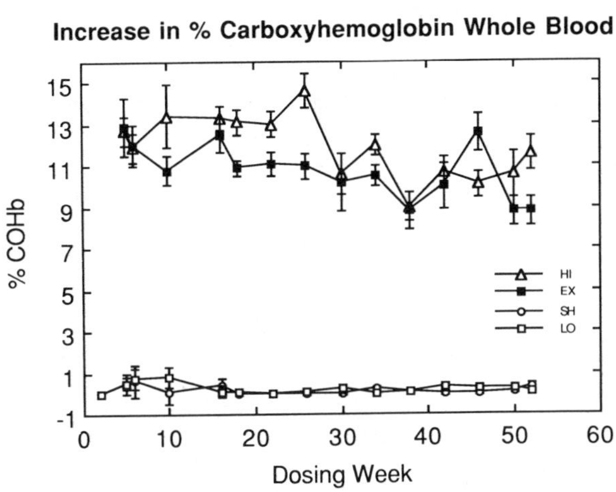

FIGURE 4. Percent whole-blood COHb 1 min after exposure (mean ± S.E. for each dose group). These values were obtained after subtraction of the preexposure baseline values from the maximally achieved % COHb for each animal within each group. No increase in COHb would be expected for the LO group because the samples were drawn during the week while they were being sham exposed.

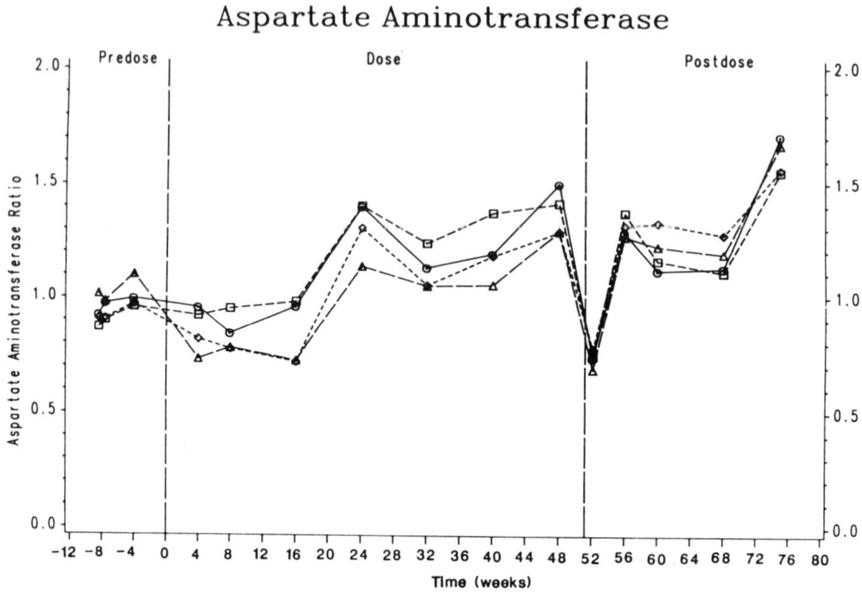

FIGURE 5. Mean group ratios for aspartate aminotransferase before, during, and after smoke exposure conditions. The abscissa represents weeks over which data were collected for each group. The vertical line at week 0 indicates the start of daily smoke exposure and the vertical line between weeks 51 and 52 indicates the end of smoke exposure conditions. Values are presented as ratios: predose values were divided by the median value for the SH group for the entire study; dose values were divided by mean predose values for the two sample periods prior to smoke exposure; and postdose values were divided by mean dose values for the last two sample periods of smoke exposure. Squares, EX group; triangles, SH group; diamonds, LO group; and circles, HI group.

Blood Chemistries

ANOVA indicated a significant group effect and subsequent contrasts indicated an increase in aspartate aminotransferase concentrations in the EX and HI groups as compared to the SH group 22 and 32 weeks after initiation of smoke exposure (Figure 5). ANOVA indicated a significant group effect and subsequent contrasts indicated a decrease in alkaline phosphatase in the HI group as compared to the SH group (11 to 12%) 8 to 16 weeks after smoke exposure (Figure 6). ANOVA indicated both a decrease in γ-glutamyl transferase in the HI group values compared to the SH and EX groups at 24 weeks after the initiation of smoke exposure (34% decrease, Figure 7). Other chemistry parameters did not exhibit any

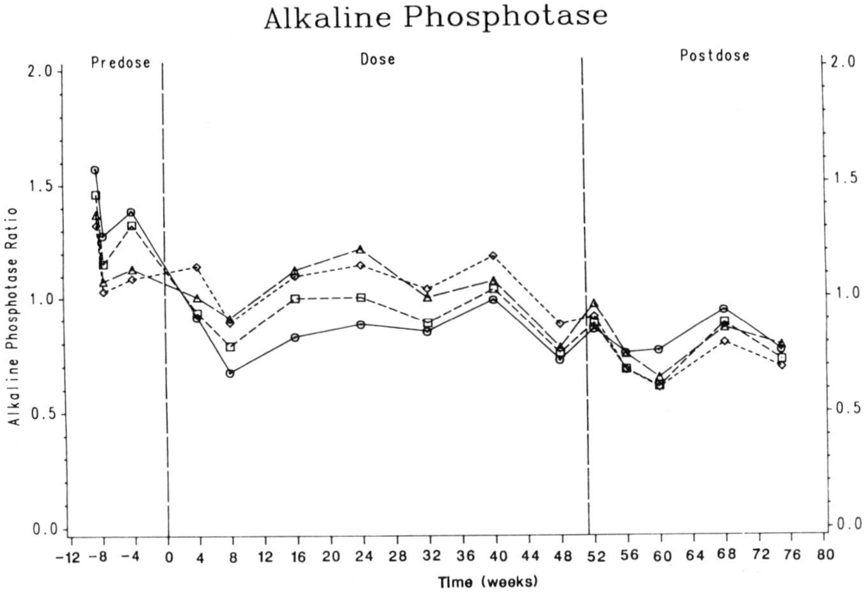

FIGURE 6. Mean group ratios for alkaline phosphatase before, during, and after smoke exposure conditions. Data presentation and symbols as in Figure 5.

significant changes throughout the study. These parameters and their mean overall values ± SD for all groups were: total serum cholesterol (190 ± 35 mg/dl); total serum protein (8.01 ± 0.64 mg/dl); β-globulin (18.08 ± 2.84%); albumin (54.38 ± 3.98%); α_1- (3.52 ± 0.94%), α_2- (12.61 ± 2.80%), and γ-globulin (11.41 ± 2.00%).

Hematologies

ANOVA indicated a significant group effect and subsequent contrasts indicated a decrease (8 to 15%) in the percent lymphocyte values of the HI group as compared to the SH group from 8 to 32 weeks after initiation of smoke exposure (Figure 8). Other hematological parameters did not exhibit any significant changes throughout the study. These parameters and their mean overall values ± SD for all groups were: red blood cell count (5.76 ± 0.47 × 10^6 cells/mm^3); hemoglobin (13.58 ± 0.96 g/dl); hematocrit (42.88 ± 3.16%); mean corpuscular volume (75.51 ± 2.93 μm^3); mean corpuscular hemoglobin (23.58 ± 1.14 pg/cell); mean corpuscular hemoglobin concentration (31.45 ± 1.03 g/dl); white blood cell count (8.09 ± 3.14 × 10^3 cells/

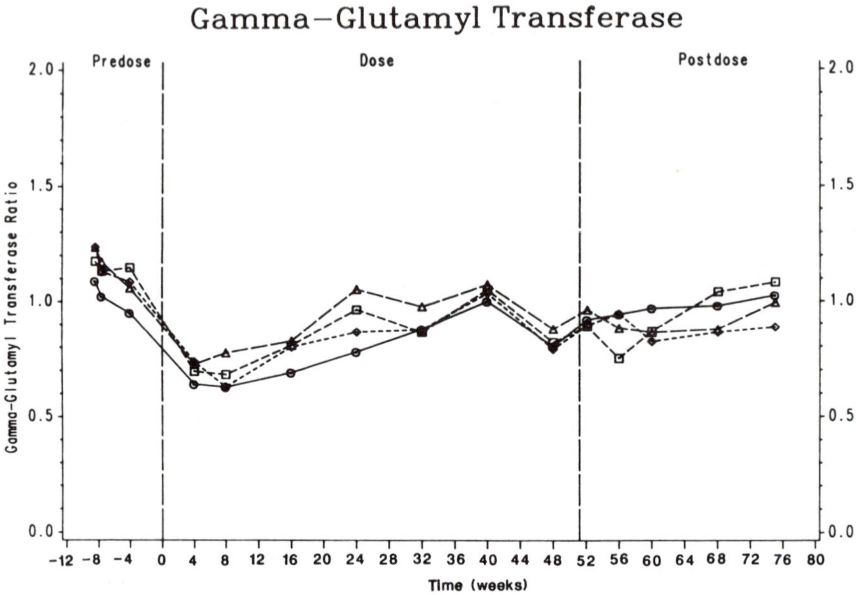

FIGURE 7. Mean group ratios for γ-glutamyl transferase before, during, and after smoke exposure conditions. Data presentation and symbols as in Figure 5.

mm^3); basophils (0.03 ± 0.18%); eosinophils (1.20 ± 1.20%); monocytes (2.21 ± 1.42%); and reticulum cells (0.55 ± 0.24 per 1000 red blood cells).

This study documents the effectiveness of the exposure procedures we utilized for evaluating the long-term effects of MJ smoke exposure in the nonhuman primate. The weight gain of these animals for the duration of the experimental period is essentially the same as those reported by others for rhesus monkeys of the same age.[15] Quarterly T.B. testing, biannual parasite assessments, and physical examinations were employed to maintain a healthy colony. The general health of the monkeys was not compromised by a year of MJ smoke exposure, as indicated by weight gain, carboxyhemoglobin, and clinical chemistry/hematology values. Although within the normal ranges of reported values, plasma lymphocyte concentrations did exhibit a dose-related decrease associated with the onset of MJ smoke exposure. Most clinical parameters, however, did not show any treatment related changes, and those few that did were of small magnitude, transient in nature, and were not different at the end of the 5-month post-dosing period.

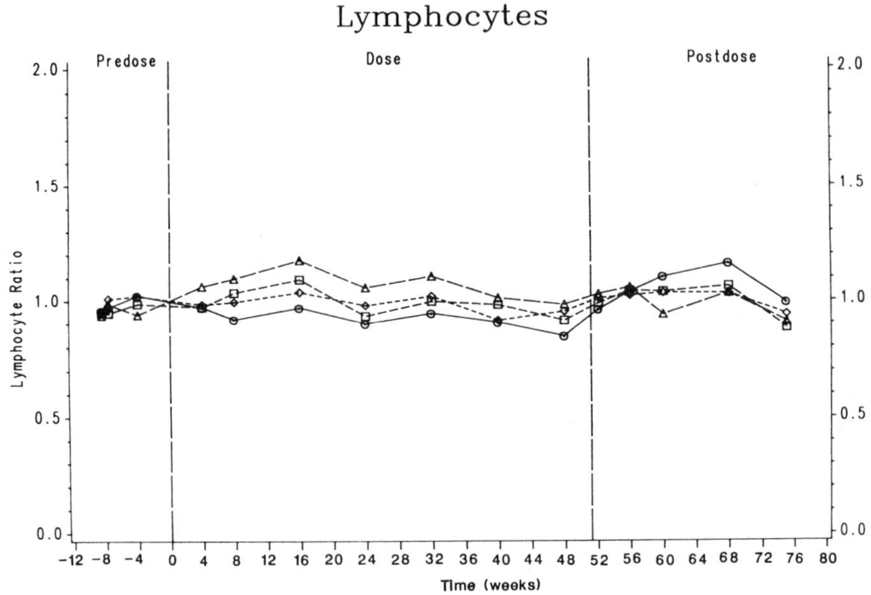

FIGURE 8. Mean group ratios for lymphocytes before, during, and after smoke exposure conditions. Data presentation and symbols as in Figure 5.

EFFECTS ON URINARY CORTISOL, PLASMA CORTISOL, AND PLASMA TESTOSTERONE

Previous studies with rats and mice involving acute and subchronic exposure to THC have demonstrated treatment-related increases in plasma corticosterone levels.[16-21] One study utilizing monkeys demonstrated no urinary cortisol increase over baseline when the animals were exposed to oral dosing.[22] Studies with humans employing MJ also did not reveal elevation of urinary cortisol excretion.[23-26] Prolonged exposure to elevated levels of glucocorticoids due to emotional activation of the adrenal cortex via the pituitary has resulted in biochemical, cellular, and tissue alterations.[27-29] In addition to the physiological changes that have been noted, psychological changes have also been reported.[30-32] Within recent years, psychoneuroendocrine influences on immunocompetence and the possibil-

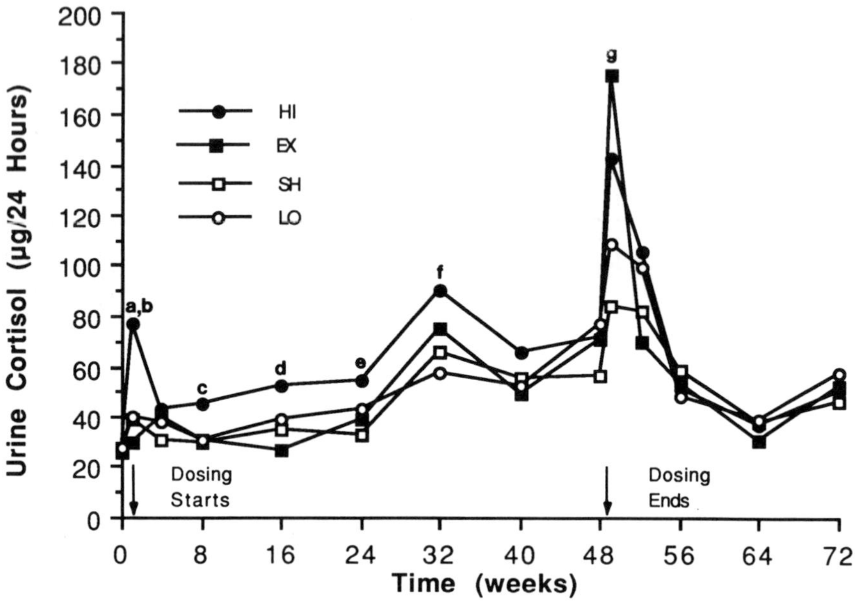

FIGURE 9. Free cortisol excreted during 24-h urine collections (n = 15 or 16). a, HI significantly different from SH, EX, and LO; b, LO and SH significantly different from EX; c, HI significantly different from LO and SH; d, HI significantly different from SH and EX; e, HI significantly different from SH and LO; f, HI significantly different from LO; g, HI and EX significantly different from SH. Values from 0 to 72 weeks were used in the Repeated Measures Analysis of Variance. Fisher's LSD tests were used to determine whether there were statistically significant (p <0.05) differences between dosing groups at specific time points.

ity of neurotoxic consequences have been reported.[20,29,33,34] Since there were no data available from a primate species on the effects of chronic MJ exposure on cortisol production/excretion prior to this study, we sought to characterize the consequences of long-term exposure to such smoke on urinary free and plasma cortisol, as an index of adrenocortical activity in the rhesus monkey. Plasma testosterone levels were also examined.

Experimental Methods

Sample Evaluation

The same monkeys, exposure regimen, and conditions were used as previously described in the previous section.[35] Urine samples (24 h) were collected on dry ice from metabolism cages at preselected time points during the course of the study. Blood samples for plasma cortisol and

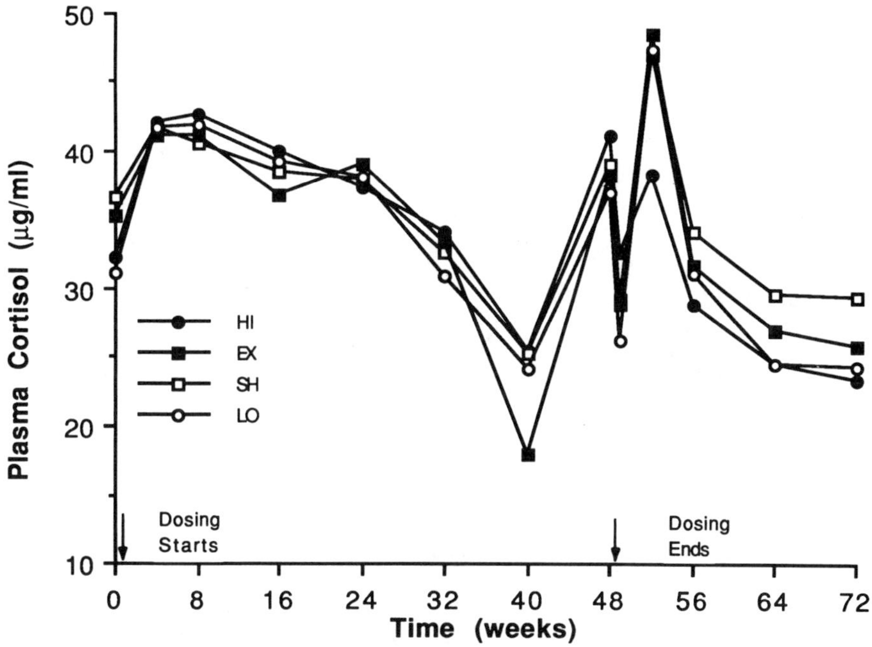

FIGURE 10. Plasma cortisol concentration in acute blood samples collected over the course of the study (n = 15 or 16). No statistically significant differences between groups were observed. Values from 0 to 72 weeks were used in the Repeated Measures Analysis of Variance. Fisher's LSD tests were used to determine whether there were statistically significant (p <0.05) differences between dosing groups at specific time points.

testosterone were collected prior to the dosing exposure so that approximately 23 h had elapsed since the previous dose. This demonstrated effects on endogenous levels as a result of chronic exposure rather than acute exposure. Samples were evaluated using a double antibody ^{125}I-radioimmunoassay designed for quantitative measurement of cortisol in serum, heparinized plasma, and urine (Diagnostic Products Corporation, Los Angeles, CA). Calculations to determine urinary free cortisol output per day were:

$$\mu g/24\ h = \frac{\mu g/dl \times \text{total volume in milliliters}}{100}$$

Plasma testosterone was also evaluated by RIA utilizing a tritiated double antibody binding assay (Diagnostic Products Corporation, Los Angeles, CA).

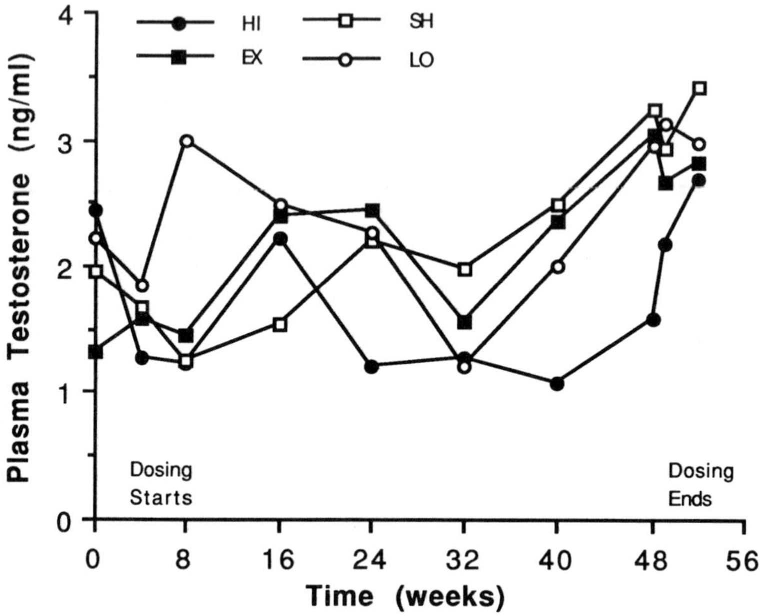

FIGURE 11. Plasma testosterone concentration in acute blood samples collected over the course of the study (n = 15 or 16). No statistically significant differences between groups were observed. Values from 0 to 56 weeks were used in the Repeated Measures Analysis of Variance. Fisher's LSD tests were used to determine whether there were statistically significant (p <0.05) differences between dosing groups at specific time points.

Results and Discussion

The data obtained in the present study are in accord with the findings of previously published rodent studies (i.e., exposure to THC does elicit elevated cortisol levels),[16-21] but do not support observations by others, using rhesus monkeys chronically exposed to THC, who reported no change in cortisol production.[22] By inspection (Figure 9), it is apparent that high dose MJ exposure over the course of the year produced elevated cortisol excretion compared to the low, placebo, and sham groups. Statistical analyses, however, indicated that this difference was not always consistent across time. The high dose group was consistently higher in urinary cortisol excretion throughout the entire exposure period. At cessation of dosing (Figure 9, 1 week after dosing ended), all groups demonstrated pronounced urinary cortisol elevation as a result of the change in routine. With cessation of THC exposure, the high dose group returned to within the range of the other three groups. Plasma cortisol (Figure 10) demonstrated an immediate increase in circulating levels with initiation of the dosing regimen. The initial surge diminished after approximately 2 months as the animals became accustomed to the exposure, and continued to decline over the dosing period. With cessation of dosing exposure, the urinary cortisol excretion again, initially, demonstrated considerable fluctuation with levels returning to or slightly below baseline values for weeks 56 to 72 of study (8 to 16 weeks after dosing ceased). Plasma testosterone profiles are more complicated (Figure 11). It would appear that exposure suppressed testosterone in the HI group compared to the SH, EX, or LO groups throughout the exposure period. The SH and LO groups were initially depressed below baseline (1st month post-exposure), then reversed to increase over the course of study. The EX group was initially stimulated and continued to increase over the course of the study. The difficulty in interpreting this data stems from the considerable within-group variability. This variability may be attributable to approaching sexual maturation. Since these animals were approximately 3 years old at initiation of dosing, they would, over the next 1.5 years, become peri- or postpubertal.

LONG-TERM EFFECTS ON BEHAVIORS USED TO ASSESS ASPECTS OF MOTIVATION AND SIMPLE DISCRIMINATION ABILITY

Many reports discussing the acute behavioral effects of MJ and THC are available.[36,37] Generally, these compounds have been shown to impair a variety of behaviors, including those associated with learning,[38] memory,[39-41] timing ability,[42,43] and choice discriminations.[44] It has been suggested that chronic MJ use results in the development of an "amotivational syndrome" in some populations.[45-47] Foltin et al.,[48] however, pointed out that there is little experimental evidence to support such an effect. It has also been suggested that the issue of whether chronic MJ and/or THC exposure results in permanent effects (i.e., brain damage) will likely only be settled using a suitable animal model since human studies are "confounded by too many other variables".[49]

There is convincing evidence in the rodent that chronic administration of relatively large doses of pure THC leads to residual changes in behaviors thought to depend upon hippocampal function[3-7] and to long-term, perhaps permanent, changes in hippocampal structure.[8,10] Earlier descriptions in monkeys of alterations in hippocampal electrophysiology and morphology after chronic treatment with THC or MJ further demonstrated the need to explore the possible long-term functional consequences of MJ use.[1,2]

The studies discussed here were carried out in an effort to examine some of the behavioral consequences associated with chronic MJ exposure.[50] To examine effects that occur during times of active chronic MJ exposure, half of the subjects were monitored during the active exposure (daily for 1 year) and withdrawal (2 months after the last treatment) phases of the study. Behavioral assessments were made 23 h after the most recent exposure; thus, the known acute effects of MJ smoke were minimized. The other half of the subjects were not behaviorally assessed during the active exposure and withdrawal phases so that they were not allowed to "practice" the behavioral tasks while under the influence of MJ.

An operant test battery (OTB) was used to monitor multiple behaviors in each test subject. The specific components of the OTB have been described previously[51-53] and consist of five different tasks performed for food reinforcers. The specific tasks to be discussed here are the Progressive Ratio (PR) and the Conditioned Position Responding (CPR) tasks. The PR task was chosen for inclusion in the OTB because of its suspected utility in generating behaviors thought to model those associated with motivation and/or reinforcer efficacy.[54] The CPR task was included because of the presumption that correct performance in similar tasks depends upon the integrity of the prefrontal cortex.[55-57]

The PR tasks, as described previously[54,58,59] and employed in the current study, have been used to examine the effects of a number of "motivational" variables.[54,58,60,61] The CPR task[51,52] was used to generate behaviors dependent primarily upon color and position discrimination.

Experimental Methods

Animals

The same monkeys, exposure regimen, and conditions were used as previously described.[35] In brief, 62 male rhesus monkeys (2 to 3 years old at the start of the study) were trained to perform the tasks in the operant battery for 1 year prior to the start of chronic MJ smoke exposure. At the end of the training phase, animals were matched according to behavioral performance and assigned to one of eight treatment groups (four dose groups X two behavioral groups).

Treatment Groups

The four dose groups were: HI, exposure once per day, 7 d/week to the smoke from one MJ cigarette (2.6% Δ^9-THC) smoked to a 10-mm butt; LO, exposure to MJ smoke only on weekends and sham exposure during the week; EX, exposure 7 d/week to the smoke from one ethanol-extracted MJ cigarette smoked to a 10-mm butt; SH = sham exposure only, 7 d/week. The two behavioral groups were: ACTIVE, operant behavior assessed throughout the 365 d of exposure and beyond; RESIDUAL, assessment of operant behavior reinstated 2 months after the last exposure.

Operant Behavioral Tasks

All behavioral testing occurred 23 h after the last exposure. The PR task was similar to that described by Hodos[54] and was performed with only the far right retractable lever (Figure 12) extended. Animals had to operate the lever a number of times (range = 1 to 7) to obtain the first reinforcer. For the second reinforcer, the response requirement was increased an amount equal to the initial response requirement and increased again the same amount for each subsequent reinforcer. Thus, if the first reinforcer "cost" three responses, the second would cost six, the third nine, and so on. Responses in the PR task (which lasted for 10 min per session) was defined in terms of response rates, breakpoints (the number of responses made for the last reinforcer in a given session), and the total number of reinforcers earned per session.

The CPR task was an automated version of a similar task described for use with the Wisconsin General Test Apparatus;[55] specific details appear elsewhere.[51-53] At the start of the CPR task, the center press-plate (Figure 12) was illuminated with either a red, yellow, blue, or green color. Each subject made "observing" responses to the center color, immediately after which it was extinguished and the two side plates were illuminated white. If the center color had been either red or yellow, responding to the left side plate resulted in reinforcer (banana-flavored food pellet) delivery. If the

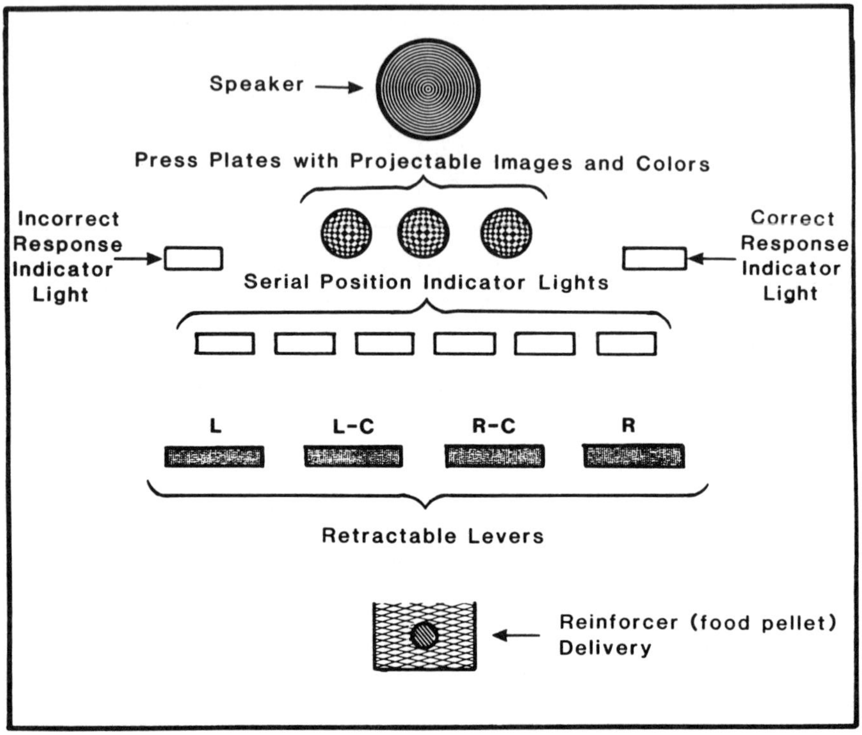

FIGURE 12. Behavioral panel used in the operant test battery. For the progressive ratio (PR) task, the far right retractable lever was used; for the conditioned position responding (CPR) task, the press plates were used.

center color had been either blue or green, then a response on the right plate resulted in reinforcer delivery. Responding in the CPR task was defined in terms of accuracy (percent correct choices), response rates, and percent task completed in the 5-min test time. The CPR task lasted until 60 reinforcers had been delivered or for a maximum of 5 min.

Results

General

There were no differences in CPR accuracy between any dose groups at any time during the study. Additionally, there were generally no significant differences between either of the control groups (SH and EX) and no significant differences between the MJ groups (LO and HI). Therefore, the groups were combined and the data are presented here simply as controls vs. MJ exposed.

Effects of MJ During Active Exposure (Active Groups)

Progressive Ratio

The control groups exhibited large and sustained increases in response rates, breakpoints, and number of reinforcers earned during the active dosing period, which became significantly greater than those for the MJ groups during the later months (see breakpoint, Figure 13, panel A). Changes in response rate and number of reinforcers earned paralleled those for breakpoint.

Conditioned Position Responding

There were generally no significant effects of chronic MJ exposure on CPR rate, or percent task completed over the entire year of exposure (Figure 13, panel B). However, the trend for the HI group was to exhibit a decrease in response rate and percent task completed during chronic exposure. Examination of individual subject data indicated that one of the eight animals in the HI group (#22) exhibited a substantial decrease in response rate beginning about 2 months after the start of daily exposure (period 6, Figure 14), which accounted for the downward trend in percent task completed for this group. The PR response rate did not show a similar effect for this animal (data not shown). Additionally, this animal did not differ from other subjects in the HI group with respect to other endpoints monitored, including plasma THC and 11-NOR-THC levels 45 min after exposure, 24-h urinary cortisol excretion, and several blood chemistry and hematological measures as discussed in an earlier section and elsewhere.[35]

Withdrawal Effects

There were no significant treatment-related alterations in any PR or CPR behavioral measures over the sessions immediately following cessation of exposure.

Progressive Ratio

Compared to their prewithdrawal baseline data, the MJ groups exhibited increases in breakpoints (Figure 15, panel A) beginning about 1 month after cessation of MJ exposure. Changes in the number of reinforcers earned and response rates (data not shown) paralleled these data, with the increases plateauing about 2 months after the last exposure. During the last observation period, raw data for these measures did not differ between groups.

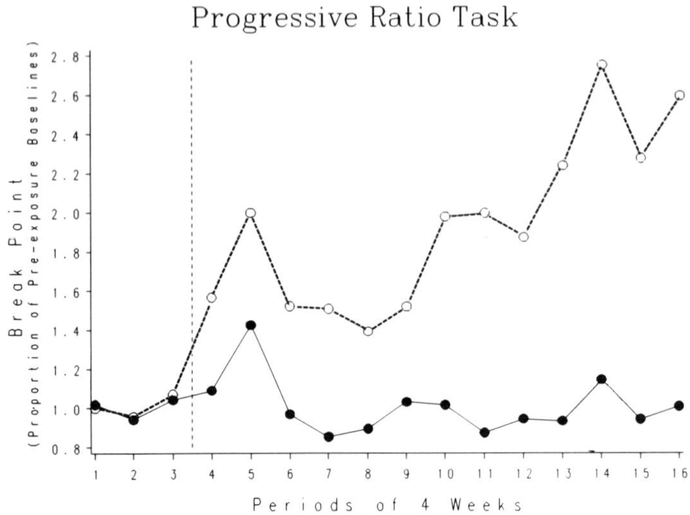

PANEL A.

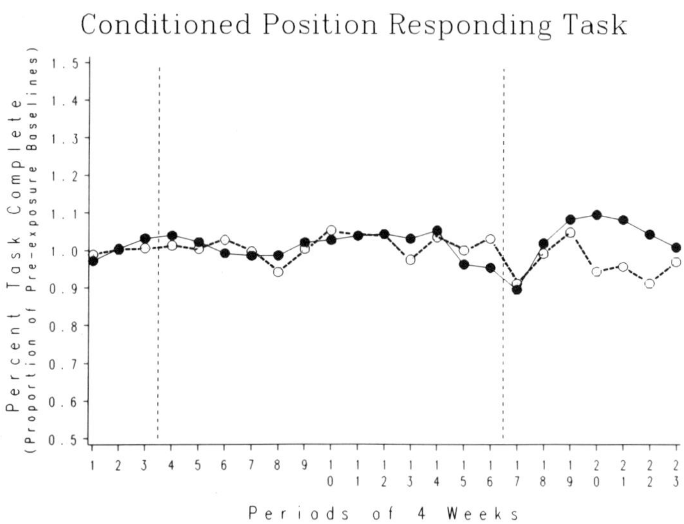

PANEL B.

FIGURE 13. ACTIVE group data, expressed as proportions of preexposure baselines for PR breakpoint (Panel A) and CPR task completed (Panel B) before and during the active exposure phase. The abscissa represents periods of 4 weeks over which data for each endpoint were averaged for each group. The vertical line between periods 3 and 4 indicates the start of daily exposure. Open symbols, controls; closed symbols, MJ groups.

Conditioned Position Responding

Compared to their prewithdrawal baseline data, the MJ groups also exhibited increases in response rates (Figure 15, panel B) beginning about 1 month after cessation of MJ exposure. The increase in response rate was due entirely to the increase exhibited by animal #22 whose response rate had been profoundly suppressed during active exposure by up to 46% (recall Figure 14). The preexposure rate of responding, however, was never reattained by this animal during the 7 months following cessation of exposure.

Residual Effects

Progressive Ratio

There were no significant group differences on any of the measures of PR responding in the animals that did not practice throughout the exposure or washout periods.

Conditioned Position Responding

Response rate and percent task completed, while less in the combined MJ groups than in the controls, were not significantly different from those of controls. When data from just the HI group were compared to that from both control groups, the differences in percent task completed did become significant during the very last observation period (data not shown).

Discussion

It is uncertain whether the effects of chronic MJ smoke exposure on PR performance resulted from an effect on the reinforcing efficacy of the specific food reinforcers used or whether they resulted from a more general "motivational" deficit. However, since performance of the CPR task by the same animals for the same food reinforcers was not affected in like fashion, it seems unlikely that the noted effects were due only to changes in the reinforcing efficacy of the banana-flavored food pellets. It is also unlikely that these effects were due to any rate-suppressing effects of MJ, since suppression of PR responding after acute exposures occurs only at doses higher than those needed to suppress CPR.[52] Additionally, mean PR rates did not at any time decrease below preexposure baseline rates. Thus, one interpretation of the data (significantly fewer reinforcers earned in the PR task) is that during chronic exposure, MJ appears to induce an "amotivational syndrome" in the rhesus monkey. This effect was noted whether exposures

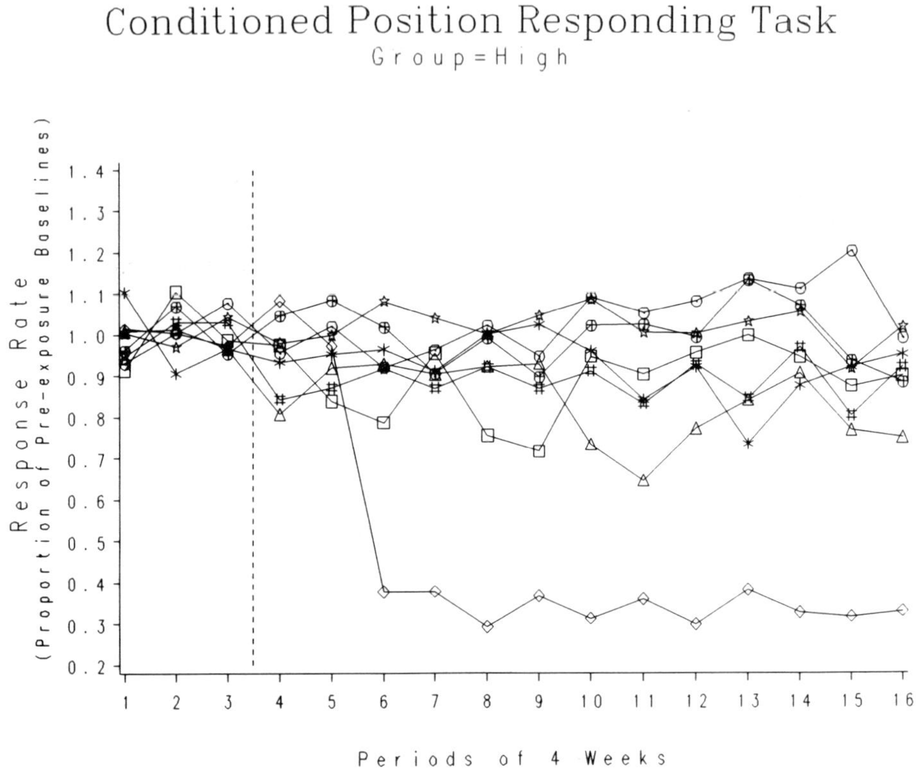

FIGURE 14. Individual animal data (HI ACTIVE dose group only), expressed as proportions of preexposure baselines, for response rate for the CPR task. Each symbol represents data for an individual subject; otherwise, data are presented as in Figure 13. Response rate for animal #22 (diamond symbol) was significantly depressed from preexposure levels beginning at Period 6.

occurred daily or on weekends only. Two to three months were required after cessation of chronic exposure for PR performance to return to control group levels. Since the effects of chronic MJ exposure on PR responding were slowly reversible and there did not appear to be any long-term "residual" behavioral effects, it is possible that persistent exposure to a compound (or compounds) that is very slowly cleared from the brain can account for these observations.

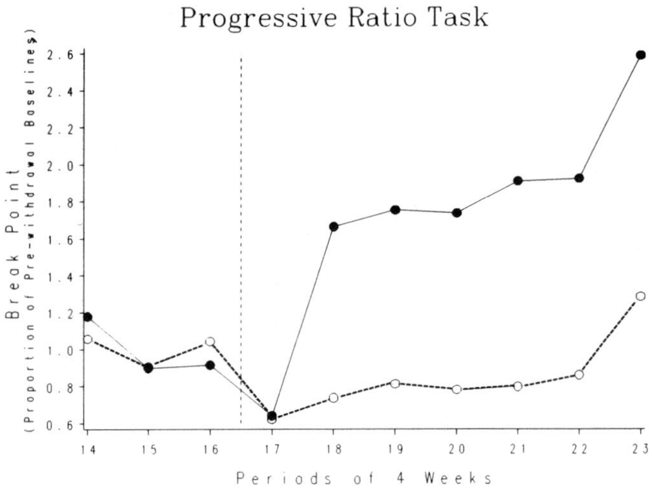

PANEL A.

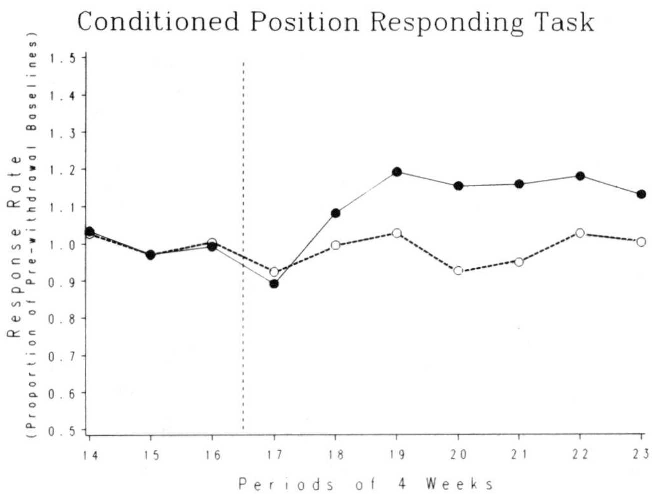

PANEL B.

FIGURE 15. ACTIVE group data, expressed as proportions of prewithdrawal baselines, for PR breakpoint (Panel A) and CPR response rate (Panel B) during the withdrawal phase of the study. The abscissa represents periods of 4 weeks over which data were averaged for each group. The vertical line between periods 16 and 17 indicates the end of daily exposure. Open symbols, control groups; closed symbols, MJ groups.

The use of animal models makes it possible to minimize or eliminate factors (i.e., socioeconomic and nutritional status, age, sex, previous drug history, concurrent drug use, etc.) known to influence research findings in most human studies.[62] The previous lack of success in experimentally demonstrating "motivational" effects of MJ in humans[63-65] may have been due to some of these factors. Epidemiological studies of chronic cannabis users have also failed to demonstrate an "amotivational" effect in chronic users.[66-68] Human epidemiologic studies are also prone to various design flaws.[69] The present findings in monkeys may have relevance to clinical observations associating chronic MJ use with an amotivational syndrome.[11,45,47,70] Detecting such effects requires several months of continued exposure (and the use of an appropriate control group) and may depend on the developmental stage of the experimental subjects at the time of treatment.

We studied these animals during their adolescence to best model adolescent humans. Our subjects were about 3.5 to 4.5 years old when PR responding in the control groups became significantly different from that of the MJ groups, and most of them became sexually mature at some point during the experiment. It is possible that the increase in PR responding we noted in the controls might not have occurred in younger or older animals. This observation is particularly interesting since, in the human literature, motivational problems have generally been described for adolescent users[46,70] and not adults.[64,71-73]

The finding that exposure to MJ only on weekends was sufficient to produce effects on PR responding suggests that this effect occurs at relatively low or "recreational" exposure levels. Additionally, the effect appears to be maximal with these intermittent exposures. There could be a gradual buildup of active compound(s) that affects performance for an entire week or more following exposure. This would not be inconsistent with the recent reports by Johansson et al.[74,75] of a long half-life for THC in blood.

During most of the chronic exposure and for several weeks afterward, one of the eight subjects in the daily MJ exposure group exhibited profound deficits in responding in the CPR task. Certain individuals may thus be more sensitive to chronic MJ exposure than are others, an observation also made in rats[76] and humans.[77,78] Monkeys exposed daily to MJ smoke, but not receiving daily "practice" in the CPR task, exhibited a reduced percent task completed during the post-treatment phase of the experiment, suggesting persistent deficits. The lack of a persistent deficit in the performance of animals that "practiced" the CPR task throughout chronic treatment suggests that such practice may have resulted in an "overlearning" situation in which performance of the CPR task became less sensitive to treatment effects.

Previous reports of a MJ abstinence syndrome, evidenced by behavioral disruptions that occur after cessation of chronic daily administration of cannabinoids,[79-82] have generally employed large doses of pure THC. We did not find any convincing evidence of treatment-related behavioral deficits in PR or CPR responding upon acute cessation of exposure. We have reported, however, findings of exposure-related differences in urinary cortisol excretion in these same subjects[83,84] (and section on "Effects on Urinary Cortisol, Plasma Cortisol, and Plasma Testosterone") and believe such a finding to be the first evidence of withdrawal-induced endocrine effects in nonhuman primates after relevant chronic MJ exposure.

It might be expected from the data presented here that, during periods of active chronic MJ use during adolescence (even on weekends only), aspects of behavior that may have relevance to motivation and/or to the effectiveness of specific rewards or reinforcers might well be altered and complete recovery from such effects may take several weeks to months. Relatively simple brain functions such as color and position discrimination will likely not be affected. Certain individuals may be much more "susceptible" to behavioral effects of chronic MJ exposure than others. It remains to be determined whether longer or greater exposures will cause more severe or additional behavioral effects.

NEUROCHEMICAL EFFECTS AND COMPARISON TO ACUTE AND CHRONIC EXPOSURE TO Δ^9-TETRAHYDROCANNABINOL (THC) IN RATS

THC is one of the major components of MJ known to produce psychopharmacological effects in animals and humans. However, the mechanism of action of THC is still unclear. It seems to affect all major neurotransmitter systems, including cholinergic,[85] dopaminergic and adrenergic,[86-94] serotonergic,[94,95] and GABAergic[96] systems in the *in vitro* studies. THC decreased dopamine release[97] and synthesis,[98] and dopamine, norepinephrine, and serotonin uptake in the mouse[98-100] and rat brain.[101] However, there are few reports on the effects of THC on neurotransmitter receptors. Recently, Bloom and associates reported that THC produced neurochemical changes *in vitro*.[102-104] In the present study, we evaluated whether acute or chronic exposure to THC produced any selective *in vitro* or *in vivo* neurochemical alterations in the rat brain. We also measured concentrations of dopamine, serotonin, and their metabolites in several brain regions in monkeys after chronic exposure to MJ smoke.

EXPERIMENTAL METHODS

Rat Treatment

Male Sprague-Dawley rats, 8 to 10 weeks old, were used in these studies. For the acute study, animals were dosed with 1, 3, 10, or 30 mg/kg THC intraperitoneally (i.p.) and sacrificed 2 h later. For the chronic study, animals were dosed by gavage (p.o.) with 10 or 20 mg/kg THC daily for 5 d/week for 90 d. Control animals received vehicle (Triton X-100/ethanol/saline; 1:10:89). After 90 d of treatment, animals were sacrificed by decapitation, brains were quickly dissected into several brain regions following the guidelines of Glowinski and Iversen,[105] frozen over dry ice and stored at –70°C until analysis.

Monkey Treatment

Male rhesus monkeys were exposed as previously described.[35] Approximately 7 months after the last exposure, animals were sacrificed by an overdose of pentobarbital, and the brains were removed and dissected on an ice-cold plate according to the method of Brown et al.,[106] using the rhesus monkey atlas of Szebenyi.[107]

Neurochemical Analysis

Receptor Binding

Membrane preparation: A crude membrane fraction was prepared from frontal cortex, hippocampus, and caudate nucleus by homogenizing in 20 volumes of 0.32 *M* sucrose followed by centrifugation (50,000 × *g* for 10 min). The precipitate from this step was then homogenized in deionized, distilled water (pH adjusted to 7.4) and centrifuged (50,000 × *g* for 10 min). The pellet was resuspended in 50 m*M* Tris-HCl buffer (pH 7.4) and centrifuged. This step was repeated one more time. The final pellet was suspended in 50 m*M* Tris-HCl buffer containing 2.5 m*M* $CaCl_2$, 1 m*M* $MgCl_2$, 5 m*M*

KCl, 120 m*M* NaCl, 0.1% ascorbate, and 10 μ*M* pargyline at a concentration of 50 mg (original weight equivalent)/ml.

[^{3}H]-Quinuclidinyl benzilate (QNB) binding: Muscarinic cholinergic receptor binding was assayed by incubating 100 μl of the membrane preparation with 1.0 n*M* [^{3}H]-QNB (33.2 Ci/mmol; New England Nuclear, Boston, MA) following the method of Ali et al.[108] For the *in vitro* experiment, membrane was prepared from different regions of the control rat brain. The effects of the addition of THC (1 to 100 μ*M*), cannabidiol (CBD; 1 to 1000 μ*M*), and cannabinol (CBN; 1 to 1000 μ*M*) on the [^{3}H]-QNB binding was evaluated following the same procedure as described for the *in vivo* experiments.[86,108]

Protein Determination

Aliquots of the membrane preparations were used for determination of protein content by the method of Lowry et al.,[109] using bovine serum albumin as the standard.

Neurotransmitter Concentration Determination

Concentrations of dopamine, serotonin and their metabolites [3,4-dihydroxyphenylacetic acid (DOPAC) and 5-hydroxyindoleacetic acid (5-HIAA)] were resolved by high performance liquid chromatography (HPLC) and quantified by electrochemical detection following the method of Ali et al.[86,108]

Statistical Analysis

Data were evaluated by one-way analysis of variance, followed, where appropriate, by Student *t*-test. A value of $p < 0.05$ was taken as significant.

Results and Discussion

In the *in vitro* study, THC produced a dose-dependent inhibition of muscarinic cholinergic receptor binding in the frontal cortex and hippocampus of the rat (Figure 16). In the *in vitro* study, THC and its two nonpsychoactive derivatives (CBD and CBN) produced a dose-dependent inhibition of muscarinic cholinergic receptor binding in the frontal cortex of the rat (Figures 17 and 18).

Acute exposure to THC in the rat, however, did not produce any significant alterations in muscarinic cholinergic receptor binding in the frontal cortex or hippocampus (Figure 19). Chronic exposure to THC in the rat did not produce any significant alterations in muscarinic cholinergic receptor binding in the hippocampus (Figure 20). Chronic MJ smoke exposure in the monkey did not produce any alterations in the concentrations of dopamine, DOPAC, serotonin, or 5-HIAA in caudate nucleus or hypothalamus (Figure 21).

In vitro exposure to THC produced dose-dependent inhibition of muscarinic cholinergic receptor binding in frontal cortex and hippocampus. We also found the same inhibition of muscarinic cholinergic receptor by CBD and CBN. However, *in vivo* studies, where rats were dosed with THC and the same brain regions were analyzed for muscarinic cholinergic receptor, we did not identify any significant changes after either acute or chronic administration of THC. There are reports in the literature describing the significant elevation of acetylcholine 30 min after large dose exposure to THC.[110] In follow-up studies, no changes were found in acetylcholine levels, but there was a significant increase of utilization of acetylcholine in hippocampus.[85,111] However, there are no previous reports available in the literature that describe the effect of THC on muscarinic cholinergic receptor binding in the brain. Our data suggest that *in vitro* exposure to THC, CBD, or CBN produced dose-dependent inhibition of muscarinic cholinergic receptor binding. However, no significant changes were found after acute or chronic exposure *in vivo*. Therefore, the *in vitro* effects of THC on the muscarinic cholinergic receptor are apparently not linked to its psychoactive effects and are not observed *in vivo*.

MORPHOMETRIC NEUROHISTOLOGICAL STUDIES

MJ is among the most frequently abused illicit drugs in the U.S. and worldwide,[112] and has often been proposed for legalization.[113] Although the presumed neurotoxic effects of MJ have an obvious relationship to the issue of legalization, surprisingly few experimental studies of MJ neurotoxicity have been published. Human MJ research has been hampered by limited sensitivity of measurements and the retrospective nature of most of the studies.[60,114] Careful neurobehavioral studies of rodents[5-7] implied potential toxicity to the hippocampus, a suggestion that has been confirmed by neurohistological studies.[8,10] However, comparability of chemical constituents, pyrolysis products, dose, route of exposure, and physiological re-

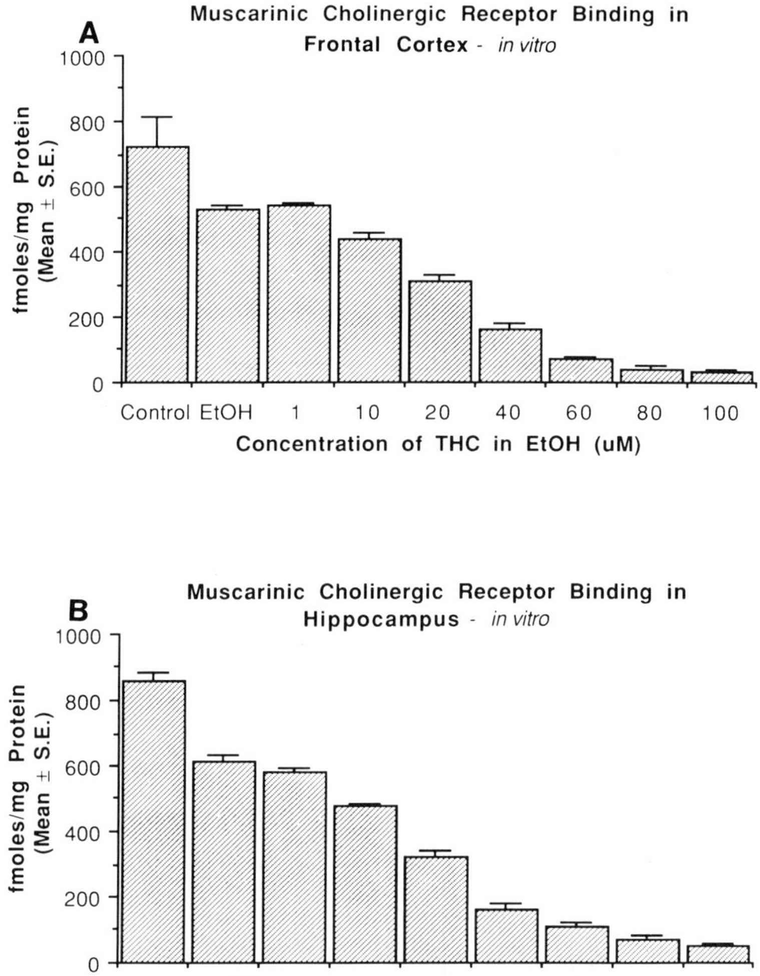

FIGURE 16. Muscarinic cholinergic receptor binding after *in vitro* exposure to THC in different regions of the rat brain. (A) frontal cortex and (B) hippocampus.

sponses to MJ between human and rodent models has not been achieved. The rhesus monkey (*M. mulatta*) has provided suggestive neurophysiological evidence, as well as neurohistological findings from two treated monkeys, of a limbic system toxicity that included the hippocampus.[2] Quantitatively, this was described as a widening of the synaptic clefts. The previ-

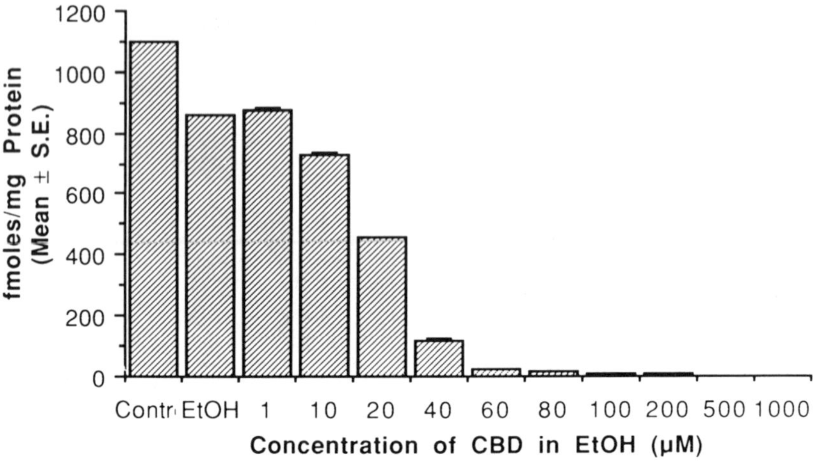

FIGURE 17. Muscarinic cholinergic receptor binding in the rat frontal cortex after *in vitro* exposure to CBD.

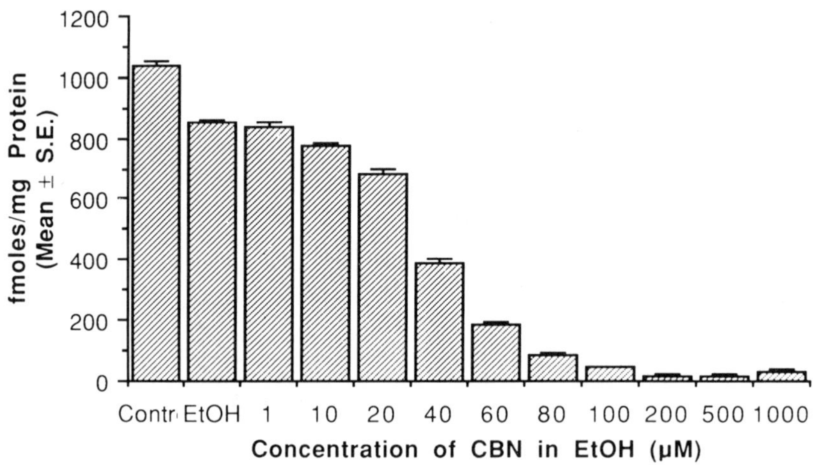

FIGURE 18. Muscarinic cholinergic receptor binding in the rat frontal cortex after *in vitro* exposure to CBN.

ous reports in this series document our overcoming the limitations of the rodent model by providing smoke exposure to the Rhesus monkey that is comparable in chemical content and dose to realistic human exposure conditions. The aim of the present study was to apply neurohistological methods to evaluate the brains, especially the hippocampal morphometry,

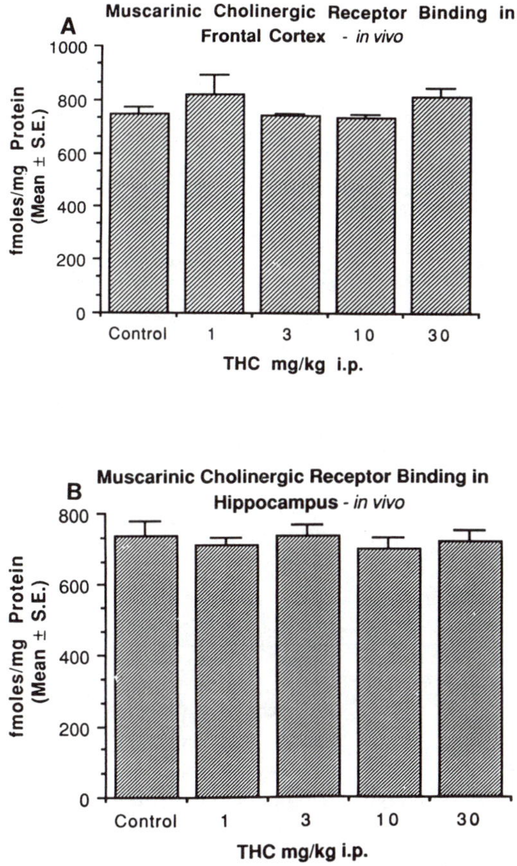

FIGURE 19. Muscarinic cholinergic receptor binding in the rat frontal cortex and hippocampus after acute exposure to THC. Animals were sacrificed 2 h after dosing. (A) frontal cortex and (B) hippocampus.

of monkeys exposed for a year to MJ smoke. Sacrifice for evaluation occurred 7 months after the last dose of MJ in order to be able to detect persistent neuropathological changes, if present.

Experimental Methods

General

The methods of treatment, behavioral testing, and experimental design have been previously described.[35] Six monkeys per drug condition, referred to here as sham, extracted, low dose, and high dose, were evaluated by

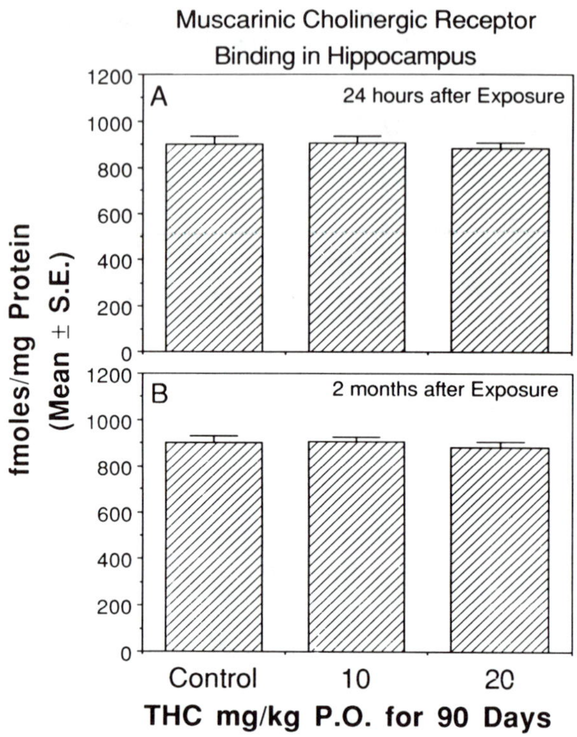

FIGURE 20. Muscarinic cholinergic receptor binding in different regions of the rat brain after chronic exposure to THC. Animals were sacrificed 24 h (A) or 2 months (B) after the last dose.

neurohistological methods applied to previous rodent studies[8] and adapted here for monkeys. Half of each drug group had been operant-tested during the year-long exposure period ("test" condition), while the other half remained in their home cages except for exposure ("rest" condition).

Macroscopic and Light Microscopic Studies

Monkeys were deeply anesthetized with i.v. pentobarbital and their brains rapidly dissected as previously described.[115] The hippocampus was removed and further divided into six approximately equal blocks and stained with the Golgi-Cox approach (cresyl-violet counterstain) to reveal hippocampal neurons and their dendrites. Blocks A to E were then serially sectioned at 150 μ for computerized morphometry (Figure 22), except for Block B which was sectioned at 200 μ for analysis of apical dendrites. Measurements were performed at the macroscopic level by outlining the

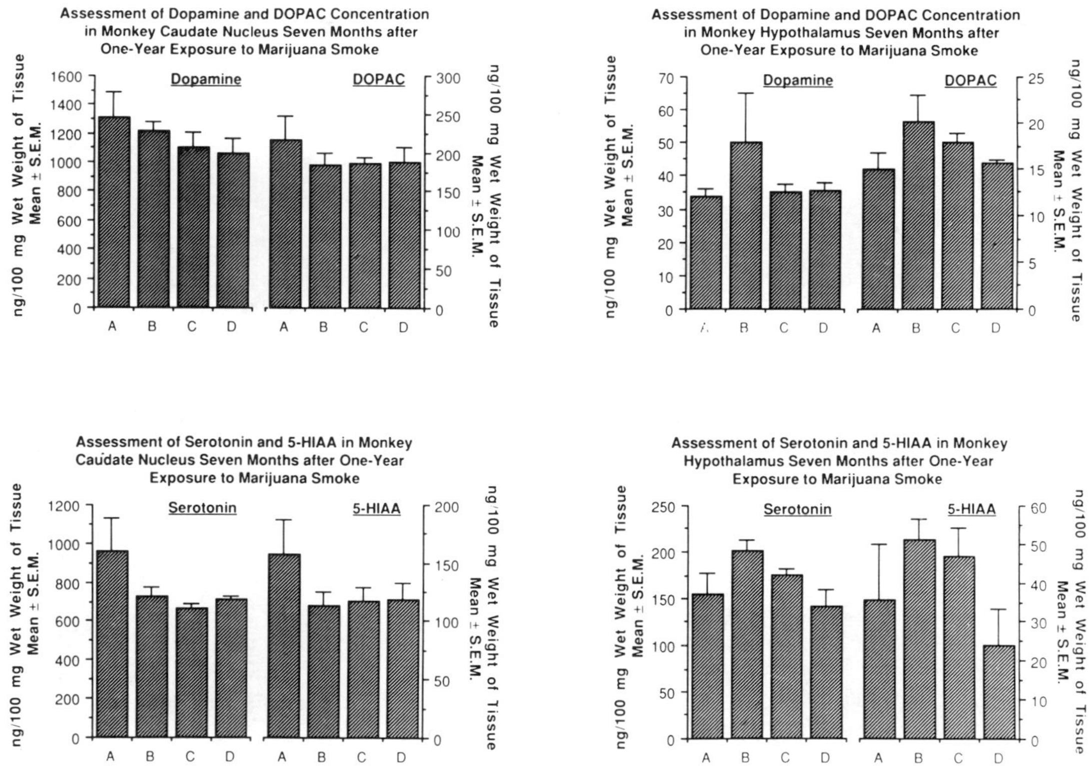

FIGURE 21. Effect of chronic MJ cigarette smoke exposure on neurotransmitter concentrations in caudate nucleus and hypothalamus of monkey brain. Animals were exposed to MJ cigarette smoke (2.6% THC) for 1 year and sacrificed 7 months after the cessation of exposure. A, sham controls; B, extracted MJ ccigarette controls; C, low dose group (weekend only exposure group); D, high dose group (every day exposure group).

total as well as regional subareas of hippocampus on a minimum of one section per block per monkey using a mouse-driven cursor and a Microcomputer Imaging Device (MCID, Imaging Research, Inc., Ontario, Canada) as illustrated in Figures 22 and 23.

Measurements at the microscopic level were performed by a Tree Analysis program on computer-aided 3-D drawings (R and M Biometrics, Nashville, TN) of the soma, dendrite lengths, branch-points, etc., as shown in Figure 24. A minimum of six neurons per regional subarea per block per monkey were averaged to derive a "prototypical neuron" reflecting that brain region for further analysis. Here, we report data from the CA3 region of blocks C and D for basilar dendrites and block B for apical dendrites. Future reports will include CA1, dentate granule cells, and cerebellar granule cells. Custom modification of the Tree Analysis program to produce data matrices for output to a standard spreadsheet program (Lotus Development Corp., Cambridge, MA), which was programmed to perform the necessary data reduction, was required to prepare the data for statistical analysis.

Electron Microscopic Studies

Deeply anesthetized monkeys were perfused with mixtures of formaldehyde and glutaraldehyde (Figure 25), replicating as closely as possible the procedures of Heath et al.[2] Samples from the anterior, middle, and posterior hippocampus of each animal, among other brain regions, were dissected, trimmed, and embedded in Epon/Araldite. Two grids from separate locations within each of the three hippocampal regions (total of 6 grids) were selected from the CA3 stratum pyramidal region, as identified in relation to the tips of the two blades of the dentate granule cells. After staining with lead citrate/uranyl acetate, 10 electron microscopic negatives (one each from the four corners plus the middle of two adjacent grid squares) were taken from each of the 6 grids at a magnification of 4000×. The resulting 60 negatives (as digitized positive images) were examined a sector at a time by means of an overlay ruled with 9-mm squares. Axo-dendritic synapses, as identified by distinct pre- and postsynaptic membranes, presynaptic vesicles, and a postsynaptic density, were counted and their lengths measured with the MCID system. Data transformations, based on the approach of DeHoff and Rhines as described by Collonier and Beaulieu,[116] were performed by means of a standard spreadsheet program (Lotus Development Corp., Cambridge, MA) prior to statistical analysis. Synaptic widths were measured at three locations along the synapse and averaged with the

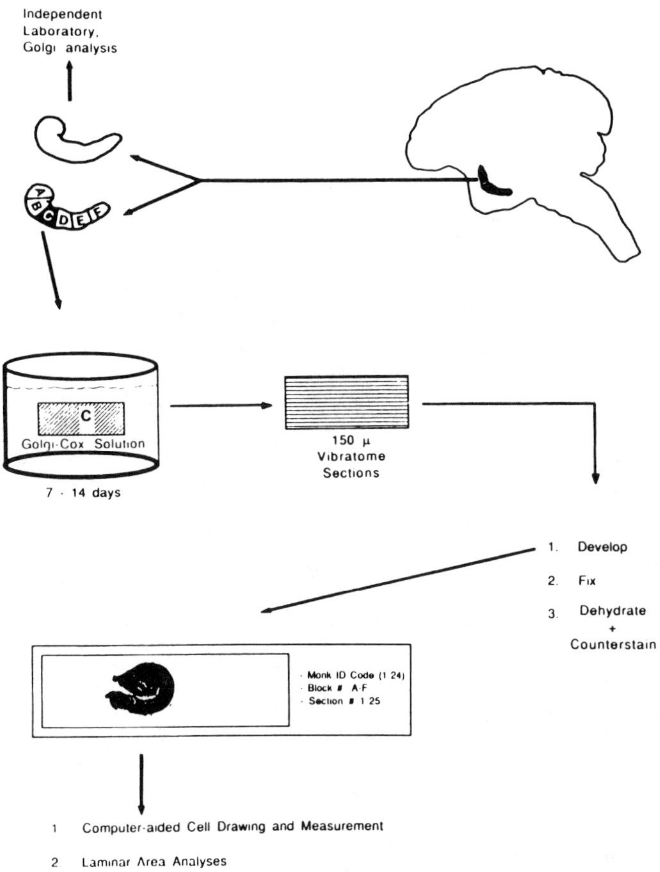

FIGURE 22. Light microscopic evaluation of THC-treated hippocampus.

MCID system. Four synapses from each of the two grids from each of the three hippocampal regions (for a total of 24 per animal) were measured for statistical analysis.

Statistical Analysis

The statistical approach taken recognized that each experimental animal contributed an "n" of 1 for comparisons based on drug or test/rest conditions and treated multiple parameters obtained from the same animal as "within" or "repeated" measures. Thus, most analyses were mixed-model factorial ANOVAs with 2 "between" factors (drug with 4 levels, test/rest with 2 levels) and several "repeated" factors chosen to allow us to examine whether there were any effects of anatomical location on the measurements (SAS

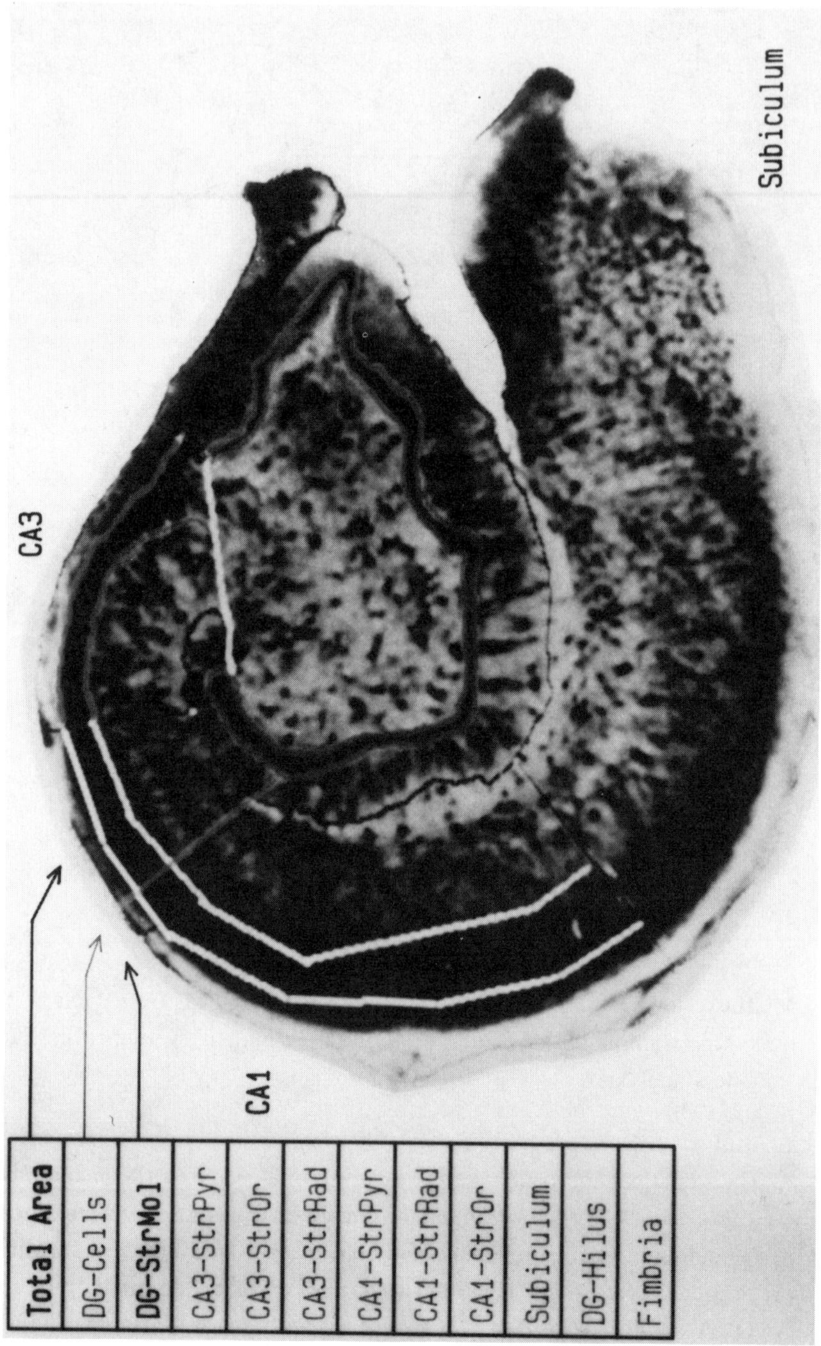

FIGURE 23. Computer-aided area measurements of the hippocampus (*M. mulatta*).

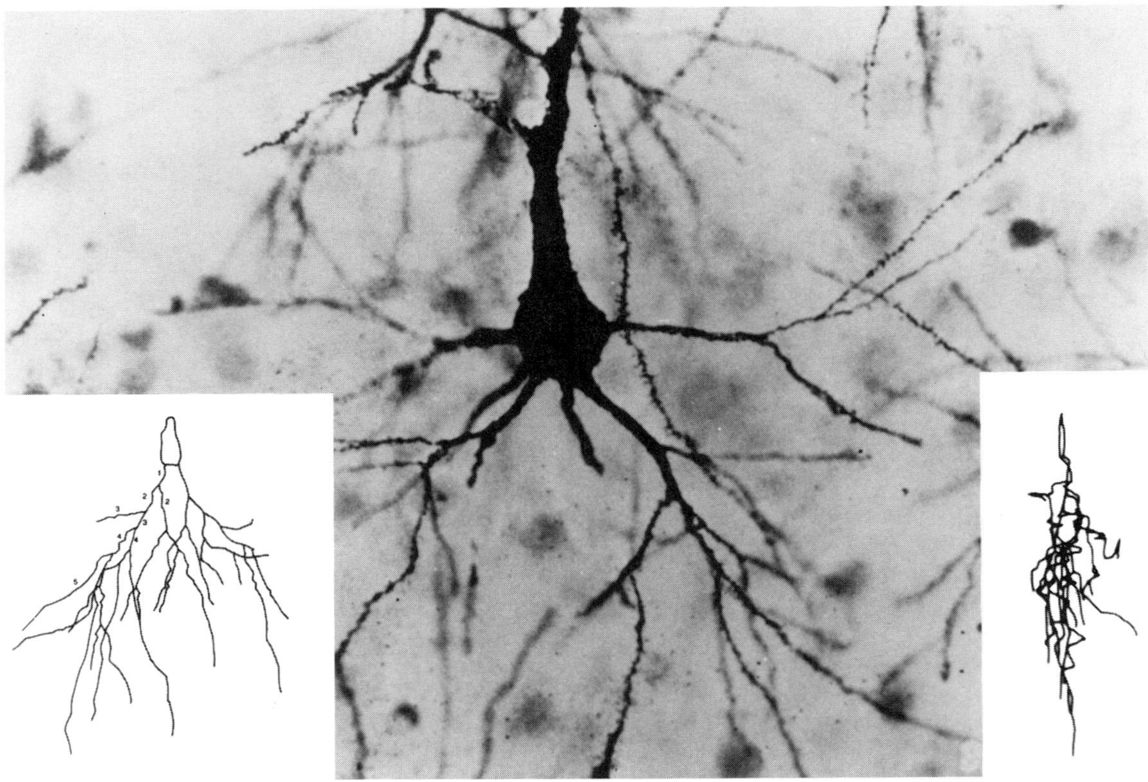

FIGURE 24. A photomicrograph of a typical Golgi-Cox stained hippocampal neuron. The inserts represent computer-aided 3-D drawings of the soma and dendrites of a typical neuron.

GLM, SAS Institute, Cary, NC, 1987). If a main effect of drug, rest/test, their interactions with each other, or with a repeated measure was significant at the 0.05 level or greater, individual mean comparisons were performed by the Fisher least significant difference (LSD) approach. Data in all the figures are presented as means ± the standard error of the mean.

Results and Discussion

Macroscopic Analyses

There were no statistically reliable effects of MJ smoke exposure on the total area of hippocampus or any of its subfields (Table 1). However, there were significant effects of behavior group on CA3 stratum oriens ($p < 0.05$). This finding indicates that daily performance of operant testing during the year's exposure period increased the total area of hippocampus, and especially the CA3 stratum oriens as a percent of the total. The areas represented by each subregion decreased from front to back of the hippocampus and their relative proportions changed as illustrated in Figure 26.

Light Microscopic Analysis

Neurons prepared as described in Figure 1 were digitized into 3-D computer-generated images stored as a matrix of X-Y-Z coordinates and displayed with a vector graphics adapter for a PC computer. An algorithm designed to quantify various dendritic parameters provided matrices of data that were then evaluated by analysis of variance. There were no effects of MJ exposure on apical or basilar dendritic length (Figures 27 and 28) nor on soma size, dendrite number, or degree of branching (Figure 29A, B, C). However, the finding of increased CA3 stratum oriens area with daily operant performance described above was echoed in the dendritic measurements since this subfield had longer, more highly branched dendrites as well (Figure 29D).

Electron Microscopic Analyses

Quantitative measurements of electron micrographs were performed, as shown in Figure 25. Analyses of variance of this data indicated no statistically reliable effects of MJ smoke on synaptic density (number of synapses per cubic millimeter), length, or width (Figures 30A, B, C), although there was a trend for an increase in synaptic density with MJ treatment (Figure 30A). Again, daily operant-tested animals differed significantly from cage-rested monkeys (Figure 30D), with synaptic density value consistently less

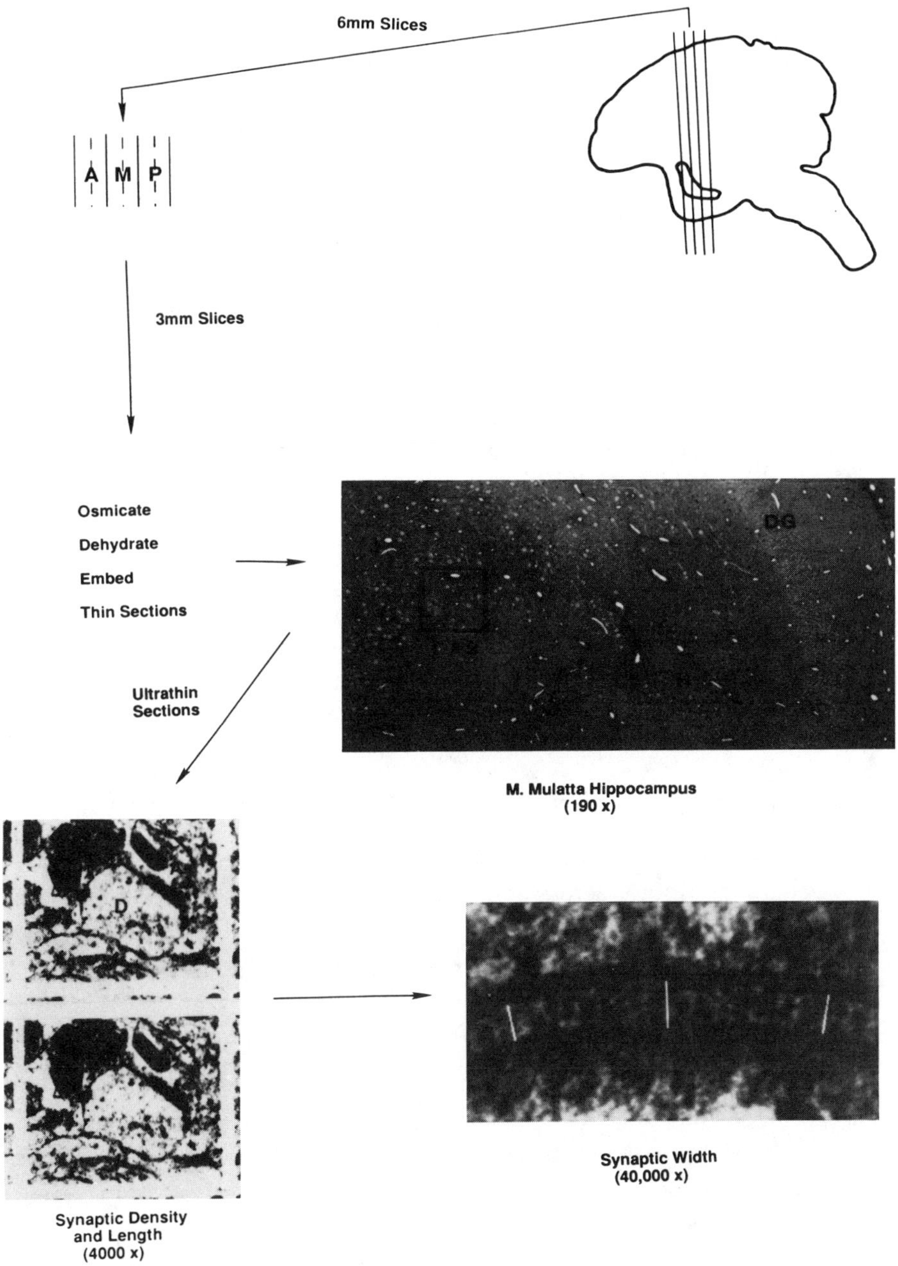

FIGURE 25. Electron microscopic evaluation of THC-treated hippocampus.

TABLE 1
F-Values for the Analyses of Variance of the Hippocampal Size Data and for the Sub-Fields as a Proportion of the Total Area

df	Drug (3,16)	Behav. (1,16)	Drug × Behav. (3,16)	Blocks (3,48)	Blocks × Behav. (3,48)	Blocks × Drug (9,48)	Blocks × Drug × Behav. (9,48)
Total area	0.2	4.8[a]	4.2[a]	56.9[b]	1.1	0.6	0.9
CA1 (S. oriens)	0.1	3.5	2.3	1.3	0.5	0.4	0.6
CA1 (S. pyramidale)	0.8	0.1	0.1	34.7[b]	0.4	1.6	0.6
CA1 (S. radiatum)	0.2	0.8	1.5	2.3	1.3	0.4	0.7
CA3 (S. oriens)	0.6	6.8[a]	0.2	19.2[b]	0.1	0.4	1.1
CA3 (S. pyramidale)	1.0	0.9	0.7	14.5[b]	0.2	0.4	1.3
CA3 (S. radiatum)	0.2	1.3	0.8	27.4[b]	0.2	0.5	1.0
Dentate (S. molec-ulare)	0.7	2.6	0.7	5.8[b]	0.6	1.1	0.5
Dentate (granule cells)	1.2	0.1	3.0	1.8	0.3	0.3	1.9
Dentate (hilus)	3.1	0.1	4.4[a]	32.4[b]	0.3	1.1	0.9
Myelinated axons	2.2	1.7	6.6[b]	36.3[b]	1.3	1.1	1.9

[a] $p < 0.05$.
[b] $p < 0.01$.

in daily-tested monkeys ($p < 0.05$). Considered as a strikingly consistent pattern of results across levels of analysis (macroscopic to electron microscopic), the effects of behavior group are quite similar to suggestive prior evidence of the effects of "enriched environments" on neuroanatomy.

The results of the present study provide some suggestive evidence in the form of trends for MJ smoke exposure to increase soma size, and decrease basilar dendrite number in the monkey hippocampal CA3 region, although dendrite length was not altered. However, these changes are dissimilar to the pattern of effects described previously by dendritic analysis of the THC-treated rat hippocampal CA3 region.[8] Rats gavaged with as little as 10 to 20 mg/kg THC had dose-related shortening of outer dendrite segments, while higher doses of THC resulted in *smaller* soma size. It is also notable that withholding daily operant behavioral trials, unlike MJ exposure, *did* produce a statistically significant shortening of CA3 stratum oriens (basilar) dendrites, a finding quite similar to previous descriptions of the effects of differential experience and training on dendritic morphology.[100] Therefore, MJ effects, if any, are of less consequence than providing differential learning environments for the monkeys. In this light, it is interesting that smoke exposure (MJ or extracted) significantly reversed the effects of depriving the monkeys of training.

Electron microscopic studies of the brains of rats treated with higher doses of THC (40 to 60 mg/kg) revealed disrupted membranes, increased extracellular space, and decreased synaptic density.[8] Lower doses (10 to 20 mg/kg) resulted in normal appearance of the CA3 neurophil with no change in its synaptic density,[8] although doses in the 4- to 10-mg/kg range were shown to produce a trend toward increased CA1 synaptic density in a study by Landfield et al.[10] The MJ smoke-exposed monkeys of the present study tended to have increased synaptic density in CA3, but it is difficult to interpret these observations without the functional behavioral data. Since withholding the daily operant behavioral trials significantly lowered synaptic density, also in similar fashion as might be expected from previous studies of differential experience effects,[117] MJ produced only a smaller and nonsignificant effect in the present study.

SUMMARY STATEMENTS

1. Subject's body weights did not differ between treatment groups at any time during the study. Additionally, the weight gain of these animals during this period was essentially the same as that reported by others for rhesus monkeys of the same age.
2. Based on comparisons to human data and data from our earlier pharmacokinetic studies in the monkey, peak plasma THC concentrations attained in our subjects were equivalent to those that would be predicted for humans after smoking four to five low to medium potency MJ cigarettes.
3. The maximum whole blood carboxyhemoglobin values obtained after smoke exposures (approximately a 13% increase over baseline) were not different between the EX "smoke-control" and the MJ groups, and have not been associated with toxicity.
4. The general health of monkeys exposed to chronic MJ smoke exposure for a year was not compromised, as indicated by weight gain, carboxyhemoglobin, clinical blood chemistry, and hematology values.

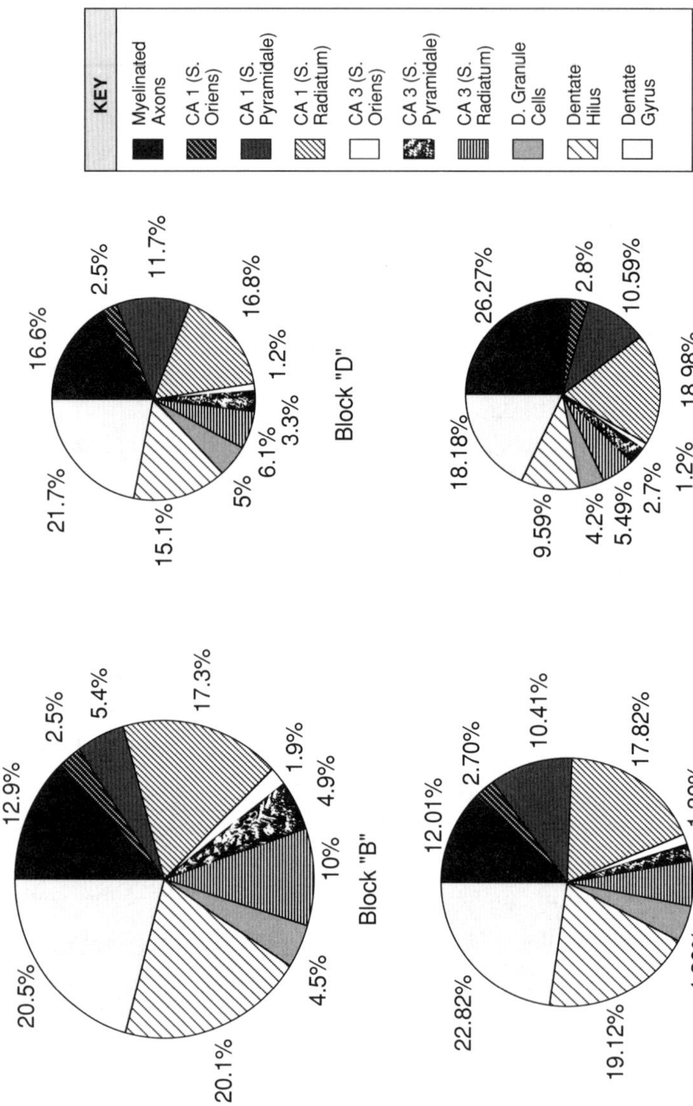

FIGURE 26. Relative areas of hippocampal subregions.

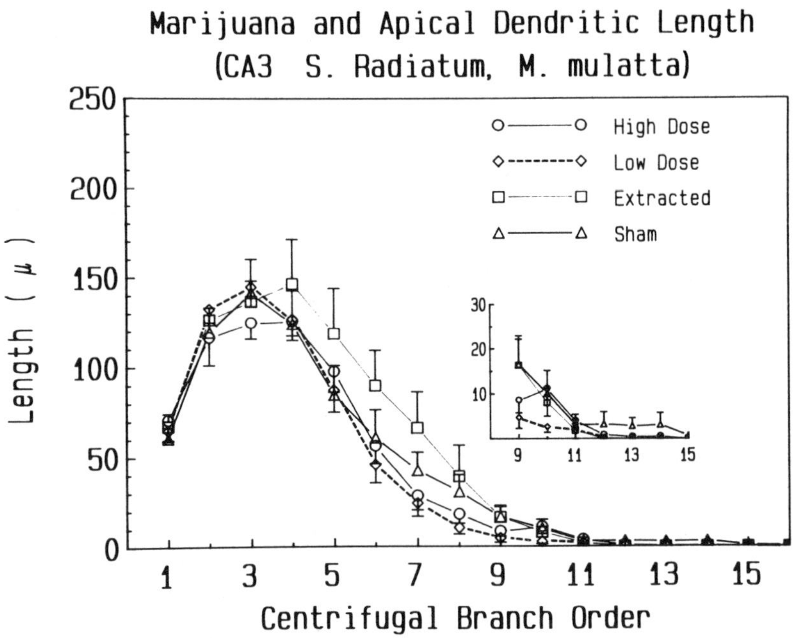

FIGURE 27. Effect of MJ smoke exposure on apical dendrite length of CA3 stratum radiatum neurons.

5. Plasma lymphocyte concentrations exhibited a dose-related decrease associated with the onset of MJ smoke exposure. However, these concentrations remained within normal ranges.
6. Initial exposure to MJ smoke caused a marked elevation in the high dose (HI) group urinary cortisol output.
7. Tolerance of cortisol output to daily high dose MJ smoke exposure is not complete since cortisol output during such treatment remains elevated throughout the year-long exposure period when compared to the LO, EX, and SH groups.
8. Acute plasma cortisol concentration differences between dosing groups are not statistically significant.
9. While there appears to be a trend toward plasma testosterone suppression by exposure to chronic administration of MJ smoke in the HI dose group, as evidenced by an initial 50% reduction in circulating levels and the continued low circulating levels over the dosing period, as well as by the dramatic rebound 1 and 4 weeks after dosing cessation, differences in testosterone concentrations between groups were not statistically significant.

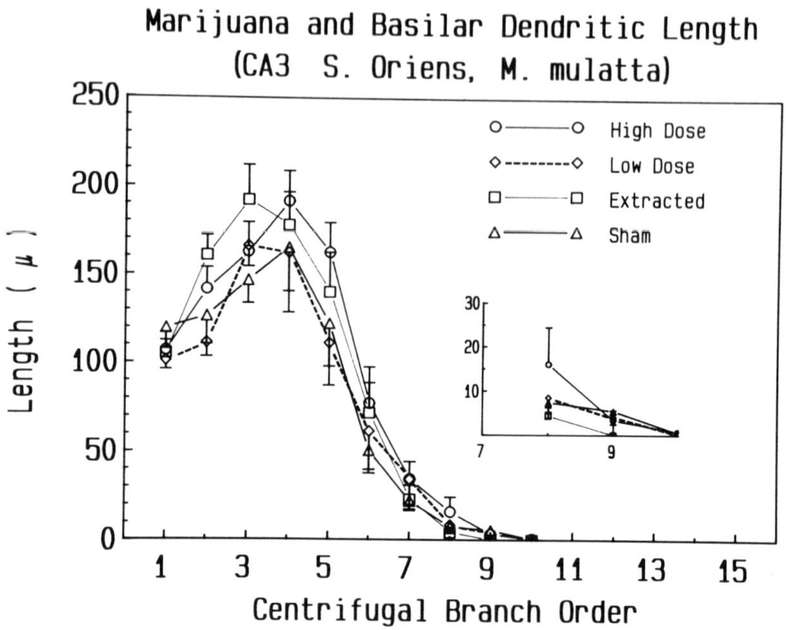

FIGURE 28. Effect of MJ smoke exposure on basilar dendrite length of CA3 stratum oriens neurons.

10. Marijuana smoking produces a physiological stress response in primates, as indicated by statistically significant increases in cortisol production in the HI dose group.
11. Abrupt cessation of chronic exposure (even to extracted MJ) also produces a physiological stress response, as indicated by large increases in urinary and plasma cortisol.
12. Presuming that measures of progressive ratio (PR) responding are related to aspects of motivation to perform for food, it seems that:

    a. Repeated performance of the PR task by "control" rhesus monkeys over the age range of about 3 to 5 years is characterized by significant increases in levels of motivation to respond for food.
    b. Chronic exposure to MJ smoke, whether occurring daily or on weekends only, prevents the occurrence of this increase in motivation.

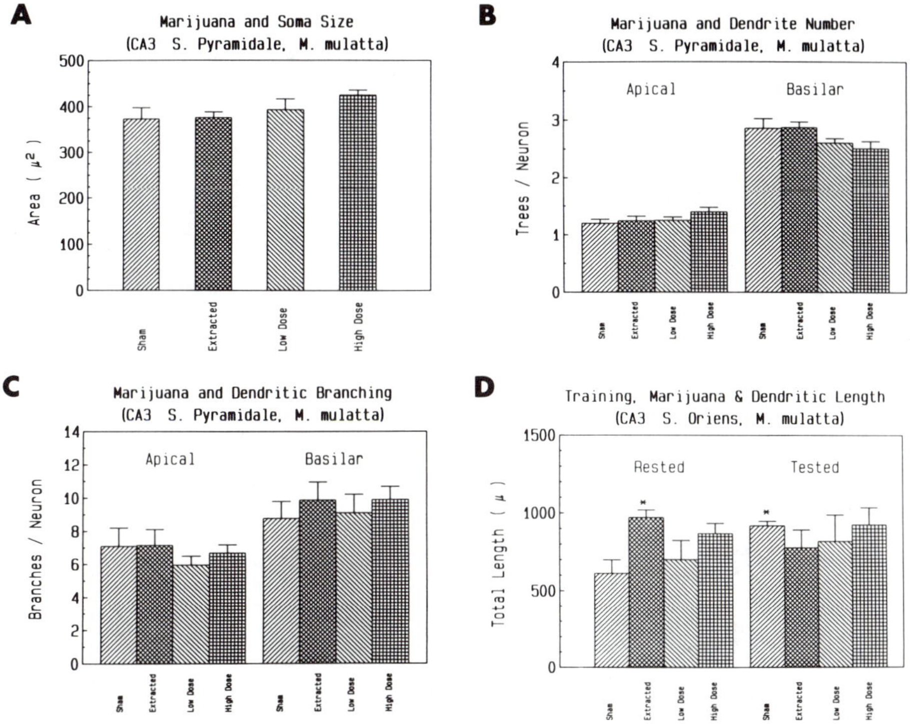

FIGURE 29. Effect of MJ smoke exposure on soma size (A), dendrite number (B), and dendrite branching (C) of CA3 stratum pyramidal neurons. Effect of behavioral training (d) on dendrite length of stratum oriens neurons.

c. Within 2 to 3 months after cessation of exposure to MJ smoke, levels of motivation to respond for food increase to match those of animals that had not been previously exposed to MJ smoke.

d. There does not appear to be any residual effect of chronic MJ smoke exposure on motivation to work for food.

13. For the task designed to monitor simple discrimination ability:

a. There are no detectable effects of chronic MJ smoke exposure on percent of the task completed in most animals that are allowed to continually practice the task during chronic exposure.

b. Upon cessation of chronic exposure to MJ smoke, the percent of the task completed increases, indicating that this behavior was suppressed during prolonged exposure.

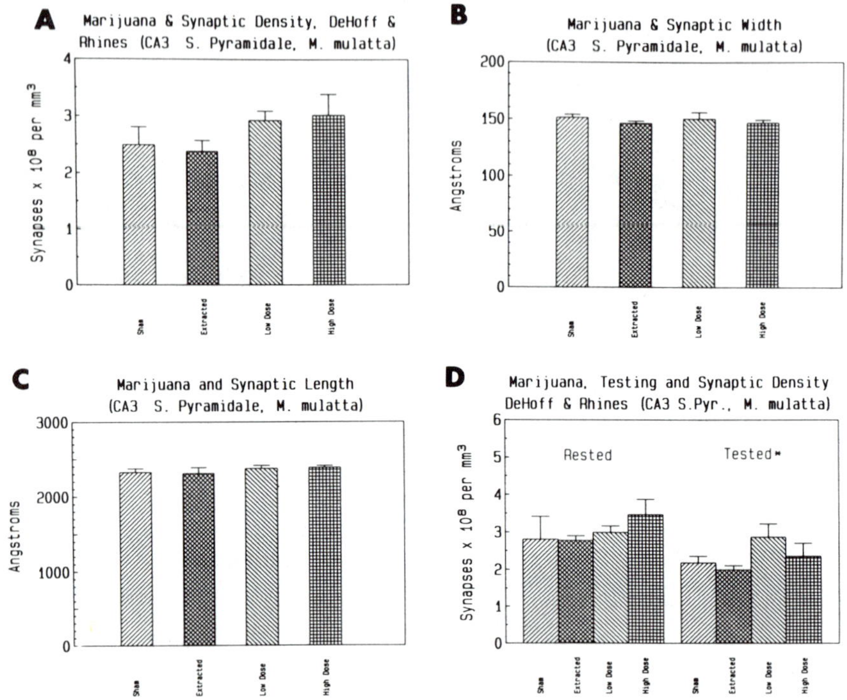

FIGURE 30. Effect of MJ smoke exposure on synaptic density (A), synaptic width (B), and synaptic length (C) of CA3 stratum pyramidal neurons. Effect of behavioral training (D) on synaptic density of CA3 stratum pyramidal neurons.

c. There appear to be residual effects (decreased percent task completed) of heavy (daily) MJ smoke exposure.

14. There are subjects that are very sensitive to the behaviorally disruptive effects of MJ smoke and there are others that show no such effects.
15. *In vitro,* a dose-dependent inhibition of muscarinic cholinergic receptor binding was seen with THC and also with CBD and CBN. No significant alterations were found after acute or chronic exposure to THC *in vivo*. Therefore, the *in vitro* effects of THC on muscarinic cholinergic receptor binding are not associated with its psychoactive effects, nor are they predictive of its *in vivo* effects in the rat or monkey.

16. MJ exposure did not significantly alter the total volume of monkey hippocampus, nor did it change the distribution of total volume among its subdivisions.
17. Despite trends toward increased soma size and fewer basilar dendrites, MJ did not significantly alter these parameters nor the mean dendritic length or branching of CA3 neurons.
18. Marijuana exposure also did not significantly affect the ultrastructural appearance of the hippocampal CA3 region, nor the width, length, or volume density of its synapses, although there was a trend toward increased synaptic density in low- and high-dose monkeys.
19. Cage-rested "sham" monkeys deprived of daily operant testing during the 1-year exposure period had significantly smaller CA3 stratum oriens volume, smaller basilar dendritic length, but greater volume density of synapses than "sham" monkeys performing the operant tasks throughout exposure. It is possible that this finding represents a kind of "isolation" or "withdrawal from stimulation" effect.
20. Smoke-exposure, whether to low, high, or extracted MJ, tended to block the deprivation effect, suggesting a possible compensatory effect of smoking on CA3 neurons.
21. Although chronic MJ smoke exposure cannot now be construed as neurotoxic to monkey hippocampal CA3 neurons, further similar analyses of CA1, dentate granule cells, and cerebellar granule cells are underway.
22. The evaluation methods employed in the present study were sufficiently sensitive to detect and describe the effects of differential training in young rhesus monkeys.
23. Withholding daily training selectively decreased the volume of the stratum oriens layer of hippocampal CA3.
24. Withholding training also selectively decreased the mean basilar dendritic length of CA3 pyramidal neurons. Since this portion of the dendritic tree, in aggregate, forms the CA3 stratum oriens, its widespread reduction in length could account for the decreased volume of CA3 stratum oriens.
25. Withholding training increased synaptic density in the vicinity of CA3 stratum pyramidale of the hippocampus.
26. Cannabis smoke, whether extracted or not, significantly reversed the shortening of CA3 basilar dendrites produced by differential training.

OVERALL CONCLUSION

These data indicate that long-term inhalation of MJ smoke does not compromise the general health of the rhesus monkey. Daily MJ smoke exposure served as a chronic physiological stressor, however, as evidenced by sustained increases in urinary free cortisol excretion. Behavioral assessment of the monkeys suggested that during daily exposure to MJ smoke, a decrease in "motivation" to respond for food occurred. Chronic MJ smoke exposure failed to result in any detectable neurochemical alterations 7 months after exposure. Neurohistological and electromicroscopy studies of the brain 7 months after the cessation of MJ smoke exposure failed to show any treatment-related differences in hippocampal volume, neuronal size, number or length of CA3 pyramidal cell dendrites, synaptic width, or density. Together, these data indicate that while behavioral and neuroendocrine effects are observed during MJ smoke exposure in the monkey, residual neuropathological and neurochemical effects of MJ exposure were not observed seven months after the year-long MJ smoke exposure regimen.

ACKNOWLEDGMENTS

The authors thank the collaborators, Drs. Etsuro Uemura and Gene Schulze and Ms. Annette Andrews, for their contributions to this study and express their appreciation for the technical assistance of Michael Gillam, Matthew Fogle, Jim Craven, Robert Rountree, and Sterling Wilson. Special thanks also go to the animal care staff at the NCTR: Vera Campbell, Cissy Freeman, James Henderson, Randy Thompson, Arnold Tripp, Gail Wagoner, O.T. Watson, Josie Watson, Betty White, and Michael Zornek. Without their efforts, these studies would not have been possible. The authors wish to thank Drs. Don McMillan, Galen Wenger, Roger Jenkins, Tom Bucci, Roger Brown, and Monique Braude for their important consultations, Richard Allen and Paul Felton for their statistical evaluations, and Barbara Jacks for her secretarial assistance. This research was supported by an Interagency Agreement between the National Institute on Drug Abuse and the National Center for Toxicological Research (NIDA IAG #224-83-0005) and NCTR (Experiment 6230).

REFERENCES

1. **Harper, J. W., Heath, R. G., and Myers, W. A.,** Effects of cannabis sativa on ultrastructure of the synapse in monkey brain, *J. Neurosci. Res.*, 3, 87, 1977.
2. **Heath, R. G., Fitzjarrell, A. T., Fontana, C. J., and Garey, R. E.,** Cannabis sativa: effects on brain function and ultrastructure in Rhesus monkeys, *Biol. Psychiatr.*, 15, 657, 1980.
3. **Stiglick, A. and Kalant, H.,** Learning impairment in the radial-arm maze following prolonged cannabis treatment in rats, *Psychopharmacology*, 77, 117, 1982a.
4. **Stiglick, A. and Kalant, H.,** Residual effects of prolonged cannabis administration on exploration and did performance in rats, *Psychopharmacology*, 77, 124, 1982b.
5. **Stiglick, A. and Kalant, H.,** Behavioral effects of prolonged administration of delta-9-tetrahydrocannabinol in the rat, *Psychopharmacology*, 80, 325, 1983.
6. **Stiglick, A. and Kalant, H.,** Residual effects of chronic cannabis treatment on behavior in mature rats, *Psychopharmacology*, 85, 436, 1985.
7. **Stiglick, A., Llewellyn, M. E., and Kalant, H.,** Residual effects of prolonged cannabis treatment on shuttle-box avoidance in the rat, *Psychopharmacology*, 84, 476, 1984.
8. **Scallet, A. C., Uemura, E., Andrews, A., Ali, S. F., McMillan, D. E., Paule, M. G., Brown, R. M., and Slikker, W., Jr.,** Morphometric studies of the rat hippocampus following chronic delta-9-tetrahydrocannabinol (THC), *Brain Res.*, 436, 193, 1987.
9. **Lorente de No, R.,** Studies on the structure of cerebral cortex, *J. Psychol. Neurol.*, 45(6), 381, 1933.
10. **Landfield, P. W., Cadwallader, L. B., and Vinsant, S.,** Quantitative changes in hippocampal structure following long-term exposure to delta-9-tetrahydrocannabinol: possible mediation by the glucocorticoid system, *Brain Res.*, 443, 47, 1988.
11. **Schwartz, R. H.,** Marijuana: an overview, *Chem. Depend. Pediatr. Clin. North Am.*, 34(2), 305, 1987.
12. **Paule, M. G., Bailey, J. R., Slikker, W., Jr., and Brown, R. M.,** Estimation of plasma delta-9-tetrahydrocannabinol (THC) levels by carboxyhemoglobin measurement in rhesus monkeys after exposure to marijuana smoke from cigarettes of known THC content, in *Marijuana: An International Research Report*, Chester, G., Consroe, P., and Musty, R., Eds., Australian Government Publication Service, Canberra, Australia, 1988, 259.
13. **Bailey, J. R., Cunny, H. C., Paule, M. G., and Slikker, W., Jr.,** Fetal disposition of delta-9-tetrahydrocannabinol (THC) during late pregnancy in the rhesus monkey, *Toxicol. Appl. Pharmacol.*, 90, 315, 1987.
14. **Barnett, G., Chiang, C.-W. N., Perez-Reyes, M., and Owens, S. M.,** Kinetic study of smoking marijuana, *J. Pharamcokinet. Biopharmaceut.*, 10(5), 494, 1982.
15. **Bourne, G. H.,** Collected anatomical and physiological data from the rhesus monkey, in *The Rhesus Monkey, Vol. I, Anatomy and Physiology*, Bourne, G. H., Ed., Academic Press, New York, 1975, 1.
16. **Neto, J. P., Junes, J. P., and Carvalho, F. V.,** The effects of chronic cannabis treatment upon brain 5-hydroxytryptamine, plasma corticosterone and aggressive behavior in female rats with different hormonal status, *Psychopharmacologia*, 42(2), 195, 1975.

17. **Kubena, R. K., Perhach, J. L., Jr., and Barry, H., III,** Corticosterone elevation mediated centrally by delta 1-tetrahydrocannabinol in rats, *Eur. J. Pharmacol.*, 14(1), 89, 1971.
18. **Mitra, G., Poddar, M. K., and Ghosh, J. J.,** Interaction of delta 9-tetrahydrocannabinol with reserpine, phenobarbital, and LSD-25 on plasma and adrenal corticosterone, *Toxicol. Appl. Pharmacol.*, 42(3), 505, 1977.
19. **Pertwee, R. G.,** Tolerance to the effect of delta 1-tetrahydrocannabinol on corticosterone levels in mouse plasma produced by repeated administration of cannabis extract or delta 1-tetrahydrocannabinol, *Br. J. Pharmacol.*, 51(3), 391, 1974.
20. **Sapolsky, R. and McEwen, B.,** Prolonged glucocorticoid exposure reduces hippocampal neuron number: implications for aging, *J. Neurosci.*, 5(5), 1221, 1985.
21. **Patel, V., Borysenko, M., Kumar, M. S., and Millard, W. J.,** Effects of acute and subchronic delta 9-tetrahydrocannabinol administration on the plasma catecholamine, beta-endorphin, and corticosterone levels and splenic natural killer cell activity in rats, *Proc. Soc. Exp. Biol. Med.*, 180(2), 400, 1985.
22. **Sassenrath, E. N. and Chapman, L. F.,** Tetrahydrocannabinol-induced manifestations of the "marihuana syndrome" in group-living macaques, *Fed. Proc. Fed. Am. Soc.*, 34, 1666, 1975.
23. **Hollister, L. E.,** Steroids and moods: correlations in schizophrenics and subjects treated with lysergic acid diethylamide (LSD), mescaline, tetrahydrocannabinol, and synhexyl, *J. Clin. Pharmacol. J. New Drugs*, 9(1), 24, 1969.
24. **Hollister, L. E., Moore, F., Kanter, S., and Noble, E.,** 1-Tetrahydrocannabinol, synhexyl and marijuana extract administered orally in man: catecholamine excretion, plasma cortisol levels and platelet serotonin content, *Psychopharmacologia*, 17(4), 354, 1970.
25. **Cruickshank, E. K.,** Physical assessment of 30 chronic cannabis users and 30 matched controls, *Ann. N.Y. Acad. Sci.*, 282, 162, 1976.
26. **Kolodny, R. C., Masters, W. H., Kolodner, R. M., and Toro, G.,** Depression of plasma testosterone levels after chronic intensive marihuana use, *N. Engl. J. Med.*, 290(16), 872, 1974.
27. **Turner, C. D. and Bagnara, J. T.,** The adrenal cortex: steroidogenic tissue, in *General Endocrinology*, 6th ed., Saunders, Philadelphia, 1971, 324.
28. **Mason, J. W.,** Specificity in the organization of neuroendocrine research profiles, in *Frontiers in Neurology and Neuroscience Research*, Seeman, P. and Brown, G. M., Eds., University of Toronto Press, Toronto, 1974, 68.
29. **Riley, V.,** Biobehavioral factors in animal work on tumorigenesis, in *Perspectives in Behavioral Medicine*, Weiss, S. M., Herd, J. A., and Fox, B. H., Eds., Academic Press, New York, 1981, 183.
30. **Riley, V.,** Psychoneuroendocrine influences on immunocompetence and neoplasia, *Science*, 212(4499), 1100, 1981a.
31. **Selye, H.,** Confusion and controversy in the stress field, *J. Hum. Stress*, 1(2), 37, 1975.
32. **Mason, J. W.,** A historical view of the stress field, *J. Hum. Stress*, 1(1), 6, 1975.
33. **Ader, R.,** *Psychoneuroimmunology*, Ader, R., Ed., Academic Press, New York, 1981.
34. **Holden, C.,** Behavioral medicine: an emergent field [news], *Science*, 209(4455), 479, 1980.
35. **Slikker, W., Jr., Paule, M. G., Ali, S. F., Scallet, A. C., and Bailey, J. R.,** Chronic marijuana smoke exposure in the rhesus monkey. I Plasma cannabinoid and blood carboxyhemoglobin concentrations and clinical chemistry parameters, *Fund. Appl. Toxicol.*, 17, 321, 1991.

36. **Paton, W. D. M. and Pertwee, R. G.,** The pharmacology of cannabis in animals, in *Marijuana: Chemistry, Pharmacology, Metabolism, and Clinical Effects*, Mechoulam, R., Ed., Academic Press, New York, 1973a, 191.
37. **Paton, W. D. M. and Pertwee, R. G.,** The action of cannabis in man, in *Marijuana: Chemistry, Pharmacology, Metabolism, and Clinical Effects*, Mechoulam, R., Ed., Academic Press, New York, 1973b, 287.
38. **Vachon, L., Sulkowski, A., and Rich, E.,** Marihuana effects on learning, attention and time-estimation, *Psychopharmacology*, 39, 1, 1974.
39. **Domino, E. F., Rennick, P., and Pearl, J. H.,** Short-term neuropsychopharmacological effects of marihuana smoking in experienced male users, in *Pharmacology of Marijuana,* Braude, M. D. and Sxara, S., Eds., Raven Press, New York, 1976, 393.
40. **Essman, E. J.,** Marijuana intoxication in rats: interruption of recent memory and effect on brain concentration of delta-9-tetrahydrocannabinol, *Psychol. Rep.*, 55, 563, 1984.
41. **Zimmerberg, B., Glick, S. D., and Jarvik, M. E.,** Impairment of recent memory by marihuana and THC in rhesus monkeys, *Nature (London)*, 333, 343, 1971.
42. **Elsmore, T. F.,** Effects of delta-9-tetrahydrocannabinol on temporal and auditory discrimination performance of monkeys, *Psychopharmacology*, 26, 62, 1972.
43. **Hicks, R. E., Gualtieri, C. T., Mayo, J. P., and Perez-Reyes, M.,** Cannabis, atropine and temporal information processing, *Neuropsychobiology*, 12, 229, 1984.
44. **Adams, P. M. and Barratt, E. S.,** The effects of marijuana extract on two-choice discrimination learning in the squirrel monkey, *Physiol. Psychol.*, 4, 155, 1976.
45. **McGlothin, W. H. and West, L. J.,** The marijuana problem: an overview, *Am. J. Psychiatr.*, 125, 370, 1968.
46. **Smith, D. E.,** Acute and chronic toxicity of marijuana, *J. Psychedel. Drugs*, 2, 37, 1968.
47. **Kolansky, H. and Moore, W. T.,** Toxic effects of chronic marijuana use, *JAMA*, 222, 35, 1972.
48. **Foltin, R. W., Fischman, M. W., Brady, J. V., Bernstein, D. J., Capriotti, R. M., Nellis, M. J., and Kelly, T. H.,** Motivational effects of smoked marijuana: behavioral contingencies and low-probability activities, *J. Exp. Anal. Behav.*, 53, 5, 1990.
49. **Hollister, L. E.,** Health aspects of cannabis, *Pharmacol. Rev.* 38(1), 1, 1986.
50. **Paule, M. G., Allen, R. R., Bailey, J. R., Scallet, A. C., Ali, S. F., Brown, R. M., and Slikker, W., Jr.,** Chronic marijuana smoke exposure in the rhesus monkey. II. Effects on progressive ratio and conditioned position responding, *J. Pharmacol. Exp. Ther.*, 280, 210, 1992.
51. **Schulze, G. E., McMillan, D. E., Bailey, J. R., Scallet, A. C., Ali, S. F., Slikker, W., Jr., and Paule, M. G.,** Acute effects of delta-9-tetrahydrocannabinol (THC) in rhesus monkeys as measured by performance in a battery of complex operant tests, *J. Pharmacol. Exp. Ther.*, 245, 178, 1988.
52. **Schulze, G. E., McMillan, D. E., Bailey, J. R., Scallet, A. C., Ali, S. F., Slikker, W., Jr., and Paule, M. G.,** Acute effects of marijuana smoke in rhesus monkeys as measured by performance in a battery of complex operant tests, *Life Sci.*, 45(6), 465, 1989.
53. **Paule, M. G., Schulze, G. E., and Slikker, W., Jr.,** Complex brain function in monkeys as a baseline for studying the effects of exogenous compounds, *Neurotoxicology*, 9, 463, 1988.
54. **Hodos, W.,** Progressive ratio as a measure of reward strength, *Science*, 134, 943, 1961.
55. **Goldman, P. S. and Rosvold, H. E.,** Localization of function within the dorsolateral prefrontal cortex of the rhesus monkey, *Exp. Neurol.*, 27, 291, 1970.

56. **Goldman, P. S., Rosvold, H. E., and Mishkin, M.,** Selective sparing of function following prefrontal lobectomy in infant monkeys, *Exp. Neurol.*, 29, 221, 1970.
57. **Kojima, S., Kojima, M., and Goldman-Rakic, P. S.,** Operant behavioral analysis of memory loss in monkeys with prefrontal lesions, *Brain Res.*, 248, 51, 1982.
58. **Hodos, W. and Kalman, G.,** Effects of increment size and reinforcer volume on progressive ratio performance, *J. Exp. Anal. Behav.*, 6, 387, 1963.
59. **Roberts, D. C. S., Loh, E. A., and Vickers, G.,** Self-administration of cocaine on a progressive ratio schedule in rats: dose-response relationship and effect of haloperidol pretreatment, *Psychopharmacology*, 97, 535, 1989.
60. **Bedford, J. A., Bailey, L. P., and Wilson, M. C.,** Cocaine-reinforced progressive ratio performance in the rhesus monkey, *Pharmacol. Biochem. Behav.*, 9, 631, 1978.
61. **Griffiths, R. R., Bradford, L. D., and Brady, J. V.,** Progressive ratio and fixed ratio schedules of cocaine-maintained responding in baboons, *Psychopharmacology*, 65, 125, 1979.
62. **Wert, R. C. and Raulin, M. L.,** The chronic cerebral effects of cannabis use. I. Methodological issues and neurological findings, *Int. J. Addict.*, 21, 605, 1986a.
63. **Mello, N. K and Mendelson, J. H.,** Operant acquisition of marijuana by women, *J. Pharmacol. Exp. Ther.*, 235, 162, 1985.
64. **Mendelson, J. H., Kuehnle, J. C., Greenberg, I., and Mello, N. K.,** Operant acquisition of marijuana in man, *J. Pharmacol. Exp. Ther.*, 198, 42, 1976.
65. **Miles, C. G., Congreve, G. R. S., Gibbins, R. J., Marshman, J., Devenyi, P., and Hicks, R. C.,** An experimental study of the effects of daily cannabis smoking on behavior patterns, *Acta Pharmacol. Toxicol.*, 34(Suppl. 7), 1, 1974.
66. **Comitas, L.,** Cannabis and work in Jamaica: a refutation of the amotivational syndrome, *Ann. N.Y. Acad. Sci.*, 282, 24, 1976.
67. **Boulougouris, J. C., Liakos, A., and Stefanis, C.,** Social traits of heavy hashish users and matched controls, *Ann. N.Y. Acad. Sci.*, 282, 17, 1976.
68. **Page, J. B.,** The amotivational syndrome hypothesis and the Costa Rica study: relationships between methods and results, *J. Psychoact. Drugs*, 15, 261, 1983.
69. **Feinstein, A. R.,** Scientific standards in epidemiologic studies of the menace of daily life, *Science*, 242, 1257, 1988.
70. **Lantner, I. L.,** Marijuana use by children and teenagers: a pediatrician's view, in *Marijuana and Youth: Clinical Observations on Motivation and Learning*, DHHS Publication No. ADM 82-1186, U.S Government Printing Office, Washington, D.C., 1982, 84.
71. **Brill, N. Q., Crumpton, E., and Grayson, H. M.,** Personality factors in marijuana use: a preliminary report, *Arch. Gen. Psychiatry*, 24, 163, 1971.
72. **Brill, N. Q. and Christie, R. L.,** Marijuana use and psychosocial adaptation: follow-up study of a collegiate population, *Arch. Gen. Psychiatry*, 31(5), 713, 1974.
73. **Carter, W. E. and Doughty, P. L.,** Social and cultural aspects of cannabis use in Costa Rica, *Ann. N.Y. Acad. Sci.*, 282, 2, 1976.
74. **Johansson, E., Agurell, S., Hollister, L. E., and Halldin, M. M.,** Prolonged apparent half-life of delta-1-tetrahydrocannabinol in plasma of chronic marijuana users, *J. Pharm. Pharmacol.*, 40, 374, 1988a.
75. **Johansson, E., Sjovall, J., Noren, K., Agurell, S., Hollister, L., and Halldin, M. M.,** Analysis of delta-1-tetrahydrocannabinol (delta-1-THC) in human plasma and fat after smoking, in *Marijuana: An International Research Report,* Chester, G., Consroe, P., and Musty, R., Eds., Australian Government Publication Service, Canberra, Australia, 1988b, 291.

76. **O'Neal, M. F., Means, L. W., Porter, J. H., Rosecrans, J. A., and Mokler, D. J.,** Rats that acquire a THC discrimination more rapidly are more sensitive to THC and faster in reaching operant criteria, *Pharmacol. Biochem. Behav.*, 29, 67, 1988.
77. **Johnson, B. A.,** Psychopharmacological effects of cannabis, *Br. J. Hosp. Med.*, 43(2), 114, 1990.
78. **Maugh, T. H.,** Marihuana (II): does it damage the brain?, *Science*, 185, 775, 1974.
79. **Branch, M. N., Dearing, M. E., and Lee, D. M.,** Acute and chronic effects of delta-9-tetrahydrocannabinol on complex behavior of squirrel monkeys, *Psychopharmacology*, 71, 247, 1980.
80. **Beardsley, P. M., Balster, R. L., and Harris, L. S. J.,** Dependence on tetrahydrocannabinol in rhesus monkeys, *Pharmacol. Exp. Ther.*, 229, 311, 1986.
81. **Fredericks, A. B. and Benowitz, N. L.,** An abstinence syndrome following chronic administration of delta-9-tetrahydrocannabinol in rhesus monkeys, *Psychopharmacology*, 71, 201, 1980.
82. **Stadnicki, S. W., Schaeppi, U., Rosenkrantz, H., and Braude, M. C.,** Crude marihuana extract EEG and behavioral effects of chronic oral administration in rhesus monkeys, *Psychopharmacologia (Berlin)*, 37, 225, 1974.
83. **Bailey, J. R., Paule, M. G., Ali, S. F., Scallet, A. C., and Slikker, W., Jr.,** Effects of chronic marijuana smoke exposure on urinary cortisol (UC) excretion in the rhesus monkey, *Pharmacologist*, 30(3), 115, 1988.
84. **Paule, M. G., Bailey, J. R., Scallet, A. C., Ali, S. F., and Slikker, W., Jr.,** Chronic marijuana (MJ) smoking and urinary cortisol (UC) excretion in the monkey, *Proc. West. Pharmacol. Soc.*, 32, 189, 1989.
85. **Domino, E. F.,** Cannabinoids and the cholinergic system, *J. Clin. Pharmacol.*, 21(Suppl.), 2495, 1981.
86. **Ali, S. F., Newport, G. D., Scallet, A. C., Gee, K. W., Paule, M. G., Brown, R. M., and Slikker, W., Jr.,** Effects of chronic delta-9-tetrahydrocannabinol (THC) administration on neurotransmitter concentrations and receptor binding in the rat brain, *Neurotoxicology*, 10, 491, 1989.
87. **Sakurai-Yamashita, Y., Kataoka, Y., Fujiwana, M., Mine, K., and Ueki, S.,** Delta-9-tetrahydrocannabinol facilitates striatal dopaminergic transmission, *Pharmacol. Biochem. Behav.*, 33, 397, 1989.
88. **Cheong Tan, J. M. N., Gerhardt, G. A., Friedemann, M., Etgen, A. M., Rose, G. M., Sharpless, N. S., and Gardner, E. L.,** The effects of delta-9-THC on potassium-evoked release of dopamine in the rat caudate nucleus: an *in vivo* electrochemical and *in vivo* microdialysis study, *Brain Res.*, 451, 59, 1988.
89. **Truitt, E. B. and Anderson, S. M.,** Biogenic amine alterations produced in brain by tetrahydrocannabinols and their metabolites, *Ann. N.Y. Acad. Sci.*, 191, 68, 1972.
90. **Holtzman, D., Lovell, R. A., Jaffe, J. H., and Freedman, D. X.,** Delta-9-tetrahydrocannabinol neurochemical and behavioral effects in mice, *Science*, 163, 1464, 1969.
91. **Sofia, R. D., Dixit, B. N., and Barry, H.,** Effects of delta-9-tetrahydrocannabinol on serotonin metabolism in the rat brain, *Life Sci.*, 10, 425, 1971.
92. **Poddar, M. K., Biswar, B., and Ghosh, J. J.,** Delta-9-tetrahydrocannabinol and brain biogenic amines, in *Drug and Central Synaptic Transmission*, Bradley, P. B. and Dhawan, B. N., Eds., McMillan Press, London, 1976, 193.
93. **Bloom, A. S. and Kiernan, C. J.,** Interaction of ambient temperature with the effects of delta-9-tetrahydrocannabinol on brain catecholamine synthesis and plasma corticosterone levels, *Psychopharmacology*, 67, 215, 1980.

94. **Miczek, K. A. and Dixit, B. N.,** Behavioral and biochemical effects of chronic delta-9-tetrahydrocannabinol in rats, *Psychopharmacology*, 67, 195, 1980.
95. **Sofia, R. D.,** A paradoxical effect for delta-1-tetrahydrocannabinol on rectal temperature in mice, *Res. Commun. Chem. Pathol. Pharmacol.*, 4, 281, 1972.
96. **Pertwee, R. G., Greentree, S. G., and Swift, P. A.,** Drugs which stimulate or facilitate central GABAergic transmission interact synergistically with delta-9-THC to produce marked catalepsy in mice, *Neuropharmacology*, 27, 1265, 1988.
97. **Poddar, M. K. and Dewey, W. L.,** Effects of cannabinoids on catecholamine uptake and release in hypothalamic striatal synaptosomes, *J. Pharmacol. Exp. Ther.*, 214, 63, 1980.
98. **Bloom, A. S.,** Effect of delta-9-tetrahydrocannabinol on the synthesis of dopamine and norepinephrine in mouse brain synaptosomes, *J. Pharmacol. Exp. Ther.*, 221, 97, 1982.
99. **Hershkowitz, M., Goldman, R., and Raz, A.,** Effect of cannabinoids on neurotransmitter uptake ATPase activity and morphology of mouse brain synaptosomes, *Biochem. Pharmacol.*, 26, 1327, 1977.
100. **Hershkowitz, M. and Szechtman, H.,** Pretreatment with delta-9-tetrahydrocannabinol and psychoactive drugs: effects on uptake of biogenic amines and on behavior, *Eur. J. Pharmacol.*, 59, 267, 1979.
101. **Banerjee, A., Snyder, S. H., and Mechoulam, R.,** Cannabinoids: influence on neurotransmitter uptake in rat brain synaptosomes, *J. Pharmacol. Exp. Ther.*, 194, 74, 1975.
102. **Bloom, A. S.,** Effects of cannabinoids on neurotransmitter receptors in the brain, in *The Cannabinoids: Chemical, Pharmacological and Therapeutic Aspects*, Agurell, S., Dewey, W., and Willette, R., Eds., Academic Press, Orlando, FL, 1984, 217.
103. **Bloom, A. S. and Hillard, C. J.,** in *Marihuana '84 Proceedings of the Oxford Symposium on Cannabis*, Harvey, D. J., Ed., IRL Press Limited, Oxford, England, 1984, 217.
104. **Hillard, C. J. and Bloom, A. S.,** Further studies of the interaction of delta-9-tetrahydrocannabinol with the beta-adrenergic receptor, in *The Cannabinoids: Chemical, Pharmacological and Therapeutic Agents*, Agurell, S., Dewey, W., and Willette, R., Eds., Academic Press, Orlando, FL, 1984, 591.
105. **Glowinski, J. and Iversen, L. L.,** Regional studies of catecholamines in the rat brain. I. The disposition of ^{3}H-norepinephrine, ^{3}H-dopamine and ^{3}H-DOPA in various regions of the brain, *J. Neurochem.*, 13, 655, 1966.
106. **Brown, R. M., Crane, A. M., and Goldman, P. S.,** Regional distribution of monoamines in the cerebral cortex and subcortical structures of the rhesus monkey: concentration and *in vivo* synthesis rates, *Brain Res.*, 168, 133, 1979.
107. **Szebenyi, E. S.,** *Atlas of Macaca mulatta*, University Press, Rutherford, NJ, 1970.
108. **Ali, S. F., Slikker, W., Jr., Newport, G. D., and Goad, P. T.,** Cholinergic and monoaminergic alteration in the mouse central nervous system following acute trimethyltin exposure, *Acta Pharmacol. Toxicol.*, 59, 179, 1986.
109. **Lowry, O. H., Rosebrough, N. J., Farr, A. L., and Randell, R. J.,** Protein measurement with the folin phenol reagent, *J. Biol. Chem.*, 193, 265, 1951.
110. **Domino, E. F.,** Neurospychopharmacological studies of marijuana: some synthetic and natural derivatives in animals and man, *Ann. N.Y. Acad. Sci.*, 191, 166, 1971.
111. **Domino, E. F., Donelson, A. C., and Tuttle, T.,** Effects of delta-9-tetrahydrocannabinol on regional brain acetylcholine, *Chol. Mech. Psychopharmacol.*, 25, 673, 1978.
112. **Schuster, C. R.,** The United States' drug abuse scene: an overview, *Clin. Chem.*, 33/11(B), 7B, 1987.

113. **Moran, T.,** Society should take the risk, *JAMA,* 261, 2443, 1989.
114. **Wert, R. C. and Raulin, M. L.,** The chronic cerebral effects of cannabis use. II. Psychological findings and conclusions, *Int. J. Addict.*, 21, 629, 1986b.
115. **Slikker, W., Jr., Ali, S. F., Scallet, A. C., Frith, C. H., Newport, G. D., and Bailey, J. R.,** Neurochemical and neurohistological alterations in the rat and monkey produced by orally administered methylenedioxymethamphetamine (MDMA), *Toxicol. Appl. Pharm.*, 94, 448, 1988.
116. **Collonier, M. and Beaulieu, C.,** An empirical assessment of stereological formulae applied to the counting of synaptic disks in the cerebral cortex, *J. Comp. Neurol.*, 231, 175, 1985.
117. **Greenough, W. T.,** Enduring brain effects of differential experience and training, in *Neural Mechanisms of Learning and Memory*, Rosenzweig, M. R. and Bennett, E. L., Eds., MIT Press, Cambridge, MA, 1976.

8 Cannabinoid Interaction with Brain Reward Systems — The Neurobiological Basis of Cannabinoid Abuse

INTRODUCTION

Cannabis (marijuana) is one of the oldest psychoactive botanicals known to mankind.[1] Among the many pharmacological properties of this fascinating plant is a dose-dependent euphorigenic potency deriving primarily from the psychoactive cannabis constituent Δ^9-tetrahydrocannabinol (Δ^9-THC). From this euphorigenic property derives marijuana's use as a recreational drug and its habit-forming and abuse potential. It is estimated that more than 20 million Americans are current marijuana users,[2] and marijuana currently serves (along with ethanol) in both North America and Europe as a "gateway drug", introducing individuals (often teenagers) to the use of euphoriant drugs, from which some then progress to more addicting and destructive drugs such as cocaine and heroin.[2,3] Outright cannabis addiction, with obsessional drug-seeking and compulsive drug-taking behavior, is relatively rare with low-dose cannabis preparations (e.g., marijuana) but common with high-dose preparations (e.g., hashish). Thus, Δ^9-THC can be categorized as an abusable and habit-forming drug, and is presently the most commonly used illicit drug in North America and Europe.[2]

COMMON PROPERTIES OF HABIT-FORMING DRUGS

Although the incidence of drug abuse in human society is large, the number of abused compounds is startlingly small. *Chemical Abstracts* lists more than 14 million different chemicals known to mankind; yet the number of chemicals voluntarily self-administered by human beings is not more than a few score compounds, primarily ethanol, cocaine, amphetamines, opiates, nicotine, caffeine, benzodiazepines, barbiturates and barbiturate-like sedative-hypnotics, hallucinogens, volatile solvents, dissociative and nondissociative anesthetics, and cannabis. Three obvious questions arise: (1) what do these habit-forming compounds have in common, (2) what distinguishes these compounds from the 14 million other compounds known to mankind, and (3) (most saliently) what makes these compounds appealing to self-administer while the vast majority of chemicals have no such appeal?

Until approximately 20 years ago, the standard medical explanation for compulsive drug use was essentially twofold.[4] First, it was widely held that compulsive drug users have personality disturbances that antedate but lead directly to their drug-taking behavior, the so-called "addict" personality that turns to drug use to self-medicate preexisting psychopathology. Second, it was equally widely held that drug users, once caught up in a vicious cycle of drug administration, tolerance, physical dependence, withdrawal, and readministration, come to self-administer habit-forming drugs primarily to ward off the unpleasant consequences of withdrawal. The inadequacy of these explanations was shown during the 1960s by demonstrations that laboratory animals, presumably devoid of personality disturbances, voluntarily (often avidly) self-administer the same habit-forming drugs that human addicts use,[5-7] and, furthermore, that they do so in the absence of tolerance, physical dependence, withdrawal, or any prior history of drug-taking. Furthermore, laboratory animals will *not* self-administer any of the millions of other chemicals not abused by humans. Obviously, the explanation for the powerful craving and drug-taking behavior induced by habit-forming drugs must reside elsewhere.

NEUROBIOLOGICAL SUBSTRATES ACTIVATED BY HABIT-FORMING DRUGS

As early as the 1940s, Donald Hebb, the brilliant pioneer of modern neuroscience, suggested that habit-forming drugs act as reinforcers by directly activating reinforcement circuits in the brain.[8] Although inherently plausible, this notion had to await empirical testing until the seminal discovery by James Olds and Peter Milner that such circuits exist and can be directly activated by electrical stimulation delivered through electrodes deep in the brain.[9] The reward elicited by such stimulation, termed "intracranial self-stimulation" or "brain-stimulation reward", and the brain substrates that support it, have been much studied and well characterized over the last 35 years.[10-12] As studied with behavioral paradigms in which animals volitionally self-administer the rewarding electrical stimulation themselves, such direct brain-stimulation reward is intensely powerful. Hungry animals ignore food to get it; thirsty animals ignore water to get it. Animals even endure pain to get it. Response rates of animals self-stimulating for electrical brain reward are extremely high, often in excess of 100 lever presses per minute. In short, electrical brain-stimulation reward is one of the most powerful rewards known to biology, rivaled only by the reward of the most intensely habit-forming drugs (e.g., cocaine). The few human studies of electrical stimulation of brain reward areas confirm this, the human experience being one of intense subjective pleasure or euphoria.

Evidence that habit-forming drugs might derive their rewarding properties by activating such brain reward circuits was presented by Killam and colleagues as early as 1957,[13] and has been amply confirmed by 3 decades of work since (see below). This acute enhancement of brain reward mechanisms now appears to be the single essential pharmacological commonality of habit-forming drugs, and the hypothesis that habit-forming drugs act on these brain mechanisms to produce the subjective reward that constitutes the "high" or "rush" or "hit" sought by drug users is, at present, the most compelling hypothesis available on the neurobiology of drug abuse.[14]

Neuroanatomy of Brain Reward Substrates

The neuroanatomical substrates of brain reward, upon which habit-forming drugs act to induce subjective reward, were initially unclear. Anatomical mapping studies of brain-stimulation reward carried out in the 1950s and 1960s demonstrated that electrical brain reward could be elicited in laboratory rats, cats, and monkeys from a large variety of brain stem, midbrain, and forebrain loci, including the ventral tegmental area, substantia nigra, hypothalamus, medial forebrain bundle, septum, amygdala, neostriatum, nucleus accumbens, ventral forebrain olfactory nuclei, and portions of the cingulate and frontal cortices.[15-21] Such a hodgepodge of sensory, motor, limbic, midbrain, diencephalic, and cortical domains made little sense, although even the early workers noted that the vast majority of brain sites positive for electrical brain-stimulation reward correspond to the aggregate of ascending and descending tracts known as the medial forebrain bundle, the projections of which extend from brain stem to cortex.

Neurochemistry of Brain Reward Substrates

In the mid-1960s, pioneering Scandinavian neuroanatomists, using the technique of histofluorescence microscopy, began to illuminate the monoaminergic anatomy of the brain;[22-25] a correspondence was noted between sites positive for brain-stimulation reward and the mesotelencephalic dopamine (DA) system, the major projections of which are carried through the medial forebrain bundle.[26] Guided by this anatomical correspondence, workers in many laboratories began to study the effects of pharmacological manipulation of DA neurotransmission on brain-stimulation reward. My colleagues and I found that selective inhibition of tyrosine hydroxylase profoundly inhibited brain reward in monkeys, but that selective inhibition of DA-β-hydroxylase did not, prompting us to postulate[27] that brain reward was critically dependent upon the functional integrity of DA neurotransmission. In 1972, Crow suggested that direct activation of the cells or axons of the mesolimbic and mesostriatal DA systems was rewarding, and that DA was the crucial neurotransmitter of at least one major reward system in the brain.[28] In 1980, Corbett and Wise, using movable electrodes to map brain reward substrates in the ventral mesencephalon, showed that brain-stimulation reward thresholds were a function of the density of DA elements surrounding the electrode tip.[29] Also compelling in this regard are the facts that DA blockade disrupts brain-stimulation reward at all brain sites adequately tested[30-32] and that, following DA denervation that abolishes neostriatal brain-stimulation reward, direct electrical brain reward can be restored in the denervated neostriatum by DA reinnervation from embryonic substantia nigra transplants.[33] From these studies, and literally hundreds of other experiments,[32,34-37] it has become abundantly clear that brain reward is, in fact, critically dependent upon the functional integrity of DA neurotransmission within the mesotelencephalic DA systems.

However, the pleasingly simple notion that electrical brain-stimulation reward directly activates DA neurons was shown to be incorrect a full decade ago by a series of elegant electrophysiological studies carried out by Gallistel, Shizgal, Yeomans, et al.[38-42] What these workers did was to demonstrate that lateral hypothalamic and ventral tegmental electrical brain-stimulation reward involve a common neural fiber system that cannot be DA because its conduction direction is caudal and its conduction velocity implicates small myelinated fibers with absolute refractory periods of 0.5 to 1.2 msec and local potential decay time constants of approximately 0.1 msec. None of these neurophysiological properties agrees with those of the ascending mesotelencephalic DA systems. Thus, the DA neurons cannot be the "first-stage" reward-relevant neurons preferentially activated by electrical brain-stimulation reward. Rather, the DA neurons appear to constitute a crucial "second-stage" anatomic convergence within the reward circuitry of the brain,[37,43,44] upon which the "first-stage" neurons (those preferentially activated by rewarding electrical stimulation) synapse to form an "in series" reward-relevant circuit.[45,46] It is upon this "second-stage" DA convergence that habit-forming drugs apparently act to enhance brain reward[32,35,37,45,47] (see discussions below). Furthermore, it seems likely that only a small subset of these DA neurons are specialized for carrying reward-relevant information[37] (see Figure 1).

Evidence that Brain Reward Mechanisms Constitute the Essential Neural Substrate Activated by Habit-Forming Drugs

In support of the hypothesis that habit-forming drugs act on these brain mechanisms to produce their euphorigenic effects, and thus their abuse potential, many lines of evidence can be adduced. First, as noted above,

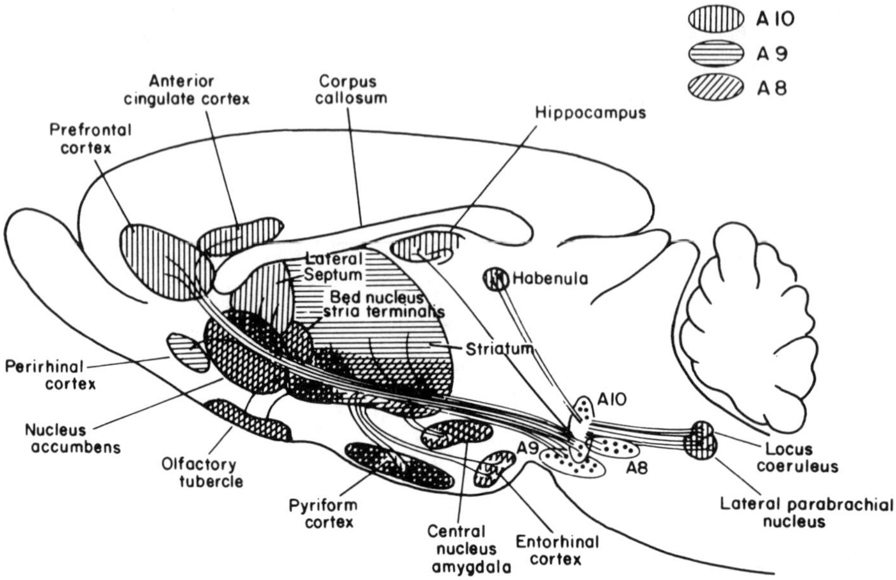

FIGURE 1. The mesotelencephalic dopamine (DA) circuitry of the mammalian (laboratory rat) brain. The primary brain-reward-relevant portion appears to be a subset of the mesolimbic projections originating in the ventral tegmental area (DA nucleus "A10") and terminating in the nucleus accumbens. (From Cooper, J. R., Bloom, F. E., and Roth, R. H., *The Biochemical Basis of Neuropharmacology,* 5th ed., Oxford University Press, New York, 1986. With permission.)

virtually all habit-forming drugs that have been adequately studied (including cocaine, amphetamines, alcohol, barbiturates, benzodiazepines, opiates, phencyclidine-like dissociative anesthetics, and cannabinoids) enhance brain-stimulation reward or lower brain reward thresholds in the mesotelencephalic DA system (see further discussion of this below). Second, most drugs with DA agonist properties are euphorigenic (and have consequential abuse potential), while most DA antagonists are dysphorigenic.[37,48] Third, virtually all habit-forming drugs that have been adequately studied have been found to enhance neuronal firing or DA release (or both) in the mesotelencephalic DA system (see further discussion of this below). Fourth, animals will work to receive microinjections of habit-forming drugs into mesotelencephalic DA loci, but not into other brain loci.[49-53] Fifth, lesions or pharmacological blockade of the mesotelencephalic DA circuitry markedly inhibit the rewarding properties of systemically administered habit-forming drugs.[54,55]

PARADIGMS FOR STUDYING HABIT-FORMING DRUG ACTION ON BRAIN REWARD MECHANISMS

For studying these brain reward mechanisms in the laboratory and the actions of habit-forming drugs thereon, a number of paradigms of electrical brain-stimulation reward have been developed.[37] Unfortunately, early studies (primarily in the 1950s, 1960s, and early 1970s) used simple response-rate measures for brain reward of fixed voltage or current and fixed electrical frequency. Such simple response-rate studies were inadequate on many grounds, not the least of which is that many habit-forming drugs are sedative and produce depressed responding for electrical brain reward when measured in a simple response-rate paradigm, even though the neural substrate for brain reward is clearly enhanced. Consequently, much of the early literature dealing with the effects of habit-forming drugs on brain reward is either extremely difficult to interpret or, worse still, leads to incorrect conclusions. For these reasons, attention in recent years has shifted toward threshold measures, response-pattern analyses, choice measures, and the use of "curve-shift" rate-frequency function measures analogous to the dose-response paradigm of traditional pharmacology.[37] In our laboratory, we have primarily used a decremental titrating-threshold brain reward paradigm[56-59] in which the animal presses a reward lever to self-administer rewarding brain stimulation that automatically decreases in intensity with successive lever presses. When this decremental stimulation passes through the threshold for activating the underlying neural substrate of reward, the animal presses a reset lever that does not deliver any brain reward, but resets the brain reward intensity back to its initial level. By analyzing reset values, we obtain a measure of the activation threshold (in microamperes of delivered current) of the neural system subserving brain reward that is independent of response rate and thus independent of the incidental sedation produced by some habit-forming drugs (e.g., barbiturates, benzodiazepines, and opiates) and the incidental motor stimulation produced by others (e.g., cocaine and amphetamines).

PLASTICITY OF (AND DRUG-INDUCED MODULATION OF) BRAIN REWARD MECHANISMS

The hypothesis that habit-forming drugs derive their euphorigenic effects, and thus their abuse potential, from acute enhancement of brain reward mechanisms carries with it a fundamental neurobiological assumption, i.e., that these brain reward mechanisms are, in fact, capable of acute functional up-regulation (to acutely enhance hedonic tone). This is by no means an irrational assumption, and is either explicit or implicit in virtually every modern theory of brain-behavior relationships and/or the neurobiology of motivation. In 1969, Pribram gave these assumptions a concrete and specific operational hypothesis; namely, that reinforcement is homeostatically regulated and that limbic circuits perform as biostats to set the range of function for reward.[60] Intrigued by Pribram's hypothesis and by the compelling arguments of Goddard[61] that one of the principal functions of the limbic circuitry of the amygdala is to modulate motivated behavior, and equally intrigued by McLardy's[62] hypothesis (based partly on cybernetic analyses) that portions of the hippocampus may act as a neural intensity-gradients detector-coder to boost and/or dampen threshold functioning of other neural systems connected to the hippocampus (such as the reinforcement circuitry coursing through the lateral hypothalamus), Jackson and I sought to determine whether brain reward thresholds are acutely modulated by concurrent stimulation of amygdaloid and/or hippocampal sites. We found[63] that brain-stimulation reward in monkeys *is* capable of rather dramatic degrees of modulation by concurrent limbic stimulation, with consistent enhancement of brain reward produced by concurrent electrical stimulation of the CA3 and CA4 regions of posterior hippocampus. Thus, my colleagues and I satisfied ourselves early on that brain reward mechanisms are capable of the modulatory plasticity required by the hypothesis that habit-forming drugs acutely enhance brain reward. That the DA substrates of the "second-stage" anatomic convergence within the reward circuitry, upon which habit-forming drugs apparently act to enhance brain reward, are also capable of this required modulatory plasticity is equally clear. Seeger and I demonstrated this to be the case in a series of experiments in which we pharmacologically induced DA receptor supersensitivity by chronic administration of DA receptor antagonists and showed that this significantly enhanced brain-stimulation reward (i.e., lowered brain reward thresholds) in monkeys.[58] We also demonstrated that this pharmacologically enhanced (up-regulated) brain reward sensitivity could be down-regulated by chronic DA agonist administration,[64] confirming that reward-relevant DA synaptic sensitivity has a two-way modulatory capability.

ACTIONS OF HABIT-FORMING DRUGS ON BRAIN REWARD

Enhancement of Electrical Brain-Stimulation Reward by Classical Habit-Forming Drugs

As studied using the electrical brain-stimulation reward threshold paradigm that we routinely use in our laboratory (described above), with electrodes in the medial forebrain bundle or ventral tegmental area, habit-forming drugs produce robust and highly replicable enhancement (i.e., threshold lowering) of brain-stimulation reward. In our laboratory, we have studied opiates,[65,66] barbiturates,[67,68] benzodiazepines, phencyclidine,[69-71] ketamine,[71] amphetamines, cocaine,[72] ethanol, and cannabis (see below). In every case, robust enhancement of brain-stimulation reward is seen. A typical example, in this case of acute brain reward enhancement by morphine, is shown in Figure 2.

Enhancement of DA Neurochemical Substrates Mediating Brain Reward by Classical Habit-Forming Drugs

If the ascending mesotelencephalic DA systems play the crucial role in brain reward that we and other researchers in the field currently ascribe to them, one would expect habit-forming drugs to act, at least indirectly, in a DA agonist-like fashion in these systems. For many years, this crucial derivative of the theory was untestable. Recently, however, two laboratory techniques have been developed that allow *in vivo* real-time determinations of neurotransmitter release in discrete brain loci of conscious, freely moving animals. These techniques are *in vivo* brain microdialysis and *in vivo* brain voltammetric electrochemistry.[73,74] For present purposes, it is not necessary to elaborate upon the technical details of these paradigms, only to appreciate that they have proven themselves to be valid and sensitive techniques for assessing real-time neurotransmitter release in discrete loci of the conscious brain. Using these techniques, as well as classical single-neuron electrophysiological recording techniques, it is now clear that all habit-

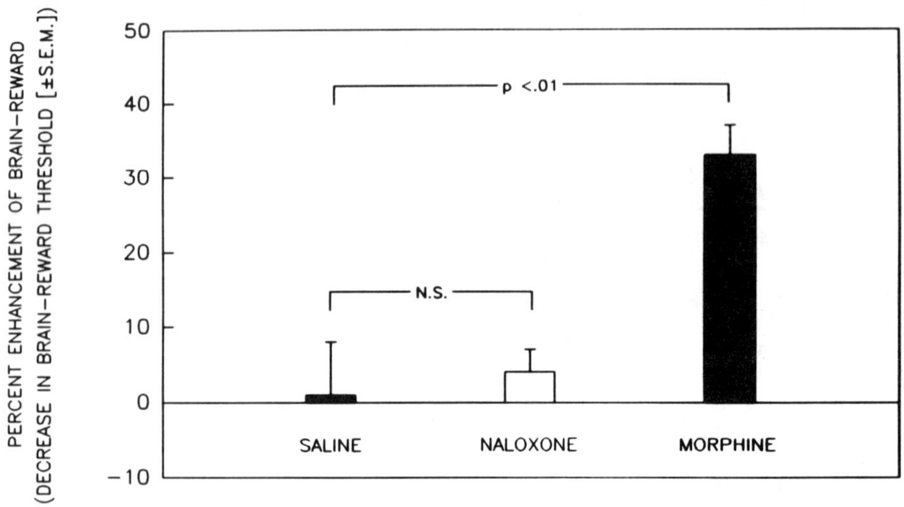

FIGURE 2. Enhanced brain reward following acute systemic administration of a prototypical habit-forming drug (5.0 mg/kg morphine, s.c. administration). Enhancement of brain reward is experimentally equivalent to a decrease in electrical brain-stimulation threshold level (see text) in the ventral tegmental area (DA nucleus "A10"). The effects of saline and naloxone (2.0 mg/kg, s.c. administration) are also shown. "N.S." ("not significant") denotes the lack of a statistically significant difference. Probability values shown are for the specific comparisons indicated on the figure. (From Seeger, T. F., Nazzaro, J. M., and Gardner, E. L., Selective inhibition of mesolimbic behavioral supersensitivity by naloxone, *Eur. J. Pharmacol.*, 65, 435, 1980; and from Nazzaro, J. M., Seeger, T. F., and Gardner, E. L., Morphine differentially affects ventral tegmental and substantia nigra brain reward thresholds, *Pharmacol. Biochem. Behav.*, 14, 325, 1981. With permission.)

forming drugs that have been adequately studied [including cocaine, amphetamines, opiates, caffeine, and cannabinoids (see below)] activate the reward-relevant mesotelencephalic DA systems.[37,75-86]

Enhancement of Brain Reward by "Atypical" Habit-Forming Drugs

The most widely used habit-forming drugs in the world are nicotine and ethanol.[2,87] For many years, it was not known whether these two ubiquitous habit-forming drugs could be fitted into this overall picture of habit-forming drug action on the crucial drug-sensitive "second-stage" DA fibers of the reward system that I have sketched above, and some workers claimed that these compounds "atypically" or "anomalously" fail to enhance brain re-

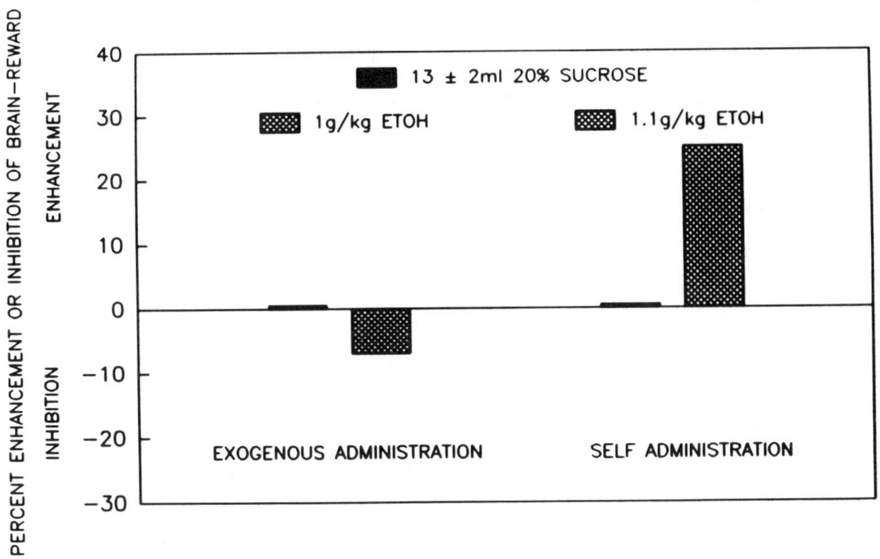

FIGURE 3. Differential effects of volitional vs. nonvolitional ethanol consumption on enhancement or nonenhancement of brain reward. Enhancement of brain reward is experimentally equivalent to a decrease in electrical brain-stimulation threshold level (see text) in the ventral tegmental area (DA nucleus "A10"). Both the volitional ("self administration") and nonvolitional ("exogenous administration") consumption was by oral route (i.e., volitional drinking in the case of the volitional consumption, and acute oral gavage in the case of the nonvolitional consumption).

ward mechanisms as do other habit-forming drugs. Recently, however, it has become clear that both nicotine and ethanol *do* fit into this conception of habit-forming drug action on brain reward mechanisms.[77,78,81] In fact, ethanol illustrates some additional interesting aspects of habit-forming drug action on brain reward. Both our laboratory and the laboratory of Kornetsky at Boston University have recently found that, when administered exogenously, ethanol has little or no facilitating effect on electrical brain-stimulation reward. However, when the very same animals are allowed to voluntarily self-administer the alcohol, brain reward is robustly and reliably enhanced (Figure 3), suggesting that there must be neural circuitry encoding volitional intent that feeds into the reward system and summates with the acute pharmacological sensitization produced by the drug. This may explain why, for many drugs of abuse, a more intense "high" is produced when self-administered than when exogenously administered. Another habit-forming drug long considered by some workers to be atypical is the dissociative anesthetic phencyclidine (PCP, "angel dust"). Much has been made in recent years of phencyclidine's interaction with a class of brain receptor that was originally claimed to be a totally unique receptor for phencyclidine,[88] that later was claimed to be identical to the sigma opiate receptor,[89,90] and that now is claimed to constitute an ion channel binding site within the *N*-methyl-D-aspartate excitatory amino acid receptor complex.[91] Regardless of the receptor site(s) to which phencyclidine binds, the obvious question arises — does phencyclidine, and the closely related dissociative anesthetic ketamine, have a final common action on the brain reward mechanisms sketched above that is similar to other habit-forming drugs? The answer appears to be an unequivocal "yes". First, we showed some years ago (Figure 4) that phencyclidine and ketamine enhance brain-stimulation reward in dose-dependent and naloxone-reversible (see further discussions of dose-dependence and naloxone-reversibility of habit-forming drug effects on brain reward below) manners,[69-71] findings that have recently been confirmed by Herberg.[92] More recently, we[82] and others[93,94] have shown that phencyclidine enhances presynaptic DA efflux in the mesotelencephalic reward-relevant circuitry, as measured by both *in vivo* voltammetric electrochemistry and *in vivo* brain microdialysis. In our hands,[82] phencyclidine's enhancement of presynaptic DA efflux appears to be action-potential independent (Figure 5). At present, then, it appears that all well-studied habit-forming drugs enhance DA-mediated brain reward mechanisms associated with the mesotelencephalic DA system to produce the subjective "high" sought by drug users.

Naloxone-Induced Attenuation of Brain Reward Enhancement Produced by Habit-Forming Drugs

Provocatively, we[65-70,95] and others[45,96,97] have found that the brain reward enhancement produced by all well-studied habit-forming drugs (belonging to very different chemical classes and pharmacological categories) is in every case significantly attenuated by the opiate antagonist naloxone, implicating an important endogenous opioid peptide (enkephalinergic or endorphinergic) link in the reward circuitry of the brain. An example, in this case of naloxone-attenuation of barbiturate-induced acute brain reward enhancement, is shown in Figure 6. Since naloxone significantly attenuates

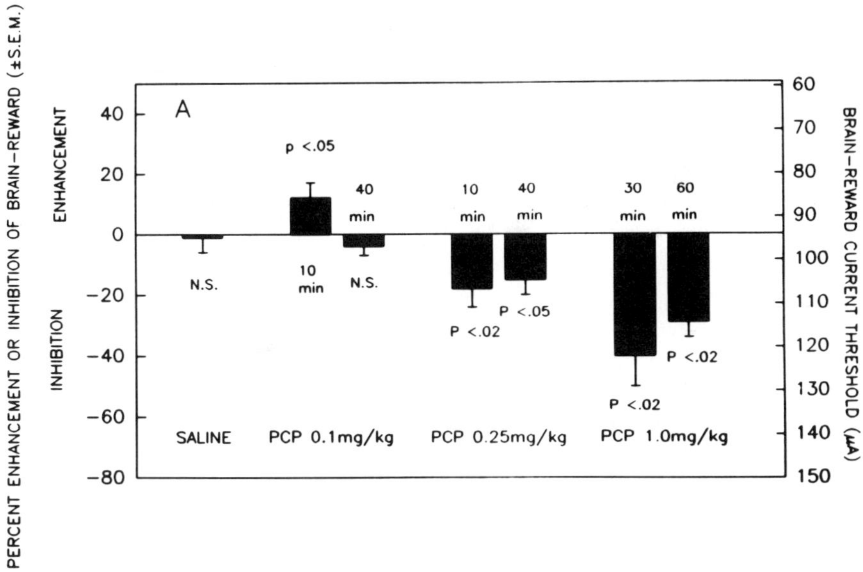

FIGURE 4. Dose-dependent enhancement or inhibition of brain reward following acute systemic administration of varying doses of phencyclidine (PCP), and naloxone-induced blockade of phencyclidine's dose-dependent enhancement and inhibition of brain reward. Enhancement of brain reward is experimentally equivalent to a decrease, and inhibition of brain reward to an increase, in electrical brain-stimulation threshold level (see text) in the ventral tegmental area (DA nucleus "A10"). (A) Effects of 0.1, 0.25, and 1.0 mg/kg PCP (i.p. administration) on brain reward at times indicated following PCP administration. Probability values given on each bar represent the degree of significance vs. saline control. "N.S." ("not significant") denotes lack of a statistically significant difference. (B) Effects of 0.1, 0.25, and 1.0 mg/kg PCP plus 2.0 mg/kg naloxone (i.p. administrations) on brain reward at times indicated following PCP administration. In every case, the naloxone was given 15 min prior to brain-stimulation reward testing. Probability values shown are for the specific comparisons indicated on the figure. (From Nazzaro, J. M., Seeger, T. F., and Gardner, E. L., Enhancing and inhibitory effects of phencyclidine ("angel dust") on mesolimbic-mesocortical intracranial self-stimulation, paper presented at meetings of the Eastern Psychological Association, Hartford, Connecticut, 1980; Nazzaro, J. M., Seeger, T. F., and Gardner, E. L., Naloxone blocks phencyclidine's dose-dependent effects on direct brain reward thresholds, in *Proc. World Conf. on Clin. Pharmacol. and Ther.*, British Pharmacological Society, London, 1980, 949; Nazzaro, J. M., Gardner, E. L., and Seeger, T. F., Effect of dissociative anesthetic drugs of abuse on direct brain reward thresholds, paper presented at 14th World Congress of the Collegium Internationale Neuro-Psychopharmacologicum, Florence, Italy, 1984.)

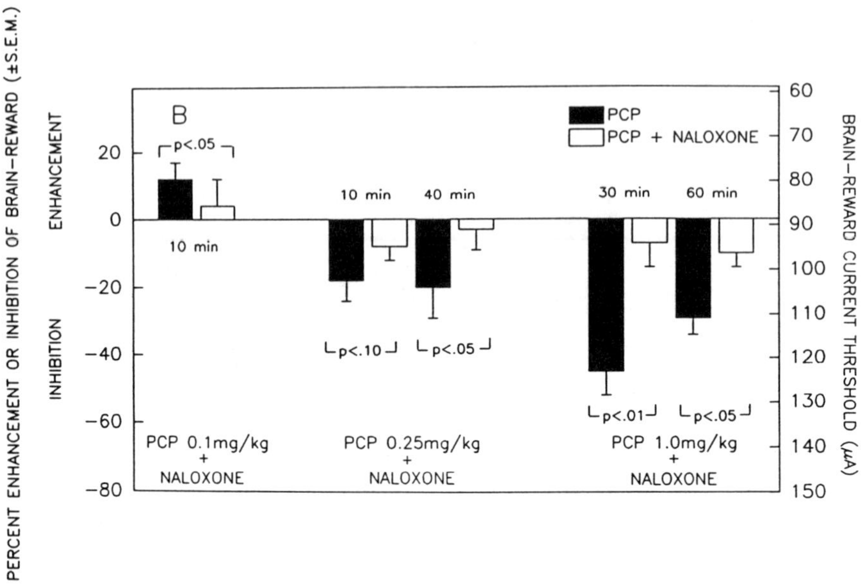

FIGURE 4B.

the brain reward enhancement produced by all well-studied drugs of abuse, it seems reasonable to speculate that endogenous opioid peptide mechanisms are involved in the modulatory machinery that allows habit-forming drugs of all classes to acutely enhance the DA mechanisms mediating brain reward, but direct proof of this is lacking (see further discussions of this point below).

Dose-Dependent Differential Effects of Habit-Forming Drugs on Brain Reward

Figure 6 also illustrates the intuitively obvious fact that for every habit-forming drug there is a threshold dose below which brain reward enhancement is not seen. For some habit-forming drugs, the dose range within which brain reward enhancement is seen is relatively narrow. Phencyclidine and ketamine, for example, produce brain reward enhancement within a comparatively narrow range of low doses, but *inhibit* brain reward at higher doses (see Figure 4A). We have suggested that this low-dose brain-reward enhancement, high-dose brain-reward inhibition phenomenon in laboratory animals is homologous with the "low-dose good-trip", "high-dose bad-trip" phenomenon reported by street users of these drugs. Interestingly, naloxone

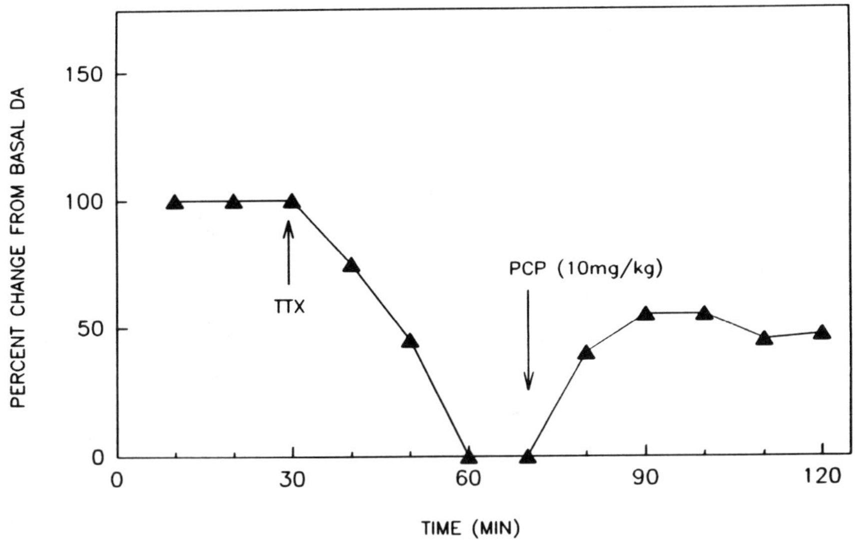

FIGURE 5. Enhancement of mesotelencephalic presynaptic DA efflux by acute systemic administration of phencyclidine (PCP) (10.0 mg/kg, i.p. administration) following abolition of action-potential-dependent presynaptic DA efflux by local intracerebral administration of tetrodotoxin (TTX). Each point represents mean amount of DA measured (shown as percent change from predrug basal DA efflux) by *in vivo* brain microdialysis for successive microdialysis samples. (From Chen, J. and Gardner, E.L., In vivo brain microdialysis study of phencyclidine on presynaptic dopamine efflux in rat caudate nucleus, *Soc. Neurosci. Abstr.*, 14, 525, 1988.)

not only attenuates the low-dose enhancement of brain reward, but also attenuates the high-dose inhibition of brain reward (Figure 4B). On this basis, we hypothesize that naloxone may be useful in the emergency room management of acute dysphoria resulting from high doses of at least some habit-forming drugs.

Differential Response of Anatomically Distinct Brain Reward Mechanisms to Habit-Forming Drugs

Although robust electrical brain-stimulation reward can be elicited from electrode placements throughout the entire medial-lateral extent of the medial forebrain bundle and the corresponding medial-lateral domains of the mesencephalic DA cell fields,[58,98] and although electrical brain-stimulation reward throughout these medial-lateral domains can be robustly modulated by *direct* pharmacological manipulation of underlying DA sub-

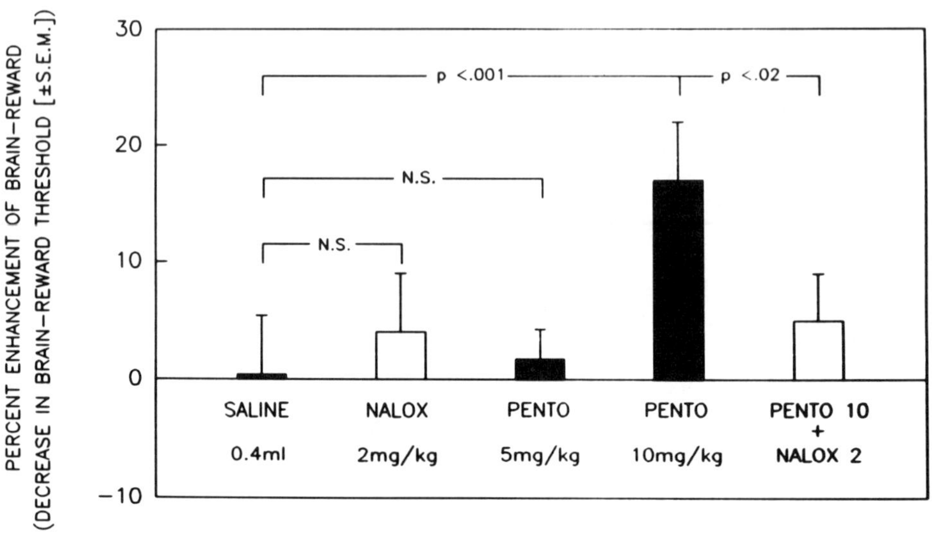

FIGURE 6. Naloxone-induced blockade of pentobarbital-enhanced brain reward. Enhancement of brain reward is experimentally equivalent to a decrease in electrical brain-stimulation threshold level (see text) in the ventral tegmental area (DA nucleus "A10"). Effects of acute systemic administration of pentobarbital (5.0 or 10.0 mg/kg, i.p. administration), naloxone (2.0 mg/kg, i.p. administration), or pentobarbital plus naloxone on brain reward thresholds are shown. Probability values shown are for the specific comparisons indicated on the figure. 'N.S.' (not significant) denotes lack of a statistically significant difference. (From Nazzaro, J. M., Gardner, E. L., Bridger, W. H., Carlson, K. R., and Seeger, T. F., Pentobarbital induces a naloxone-reversible decrease in mesolimbic self-stimulation thresholds, *Soc. Neurosci. Abstr.*, 7, 262, 1981; Seeger, T. F., Carlson, K. R., and Nazzaro, J. M., Pentobarbital induces a naloxone-reversible decrease in mesolimbic self-stimulation threshold, *Pharmacol. Biochem. Behav.*, 15, 583, 1981. With permission.)

strates (i.e., by chronic administration of DA antagonists to pharmacologically upregulate the sensitivity of postsynaptic DA receptors),[58,98-100] habit-forming drugs do *not* appear to uniformly enhance brain reward substrates throughout these medial-lateral domains. Quite to the contrary, we have consistently found that the medially lying brain-reward-relevant DA circuitry, originating in the ventral tegmental area and projecting through the medial portions of the medial forebrain bundle[26] to terminate in (among other areas) the nucleus accumbens, appears uniquely sensitive to both the acute enhancing effects of habit-forming drugs and to the attenuating effect

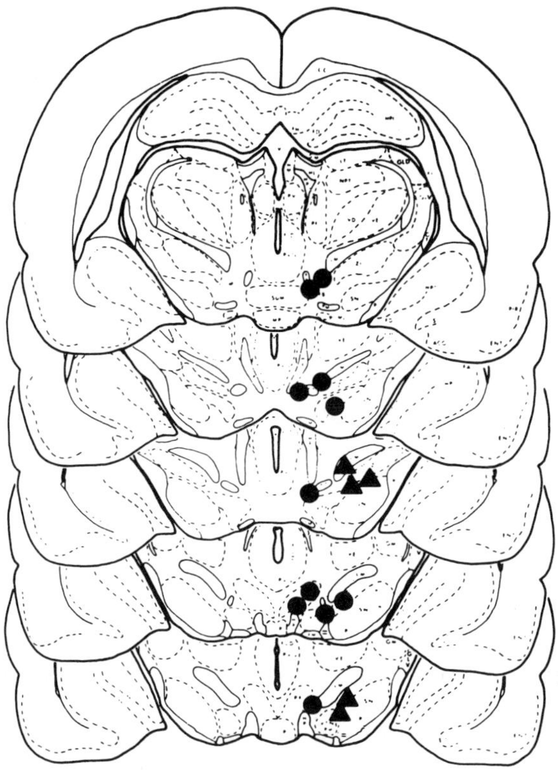

FIGURE 7. Brain reward sites showing differential susceptibility to naloxone blockade of habit-forming drug-induced enhancement of electrical brain-stimulation reward. Sites showing a naloxone reversal of pentobarbital-enhanced brain reward [clustered in the medial (mesolimbic) domains of the mesencephalic DA cell fields] are indicated by circles (●). Sites that did not show enhanced brain reward after pentobarbital or did not show naloxone blockade of pentobarbital-induced enhancement of brain reward [clustered in the lateral (mesostriatal) domains of the mesencephalic DA cell fields] are indicated by triangles (▲). The numerals adjacent to the brain sections indicate the vertical plane coordinates in millimeters anterior to the interaural line, according to the system of rat brain stereotaxic coordinates used in Pellegrino, L. J., Pellegrino, A. S., and Cushman, A. J., *A Stereotaxic Atlas of the Rat Brain,* Plenum Press, New York, 1979. (From Nazzaro, J. M., Gardner, E. L., Bridger, W. H., Carlson, K. R., and Seeger, T. F., Pentobarbital induces a naloxone-reversible decrease in mesolimbic self-stimulation thresholds, *Soc. Neurosci. Abstr.*, 7, 262, 1981; Seeger, T. F., Carlson, K. R., and Nazzaro, J. M., Pentobarbital induces a naloxone-reversible decrease in mesolimbic self-stimulation threshold, *Pharmacol. Biochem. Behav.*, 15, 583, 1981. With permission.)

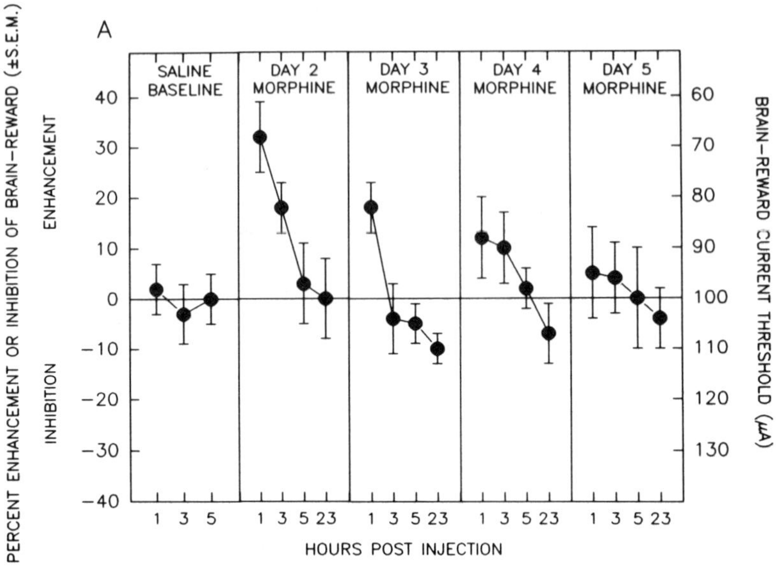

FIGURE 8. Differential effects of morphine on brain-stimulation reward in mesolimbic vs. mesostriatal DA loci. Enhancement of brain reward is experimentally equivalent to a decrease, and inhibition of brain reward to an increase, in electrical brain-stimulation threshold level (see text). (A) Effect of repeated once-daily acute systemic administrations of morphine (5.0 mg/kg/d, s.c. administration) on brain reward in the ventral tegmental area (mesolimbic DA nucleus "A10"). (B) Effect of repeated once-daily acute systemic administrations of morphine (5.0 mg/kg/d, s.c. administration) on brain reward in the substantia nigra (mesostriatal DA nucleus "A9"). (From Nazzaro, J. M., Seeger, T. F., and Gardner, E. L., Morphine differenially affects ventral tegmental and substantia nigra brain reward thresholds, *Pharmacol. Biochem. Behav.*, 14, 325, 1981. With permission.)

of naloxone on drug-enhanced brain reward.[66-68] Typical anatomic placements yielding these differential sensitivities to habit-forming drugs and to naloxone are shown in Figure 7. The apparently crucial role in the drug-enhanced brain reward of these medially lying (mesolimbic) DA fibers that terminate in the nucleus accumbens will also be discussed below in the context of the fact that the nucleus accumbens appears more sensitive than any other brain-reward-relevant locus to the enhancing effect of cannabinoids on brain reward.

In 1979, Nazzaro, Seeger, and I found[66,101] that this medial-lateral anatomic distinction within the reward circuitry, and the effects of habit-forming drugs thereon, was even more complex than we had imagined. With electrodes in the mesolimbic (medial) portions of the reward circuitry,

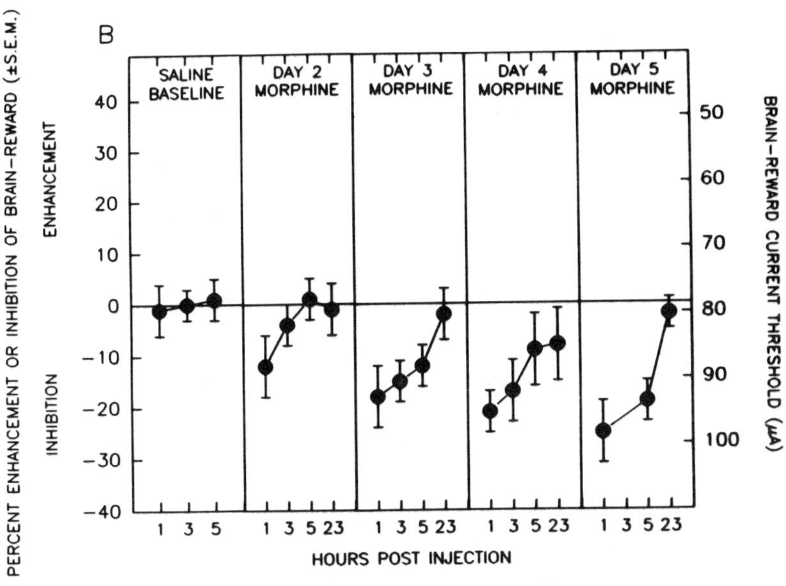

FIGURE 8B.

morphine produces a clear enhancement of brain reward, with development of partial tolerance over days of repeated morphine administrations (Figure 8A). However, with electrodes in the mesostriatal (lateral) portions of the reward circuitry, morphine is initially without effect and then grows increasingly *dys*phoric with repeated morphine administration over days (Figure 8B). In order to explain these behavioral data, we postulated[66] that morphine exerts a straight-forward DA agonist-like action on the medially lying mesolimbic reward circuitry, but exerts a complex combination of DA agonist-like and DA antagonist-like actions on the laterally lying mesostriatal reward circuitry. With the development of *in vivo* electrochemical and *in vivo* brain microdialysis techniques for measuring presynaptic neurotransmitter efflux within highly specific anatomic loci in living animals (see above), it became possible for us to directly test these notions.[102-105] We found that, in the DA terminal regions of the mesolimbic DA system, morphine produces a straightforward enhancement of presynaptic DA efflux (Figure 9), whereas in the mesostriatal DA terminal regions, morphine produces both an enhancement *and inhibition* of presynaptic DA efflux (Figure 10). Thus, the biochemical data comport well with the inferences previously drawn from our prior studies of morphine's effect on electrical

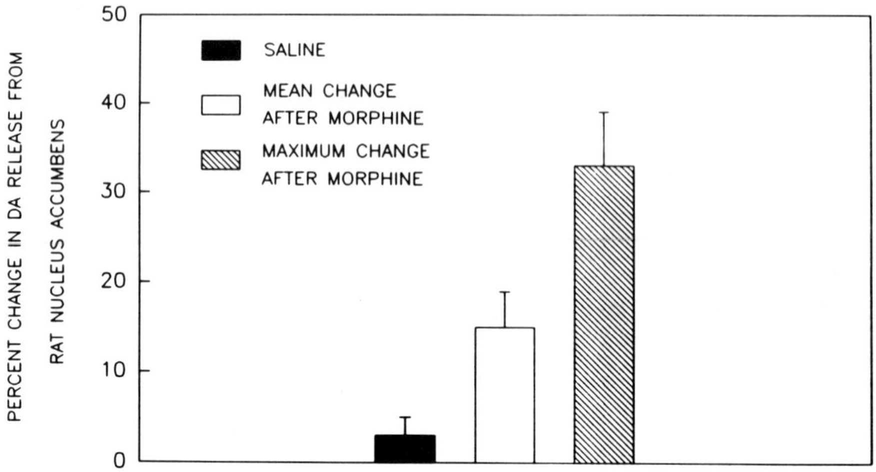

FIGURE 9. Enhanced mesolimbic DA efflux produced by morphine. The effect of acute systemic administration of morphine (5.0 mg/kg, i.p. administration) on voltammetric electrochemical signals corresponding to presynaptic DA efflux is shown for electrochemical recordings taken in the nucleus accumbens, terminal region of the "A10" mesolimbic DA system. (From Broderick, P. A. and Gardner, E. L., unpublished data, 1984; Broderick, P. A., Gardner, E. L., and van Praag, H. M., Enkephalinergic action on dopaminergic transmission: behavioral and *in vivo* electrochemical studies, paper presented at meetings of the Soc. Biol. Psychiatry, New York, 1983; Broderick, P. A., Gardner, E. L., and van Praag, H. M., *In vivo* electrochemical evidence for an opiate-induced modulation of dopaminergic and serotonergic systems in rat striatum, paper presented at meetings of the American College of Neuropsychopharmacology, San Juan, Puerto Rico, 1983.)

brain-stimulation reward thresholds in these two anatomically distinct domains of the reward circuitry. These data also agree with evidence from more traditional single-neuron electrophysiological recording studies showing the existence of two classes of DA neurons in the ventral mesencephalic cell fields of the mesotelencephalic DA system, one class that is activated by opiates and a second class that is inhibited by opiates.[106,107] Given these kinds of data, and given the evidence reviewed above that the DA "second-stage" neurons of the mesotelencephalic reward circuitry constitute the crucial convergence upon which habit-forming drugs act to acutely enhance brain reward, it does not seem unreasonable to suppose that the mesolimbic portion of the mesotelencephalic reward circuitry is uniquely specialized for brain reward enhancement by habit-forming drugs because these drugs act on mesolimbic DA neurons to produce pure DA agonist-like effects, whereas these same drugs act on mesostriatal DA neurons in a more complex fashion in which DA antagonist-like effects cancel out or override the DA agonist-

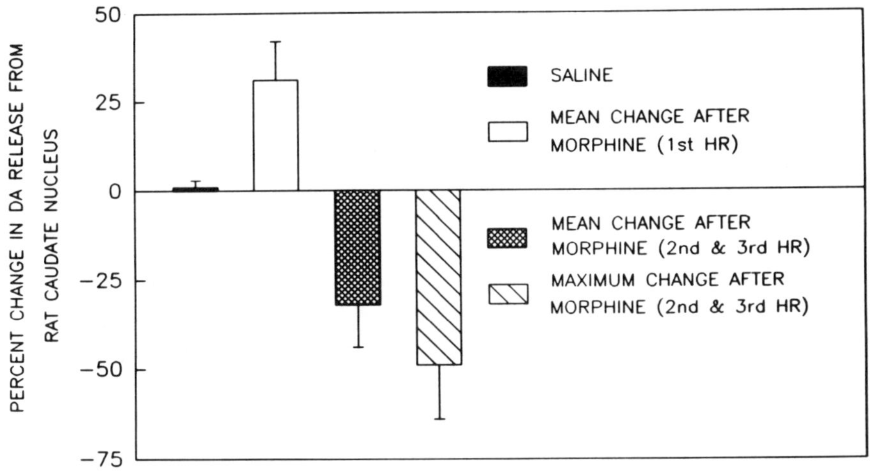

FIGURE 10. Biphasic enhancement *and* inhibition of mesostriatal DA efflux by morphine. The effect of acute systemic administration of morphine (5.0 mg/kg, i.p. administration) on voltammetric electrochemical signals corresponding to presynaptic DA efflux is shown for electrochemical recordings taken in the neostriatum, terminal region of the "A9" mesostriatal DA system. (From Broderick, P. A. and Gardner, E. L., unpublished data, 1984; Broderick, P. A., Gardner, E. L., and van Praag, H. M., Enkephalinergic action on dopaminergic transmission: behavioral and *in vivo* electrochemical studies, paper presented at meetings of the Soc. Biol. Psychiatry, New York, 1983; Broderick, P. A., Gardner, E. L. and van Praag, H. M., *In vivo* electrochemical evidence for an opiate-induced modulation of dopaminergic and serotonergic systems in rat striatum, paper presented at meetings of the Am. College of Neuropsychopharmacology, San Juan, Puerto Rico, 1983; Broderick, P. A., *In vivo* electrochemical studies of rat striatal dopamine and serotonin release after morphine, *Life Sci.*, 36, 2269, 1985. With permission.)

like effects, perhaps because of a preferential anatomical localization of DA neurons inhibited by opiates in the more lateral (mesostriatal) domains of the ventral mesencephalic DA cell fields. Many habit-forming drugs are clearly more potent as presynaptic DA efflux enhancers in the nucleus accumbens than in other DA brain loci.[82] This is illustrated for the case of phencyclidine in Figure 11. We have observed a similar differential sensitivity of the nucleus accumbens to the DA-efflux-enhancing effects of all habit-forming drugs that we have studied, as has the Italian research team of Imperato and Di Chiara.[77,78,81] We believe that these data speak directly to the accuracy of Wise's conception that the nucleus accumbens constitutes a crucial synaptic convergence of the "second-stage" DA reward neurons.[37,46]

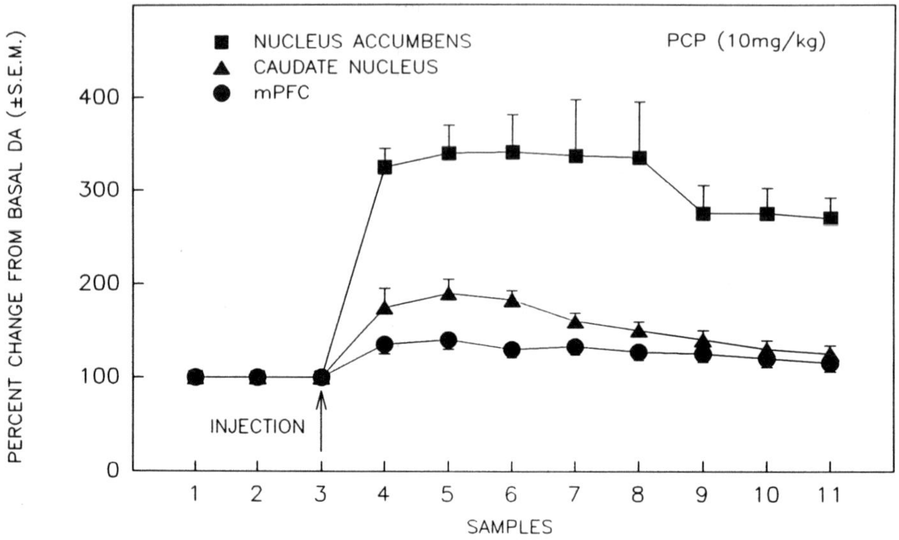

FIGURE 11. Differential sensitivities of three different mesotelencephalic DA terminal regions to enhancement of presynaptic DA efflux by acute systemic administration of phencyclidine (PCP) (10.0 mg/kg, i.p. administration). Each point represents mean amount of DA measured (shown as percent change from predrug basal DA efflux) by *in vivo* brain microdialysis for successive microdialysis samples. "mPFC" indicates the DA terminal field of the medial prefrontal cortex. (From Chen, J. and Gardner, E. L., *In vivo* brain microdialysis study of phencyclidine on presynaptic dopamine efflux in rat caudate nucleus, *Soc. Neurosci. Abstr.*, 14, 525, 1988.)

BRAIN MECHANISMS MODULATING BRAIN REWARD SUBSTRATES

Enkephalinergic/Endorphinergic Modulatory Mechanisms

In view of the fact that naloxone significantly attenuates the enhanced brain-stimulation reward induced by all known classes of habit-forming drugs (see above), it would appear that there exists an important functional interrelationship between the crucial drug-sensitive "second-stage" DA fibers of the reward system and endogenous opioid peptide circuitry; furthermore, this interrelationship is important for the brain reward enhancement produced by *all* habit-forming drugs, not just opiates. Certainly, there is no dearth of anatomical sites where such a functional interaction between reward-relevant DA neurons and endogenous opioid peptide neurons could

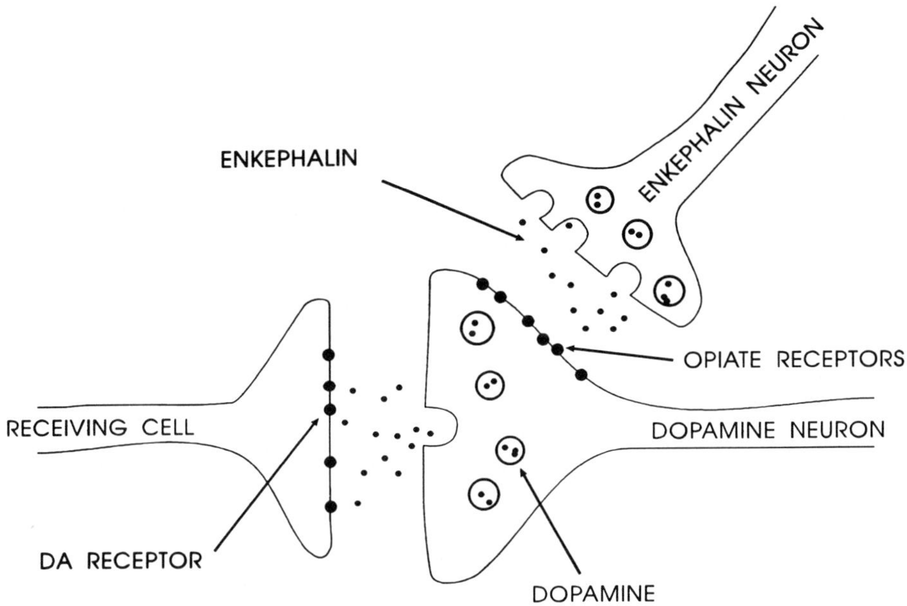

FIGURE 12. Schematic representation of axo-axonic enkephalin-DA synaptic interconnections believed to exist in terminal regions of the "A10" mesolimbic DA system, capable of modulating flow of reward-relevant neural signals through the DA circuitry.

take place. Cell bodies, axons, and synaptic terminals of enkephalinergic, endorphinergic, and dynorphinergic neurons are found in veritable profusion throughout the extent of the reward-relevant mesotelencephalic DA circuitry.[108,109] Makman, Zukin, and I have shown,[110] as have others,[111,112] that endogenous opioid peptide neurons synapse onto mesotelencephalic DA axon terminals, forming the type of axo-axonic synapses one would expect of a system designed to modulate the flow of reward-relevant neural signals through the DA circuitry. Such an axo-axonic enkephalin-DA synaptic arrangement is shown schematically in Figure 12. In addition to the DA axon terminal regions, other possible sites of enkephalinergic-DA functional interaction include the DA cell body region of the ventral mesencephalon[36] and transsynaptic modulation via afferents to the ventral mesencephalon from the locus coeruleus.[45] We also believe that some DA neurons of the reward system may synapse directly onto endogenous opioid peptide neurons located postsynaptically in the DA terminal regions, which may then carry the reward signal one synapse further. Our reasons for so believing are twofold. First, we have found[65] that naloxone attenuates the

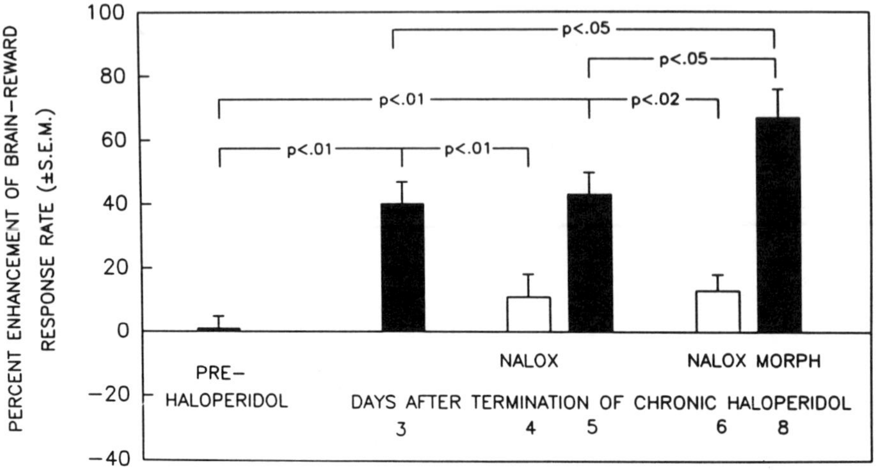

FIGURE 13. Enhanced mesolimbic brain-stimulation reward seen with pharmacological up-regulation of DA receptors produced by 3 weeks of daily haloperidol treatment (1.0 mg/kg/d, i.p. administration), and attenuation of resulting enhancement of brain-stimulation reward by acute administration of naloxone (2.0 mg/kg, i.p. administration). Probability values shown are for the specific comparisons indicated on the figure. (From Seeger, T. F., Nazzaro, J. M., and Gardner, E. L., Selective inhibition of mesolimbic behavioral supersensitivity by naloxone, *Eur. J. Pharmacol.*, 65, 435, 1980. With permission.)

enhanced brain reward induced by pharmacological up-regulation of DA receptors in the mesolimbic DA system (see Figure 13), suggesting that a crucial naloxone-blockable endogenous opioid peptide link lies *efferent* to the up-regulated DA receptors. Second, we have found[113] that naloxone significantly modulates behavioral responses induced by direct postsynaptic DA-receptor agonists in animals in whom the presynaptic DA fiber system has been destroyed by selective lesions of the DA mesotelencephalic system, again implicating a crucial naloxone-blockable endogenous opioid peptide link *efferent* to the ascending DA mesotelencephalic system. We also find that naloxone-reversibility of habit-forming drug-induced enhancement of brain reward varies sharply with anatomic placement of the stimulating electrode within the reward system. In our hands, naloxone-reversibility is invariably seen with electrodes in the medial portions of the mesencephalic DA cell fields (i.e., within the ventral tegmental area), but never with electrodes in the lateral portions of the mesencephalic DA cell fields (see also above discussion of medial-lateral differences in vulnerabil-

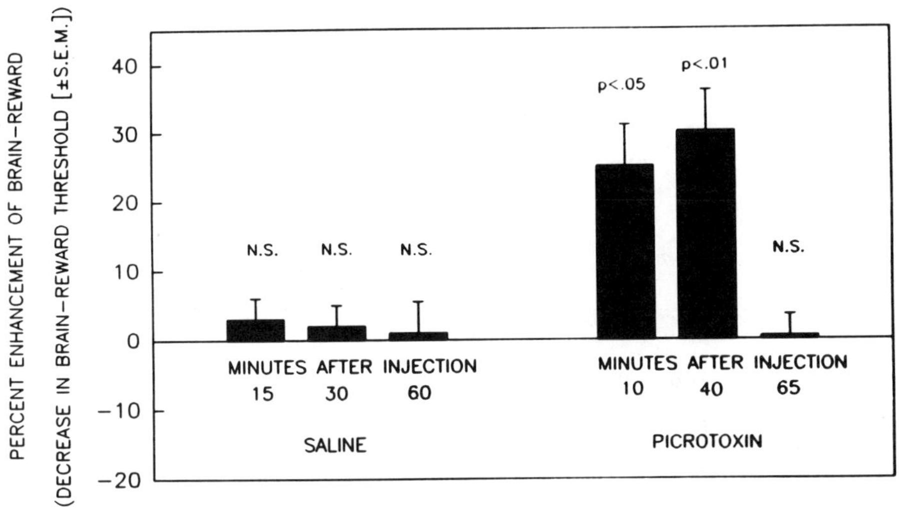

FIGURE 14. Enhanced brain-stimulation reward following acute systemic administration of the GABA antagonist picrotoxin (1.0 mg/kg, i.p. administration). Enhancement of brain reward is experimentally equivalent to a decrease in electrical brain-stimulation threshold level (see text) in the ventral tegmental area (DA nucleus "A10"). Probability values given on each bar represent the degree of significance vs. predrug baseline control. "N.S." (not significant) denotes lack of a statistically significant difference. (From Nazzaro, J. M. and Gardner, E. L., GABA antagonism lowers self-stimulation thresholds in the ventral tegmental area, *Brain Res.*, 189, 279, 1980. With permission.)

ity of reward loci to enhancement by habit-forming drugs), even in those instances where a lateral stimulation site does show acute enhancement by habit-forming drug administration. We surmise, therefore, that the medial portions of the reward-relevant ascending mesotelencephalic DA projections are modulated enkephalinergically, whereas the lateral portions are not. The implications of this for discerning cannabinoid sites of action within the reward circuitry will be discussed below.

GABAergic Modulatory Mechanisms

We have also shown[59] that additional important modulatory input to the brain reward system is received from GABAergic circuitry (see Figure 14), and since the CNS actions of barbiturates and benzodiazepines are mediated through GABAergic synapses,[114] it is not unlikely that this GABAergic modulatory synaptic input to the reward circuitry may be the site at which barbiturates and benzodiazepines enhance brain reward, producing their euphorigenic effects and abuse potential. Another possible site of barbiturate and benzodiazepine action on brain reward is the dense GABAergic

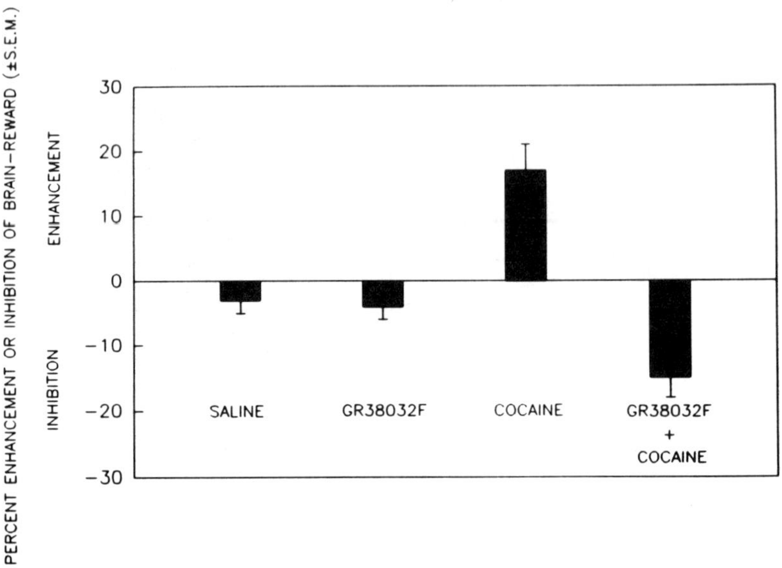

FIGURE 15. Enhanced brain-stimulation reward following acute systemic administration of cocaine (20.0 mg/kg, i.p. administration), and reversal of cocaine-enhanced brain reward by administration of the selective 5-HT_3 receptor antagonist GR38032F (0.5 mg/kg, i.p. administration). Enhancement of brain reward is experimentally equivalent to a decrease in electrical brain-stimulation threshold level (see text) in the medial forebrain bundle. (From Gardner, E. L., Paredes, W., Seeger, T. F., Smith, D., and van Praag, H.M., Mesolimbic DA antagonism by 5HT_3 receptor blockade, in *Book of Abstracts, World Psychiatric Association Regional Symposium, The Research and Clinical Interface for Psychiatric Disorders,* October 13–16, 1988, Washington, D.C., American Psychiatric Association, Washington, D.C., 1988, 222.)

innervation of the locus coeruleus, with transsynaptic modulation of brain reward being mediated via the noradrenergic connections from locus coeruleus to the general region of the ventral mesencephalic DA cell fields.[45]

Serotonergic Modulatory Mechanisms

Yet another important modulatory influence upon the mesotelencephalic DA reward circuitry is that derived from serotonergic inputs;[115-119] we have recently shown[72] that the brain reward enhancing effects of cocaine are dramatically attenuated by selective 5-HT_3 serotonin receptor blockade (see Figure 15). Also, compounds that block presynaptic 5-HT reuptake appear to significantly attenuate voluntary drug self-administration in laboratory animals.[120-128]

Neuropeptide (Nonopioid) Modulatory Mechanisms

Still other modulatory synaptic inputs onto these brain reward circuits are known to exist, including a variety of neuropeptides (neurotensin, cholecystokinin, and substance P),[129-142] and theoretically an even wider variety of transsynaptic modulatory influences on the brain reward systems are possible. With such a myriad of modulatory mechanisms and influences on the reward circuitry of the forebrain, the potential mechanisms by which drugs such as marijuana can influence brain reward are virtually legion.

BRAIN SITES AT WHICH THE INITIATION OF BRAIN REWARD ENHANCEMENT BY HABIT-FORMING DRUGS TAKES PLACE

Although, as indicated above, it is now abundantly clear that habit-forming drugs enhance direct electrical brain reward and that this enhancement involves a DA agonist-like final common pharmacological action on the "second-stage" DA fibers of the reward system, the above-summarized research does not speak to *where* the neuropharmacological enhancement takes place. For this, one turns to yet another research paradigm, that of voluntary direct intracranial self-administration of habit-forming drugs. Rats will work for direct microinjections of amphetamine and opiates into the nucleus accumbens,[37,51] noted above as perhaps *the* crucial reward-relevant terminal projection area of the mesotelencephalic DA circuitry. Compellingly, rats will also work for microinjections of DA itself into the nucleus accumbens.[143] Rats and monkeys will also self-inject cocaine into the frontal cortex,[37,52] another terminal field of the mesotelencephalic DA reward circuitry. Opioids are self-administered into both the nucleus accumbens[53] and the ventral tegmental area.[49,50] By and large, in these types of experiments, drug self-administration drops off with distance from the DA fibers. Although diffusion to distal loci can yield misleading conclusions if such intracranial self-administration experiments are not carefully done, it appears that, at least for stimulants and opiates, the crucial locus of action on the DA reward system may be on or synaptically close to the DA reward fibers themselves, although it is of course unlikely that identical receptor-mediated or membrane-mediated sites would be involved for drugs of different chemical classes. Unfortunately, comparable intracranial self-administration data are unavailable for habit-forming drugs other than stimulants and opiates.

DRUG CRAVING AND BRAIN CODING OF HEDONIC VALUE IMPARTED BY HABIT-FORMING DRUGS

Although, as also indicated above, it is clear that i.v. drug self-administration in laboratory animals can be markedly attenuated by inhibiting the function of the mesotelencephalic DA reward circuitry, it has not until recently been clear that the actual hedonic value (as coded in the brain) of habit-forming drugs also depends upon a DA substrate. Hedonic value of drugs can be inferred from a number of paradigms, but it is probably best assessed directly using conditioned place preference.[144] By using place conditioning to the injection of habit-forming drugs in conjunction with selective neurochemical blockade of the DA systems, White and Hiroi have recently demonstrated that hedonic value of habit-forming drugs appears not only to depend on a DA substrate (which had been demonstrated previously by several research groups using the conditioned place preference paradigm), but that different aspects of conditioned hedonic value depend upon different neurochemically specific DA substrates. Specifically, the newly synthesized DA pool appears crucial to the neural *en*coding of hedonic value, while the vesicular DA pool appears crucial for the behavioral expression/readout of that previously encoded hedonic value.[145-147] Thus, we may well be close with our animal models to understanding the neurochemical substrates for the craving that drug addicts feel when confronted visually with syringes, needles, crack vials, smoking pipes, or even the street corner where they normally buy their illicit drugs.

What, then, is the actual role of the ascending DA reward-relevant neuron and the seemingly crucial DA synapse to which it feeds? In this author's opinion, one can do no better, even after a full decade, than Wise's hypothesis that "the dopamine junction represents a synaptic way station for messages signaling the rewarding impact of a variety of normally powerful rewarding events. It seems likely that this synapse lies at a critical junction between branches of the sensory pathways that carry signals of the intensity, duration, and quality of the stimulus, and the motivational pathways where these sensory inputs are translated into the hedonic messages we experience as pleasure, euphoria or, 'yumminess'. The dopamine synapse qualifies as a reasonable candidate for the site in the brain where the hedonic impact of the sensory message is first associated with the sense impressions of external events that constitute natural rewards."[32]

Although none of the data or theoretical conceptions presented above require *ipso facto* that craving be coded in the same neural circuitry as reward, most workers, in the absence of hard data one way or the other, have adopted this stance out of regard for parsimony. Some have even speculated that drug craving results directly from a functional deficiency of DA in the reward-relevant DA circuitry.[148]

CANNABINOID ENHANCEMENT OF BRAIN REWARD

Like ethanol, nicotine, and the phencyclidine-like drugs, cannabis was long considered by some to not interact with the brain reward mechanisms described above, and, to that degree, to constitute an "atypical" habit-forming drug. Given, however, the facts (reviewed above) that ethanol, nicotine, and the phencyclidine-like drugs are not "atypical" at all, but rather interact with brain reward mechanisms in a rather classical fashion; and given the striking common enhancement of brain reward (also reviewed above) produced by *all* well-studied recreational and habit-forming drugs; and given that marijuana is the most widely used illicit recreational drug in North America and Europe, with well-described and well-characterized euphorigenic properties;[2,87,149,150] my colleagues and I felt that careful, well-controlled studies of cannabinoid interaction with brain reward mechanisms were called for. In recent years, these studies have been rewarded and we have found that Δ^9-THC, the principal psychoactive constituent of cannabis, acts on the brain reward systems outlined above in a fashion remarkably similar to that of "classical" habit-forming drugs.

Cannabinoid Enhancement of Electrical Brain-Stimulation Reward

As studied using the electrical brain-stimulation reward threshold paradigm described above, with electrodes in the medial forebrain bundle of

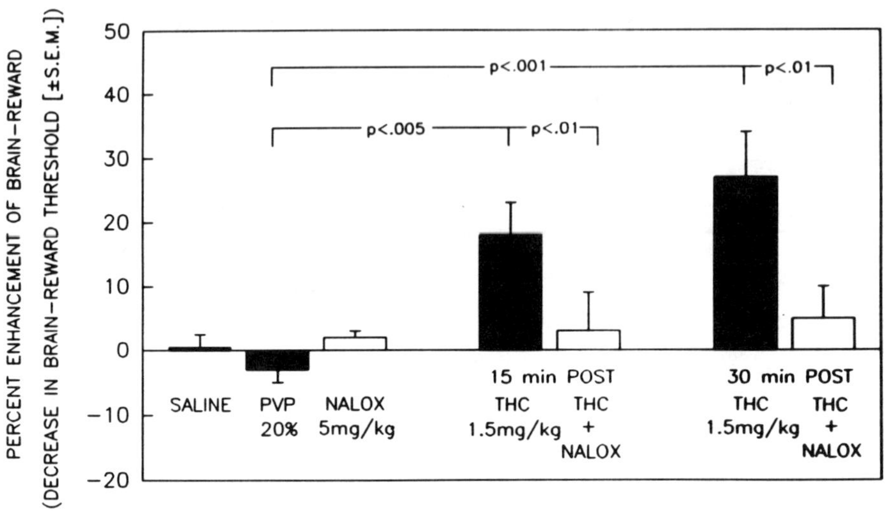

FIGURE 16. Enhanced brain reward following acute systemic administration of marijuana's psychoactive ingredient, Δ^9-tetrahydrocannabinol (THC) (1.5 mg/kg, i.p. administration), and attenuation of resulting enhancement of brain-stimulation reward by acute administration of naloxone (5.0 mg/kg, i.p. administration). Enhancement of brain reward is experimentally equivalent to a decrease in electrical brain-stimulation threshold level (see text) in the medial forebrain bundle. "PVP 20%" is the 20% polyvinylpyrrolidone vehicle used for suspending the Δ^9-THC into an injectable form. The effect of Δ^9-tetrahydrocannabinol on brain reward is shown at both 15 and 30 min post-Δ^9-THC. Probability values shown are for the specific comparisons indicated on the figure. (From Gardner, E. L., Paredes, W., Smith, D., Donner, A., Milling, C., Cohen, D., and Morrison, D., Facilitation of brain stimulation reward by Δ^9-tetrahydrocannabinol, *Psychopharmacology,* 96, 142, 1988; Gardner, E. L., Paredes, W., Smith, D., and Zukin, R. S., Facilitation of brain stimulation reward by Δ^9-tetrahydrocannabinol is mediated by an endogenous opioid mechanism, *Adv. Biosci.,* 75, 671, 1989. With permission.)

Lewis strain laboratory rats, Δ^9-THC produces robust and highly replicable enhancement (i.e., threshold lowering) of brain-stimulation reward.[57] This effect is shown in Figure 16. Strikingly, this effect is seen at low doses (e.g., 1.5 mg/kg Δ^9-THC injected i.p.). This is much lower than the dose range used in most previous studies of Δ^9-THCs effects on brain mechanisms *in vivo* and *in vitro*. The doses used in most previous studies have been severely criticized as being so high as to preclude pharmacological and physiological relevance to human recreational use of marijuana.[151] In contrast, our studies appear to be the first that can conservatively be judged as within the range of relevance to human use and abuse. Rosenkrantz and colleagues[152] translated oral doses of Δ^9-THC in rats to inhalation doses in a 50-kg human by correcting for 7:1 differences in body surface area, 50% pyrolysis, and a 3:1 difference in oral:inhalation dose. Taking these assumptions, one marijuana cigarette per day weighing 1 g with 2% Δ^9-THC would result in absorption of 0.2 mg/kg Δ^9-THC. Smoking of one to three marijuana cigarettes per day is well within the range of what is seen in human recreational marijuana use, and the equivalent Δ^9-THC content of five to ten marijuana cigarettes is well within the range of what is seen in marijuana abuse or hashish use. Making additional modest assumptions about the conversion of oral to i.p. doses in laboratory rats, these calculations indicate that the robust effects we have consistently seen with 1.5 mg/kg Δ^9-THC may well be pharmacologically and physiologically relevant to human patterns of cannabis use.

Cannabinoid Enhancement of DA Neurochemical Substrates Mediating Brain Reward

Given the enhancing action of all other well-characterized recreational and habit-forming drugs on extracellular DA at axon terminals in reward-relevant brain circuits, and given that the DA link in the reward circuitry appears to be the crucial drug-sensitive link (see review of these points above), the question arises: does cannabis also enhance extracellular DA in the reward circuitry? We have explored this question with both *in vivo* brain microdialysis and *in vivo* brain voltammetric electrochemistry. Using both techniques, we have shown that Δ^9-THC enhances both basal and potassium-stimulated extracellular DA efflux in reward-relevant brain loci, including striatum, nucleus accumbens, and medial prefrontal cortex.[84,153-155] This effect is shown, as measured by *in vivo* brain microdialysis in the nucleus accumbens, in Figure 17. Corroborating experiments, showing a similar DA-enhancing effect of Δ^9-THC in neostriatum have been recently reported by Taylor and colleagues, also using *in vivo* brain microdialysis.[156,157] In our hands, robust and highly replicable enhancements of extracellular DA in nucleus accumbens, neostriatum, and medial prefrontal cortex are seen in the dose range of 0.5 to 2.0 mg/kg Δ^9-THC, which, as noted above, may well correspond to recreational human use of as little as 1 or 2 marijuana cigarettes of moderate Δ^9-THC content, and thus pharmacologically and physiologically relevant to human patterns of marijuana use.

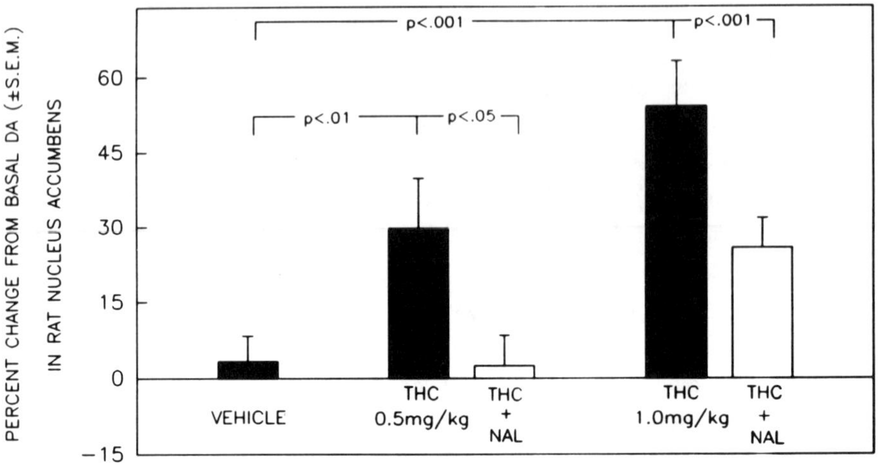

FIGURE 17. Enhanced presynaptic DA efflux produced by Δ[9]-tetrahydrocannabinol (Δ[9]-THC) as measured by *in vivo* brain microdialysis in nucleus accumbens (brain-reward-relevant mesolimbic DA terminal region), and naloxone-induced blockade of same. Each bar represents the peak amount of DA measured (shown as percent change from predrug basal DA efflux) following acute systemic administration of Δ[9]-tetrahydrocannabinol (THC) (0.5 and 1.0 mg/kg, i.p. administration) or vehicle. Probability values shown are for the specific comparisons indicated on the figure. (From Chen, J., Paredes, W., Li, J., Smith, D., and Gardner, E. L., *In vivo* brain microdialysis studies of Δ[9]-tetrahydrocannabinol on presynaptic dopamine efflux in nucleus accumbens of the Lewis rat, *Soc. Neurosci. Abstr.*, 15, 1096, 1989.)

Naloxone-Induced Attenuation of Brain Reward Enhancement Produced by Cannabis

As noted previously in this chapter, acute enhancement of brain reward mechanisms appears to be the single essential commonality of habit-forming drugs, and this facilitation of brain reward is naloxone-reversible for all well-studied drugs of abuse, implicating an endogenous opioid mechanism in mediating the euphorigenic action of such drugs. Given our findings that Δ[9]-THC resembles other recreational and abuse-prone drugs in enhancing electrical brain-stimulation reward and enhancing extracellular DA efflux in brain reward loci, the question obviously arises as to whether Δ[9]-THCs effects on these mechanisms are also naloxone reversible. We addressed this question in a series of experiments using both the electrical brain-stimulation reward and *in vivo* brain microdialysis paradigms outlined above. As shown in Figure 16, we found that Δ^9-THCs enhancing effects on electrical brain-stimulation reward are indeed naloxone blockable, at doses of naloxone which themselves have no effect on brain reward.[95] Furthermore, using *in vivo* brain microdialysis in the nucleus accumbens, convergence of the drug-sensitive forebrain reward circuitry, we found that naloxone, at doses as low as 0.1 mg/kg, significantly blocked Δ^9-THCs enhancing effects on basal extracellular DA efflux.[155,158] This naloxone blockade of Δ^9-THC-induced enhancement of basal extracellular DA efflux is shown, for the case of the nucleus accumbens, in Figure 17.

GENETIC VARIATION IN VULNERABILITY OF BRAIN REWARD MECHANISMS TO CANNABINOID ACTION

Interestingly, the brain reward enhancement induced by Δ^9-THC appears to be genetically variable.[159] Laboratory rats of one genetically inbred strain (Lewis strain) show robust and highly consistent enhancement of electrical brain-stimulation reward when given Δ^9-THC, but Δ^9-THC utterly fails to enhance electrical brain-stimulation reward in other inbred strains (Sprague-Dawley, Long-Evans, and Fischer 344 strains). These strain differences in Δ^9-THCs effect on brain-stimulation reward were striking — in the Lewis strain animals, Δ^9-THC facilitated brain-stimulation reward in *every* animal and with *every* Δ^9-THC administration; in the Sprague-Dawley, Long-Evans, and Fischer 344 rats, Δ^9-THC *never* facilitated brain reward in any animal or at any dose. In order to determine whether these striking strain differences in susceptibility to Δ^9-THCs enhancement of electrical brain-stimulation reward are mirrored by similar strain differences in susceptibility to Δ^9-THCs enhancement of extracellular DA efflux in brain reward loci, we studied Δ^9-THCs effects on basal DA efflux in the nucleus accumbens by *in vivo* brain microdialysis in awake behaving animals of the four inbred rat strains that we had previously used in our studies of Δ^9-THCs effect on electrical brain-stimulation reward: Lewis, Sprague-Dawley, Long-Evans, and Fischer 344. We found that while basal (nondrugged) extracellular DA efflux in nucleus accumbens was similar across all strains, clear strain differences emerged in vulnerability to Δ^9-THCs enhancement of extracellular DA efflux.[160] In the Lewis strain rats, Δ^9-THC at both 0.5 mg/kg and 1.0 mg/kg produced robust and long-lasting enhancement of extracellular DA efflux; the other strains tested showed no enhancement of extracellular DA efflux in nucleus accumbens at either dose.[160] Thus, the strain-specific facilitation of electrical brain stimulation reward by Δ^9-THC is mirrored by strain-specific facilitation of extracellular DA efflux by Δ^9-THC in reward-relevant neurons of the nucleus accumbens. Provocatively, the inbred rat strain that we have found to be so vulnerable to the brain reward enhancing effects of Δ^9-THC (Lewis strain) also shows increased vulnerability to the effects of cocaine, and increased vulnerability to ethanol self-administration,[161] suggesting the possibility that this strain of rat may possess a generalized vulnerability to the rewarding effects of habit-forming drugs similar to that observed in some genetically inbred mouse strains such as the C57BL/6J.[162] We believe that these data suggest a genetic variation in the vulnerability of brain reward systems to cannabis. Thus, genetic variations in proclivity to self-administer habit-forming drugs, well documented in both man and laboratory animals for ethanol,[161,163-165] but long suspected for other drugs as well, may result directly from genetic variations in the vulnerability of brain reward systems to habit-forming drugs. With specific respect to cannabis, we believe that these data raise the provocative possibility that significant genetic variation in vulnerability to the brain-reward-enhancing effects (and thus, presumably, to the euphorigenic effects and abuse potential) of marijuana and hashish may likewise exist in the human, and, furthermore, that this neurobiological vulnerability may generalize to noncannabinoid recreational and habit-forming drugs.

CELLULAR MECHANISMS OF CANNABINOID ACTION ON BRAIN REWARD CIRCUITS

Cannabinoid Action on DA Reuptake in Brain Reward Loci

As noted above, all adequately studied recreational and habit-forming drugs enhance basal DA neuronal firing and/or basal DA neurotransmitter release in brain reward circuits. However, the specific cellular mechanism(s) on which each individual drug acts to achieve these ends appear to differ widely. Thus, amphetamine appears to act as a presynaptic DA releaser,

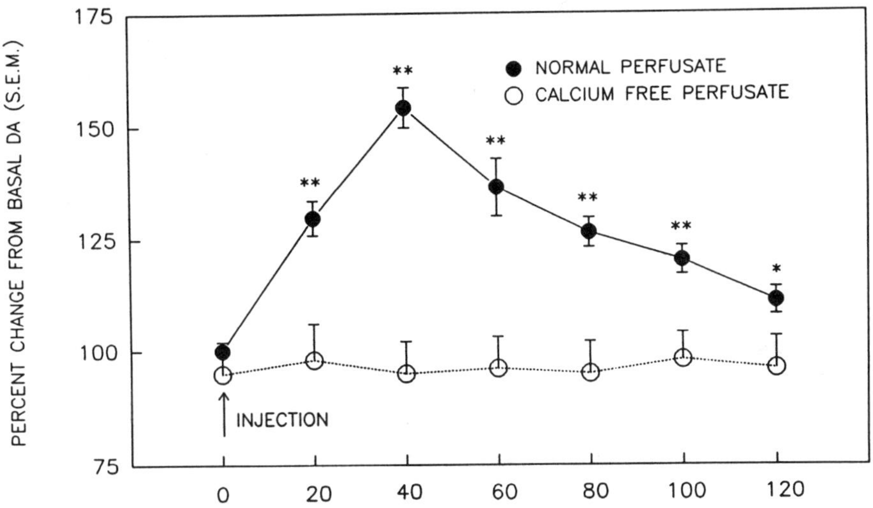

FIGURE 18. Calcium dependence of enhancing effect of Δ^9-THC (1.0 mg/kg, i.p.) on basal DA efflux in nucleus accumbens of conscious, freely moving rats as measured by *in vivo* brain microdialysis. The Δ^9-THC injections were given at the time indicated by the arrow. Asterisks indicate degree of statistical significance of each post-Δ^9-THC data point with normal perfusate as compared to the corresponding post-Δ^9-THC data point with Ca^{2+}-free perfusate (* $p < 0.05$; ** $p < 0.01$). (From Chen, J., Paredes, W., Li, J., Smith, D., Lowinson, J., and Gardner, E. L., Δ^9-tetrahydrocannabinol produces naloxone-blockable enhancement of presynaptic dopamine efflux in nucleus accumbens of conscious, freely-moving rats as measured by intracerebral microdialysis, *Psychopharmacology,* 102, 156, 1990. With permission.)

cocaine as a presynaptic DA reuptake blocker, opiates as transsynaptic enhancers of DA neuronal firing, and other recreational and habit-forming drugs by yet other means and mechanisms.[37,75-81,83,85,86,93] The question arises: how do cannabis and habit-forming cannabinoid preparations act? While not yet completely definitive, we believe that we have gathered sufficient data to strongly suggest that Δ^9-THC action on DA reward neurons of the mesotelencephalic DA system resembles that of a DA reuptake blocker.[84,155,166] First, we have shown that the extracellular DA efflux is tetrodotoxin sensitive,[155] indicating that the Δ^9-THC-enhanced DA signal is dependent upon active neuronal volleys of action potentials.[85] This is in contrast, of course, to the tetrodotoxin-*insensitivity* of the DA-enhancing properties of some other habit-forming drugs (e.g., phencyclidine, see Figure 5). Second, we have shown that the enhancing effect of Δ^9-THC on presynaptic DA efflux in brain reward loci is calcium dependent.[155] Basal

DA efflux in nucleus accumbens decreases approximately 50% in the presence of calcium-free dialysate. Challenge with 1.0 mg/kg Δ^9-THC does not enhance DA efflux at all when calcium-free dialysate is used, in contrast to the robust enhancement (to approximately 150% of pre-Δ^9-THC baseline) of DA efflux by the same Δ^9-THC dose when calcium-containing dialysate is used.[155] This calcium dependence of the Δ^9-THC enhancing effect on extracellular DA efflux in brain reward loci is shown in Figure 18. Third, we find that pretreatment (1 h prior to Δ^9-THC) with the DA receptor blocker haloperidol (0.1 mg/kg) has a synergistic effect on the Δ^9-THC enhancement of presynaptic DA efflux in the nucleus accumbens;[166] further, Δ^9-THC (1.0 mg/kg) pretreatment 1 h before haloperidol has a similar synergistic effect on haloperidol's enhancement of presynaptic DA efflux in the nucleus accumbens.[166] Tetrodotoxin, perfused locally into the nucleus accumbens, abolishes the synergism between Δ^9-THC and haloperidol. Since impulse-induced facilitation of DA release underlies the synergistic effect between DA receptor blockers and DA reuptake inhibitors,[167] these data are congruent with the concept that Δ^9-THC acts as a DA reuptake blocker. Finally, using high-speed *in vivo* voltammetric electrochemical techniques,[168] we have shown that the high-speed time dynamics of the Δ^9-THC enhancement of the potassium-evoked electrochemical signal corresponding to released extracellular DA are identical to the high-speed time dynamics of nomifensine's enhancement of the same electrochemical signal.[84] These findings are shown in Figure 19. Inasmuch as nomifensine has been shown to be a relatively pure reuptake inhibitor,[169] these findings are also congruent with the conception that Δ^9-THC augments presynaptic DA efflux by ultimately acting (probably indirectly) as a DA reuptake blocker.[84]

Cannabinoid Modulation on Brain Opioid Receptors

As noted above, the opiate antagonist naloxone antagonizes the Δ^9-THC enhancement of both electrical brain-stimulation reward and extracellular DA efflux in brain reward loci. On these grounds, we have suggested[95,158] that a profoundly important functional and anatomic interaction exists between endogenous brain opioid mechanisms and the neurobiological reward substrates on which cannabis acts to produce its euphorigenic actions. In addition, other findings strongly suggest that at least some cannabinoid effects on the brain are mediated by endogenous opioid systems. Thus, Δ^9-THC ameliorates naloxone-precipitated morphine abstinence,[170,171] and,

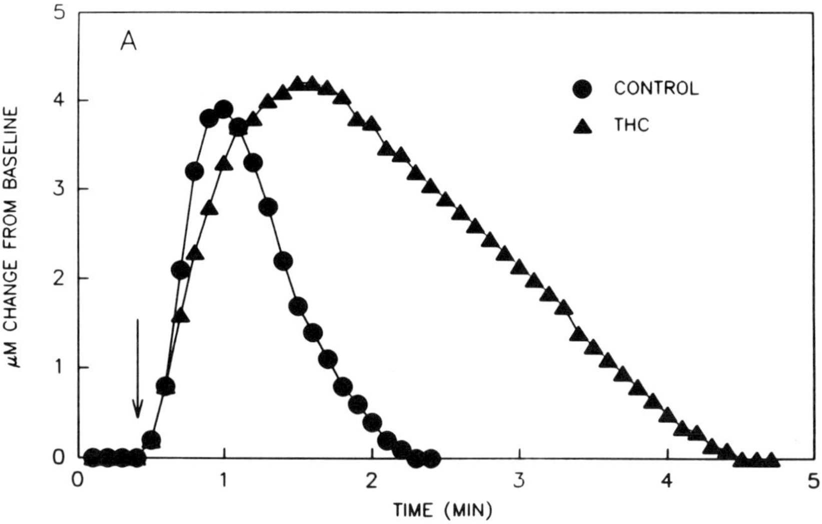

FIGURE 19. Enhanced presynaptic DA efflux produced by Δ^9-tetrahydrocannabinol and by the presynaptic DA reuptake blocker nomifensine, as measured by *in vivo* voltammetric brain electrochemistry. (A) Effects of Δ^9-tetrahydrocannabinol (THC) (0.5 mg/kg, i.p. administration) on K^+-evoked voltammetric electrochemical signals corresponding to presynaptic DA efflux. The arrow indicates the timing of the localized intracerebral micropressure K^+ applications. (B) Effects of nomifensine (5.0 mg/kg, i.p. administration) on K^+-evoked voltammetric electrochemical signals corresponding to presynaptic DA efflux. The arrow indicates the timing of the localized intracerebral micropressure K^+ applications. The time dynamics of the Δ^9-THC-enhanced presynaptic DA efflux signal are identical to those of the nomifensine-enhanced presynaptic DA efflux signal, raising the possibility that Δ^9-THC may act on reward-relevant DA neurons by inhibiting presynaptic DA reuptake. (From Ng Cheong Ton, J. M., Gerhardt, G. A., Friedemann, M., Etgen, A. M., Rose, G. M., Sharpless, N. S., and Gardner, E. L., The effects of Δ^9-tetrahydrocannabinol on potassium-evoked release of dopamine in the rat caudate nucleus: an *in vivo* electrochemical and *in vivo* microdialysis study, *Brain Res.*, 451, 59, 1988. With permission.)

conversely, a variety of cannabinoid-induced pharmacological effects are significantly attenuated by opiate antagonists.[172,173] Of specific relevance to brain DA systems, modulating effects of Δ^9-THC on both luteinizing hormone releasing hormone and prolactin are blocked by naloxone.[174] For these reasons, we undertook *in vitro* studies of cannabinoid effects on opiate binding to both membrane-bound and solubilized, partially purified brain opioid receptors as well as on muscarinic and DA receptor binding.[175-177] For these studies, rat brains were removed, dissected, and prepared for receptor binding using our standard laboratory receptor binding assay procedures.[178] Binding assays were carried out for mu (μ), delta (δ), kappa (κ),

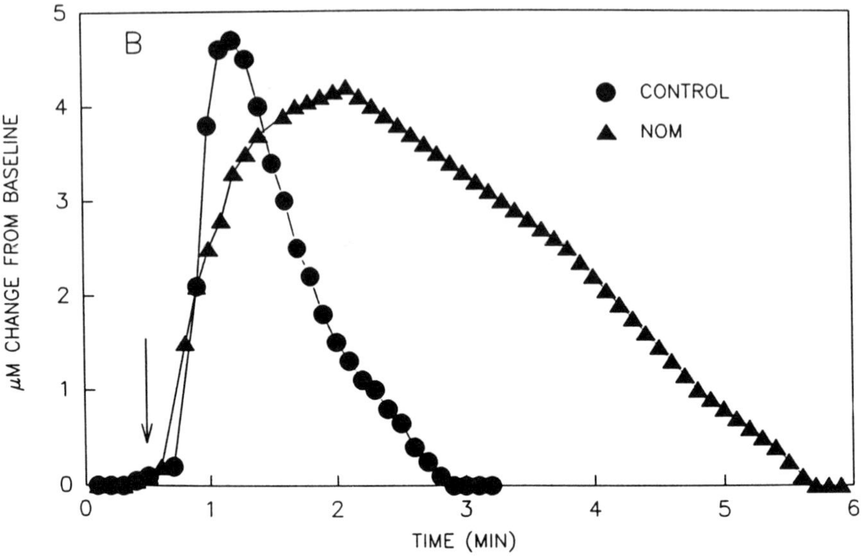

FIGURE 19B.

and sigma (σ) opioid receptors, muscarinic receptors, and DA receptors. Solubilized, partially purified opioid receptors were prepared according to previously published procedures,[179] and opiate binding studies were carried out with them according to the same previously published procedures.[179] We found that Δ^9-THC produced a dose-dependent inhibition of μ and δ opioid receptor binding, but failed to alter κ, σ, DA, or muscarinic binding.[176] Scatchard analyses of the μ receptor binding indicated that Δ^9-THC produced a significant decrease in receptor density with no significant change in receptor affinity, consistent with a noncompetitive mechanism for Δ^9-THC-induced inhibition of μ opioid receptors. The effect of Δ^9-THC on solubilized, partially purified opioid receptors was qualitatively and quantitatively similar to the effect observed on membrane-bound receptors, prompting us to suggest as early as 1985[175,177] that a cannabinoid-induced allosteric modulation of the opioid receptor results from direct interaction with the receptor protein or with a specific protein-lipid complex and not merely from nonspecific perturbation of the lipid bilayer of the membrane. We also compared the potencies of a large series of cannabinoids and related compounds in their ability to inhibit μ receptor binding, and found a general correlation with psychoactive potencies in humans, with the

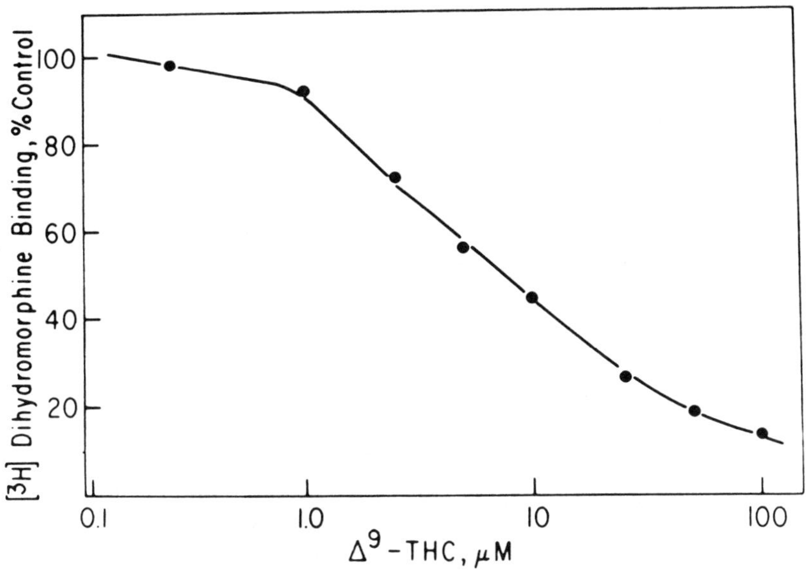

FIGURE 20. Specific binding of [^{3}H]dihydromorphine to rat brain membranes as a function of Δ^9-tetrahydrocannabinol (Δ^9-THC) concentration. Samples (1.0 ml, 0.8 mg of protein) were incubated in triplicate in 50 m*M* Tris HCl buffer pH 7.4 at 4°C for 45 min. Incubation samples included [^{3}H]dihydromorphine (0.5 n*M*) and Δ^9-THC at indicated concentrations in the presence or absence of levorphanol (5 μ*M*). The mean Δ^9-THC concentration required to inhibit [^{3}H]dihydromorphine binding to brain by 50% was 7 ± 1 μ*M*. (From Vaysse, P. J.-J., Gardner, E. L., and Zukin, R. S., Modulation of rat brain opioid receptors by cannabinoids, *J. Pharmacol. Exp. Ther.*, 241, 534, 1987. With permission.)

exception of a lower-than-expected *in vitro* opioid receptor modulatory potency for 11-hydroxy-THC and a greater-than-expected *in vitro* opioid receptor modulatory potency for cannabidiol.[176] These findings are shown in Figures 20 and 21 and Tables 1 and 2.

While we believe that these *in vitro* binding studies are suggestive of a negative allosteric modulation of cannabinoids on endogenous opioid receptor function, it must be readily conceded that other data (primarily from *in vivo* behavioral and pharmacological studies)[180-184] are highly suggestive of a *positive* modulation by cannabinoids on endogenous opioid function. Of course, the nature of cannabinoid-opioid functional interaction may differ in brain circuits mediating reward and in brain circuits mediating other pharmacological properties such as antinociception. In our current thinking, the nature of this cannabinoid-opioid functional interaction in reward-relevant forebrain circuits is very important because a positive

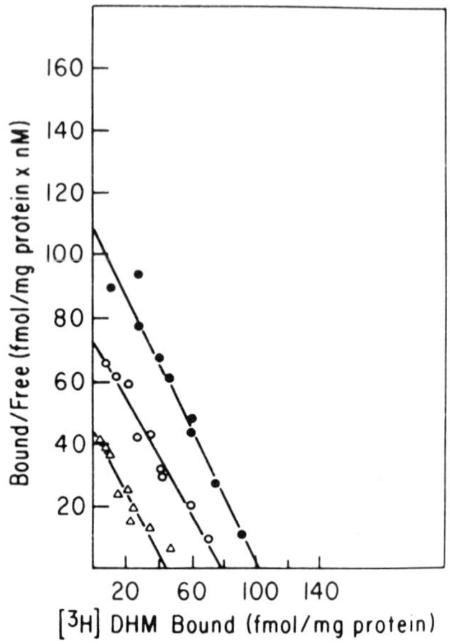

FIGURE 21. Scatchard plot of [^{3}H]dihydromorphine binding to rat brain membranes in the absence (●) or presence of Δ^9-tetrahydrocannabinol (Δ^9-THC) at indicated concentration (○, 5 μ*M*; Δ, 10 μ*M*). Aliquots of brain homogenate (1.0 ml) in 50 m*M* Tris HCl buffer, pH 7.4, were incubated with [^{3}H]dihydromorphine (0.05 to 10.0 n*M*) in the absence or presence of levorphanol (5 μ*M*) and Δ^9-THC at the indicated concentration. Binding plots were analyzed by computer-assisted nonlinear regression analysis. (From Vaysse, P. J.-J., Gardner, E. L., and Zukin, R. S., Modulation of rat brain opioid receptors by cannabinoids, *J. Pharmacol. Exp. Ther.*, 241, 534, 1987. With permission.)

interaction would imply one hypothetical model for cannabinoid action on reward synapses while a negative interaction would imply another hypothetical model (see below).

Cannabinoid Interaction with Specific Brain Cannabis Binding Sites

In the opinion of many, the most seminal and important discovery in cannabinoid pharmacology in recent years has been the discovery that cannabinoids interact with specific brain binding sites.[185-190] These sites, apparently allosterically regulated by G proteins, appear to fulfill standard criteria for high-affinity, stereoselective, and pharmacologically distinct brain receptors.[186] The relative potencies of natural and synthetic cannabinoids as competitors for these binding sites correlate closely with their

TABLE 1
Potencies of Δ⁹-THC in Competing for Radioligand Binding to Rat Brain Receptors

Receptor	Radioligand[a]	Δ⁹-THC IC_{50} (m*M*)
μ, δ, and κ opioid	[³H]Etorphine	10 ± 1.2
		14[b] ± 3.0
	[³H]Naloxone	
	100 m*M* NaCl	19 ± 1.0
	No NaCl	10 ± 1.5
μ opioid	[³H]Dihydromorphine	7 ± 1.0
δ opioid	[³H]D-Pen²,D-Pen⁵-enkephalin	16 ± 3.0
κ opioid (μ, δ blockers present)	[³H]Ethylketocyclazocine	>100
σ opioid/phencyclidine	[³H]N-(1-[2-thienyl]cyclohexyl)piperidine	>100
Dopamine	[³H]Spiroperidol	>100
Muscarinic	[³H]Quinuclidinyl benzilate	>100

Note: Whole rat brain was prepared and binding was carried out as described in Reference 176. Aliquots (1.0 ml) of rat brain homogenate in either (for opioid and dopamine receptor assays) 50 m*M* Tris HCl, pH 7.4, or (for muscarinic receptor assays) 50 m*M* sodium potassium phosphate buffer, pH 7.4, with radioligand in the absence or presence of 5 μ*M* levorphanol (μ opioid receptor assay), D-Pen²,D-Pen⁵-enkephalin (δ opioid receptor assay), ethylketocyclazocine (κ opioid receptor assay), 10 μ*M* phencyclidine (σ opioid/phencyclidine receptor assay), naltrexone (opioid antagonist assay), 1 μ*M* atropine (muscarinic receptor assay) or 10 μ*M* (+)-butaclamol (dopamine receptor assay), and Δ⁹-THC (at eight concentrations). Incubation with [³H]spiroperidol was carried out at 25°C for 20 min; all other radioligand incubations were at 4°C for 45 min. Log-probit plots were constructed, and the IC_{50} values were determined graphically. Each determination represents the mean ± S.E.M. of three independent experiments.

[a] Radioligand concentration was 0.5 n*M* in each assay, except in the case of [³H]Ethylketocyclazocine and [³H]N-(1-[2-thienyl]cyclohexyl)piperidine, which were 2.5 n*M*, and [³H]Quinuclidinyl benzilate, which was 0.05 n*M*.

[b] IC_{50}value was obtained for solubilized, partially purified opioid receptors.

From Vaysse, P. J.-J., Gardner, E. L., and Zukin, R. S., *J. Pharmacol. Exp. Ther.*, 241, 534, 1987. With permission.

relative potencies in biological and behavioral assays,[186,188] suggesting that these specific brain cannabinoid binding sites are the same brain receptors that mediate the behavioral and pharmacological effects of cannabis. As determined autoradiographically, the anatomic distribution of these binding sites is selective.[188] Dense binding is seen in the striatum, cerebral cortex,

TABLE 2
Relative Potencies of Cannabinoids in Inhibiting [^{3}H]Dihydromorphine Binding to μ Opioid Receptors of Rat Brain Membranes

Cannabinoid	IC_{50} (K_i)	Relative potency[a]
Δ^9-THC	7 ± 1[b] μ*M*	1
(+)Δ^9-THC	45%[c]	<0.07[c]
11-Hydroxy-Δ^9-THC	20%[c]	<0.07[c]
Δ^8-THC	20 ± 5 μ*M*	0.35
8α,11-Dihydroxy-Δ^9-THC	40%[c]	<0.07[c]
Equatorial hexahydrocannabinol	10 ± 1 μ*M*	0.70
Cannabidiol	7 ± 1 m*M*	1.00
Cannabinol	35 ± 5 μ*M*	0.20
Cannabinol acetate	20%[c]	<0.07[c]
1′-Oxocannabinol	48 ± 7 μ*M*	0.15
1′,2′,3′,4′,5′-Pentanor-Δ^9-THC 3-carboxylic acid	20 ± 4 μ*M*	0.35
Levonantradol	70 ± 10 μ*M*	0.10
Dextronantradol	70 ± 10 μ*M*	0.10

Note: Whole rat brain was prepared and binding was carried out as described in Reference 176. 1.0-ml samples (0.8 mg of protein) were incubated with [^{3}H]Dihydromorphine (0.5 n*M*, 87.7 Ci/mmol) at 4°C for 45 min in the absence or presence of levorphanol (5 μ*M*) or other indicated drugs at eight concentrations. The incubation buffer was 50 m*M* Tris HCl, pH 7.4. The control binding value for 0.5 n*M* [^{3}H]dihydromorphine was 3000 cpm specifically bound per sample. The IC_{50} value is defined as the concentration of drug required to half-maximally displace [^{3}H]Dihydromorphine (0.5 n*M*) binding to rat brain membranes. In the case of cannabinol acetate and 8α,11-dihydroxy-Δ^9-THC, the percentage of inhibition is reported at 0.1 m*M*.

[a] Relative potencies are expressed relative to Δ^9-THC.
[b] IC_{50} value as reported is for a colloidal suspension of Δ^9-THC. IC_{50} value determined for a fully soluble Δ^9-THC solution is 3 ± 1 μ*M*.
[c] Since many of these cannabinoids were available in only limited quantities, the approximate range of concentrations necessary to produce 50% inhibition was estimated by comparing the inhibition produced by each drug at 25 and 50 μ*M* with that produced by Δ^9-THC.

From Vaysse, P. J.-J., Gardner, E. L. and Zukin, R. S., *J. Pharmacol. Exp. Ther.*, 241, 534, 1987. With permission.

substantia nigra pars reticulata, globus pallidus, hippocampus, and the molecular layers of the cerebellum. Sparse but significant binding is seen in the hypothalamus, basal amygdala, central gray, nucleus of the solitary tract, and certain spinal laminae. Very low and essentially nonspecific binding is seen in the thalamus, colliculi, and brain stem. Provocatively, very dense specific cannabinoid binding is seen in the ventral mesencephalic regions from which the DA reward-relevant neurons arise, and dense binding is seen in the striatal regions to which these DA reward-relevant neurons project. Very dense binding is also seen in the globus pallidus, which receives an enkephalinergic projection from the reward-relevant DA terminal projection fields of the mesotelencephalic DA system.[191]

ANATOMIC SITES OF CANNABINOID ACTION ON BRAIN REWARD CIRCUITS

From many of the separate strands of data reviewed above, preliminary inferences may be made regarding the specific brain loci at which cannabinoids interact with the reward circuitry of the forebrain to produce the euphorigenic "high" that constitutes the neurobiological basis for cannabinoid recreational use and abuse. First, by analogy with all other adequately studied classes of abused substances, and from electrical brain-stimulation reward experiments and *in vivo* brain microdialysis experiments with Δ^9-THC, we infer that cannabis interacts with the crucial drug-sensitive "second-stage" DA fibers of the reward system. Second, because the enhancing effect of Δ^9-THC on the DA reward neurons is naloxone blockable, we infer that cannabis interacts with these "second-stage" DA reward neurons in a synaptic plexus containing opioid peptide neurons. Third, on the basis of data from our laboratory and elsewhere (reviewed above) showing that endogenous opioid peptide neurons synapse directly onto the "second-stage" reward-relevant DA fibers, forming axo-axonic modulatory synapses regulating the flow of reward-relevant neural signals through the DA circuitry, and additional data (also reviewed above) suggesting that an additional opioid peptide neuron lies efferent to the DA reward neuron, we infer that the most likely site for cannabinoid interaction with reward-modulating endogenous brain opioid mechanisms is in the DA terminal regions of the reward-relevant ascending DA mesotelencephalic system, in reward-relevant synapses containing both an opioid-DA axo-axonic link and a DA-opioid presynaptic-postsynaptic link. Fourth, in view of the evidence reviewed above that Δ^9-THC acts in a manner analogous to that of a presynaptic DA reuptake blocker, and arguing on the basis of parsimony that the specific locus of action of Δ^9-THC would not be far removed from the presynaptic DA reuptake pumps that it affects, we infer that cannabis acts within the region of the terminal projections of the ascending DA mesotelencephalic system. Fifth, in view of the fact (reviewed above) that electrical brain-stimulation reward enhancement by habit-forming drugs is naloxone blockable only for brain-stimulation reward loci lying within the medial portions of the ascending DA mesotelencephalic projections, and in view of the naloxone-blockable nature of Δ^9-THC enhancement of brain reward, we infer that cannabis acts primarily within the medial portions of the reward-relevant terminal projections of the ascending DA mesotelencephalic system. Sixth, in view of the focal involvement of the nucleus accumbens in the neurobiological substrates of brain reward, and in view of the fact that Δ^9-THCs enhancing effect on extracellular DA in the reward-relevant fiber systems of the forebrain is markedly more pronounced in the nucleus accumbens than in other reward-relevant forebrain loci,[154,155,192] we infer that cannabinoid enhancement of brain reward is mediated primarily at synapses within the nucleus accumbens.

HYPOTHETICAL MODELS OF CANNABINOID ACTION ON REWARD SYNAPSES

From all of the data reviewed in this chapter, it is possible to develop hypothetical models of psychoactive cannabinoid interaction with reward-relevant synapses of the mesotelencephalic DA system. Two such models are presented in Figure 22. Both models presume that, for reasons outlined above, the principal cannabis-sensitive link in the mesotelencephalic DA reward pathway is in the region of the DA terminal projections of the mesotelencephalic DA reward pathway, most likely within the medial nucleus accumbens and associated cell fields of the medial ventral striatum. Both models presume, again for reasons outlined above, that the reward-relevant DA terminals of this system synapse upon endogenous opioid peptidergic neurons that carry the reward signal one synapse further, and that other modulatory endogenous opioid peptidergic neurons synapse, in axo-axonic fashion, onto the DA terminals of the reward-relevant DA neurons. They further presume that this second class of opioid peptidergic neurons actively modulates the flow of reward-relevant neural signals through the DA circuitry, perhaps by classical presynaptic excitation and inhibition, but also perhaps by a modulatory interaction between the opioid receptors located on the DA terminals and the DA reuptake mechanism within those terminals.

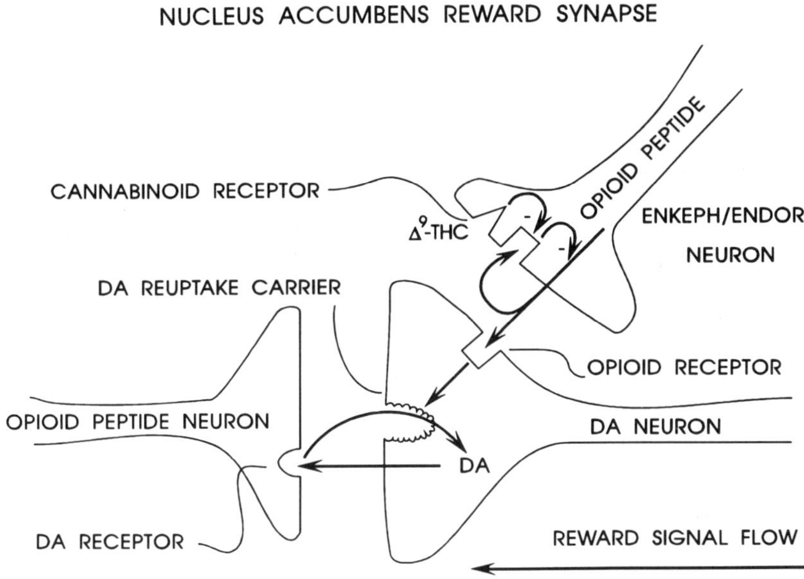

FIGURE 22. Hypothetical models of cannabinoid interaction with nucleus accumbens reward-relevant synapses (see text). (A) Model assuming negative allosteric modulation of endogenous opioid peptide receptors by cannabinoid receptor activation. (B) Model assuming positive modulation of endogenous opioid peptide receptors by cannabinoid receptor activation.

This last possibility is admittedly highly speculative, but finds support in previous work suggesting the existence within nerve endings of functional coupling between mechanisms of neurotransmitter uptake and presynaptic receptor activation.[193-198] Both models presume that specific cannabinoid receptors are found within these postulated reward synapses of the medial nucleus accumbens. The models speculate that, for reasons outlined above, there is a tight functional link, perhaps allosteric in nature, between the cannabinoid receptors and adjacent opioid receptors. They further presume, again for reasons outlined above, that the opioid receptors allosterically linked to the cannabinoid receptors are of the δ and/or μ (but not κ) type. The two models differ primarily in the nature of the postulated functional link between cannabinoid receptor activation and opioid receptor function, and in derivative constraints therefrom. The model depicted in Figure 22A postulates that this interaction is negative, thus constraining the location of the cannabinoid receptor and its allosterically linked opioid receptor to the axon terminal of the enkephalinergic or endorphinergic neuron that syn-

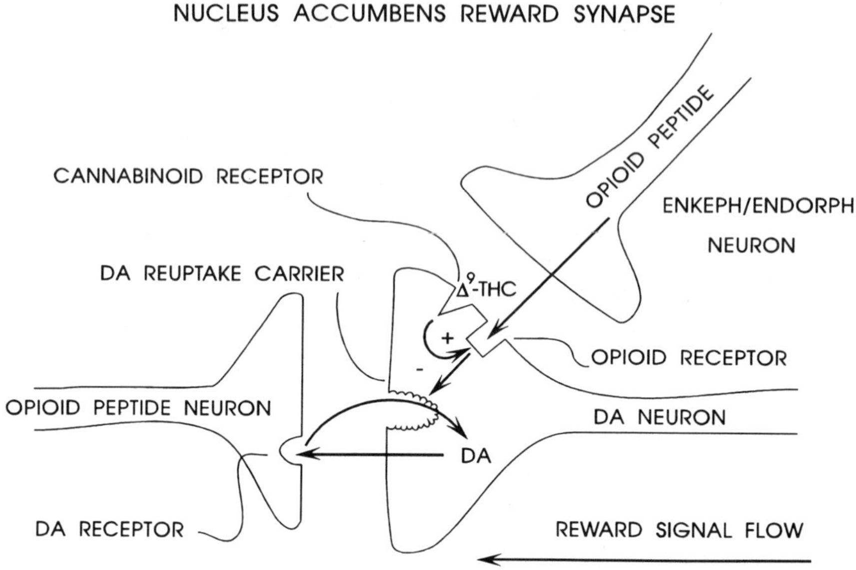

FIGURE 22B.

apses axo-axonically onto the axon terminal of the "second-stage" DA reward neuron of the mesotelencephalic reward circuitry. This also requires postulating a class of opioid autoreceptor[199] within this synaptic plexus. The model depicted in Figure 22B postulates that the interaction between cannabinoid receptor activation and opioid receptor function is positive, thus allowing the location of the cannabinoid receptor and its allosterically linked opioid receptor to be directly on the axon terminal of the "second-stage" DA reward neuron of the mesotelencephalic reward circuitry. The model depicted in Figure 22B has the appeal of simplicity and parsimony, but seems counter to our own receptor-binding evidence for a negative allosteric modulation of endogenous brain δ and μ opioid receptors by cannabinoids. Other, more complicated models than the two presented here are, of course, possible. Such models might invoke more neural elements and neuron types, more complex transsynaptic relationships, and/or interneuron modulatory links. Whether simple or complex, hypothetical models of cannabinoid interaction with brain reward substrates generate testable hypotheses, and the testing of such hypotheses, both in our laboratory and elsewhere, can reasonably be expected to significantly augment our knowledge of cannabinoid interaction with brain reward substrates, and our understanding of the neurobiological basis of cannabinoid abuse, over the coming decade.

CONCLUSIONS

Given the above description of cannabinoid interaction with brain reward mechanisms, the question may well be asked: "why are cannabinoid preparations not avidly self-administered by infrahuman species?" Indeed, the published literature on cannabis self-administration in animals would suggest, on its face, that cannabis has little appetitive or rewarding value. A number of workers have reported that marijuana, hashish, and other cannabinoid preparations are either not self-administered or only poorly self-administered in laboratory animals.[200-202] Even in those studies where successful cannabis self-administration has been seen, most environmental contingencies have been unusual and self-administration response rates low. Thus, Pickens et al.[203] obtained i.v. cannabinoid self-administration in monkeys only after exposing the animals to phencyclidine; Kaymakçalan and Deneau[204,205] obtained i.v. Δ^9-THC self-administration in monkeys only after cannabinoid physical dependence had been established. Takahashi and Singer[206] found that drug-naive rats will intravenously self-administer Δ^9-THC at rates significantly above control only if kept at 80% reduced body weight. On the other hand, Pickens et al.[203] obtained small but significantly elevated self-administration of hashish smoke (oral smoke-sucking with frequent inhalation) in monkeys as compared to oral self-administration of heated air. Several points need to be made with respect to such studies. First, in most studies, the number of animals tested was exceedingly low. Second, the insolubility of Δ^9-THC in water, its delayed onset of action, and its induction of taste aversion all militate against successful self-administration in laboratory settings. Third, and probably most germane of all, no attention was paid in previous self-administration work to strain differences or genetic factors. As delineated above, our findings of Δ^9-THC-induced enhancement of electrical brain-stimulation reward and of extracellular DA efflux in reward-relevant forebrain DA loci have been exquisitely specific to Lewis strain rats. In our hands, the Lewis strain animals become slightly quiescent when given low doses of Δ^9-THC, while rats of other strains exhibit a strange hyperreactivity characterized by vocalizations when approached or held. Tellingly, this hyperreactivity and vocalization, well described in the literature,[207,208] is a prominent feature of the behavior shown by rats in previous Δ^9-THC self-administration trials,[206] suggesting that the animals (of various non-Lewis strains) were indeed responding to Δ^9-THC in a "non-Lewis" manner.

Another question that may well be asked is: "what is the biological function of cannabinoid-sensitive sites within the neural reward substrates of the forebrain?" This is really a subset of a larger question dealing with the biological function of the reward circuits of the brain. After all, no one seriously believes that these brain reward systems and their sensitivity to cannabis emerged during the eons of mammalian evolution simply so that *Homo sapiens* can smoke marijuana and consume hashish-laced brownies. What purposes, then, did these brain reward systems evolve to serve? As suggested earlier in this chapter in the quotation from Wise,[32] many workers believe that habit-forming drugs act on brain circuits that evolved to subserve the normal reinforcement functions of the CNS, for example, in reinforcing such biologically essential behaviors as feeding, as postulated more than 40 years ago in the pioneering ideas of Hebb.[8] In fact, one of the oldest observations in the research literature on electrical stimulation of the brain is that stimulation in brain reward areas can also evoke natural consummatory behaviors such as feeding and other species-typical biologically essential behaviors.[200-212] Furthermore, the directly activated neural substrates of stimulation-induced feeding and brain-stimulation reward in the lateral hypothalamus and ventral tegmental area appear identical in terms of refractory periods, conduction velocities, and medial-lateral and dorsal-ventral alignment of the nerve fibers subserving the two stimulation-induced effects (feeding and reward), suggesting strongly that the same fibers mediate both effects.[213,214] It is intriguing in this regard to note that the same mesotelencephalic DA circuits that appear to subserve the reward induced by habit-forming drugs have recently been shown by Hernandez and Hoebel to also be biochemically activated, in a manner seemingly identical to the DA activation produced by habit-forming drugs, by natural rewards.[83] If all this is true, then presumably the cannabinoid-sensitive sites in the reward circuitry of the mammalian forebrain are there to subserve some biologically relevant natural reward or reinforcement function. Presumably, this also implies that there is a natural endogenous cannabinoid ligand in mammalian brains that serves some as yet unimagined biologically essential reward or reinforcement function. Many laboratories are presently searching for just such an endogenous "endocannabin".[215-217] In truth, these are very exciting times in cannabinoid neurobiology.

ACKNOWLEDGMENTS

Work from the author's laboratory mentioned in this manuscript was supported, in part, by the U.S. Public Health Service under research grants DA-01560, DA-02089, and DA-03622 from the National Institute on Drug Abuse, research grant NS-09649 from the National Institute of Neurological Disorders and Stroke, research grant RR-05397 (Biomedical Research Support Grant) from the National Institutes of Health, and training grants GM-00065 and GM-07260 from the National Institute of General Medical Sciences; by U.S. National Science Foundation research grant BNS 86-09351; by the U.S. Air Force Aeromedical Division, under research projects 6893-02-005 and 6893-02-039; by the Natural Sciences and Engineering Research Council of Canada; and by funds from the Sue Golding Graduate Division of Medical Sciences, Albert Einstein College of Medicine. Preparation of this manuscript was supported, in part, by the New York State Department of Substance Abuse Services. The author is indebted to colleagues too numerous to mention for their generous collaborations, inspirations, and criticisms, and to Dr. Herman M. van Praag, Chairman, Department of Psychiatry, Albert Einstein College of Medicine, for his support. Specific thanks are due to Dr. Roger Pertwee of the University of Aberdeen, Scotland, and to Drs. Suanne Garber, Amos Fleischmann, and Maynard Makman of the Albert Einstein College of Medicine for generous criticisms and helpful suggestions regarding the functional interaction of cannabinoids with endogenous opioid systems and regarding the functional interaction of receptors with membrane carrier systems. The tireless work of the following former and present students and laboratory assistants is gratefully acknowledged: Thomas Seeger, Jules Nazzaro, Pierre Vaysse, Jean Max Ng Cheong Ton, Cassandra Milling, Amy Donner, David Cohen, David Morrison, Jianping Chen, Diane Smith, Ronen Marmur, Eric Rosenbaum, Addy Pulles, and William Paredes. The very generous support and encouragement of Dr. Joyce Lowinson, Director of the Division of Substance Abuse, Albert Einstein College of Medicine, is most especially and gratefully acknowledged.

REFERENCES

1. **Mechoulam, R.,** The pharmacohistory of cannabis sativa, in *Cannabinoids as Therapeutic Agents*, Mechoulam, R., Ed., CRC Press, Boca Raton, FL, 1986, 1.
2. **Kozel, N. J. and Adams, E. H.,** Epidemiology of drug abuse: an overview, *Science*, 234, 70, 1986.
3. **Kleber, H. D.,** Introduction — cocaine abuse: historical, epidemiological, and psychosocial perspectives, *J. Clin. Psychiatry*, 4(Suppl.), 3, 1988.
4. **Jaffe, J. H.,** Drug addiction and drug abuse, in *The Pharmacological Basis of Therapeutics*, 3rd ed., Goodman, L. S. and Gilman, A., Eds., Macmillan, New York, 1965, 285.
5. **Schuster, C. R. and Thompson, T.,** Self administration of and behavioral dependence on drugs, *Annu. Rev. Pharmacol.*, 9, 483, 1969.
6. **Griffiths, R. R., Bigelow, G. E., and Liebson, I.,** Experimental drug self-administration: generality across species and type of drug, *Natl Inst. Drug Abuse Res. Monogr. Ser.*, 20, 24, 1978.
7. **Brady, J. V. and Lukas, S. E.,** *Testing Drugs for Physical Dependence Potential and Abuse Liability, Natl. Inst. on Drug Abuse Res. Monogr. Ser., No. 52,* U.S. Government Printing Office, Washington, D.C., 1984.
8. **Hebb, D. O.,** *The Organization of Behavior: A Neuropsychological Theory*, John Wiley & Sons, New York, 1949.
9. **Olds, J. and Milner, P.,** Positive reinforcement produced by electrical stimulation of septal area and other regions of rat brain, *J. Comparative Physiol. Psychol.*, 47, 419, 1954.
10. **Rolls, E. T.,** *The Brain and Reward*, Pergamon Press, Oxford, 1975.
11. **Routtenberg, A.,** The reward system of the brain, *Sci. Am.*, 239(5), 154, 1978.
12. **Olds, M. E. and Fobes, J. L.,** The central basis of motivation: intracranial self-stimulation studies, *Annu. Rev. Psychol.*, 32, 523, 1981.
13. **Killam, K. F., Olds, J., and Sinclair, J.,** Further studies on the effects of centrally acting drugs on self-stimulation, *J. Pharmacol. Exp. Ther.*, 119, 157, 1957.
14. **Kornetsky, C.,** Brain-stimulation reward: a model for the neuronal bases for drug-induced euphoria, *Natl. Inst. Drug Abuse Res. Monogr. Ser.*, 62, 30, 1985.
15. **Bursten, B. and Delgado, J. M. R.,** Positive reinforcement induced by intracerebral stimulation in the monkey, *J. Comparative Physiol. Psychol.*, 51, 6, 1958.
16. **Olds, M. E. and Olds, J.,** Approach-avoidance analysis of rat diencephalon, *J. Comparative Neurol.*, 120, 259, 1963.
17. **Wilkinson, H. A. and Peele, T. L.,** Intracranial self-stimulation in cats, *J. Comparative Neurol.*, 121, 425, 1963.
18. **Briese, E. and Olds, J.,** Reinforcing brain stimulation and memory in monkeys, *Exp. Neurol.*, 10, 493, 1964.
19. **Valenstein, E. S.,** The anatomical locus of reinforcement, *Progr. Physiol. Psychol.*, 1, 149, 1966.
20. **Wetzel, M. C.,** Self-stimulation's anatomy: data needs, *Brain Res.*, 10, 287, 1968.
21. **Routtenberg, A. and Malsbury, C.,** Brainstem pathways of reward, *J. Comparative Physiol. Psychol.*, 68, 22, 1969.
22. **Dahlström, A. and Fuxe, K.,** Evidence for the existence of monoamine-containing neurons in the central nervous system. I. Demonstration of monoamines in the cell bodies of brainstem neurons, *Acta Physiol. Scand.*, 62(Suppl. 232), 1, 1964.

23. **Fuxe, K.,** Evidence for the existence of monoamine neurons in the central nervous system. IV. Distribution of monoamine nerve terminals in the central nervous system, *Acta Physiol. Scand.*, 64(Suppl. 247), 37, 1965.
24. **Andén, N.- E., Dahlström, A., Fuxe, K., Larsson, K., Olson, L., and Ungerstedt, U.,** Ascending monoamine neurons to the telencephalon and diencephalon, *Acta Physiol. Scand.*, 67, 313, 1966.
25. **Ungerstedt, U.,** Stereotaxic mapping of the monoamine pathways in the rat brain, *Acta Physiol. Scand.*, 82(Suppl. 367), 1, 1971.
26. **Moore, R. Y. and Bloom, F. E.,** Central catecholamine neuron systems: anatomy and physiology of the dopamine systems, *Annu. Rev. Neurosci.*, 1, 129, 1978.
27. **Gardner, E. L.,** Pharmacological aspects of intracranial self-stimulation in Rhesus monkeys, paper presented at Winter Conference on Brain Research, Aspen, Colorado, 1969.
28. **Crow, T. J.,** A map of the rat mesencephalon for electrical self-stimulation, *Brain Res.*, 36, 265, 1972.
29. **Corbett, D. and Wise, R. A.,** Intracranial self-stimulation in relation to the ascending dopaminergic systems of the midbrain: a moveable electrode mapping study, *Brain Res.*, 185, 1, 1980.
30. **Mogenson, G. J., Takigawa, M., Robertson, A., and Wu, M.,** Self-stimulation of the nucleus accumbens and ventral tegmental area of Tsai attenuated by microinjections of spiroperidol into the nucleus accumbens, *Brain Res.*, 171, 247, 1979.
31. **Zarevics, P. and Setler, P. E.,** Simultaneous rate-independent and rate-dependent assessment of intracranial self-stimulation: evidence for the direct involvement of dopamine in brain reinforcement mechanisms, *Brain Res.*, 169, 499, 1979.
32. **Wise, R. A.,** The dopamine synapse and the notion of "pleasure centers" in the brain, *Trends Neurosci.*, 3, 91, 1980.
33. **Fray, P. J., Dunnett, S. B., Iversen, S. D., Björklund, A., and Stenevi, U.,** Nigral transplants reinnervating the dopamine-depleted neostriatum can sustain intracranial self-stimulation, *Science*, 219, 416, 1983.
34. **Wise, R. A.,** Brain dopamine and reward, in *Theory in Psychopharmacology*, Vol. 1, Cooper, S. J., Ed., Academic Press, New York, 1981, 103.
35. **Wise, R. A. and Bozarth, M. A.,** Brain substrates for reinforcement and drug self-administration, *Progr. Neuro-Psychopharmacol.*, 5, 467, 1981.
36. **Fibiger, H. C. and Phillips, A. G.,** Mesocorticolimbic dopamine systems and reward, *Ann. N.Y. Acad. Sci.*, 537, 206, 1988.
37. **Wise, R. A. and Rompre, P.- P.,** Brain dopamine and reward, *Annu. Rev. Psychol.*, 40, 191, 1989.
38. **Gallistel, C. R.,** Self-stimulation in the rat: quantitative characteristics of the reward pathway, *J. Comparative Physiol. Psychol.*, 92, 977, 1978.
39. **Yeomans, J. S.,** The absolute refractory periods of self-stimulation neurons, *Physiol. Behav.*, 22, 911, 1979.
40. **Yeomans, J. S., Matthews, G. G., Hawkins, R. D., Bellman, K., and Doppelt, H.,** Characterization of self-stimulation neurons by their local potential summation properties, *Physiol. Behav.*, 22, 921, 1979.
41. **Shizgal, P., Bielajew, C., Corbett, D., Skelton, R., and Yeomans, J.,** Behavioral methods for inferring anatomical linkage between rewarding brain stimulation sites, *J. Comparative Physiol. Psychol.*, 94, 227, 1980.
42. **Gallistel, C. R., Shizgal, P., and Yeomans, J.,** A portrait of the substrate for self-stimulation, *Psychol. Rev.*, 88, 228, 1981.

43. **Wise, R. A.,** "Yes, but!..." a response to Arbuthnott from Roy Wise, *Trends Neurosci.*, 3, 200, 1980.
44. **Gardner, E. L.,** Brain reward mechanisms, in *Comprehensive Textbook of Substance Abuse*, Lowinson, J. H., Ruiz, P., and Millman, R. B., Eds., Williams & Wilkins, Baltimore, 1992.
45. **Wise, R. A.,** Action of drugs of abuse on brain reward systems, *Pharmacol. Biochem. Behav.*, 13(Suppl. 1), 213, 1980.
46. **Wise, R. A. and Bozarth, M. A.,** Brain reward circuitry: four circuit elements "wired" in apparent series, *Brain Res. Bull.*, 12, 203, 1984.
47. **Wise, R. A.,** Neural mechanisms of the reinforcing action of cocaine, *Natl. Inst. Drug Abuse Res. Monogr. Ser.*, 50, 15, 1984.
48. **Wise, R. A.,** Neuroleptics and operant behavior: the anhedonia hypothesis, *Behav. Brain Sci.*, 5, 39, 1982.
49. **Phillips, A. G. and LePiane, F. G.,** Reinforcing effects of morphine microinjection into the ventral tegmental area, *Pharmacol. Biochem. Behav.*, 12, 965, 1980.
50. **Bozarth, M. A. and Wise, R. A.,** Intracranial self-administration of morphine into the ventral tegmental area of rats, *Life Sci.*, 28, 551, 1981.
51. **Hoebel, B. G., Monaco, A. P., Hernandez, L., Aulisi, E. F., Stanley, B. G., and Lenard, L.,** Self-injection of amphetamine directly into the brain, *Psychopharmacology*, 81, 158, 1983.
52. **Goeders, N. E. and Smith, J. E.,** Cortical dopaminergic involvement in cocaine reinforcement, *Science*, 221, 773, 1983.
53. **Goeders, N. E., Lane, J. D., and Smith, J. E.,** Self-administration of methionine enkephalin into the nucleus accumbens, *Pharmacol. Biochem. Behav.*, 20, 451, 1984.
54. **Bozarth, M. A. and Wise, R. A.,** Heroin reward is dependent on a dopaminergic substrate, *Life Sci.*, 29, 1881, 1981.
55. **Spyraki, C., Fibiger, H. C., and Phillips, A. G.,** Attenuation of heroin reward in rats by disruption of the mesolimbic dopamine system, *Psychopharmacology*, 79, 278, 1983.
56. **Gardner, E. L.,** An improved technique for determining brain reward thresholds in primates, *Behav. Res. Meth. Instrument.*, 3, 273, 1971.
57. **Gardner, E. L., Paredes, W., Smith, D., Donner, A., Milling, C., Cohen, D., and Morrison, D.,** Facilitation of brain stimulation reward by Δ^9-tetrahydrocannabinol, *Psychopharmacology*, 96, 142, 1988.
58. **Seeger, T. F. and Gardner, E. L.,** Enhancement of self-stimulation behavior in rats and monkeys after chronic neuroleptic treatment: evidence for mesolimbic supersensitivity, *Brain Res.*, 175, 49, 1979.
59. **Nazzaro, J. M. and Gardner, E. L.,** GABA antagonism lowers self-stimulation thresholds in the ventral tegmental area, *Brain Res.*, 189, 279, 1980.
60. **Pribram, K. H.,** The neurobehavioral analysis of limbic forebrain mechanisms: revision and progress report, in *Advances in the Study of Behavior*, Vol. 2, Lehrman, D. S., Hinde, R. A., and Shaw, E., Eds., Academic Press, New York, 1969, 297.
61. **Goddard, G. V.,** Functions of the amygdala, *Psychol. Bull.*, 62, 89, 1964.
62. **McLardy, T.,** Hippocampal formation as a detector-coder of temporal patterns, *Perspect.Biol. Med.*, 2, 443, 1959.
63. **Jackson, W. J. and Gardner, E. L.,** Modulation of hypothalamic ICSS by concurrent limbic stimulation, *Physiol. Behav.*, 12, 177, 1974.
64. **Seeger, T. F., Gardner, E. L., and Bridger, W. F.,** Increase in mesolimbic electrical self-stimulation after chronic haloperidol: reversal by L-dopa or lithium, *Brain Res.*, 215, 404, 1981.

65. **Seeger, T. F., Nazzaro, J. M., and Gardner, E. L.,** Selective inhibition of mesolimbic behavioral supersensitivity by naloxone, *Eur. J. Pharmacol.*, 65, 435, 1980.
66. **Nazzaro, J. M., Seeger, T. F., and Gardner, E. L.,** Morphine differentially affects ventral tegmental and substantia nigra brain reward thresholds, *Pharmacol. Biochem. Behav.*, 14, 325, 1981.
67. **Nazzaro, J. M., Gardner, E. L., Bridger, W. H., Carlson, K. R., and Seeger, T. F.,** Pentobarbital induces a naloxone-reversible decrease in mesolimbic self-stimulation thresholds, *Soc. Neurosci. Abstr.*, 7, 262, 1981.
68. **Seeger, T. F., Carlson, K. R., and Nazzaro, J. M.,** Pentobarbital induces a naloxone-reversible decrease in mesolimbic self-stimulation threshold, *Pharmacol. Biochem. Behav.*, 15, 583, 1981.
69. **Nazzaro, J. M., Seeger, T. F., and Gardner, E. L.,** Enhancing and inhibitory effects of phencyclidine ("angel dust") on mesolimbic-mesocortical intracranial self-stimulation, paper presented at meetings of the Eastern Psychological Association, Hartford, Connecticut, 1980.
70. **Nazzaro, J. M., Seeger, T. F., and Gardner, E. L.,** Naloxone blocks phencyclidine's dose-dependent effects on direct brain reward thresholds, in *Proceedings of World Conference on Clinical Pharmacology and Therapeutics*, British Pharmacological Society, London, 1980, 949.
71. **Nazzaro, J. M., Gardner, E. L., and Seeger, T. F.,** Effect of dissociative anesthetic drugs of abuse on direct brain reward thresholds, paper presented at 14th World Congress of the Collegium Internationale Neuro-Psychopharmacologicum, Florence, Italy, 1984.
72. **Gardner, E. L., Paredes, W., Seeger, T. F., Smith, D., and van Praag, H. M.,** Mesolimbic DA antagonism by $5HT_3$ receptor blockade, in *Book of Abstracts, World Psychiatric Association Regional Symposium, The Research and Clinical Interface for Psychiatric Disorders,* October 13–16, 1988, Washington, D.C., American Psychiatric Association, Washington, D.C., 1988, 222.
73. **Marsden, C. A., Ed.,** *Measurement of Neurotransmitter Release In Vivo (IBRO Handbook Series: Methods in the Neurosciences,* Vol. 6), John Wiley & Sons, New York, 1984.
74. **Myers, R. D. and Knott, P. J., Eds.,** Neurochemical Analysis of the Conscious Brain: Voltammetry and Push-Pull Perfusion, *Ann. N.Y. Acad. Sci.*, 473, 1986.
75. **Gysling, K. and Wang, R. Y.,** Morphine-induced activation of A10 dopamine neurons in rats brain, *Brain Res.*, 277, 119, 1983.
76. **Hommer, D. W. and Pert, A.,** The action of opiates in the rat substantia nigra: an electrophysiological analysis, *Peptides*, 4, 603, 1983.
77. **Imperato, A. and Di Chiara, G.,** Preferential stimulation of dopamine release in the nucleus accumbens of freely moving rats by ethanol, *J. Pharmacol. Exp. Ther.*, 239, 219, 1986.
78. **Imperato, A., Mulas, A., and Di Chiara, G.,** Nicotine preferentially stimulates dopamine release in the limbic system of freely moving rats, *Eur. J. Pharmacol.*, 132, 337, 1986.
79. **Kalivas, P. W., Widerlov, E., Stanley, D., Breese, G., and Prange, A. J., Jr.,** Enkephalin action on the mesolimbic system: a dopamine-dependent and a dopamine-independent increase in locomotor activity, *J. Pharmacol. Exp. Ther.*, 227, 229, 1983.
80. **Kalivas, P. W., Duffy, P., Dilts, R., and Abhold, R.,** Enkephalin modulation of A10 dopamine neurons: a role in dopamine sensitization, *Ann. N.Y. Acad. Sci.*, 537, 405, 1988.

81. **Di Chiara, G. and Imperato, A.,** Preferential stimulation of dopamine release in the nucleus accumbens by opiates, alcohol, and barbiturates: studies with transcerebral dialysis in freely moving rats, *Ann. N.Y. Acad. Sci.*, 473, 367, 1986.
82. **Chen, J. and Gardner, E. L.,** *In vivo* brain microdialysis study of phencyclidine on presynaptic dopamine efflux in rat caudate nucleus, *Soc. Neurosci. Abstr.*, 14, 525, 1988.
83. **Hernandez, L. and Hoebel, B. G.,** Food reward and cocaine increase extracellular dopamine in the nucleus accumbens as measured by microdialysis, *Life Sci.*, 42, 1705, 1988.
84. **Ng Cheong Ton, J. M., Gerhardt, G. A., Friedemann, M., Etgen, A. M., Rose, G. M., Sharpless, N. S., and Gardner, E. L.,** The effects of Δ^9-tetrahydrocannabinol on potassium-evoked release of dopamine in the rat caudate nucleus: an *in vivo* electrochemical and *in vivo* microdialysis study, *Brain Res.*, 451, 59, 1988.
85. **Westerink, B. H., Tuntler, J., Damsma, G., Rollema, H., and de Vries, J. B.,** The use of tetrodotoxin for the characterization of drug-enhanced dopamine release in conscious rats studied by brain dialysis, *Naunyn-Schmiedeberg's Arch. Pharmacol.*, 336, 502, 1987.
86. **Wise, R. A. and Bozarth, M. A.,** A psychomotor stimulant theory of addiction, *Psychol. Rev.*, 94, 469, 1987.
87. **Lierman, T. L., Ed.,** *Building a Healthy America*, Mary Ann Liebert, Inc., New York, 1986.
88. **Zukin, S. R. and Zukin, R. S.,** Specific [^{3}H]phencyclidine binding in rat central nervous system, *Proc. Natl. Acad. Sci. U.S.A.*, 76, 5372, 1979.
89. **Sircar, R. and Zukin, S. R.,** Further evidence of phencyclidine/sigma opioid receptor commonality, *Natl. Inst. Drug Abuse Res. Monogr. Ser.*, 64, 14, 1986.
90. **Sircar, R., Nichtenhauser, R., Ieni, J. R., and Zukin, S. R.,** Characterization and autoradiographic visualization of (+)-[^{3}H]SKF10,047 binding in rat and mouse brain: further evidence for phencyclidine/"sigma opiate" receptor commonality, *J. Pharmacol. Exp. Ther.*, 237, 681, 1986.
91. **Sonders, M. S., Keana, F. W., and Weber, E.,** Phencyclidine and psychotomimetic sigma opiates: recent insights into their biochemical and physiological sites of action, *Trends Neurosci.*, 11, 37, 1988.
92. **Herberg, L. J. and Rose, I. C.,** The effect of MK-801 and other antagonists of NMDA-type glutamate receptors on brain-stimulation reward, *Psychopharmacology*, 99, 87, 1989.
93. **Gerhardt, G. A., Pang, K., and Rose, G. M.,** *In vivo* electrochemical demonstration of the presynaptic actions of PCP in rat caudate nucleus, *J. Pharmacol. Exp. Ther.*, 241, 714, 1987.
94. **Hernandez, L., Auerbach, S., and Hoebel, B. G.,** Phencyclidine (PCP) injected in the nucleus accumbens increases extracellular dopamine and serotonin as measured by microdialysis, *Life Sci.*, 42, 1713, 1988.
95. **Gardner, E. L., Paredes, W., Smith, D., and Zukin, R. S.,** Facilitation of brain stimulation reward by Δ^9-tetrahydrocannabinol is mediated by an endogenous opioid mechanism, *Adv. Biosci.*, 75, 671, 1989.
96. **Lorens, S. A. and Sainati, S. M.,** Naloxone blocks the excitatory effect of ethanol and chlordiazepoxide on lateral hypothalamic self-stimulation behavior, *Life Sci.*, 23, 1359, 1978.
97. **Esposito, R. U., Perry, W., and Kornetsky, C.,** Effects of *d*-amphetamine and naloxone on brain stimulation reward, *Psychopharmacology*, 69, 187, 1980.

98. **Gardner, E. L. and Seeger, T. F.,** Anatomically selective action of atypical neuroleptics on the mesocorticolimbic dopamine system, *Ann. N.Y. Acad. Sci.*, 537, 502, 1988.
99. **Eichler, A. J., Antelman, S. M., and Fisher, A. E.,** Self-stimulation: site-specific tolerance to chronic dopamine receptor blockade, *Soc. Neurosci. Abstr.*, 2, 440, 1976.
100. **Ettenberg, A. and Wise, R. A.,** Non-selective enhancement of locus coeruleus and substantia nigra self-stimulation after termination of chronic dopaminergic receptor blockade with pimozide in rats, *Psychopharmacol. Commun.*, 7, 507, 1976.
101. **Nazzaro, J. M., Seeger, T. F., and Gardner, E. L.,** Differential opiate effects on intracranial self-stimulation thresholds in substantia nigra (A-9) and ventral tegmentum (A-10), *Soc. Neurosci. Abstr.*, 5, 658, 1979.
102. **Broderick, P. A., Gardner, E. L., and van Praag, H. M.,** Enkephalinergic action on dopaminergic transmission: behavioral and *in vivo* electrochemical studies, paper presented at meetings of the Society of Biological Psychiatry, New York, 1983.
103. **Broderick, P. A., Gardner, E. L., and van Praag, H. M.,** *In vivo* electrochemical evidence for an opiate-induced modulation of dopaminergic and serotonergic systems in rat striatum, paper presented at meetings of the American College of Neuropsychopharmacology, San Juan, Puerto Rico, 1983.
104. **Broderick, P. A. and Gardner, E. L.,** unpublished data, 1984.
105. **Broderick, P. A.,** *In vivo* electrochemical studies of rat striatal dopamine and serotonin release after morphine, *Life Sci.*, 36, 2269, 1985.
106. **Ostrowski, N. L., Paul, I. A., Drnach, M., and Caggiula, A. R.,** Morphine produces time-dependent changes in locomotor activity and in discharge rates of mesencephalic dopamine neurons, *Soc. Neurosci. Abstr.*, 8, 230, 1982.
107. **Ostrowski, N. L., Hatfield, C. B., and Caggiula, A. R.,** The effects of low doses of morphine on the activity of dopamine-containing cells and on behavior, *Life Sci.*, 31, 2347, 1982.
108. **Khachaturian, H., Lewis, M. E., Schäfer, M. K.-H., and Watson, S. J.,** Anatomy of the CNS opioid systems, *Trends Neurosci.*, 8, 111, 1985.
109. **Mansour, A., Khachaturian, H., Lewis, M., Akil, H., and Watson, S.,** Anatomy of CNS opioid receptors, *Trends Neurosci.*, 11, 308, 1988.
110. **Gardner, E. L., Zukin, R. S., and Makman, M. H.,** Modulation of opiate receptor binding in striatum and amygdala by selective mesencephalic lesions, *Brain Res.*, 194, 232, 1980.
111. **Pollard, H., Llorens, C., Bonnet, J. J., Costentin, J., and Schwartz, J. C.,** Opiate receptors on mesolimbic dopaminergic neurons, *Neurosci. Lett.*, 7, 295, 1977.
112. **Pollard, H., Llorens, C., Schwartz, J. C., Gros, C., and Dray, F.,** Localization of opiate receptors and enkephalins in the rat striatum in relationship with the nigrostriatal dopaminergic system: lesion studies, *Brain Res.*, 151, 392, 1978.
113. **Hirschhorn, I. D., Hittner, D., Gardner, E. L., Cubells, J., and Makman, M. H.,** Evidence for a role of endogenous opioids in the nigrostriatal system: influence of naloxone and morphine on nigrostriatal dopaminergic supersensitivity, *Brain Res.*, 270, 109, 1983.
114. **Tallman, J. F. and Gallagher, D. W.,** The GABA-ergic systems: a locus of benzodiazepine action, *Annu. Rev. Neurosci.*, 8, 21, 1985.
115. **de Belleroche, J. S. and Bradford, H. F.,** Presynaptic control of the synthesis and release of dopamine from striatal synaptosomes: a comparison between the effects of 5-hydroxytryptamine, acetylcholine, and glutamate, *J. Neurochemistry*, 35, 1227, 1980.
116. **Westfall, T. C. and Titternary, V.,** Inhibition of the electrically induced release of [^{3}H]dopamine by serotonin from superfused rat striatal slices, *Neurosci. Lett.*, 28, 205, 1982.

117. **Chesselet, M. F.,** Presynaptic regulation of neurotransmitter release in the brain: facts and hypothesis, *Neuroscience*, 12, 347, 1984.
118. **Hetey, L. and Drescher, K.,** Influence of antipsychotics on presynaptic receptors modulating the release of dopamine in synaptosomes of the nucleus accumbens of rats, *Neuropharmacology*, 25, 1103, 1986.
119. **Tricklebank, M. D.,** Interactions between dopamine and $5HT_3$ receptors suggest new treatment for psychosis and drug addiction, *Trends Pharmacol. Sci.*, 10, 127, 1989.
120. **Amit, Z., Sutherland, E. A., Gill, K., and Ögren, S. O.,** Zimeldine: a review of its effects on ethanol consumption, *Neurosci. Biobehav. Rev.*, 8, 35, 1984.
121. **Gill, K., Amit, Z., and Ögren, S. O.,** The effects of zimelidine on voluntary ethanol consumption: studies on the mechanism of action, *Alcohol*, 2, 343, 1985.
122. **Le Bourhis, B., Uzan, A., Aufrere, G., and Lefur, G.,** Effets de l'indalpine, inhibiteur spécifique de la récapture de la sérotonine sur la dépendance comportementale à l'éthanol et sur la prise volontaire d'alcool chez le rat [Effects of indalpine, a specific serotonin reuptake inhibitor, on ethanol behavioral dependence and voluntary alcohol consumption in the rat], *Ann. Pharmaceut. Fr.*, 39, 11, 1981.
123. **Leccese, A. P. and Lyness, W. H.,** The effects of putative 5-hydroxytryptamine receptor active agents on *d*-amphetamine self-administration in controls and rats with 5,7-dihydroxytryptamine medial forebrain bundle lesions, *Brain Res.*, 303, 153, 1984.
124. **Murphy, J. M., Waller, M. B., Gatto, G. J., McBride, W. J., Lumeng, L., and Li, T.-K.,** Monoamine uptake inhibitors attenuate ethanol intake in alcohol-preferring (P) rats, *Alcohol*, 2, 349, 1985.
125. **Naranjo, C. A., Sellers, E. M., and Lawrin, M. O.,** Modulation of ethanol intake by serotonin uptake inhibitors, *J. Clin. Psychiatry*, 47(Suppl.), 16, 1986.
126. **Rockman, G. E., Amit, Z., Carr, G., Brown, Z. W., and Ögren, S. O.,** Attenuation of ethanol intake by 5-hydroxytryptamine uptake blockade in laboratory rats. I. Involvement of brain 5-hydroxytryptamine in the mediation of the positive reinforcing properties of ethanol, *Arch. Int. Pharmacodyn. Ther.*, 241, 245, 1979.
127. **Yu, D. S. L., Smith, F. L., Smith, D. G., and Lyness, W. H.,** Fluoxetine-induced attenuation of amphetamine self-administration in rats, *Life Sci.*, 39, 1383, 1986.
128. **Zabik, J. E., Roache, J. D., Sidor, R., and Nash, J. F., Jr.,** The effects of fluoxetine on ethanol preference in the rat, *Pharmacologist*, 24, 204, 1982.
129. **Kelley, A. E. and Cador, M.,** Behavioral evidence for differential neuropeptide modulation of the mesolimbic dopamine system, *Ann. N.Y. Acad. Sci.*, 537, 415, 1988.
130. **Kelley, A. E., Stinus, L., and Iversen, S. D.,** Behavioural activation induced in the rat by substance P infusion into ventral tegmental area: implication of dopaminergic A10 neurones, *Neurosci. Lett.*, 11, 335, 1979.
131. **Palacios, J. M. and Kuhar, M. J.,** Neurotensin receptors are located on dopamine-containing neurones in rat midbrain, *Nature (London)*, 294, 587, 1981.
132. **Pernow, B.,** Substance P., *Pharmacol. Rev.*, 35, 85, 1983.
133. **Phillips, A. G., Blaha, C. D., Fibiger, H. C., and Lane, R. F.,** Interactions between mesolimbic dopamine neurons, cholecystokinin, and neurotensin: evidence using *in vivo* voltammetry, *Ann. N.Y. Acad. Sci.*, 537, 347, 1988.
134. **Pickel, V. M., Joh, T. H., and Chan, J.,** Substance P in the rat nucleus accumbens: ultrastructural localization in axon terminals and their relation to dopaminergic afferents, *Brain Res.*, 444, 247, 1988.
135. **Bissette, G. and Nemeroff, C. B.,** Neurotensin and the mesocorticolimbic dopamine system, *Ann. N.Y. Acad. Sci.*, 537, 397, 1988.

136. **Hökfelt, T., Everitt, B. S., Theodorsson-Norheim, E., and Goldstein, M.,** Occurrence of neurotensinlike immunoreactivity in subpopulations of hypothalamic, mesencephalic, and medullary catecholamine neurons, *J. Comparative Neurol.*, 222, 543, 1984.
137. **Kalivas, P. W., Jennes, L., and Miller, J. S.,** A catecholaminergic projection from the ventral tegmental area to the diagonal band of Broca: modulation by neurotensin, *Brain Res.*, 326, 229, 1985.
138. **Hervé, D., Tassin, J. P., Studler, J. M., Dana, C., Kitagbi, P., Vincent, J. P., and Glowinski, J.,** Dopaminergic control of ^{125}I-labelled neurotensin binding site density in corticolimbic structures of the rat brain, *Proc. Natl. Acad. Sci. U.S.A.*, 83, 6203, 1986.
139. **Nemeroff, C. C. and Cain, S. T.,** Neurotensin-dopamine interactions in the CNS, *Trends Pharmacol. Sci.*, 6, 201, 1985.
140. **Quirion, R., Chiueh, C. C., Everist, H. D., and Pert, A.,** Comparative localization of neurotensin receptors on nigrostriatal and mesolimbic dopaminergic terminals, *Brain Res.*, 327, 835, 1985.
141. **Studler, J.-M., Kitagbi, P., Tramu, G., Hervé, D., Glowinski, J., and Tassin, J.-P.,** Extensive co-localization of neurotensin with dopamine in rat mesocortico-frontal dopaminergic neurons, *Neuropeptides*, 11, 95, 1988.
142. **Voigt, M. M., Wang, R. Y., and Westfall, T. C.,** The effects of cholecystokinin on the *in vivo* release of newly synthesized [^{3}H]dopamine from the nucleus accumbens of the rat, *J. Neurosci.*, 5, 2744, 1985.
143. **Dworkin, S. I., Goeders, N. E., and Smith, J. E.,** The reinforcing and rate effects of intracranial dopamine administration, *Natl. Inst. Drug Abuse Res. Monogr. Ser.*, 67, 242, 1986.
144. **Bozarth, M. A., Ed.,** *Methods of Assessing the Reinforcing Properties of Abused Drugs*, Springer-Verlag, New York, 1987.
145. **Hiroi, N. and White, N. M.,** The reserpine-sensitive dopamine pool mediates (+)-amphetamine-conditioned reward in the place preference paradigm, *Brain Res.*, 510, 33, 1990.
146. **Hiroi, N., McDonald, R. J., and White, N. M.,** Involvement of the lateral nucleus of the amygdala in amphetamine and food conditioned place preferences (CPP), *Soc. Neurosci. Abstr.*, 16, 605, 1990.
147. **Hiroi, N.,** A Pharmacological and Neuroanatomical Investigation of the Conditioned Place Preference Produced by Amphetamine, Ph.D. thesis, McGill University, Montreal, 1990.
148. **Dackis, C. A. and Gold, M. S.,** New concepts in cocaine addiction: the dopamine depletion hypothesis, *Neurosci. Biobehav. Rev.*, 9, 469, 1985.
149. **Glantz, M. D., Ed.,** *Correlates and Consequences of Marijuana Use ADAMHA Research. Issues,* No. 34, Alcohol, Drug Abuse, and Mental Health Administration, U.S. Public Health Service, Rockville, MD, 1984.
150. **National Institute on Drug Abuse,** Marijuana Res. Announcement, February 1983 (*Catalog of Federal Domestic Assistance* No. 13.279), National Institute on Drug Abuse, U.S. Public Health Service, Rockville, MD, 1983.
151. **Harris, A. R. and Stokes, J. A.,** Cannabinoids inhibit calcium uptake by brain synaptosomes, *J. Neurochemistry*, 2, 443, 1982.
152. **Rosenkrantz, H., Sprague, R. A., Fleischman, R. W., and Braude, M. C.,** Oral Δ^9-tetrahydrocannabinol toxicity in rats treated for periods up to six months, *Toxicol. Appl. Pharmacol.*, 32, 399, 1975.

153. **Ng Cheong Ton, J. M. and Gardner, E. L.,** Effects of delta-9-tetrahydrocannabinol on dopamine release in the brain: intracranial microdialysis experiments, *Soc. Neurosci. Abstr.*, 12, 135, 1986.
154. **Chen, J., Paredes, W., Li, J., Smith, D., and Gardner, E. L.,** *In vivo* brain microdialysis studies of Δ^9-tetrahydrocannabinol on presynaptic dopamine efflux in nucleus accumbens of the Lewis rat, *Soc. Neurosci. Abstr.*, 15, 1096, 1989.
155. **Chen, J., Paredes, W., Li, J., Smith, D., Lowinson, J., and Gardner, E. L.,** Δ^9-tetrahydrocannabinol produces naloxone-blockable enhancement of presynaptic dopamine efflux in nucleus accumbens of conscious, freely-moving rats as measured by intracerebral microdialysis, *Psychopharmacology*, 102, 156, 1990.
156. **Taylor, D. A., Sitaram, B. R., and Elliot-Baker, S.,** Effect of Δ-9-tetrahydrocannabinol on the release of dopamine in the corpus striatum of the rat, in *Marijuana: An International Research Report*, Chesher, G., Consroe, P., and Musty, R., Eds., Australian Government Publishing Service, Canberra, Australia, 1988.
157. **Taylor, D. A., Sitaram, B., and Nyoni, E.,** Effect of repeated administration of Δ^9-tetrahydrocannabinol (Δ^9-THC) on dopaminergic transmission and 5-HIAA levels in the corpus striatum of rats, paper presented at Marijuana '90 International Conference on Cannabis and Cannabinoids, Satellite meeting of the XIth International Congress of Pharmacology, Kolympari, Chania, Greece, 1990.
158. **Gardner, E. L., Chen, J., Paredes, W., Smith, D., Li, J., and Lowinson, J.,** Enhancement of presynaptic dopamine efflux in brain by Δ^9-tetrahydrocannabinol is mediated by an endogenous opioid mechanism, in *New Leads in Opioid Research*, van Ree, J. M., Mulder, A. H., Wiegant, V. M., and van Wimersma Greidanus, T. B., Eds., Elsevier, Amsterdam, 1990, 243.
159. **Gardner, E. L., Paredes, W., Smith, D., Seeger, T., Donner, A., Milling, C., Cohen, D., and Morrison, D.,** Strain-specific sensitization of brain stimulation reward by Δ^9-tetrahydrocannabinol in laboratory rats, *Psychopharmacology*, 96(Suppl.), 365, 1988.
160. **Gardner, E. L., Chen, J., Paredes, W., Li, J., and Smith, D.,** Strain-specific facilitation of brain stimulation reward by Δ^9-tetrahydrocannabinol in laboratory rats is mirrored by strain-specific facilitation of presynaptic dopamine efflux in nucleus accumbens, *Soc. Neurosci. Abstr.*, 15, 638, 1989.
161. **George, F. R.,** Genetic and environmental factors in ethanol self-administration, *Pharmacol. Biochem. Behav.*, 27, 379, 1987.
162. **Khodzhagel'diev, T.,** Formirovanie vlecheniia k nikotinu u mysheĭ liniĭ C57BL/6 i CBA [Development of nicotine preference in C57BL/6 and CBA mice], *Biulleten Eksperimentalnoi Biologii i Meditsiny*, 101, 48, 1986.
163. **Li, T. K. and Lumeng, L.,** Alcohol preference and voluntary alcohol intakes of inbred rat strains and the National Institutes of Health heterogenous stock of rats, *Alcoholism*, 8, 485, 1984.
164. **Ritz, M. C., George, F. R., deFiebre, C. M., and Meisch, R. A.,** Genetic differences in the establishment of ethanol as a reinforcer, *Pharmacol. Biochem. Behav.*, 24, 1089, 1986.
165. **Cannon, D. S. and Carrell, L. E.,** Rat strain differences in ethanol self-administration and taste aversion, *Pharmacol. Biochem. Behav.*, 28, 57, 1987.
166. **Gardner, E. L., Paredes, W., and Chen, J.,** Further evidence for Δ^9-tetrahydrocannabinol as a dopamine reuptake blocker: brain microdialysis studies, *Soc. Neurosci. Abstr.*, 16, 1100, 1990.
167. **Westerink, B. H. C., Damsma, G., de Vries, J. B., and Koning, H.,** Dopamine reuptake inhibitors show inconsistent effects on the *in vivo* release of dopamine as measured by intracerebral dialysis in the rat, *Eur. J. Pharmacol.*, 135, 123, 1987.

168. **Gerhardt, G. A., Rose, G. M., and Hoffer, B. J.,** Release of monoamines from striatum of rat and mouse evoked by local application of potassium: evaluation of a new *in vivo* electrochemical technique, *J. Neurochemistry*, 46, 842, 1986.
169. **Hunt, P., Reynaud, J.-P., Leven, M., and Schacht, U.,** Dopamine reuptake inhibitors and releasing agents differentiated by the use of synaptosomes and field-stimulated brain slices *in vitro, Biochemical Pharmacol.*, 28, 2011, 1979.
170. **Bhargava, N. M.,** Inhibition of naloxone-induced withdrawal in morphine dependent mice by 1-*trans*-Δ^9-tetrahydrocannabinol, *Eur. J. Pharmacol.*, 36, 259, 1976.
171. **Hine, B., Friedman, E., Torrelio, M., and Gershon, S.,** Morphine-dependent rats: blockage of precipitated abstinence by tetrahydrocannabinol, *Science*, 187, 443, 1975.
172. **Tulunay, F. C., Ayman, I. H., Portoghese, P. S., and Takemori, A. E.,** Antagonism by chlornaltrexamine of some effects of Δ^9-tetrahydrocannabinol in rats, *Eur. J. Pharmacol.*, 70, 219, 1981.
173. **Wilson, R. S. and May, E. L.,** Analgesic properties of the tetrahydrocannabinols, their metabolites, and analogs, *J. Med. Chem.*, 18, 700, 1975.
174. **Kumar, M. S. and Chen, C. L.,** Effect of an acute dose of delta-9-THC on hypothalamic luteinizing hormone releasing hormone and met-enkephalin content and serum levels of testosterone and corticosterone in rats, *Subst. Alcohol Actions Misuse*, 4, 37, 1983.
175. **Vaysse, P. J.-J., Gardner, E. L., and Zukin, R. S.,** Modulation of rat brain opiate receptors by cannabinoids, paper presented at Int. Narcotics Res. Conf., Falmouth, MA, 1985.
176. **Vaysse, P. J.-J., Gardner, E. L., and Zukin, R. S.,** Modulation of rat brain opioid receptors by cannabinoids, *J. Pharmacol. Exp. Ther.*, 241, 534, 1987.
177. **Zukin, R. S., Vaysse, P., and Gardner, E. L.,** Interaction of THC derivatives with brain opioid receptors, paper presented at meetings of Am. Soc. for Pharmacol. and Exp. Therapeutics (Symposium on Mechanisms of Central Action of Cannabinoids), Boston, 1985.
178. **Tempel, A., Gardner, E. L., and Zukin, R. S.,** Neurochemical and functional correlates of naltrexone-induced opiate receptor upregulation, *J. Pharmacol. Exp. Ther.*, 232, 439, 1985.
179. **Chow, T. and Zukin, R. S.,** Solubilization and preliminary characterization of mu and kappa opiate receptor subtypes from rat brain, *Mol. Pharmacol.*, 24, 203, 1983.
180. **Pertwee, R. G.,** *In vivo* interactions between psychotropic cannabinoids and other drugs involving central and peripheral neurochemical mediators, in *Marijuana/Cannabinoids: Neurobiology and Neurophysiology*, Murphy, L. L. and Bartke, A., Eds., CRC Press, Boca Raton, FL, 1992.
181. **Kaymakçalan, S. and Deneau, G. A.,** Some pharmacologic properties of synthetic Δ^9-tetrahydrocannabinol, *Acta Med. Turcica*, Suppl. 1, 1, 1972.
182. **Ghosh, P. and Bhattacharya, S. K.,** Cannabis-induced potentiation of morphine analgesia in rat — role of monoamines, *Indian J. Med. Res.*, 70, 275, 1979.
183. **Welch, S. P.,** Interactions of morphine and cannabinoids in the production of antinociception at the spinal level, paper presented at meetings of the Int. Cannabinoid Study Group, Palm Beach, FL, 1991.
184. **Welch, S. P. and Stevens, D. L.,** Antinociceptive activity of intrathecally administered cannabinoids alone, and in combination with morphine, in mice, *J. Pharmacol. Exp. Ther.*, 1992, in press.
185. **Bidaut-Russell, M., Devane, W. A., and Howlett, A. C.,** Cannabinoid receptors and modulation of cyclic AMP accumulation in the rat brain, *J. Neurochem.*, 55, 21, 1990.

186. **Devane, W. A., Dysarz, F. A., Johnson, M. R., Melvin, L. S., and Howlett, A. C.,** Determination and characterization of a cannabinoid receptor in rat brain, *Mol. Pharmacol.*, 34, 605, 1988.
187. **Herkenham, M., Little, M. D., Johnson, M. R., Melvin, L. S., Howlett, A. C., Rothman, R. B., de Costa, B., and Rice, K. C.,** Localization of cannabinoid receptors in brain, *Soc. Neurosci. Abstr.*, 14, 104, 1988.
188. **Herkenham, M., Lynn, A. B., Little, M. D., Johnson, M. R., Melvin, L. S., de Costa, B. R., and Rice, K. C.,** Cannabinoid receptor localization in brain. *Proc. Natl. Acad. Sci. U.S.A.*, 87, 1932, 1990.
189. **Howlett, A. C., Johnson, M. R., Melvin, L. S., and Milne, G. M.,** Nonclassical cannabinoid analgetics inhibit adenylate cyclase: development of a cannabinoid receptor model, *Mol. Pharmacol.*, 33, 297, 1988.
190. **Johnson, M., Devane, W., Howlett, A., Melvin, L., and Milne, G.,** Structural studies leading to the discovery of a cannabinoid binding site, *Natl. Inst. Drug Abuse Res. Monogr. Ser.*, 90, 129, 1988.
191. **Cuello, A. C. and Paxinos, G.,** Evidence for a long leu-enkephalin striopallidal pathway in rat brain, *Nature (London)*, 271, 178, 1978.
192. **Gardner, E. L. and Chen, J.,** unpublished data, 1991.
193. **Cubeddu, L. X., Hoffmann, I. S., and James, M. K.,** Frequency-dependent effects of neuronal uptake inhibitors on the autoreceptor-mediated modulation of dopamine and acetylcholine release from the rabbit striatum, *J. Pharmacol. Exp. Ther.*, 226, 88, 1983.
194. **Galzin, A. M., Dubocovich, M. L., and Langer, S. Z.,** Presynaptic inhibition of dopamine receptor agonists of noradrenergic neurotransmission in the rabbit hypothalamus, *J. Pharmacol. Exp. Ther.*, 221, 461, 1982.
195. **Galzin, A. M., Moret, C., Verzier, B., and Langer, S. Z.,** Interaction between tricyclic and nontricyclic 5-hydroxytryptamine uptake inhibitor and the presynaptic 5-hydroxytryptamine inhibitory autoreceptors in the rat hypothalamus, *J. Pharmacol. Exp. Ther.*, 235, 200, 1985.
196. **Göthert, M., Schliker, E., and Kostermann, F.,** Relationship between transmitter uptake inhibition and effects of α-adrenoreceptor agonists on serotonin and noradrenaline release in the rat brain cortex. *Naunyn-Schmiedeberg's Arch. Pharmacol.*, 322, 121, 1983.
197. **Langer, S. Z. and Moret, C.,** Citalopram antagonizes the stimulation by lysergic acid diethylamide of presynaptic inhibitory serotonin autoreceptors in the rat hypothalamus, *J. Pharmacol. Exp. Ther.*, 222, 220, 1982.
198. **Moret, C. and Briley, M.,** Sensitivity of the response of 5-HT autoreceptors to drugs modifying synaptic availability of 5-HT, *Neuropharmacology*, 27, 43, 1988.
199. **Gintzler, A. R. and Xu, H.,** Differential G proteins mediate the opioid inhibition or enhancement of evoked [5-methionine]enkephalin release, *Proc. Natl. Acad. Sci. U.S. A.*, 88, 4741, 1991.
200. **Harris, R. T., Waters, W., and McLendon, D.,** Evaluation of reinforcing capability of Δ^9-tetrahydrocannabinol in rhesus monkeys, *Psychopharmacologia*, 37, 23, 1974.
201. **Leite, J. R. and Carlini, E. A.,** Failure to obtain "cannabis directed behavior" and abstinence syndrome in rats chronically treated with cannabis sativa extracts, *Psychopharmacologia*, 36, 133, 1974.
202. **Corcoran, M. E. and Amit, Z.,** Reluctance of rats to drink hashish suspensions: free-choice and forced consumption, and the effects of hypothalamic stimulation, *Psychopharmacologia*, 35, 129, 1974.

203. **Pickens, R., Thompson, T., and Muchow, D. C.,** Cannabis and phencyclidine self-administered by animals, in *Psychic Dependence (Bayer-Symposium IV)*, Goldfarb, L. and Hoffmeister, F., Eds., Springer-Verlag, Berlin, 1973, 78.
204. **Deneau, G. A. and Kaymakçalan, S.,** Physiological and psychological dependence to synthetic Δ^9-tetrahydrocannabinol (THC) in rhesus monkeys, *Pharmacologist*, 13, 246, 1971.
205. **Kaymakçalan, S.,** Physiological and psychological dependence on THC in rhesus monkeys, in *Cannabis and Its Derivatives*, Paton, W.D. M. and Crown, J., Eds., Oxford University Press, London, 1972, 142.
206. **Takahashi, R. N. and Singer, G.,** Self-administration of Δ^9-tetrahydrocannabinol by rats, *Pharmacol. Biochem. Behav.*, 11, 737, 1979.
207. **Carlini, E. A. and Kramer, C.,** Effects of Cannabis sativa (marihuana) on maze performance of the rat, *Psychopharmacologia*, 7, 175, 1965.
208. **Schildkraut, J. J. and Efron, D. H.,** The effects of Δ^9-tetrahydrocannabinol on the metabolism of norepinephrine in rat brain, *Psychopharmacologia*, 20, 191, 1971.
209. **Hoebel, B. G. and Teitelbaum, P.,** Hypothalamic control of feeding and self-stimulation, *Science*, 135, 375, 1962.
210. **Margules, D. L. and Olds, J.,** Identical "feeding" and "rewarding" systems in the lateral hypothalamus of rats, *Science*, 135, 376, 1962.
211. **Miller, N. E.,** Motivational effects of brain stimulation and drugs, *Fed. Proc.*, 19, 846, 1960.
212. **Coons, E. E., Levak, M., and Miller, N. E.,** Lateral hypothalamus: learning of food-seeking response motivated by electrical stimulation, *Science*, 150, 1320, 1965.
213. **Gratton, A. and Wise, R. A.,** Comparisons of refractory periods for medial forebrain bundle fibers subserving stimulation-induced feeding and brain stimulation reward: a psychophysical study, *Brain Res.*, 438, 256, 1988.
214. **Gratton, A. and Wise, R. A.,** Comparisons of connectivity and conduction velocities for medial forebrain bundle fibers subserving stimulation-induced feeding and brain stimulation reward, *Brain Res.*, 438, 264, 1988.
215. **Mechoulam, R.,** Recent progress towards the identification of an endogenous brain constituent binding to the *cannabis* receptor, paper presented at meetings of the Int. Cannabinoid Study Group, Palm Beach, FL, 1991.
216. **Howlett, A.,** Cannabinoid receptors in brain, paper presented at meetings of the Int. Cannabinoid Study Group, Palm Beach, FL, 1991.
217. **Childers, S.,** Cannabinoid receptors in rat cerebellar granule cells and NG-108-15 cells, paper presented at meetings of the Int. Cannabinoid Study Group, Palm Beach, FL, 1991.

9 The Effects of Marijuana on Cerebral Blood Flow and Metabolism

INTRODUCTION

Marijuana is remarkable for its wide ranging effects on human behavior. It is known to induce euphoria, anxiety, lethargy, drowsiness, confusion, memory defects, altered time sense, depersonalization, impaired performance, cognitive dysfunctions, and psychotic symptoms.[15] These behavioral effects tend to vary considerably from subject to subject and seem to be influenced by a variety of nonspecific factors, including the personality of the subject, preexisting mood state, expectancy, social setting in which the drug is consumed, previous exposure to the drug, gender, mental health, etc.[1-10]

The neurobiologic mechanisms that underlie the behavioral manifestations of marijuana would seem to be of considerable interest. Indeed, an impressive body of literature is available on the neurochemical changes that accompany marijuana intoxication.[11-13] However, relationships between the neurochemical changes and changes in behavior are complicated and not easy to comprehend. Jacques-Joseph Moreau, the 19th century French psychiatrist noted for his pioneering work on the acute behavioral effects of marijuana, stated, "Yes, unquestionably there are modifications (I do not dare use the word lesion) in the organ that is in charge of mental functions, but these modifications are not those one would generally expect. They will always escape the investigations of the researchers seeking alleged or imagined structural changes. One must not look for particular, abnormal changes in either the gross anatomical or defined histological structure of the brain; but one must look for an alterations of its sensibility. That is to say for an irregular, enhanced, diminished, or distorted activity of the specific mechanism upon which depends the performance of mental functions."[14]

BRAIN FUNCTION AND BEHAVIOR

In recent years we have seen an increased interest in the study of the biological basis for mental functions in human subjects.[15-18] In the normal brain, cerebral blood flow (CBF) and cerebral metabolism rate (CMR) are closely coupled with brain function. An extensive body of research conducted in animals and human subjects and in normal and neuropsychiatric patients confirmed this hypothesis.[15-18] Since Kety and Schmidt developed the first quantitative measurement technique for CBF and CMR,[19] investigators utilized the technique to examine the CNS effects of psychopharmacological agents.[20] However, these earlier studies were limited because of the invasive nature of the available measurement techniques and poor spatial resolution. With the development of totally noninvasive techniques with improved spatial resolution, study of the effects of CNS acting drugs on CBF and CMR became easier and more meaningful. Acute and chronic effects of a variety of psychopharmacological agents, especially ethyl alcohol, were studied extensively.[21] In spite of the high prevalence of marijuana smoking, very little information was available on its effects on CBF and CMR until recently.

MULTIPLE EFFECTS OF MARIJUANA ON CBF

As mentioned above, CBF measurement is important to marijuana research as it might identify brain regions responsible for the behavioral changes. Unfortunately, marijuana (and several other CNS acting drugs) induce CBF changes through mechanisms other than alterations in brain function[21] (Table 1).

Effects on Vascular Smooth Muscle

A number of CNS acting drugs are vasoactive, i.e., they act upon the smooth muscle in the blood vessel wall. Caffeine-induced cerebral vasocon-

TABLE 1
Acute Effects of Drugs on CBF — Some Mechanisms

Changes in brain function
Changes in blood gases (CO_2)
Vasoactive properties
Effects on blood viscosity
Autonomic activation or inhibition
Activation or inhibition of intraparenchymal innervation
Effects on vasoactive neuropeptides

striction and increases in CBF induced by amyl nitrite are examples of this.[22,23] Although marijuana has not been demonstrated to have a vascular effect in the brain, it does influence blood vessels elsewhere in the body. For example, its dilatory effect on the congenital and muscle blood vessels are well known.[24-26] The "red eye" seen after marijuana smoking is known to last for several hours.[24,25] It is conceivable that marijuana might have a similar dilatory effect on cerebral blood vessels.

Changes in General Circulation and Respiration

A number of physiological changes associated with marijuana are relevant to CBF. Marijuana may increase respiratory rate and alter blood carbon dioxide levels. CBF is acutely sensitive to alterations in carbon dioxide. Even modest changes in carbon dioxide are associated with large changes in CBF, i.e., an increase in arterial carbon dioxide levels by 1 mm of mercury partial pressure will be accompanied by a 3 to 4% increase in CBF.[27,28]

Increase in pulse rate is one of the most consistent physiological changes seen after marijuana. Fortunately, changes in pulse and blood pressure are unlikely to influence CBF substantially. In the normal brain, CBF is insulated from modest changes in blood pressure through autoregulatory mechanisms.[29]

Autonomic Changes

The drug clearly has autonomic effects. Cerebral blood vessels receive vasoconstrictive sympathetic fibers from the superior cervical ganglion.[30,31] Sympathetic stimulation can produce CBF reduction. However, cerebral blood vessels in human subjects are believed to be less sensitive to auto-

TABLE 2
Long-Term Effects of Drugs on CBF — Some Mechanisms

Functional disorders
Brain atrophy
Endocarditis with embolism
Foreign material embolization
Injury to blood vessels (vasculitis)
Infections of the brain and appendages (HIV)
Presence of toxins in blood (ammonia in liver disease)
Changes in blood viscosity (anemia)
Changes in blood gases (emphysema)

nomic activation as compared to several animal species. It also needs to be pointed out that the argument that marijuana induces tachycardia via vagal blockade (and not sympathetic stimulation) has been put forward.

Carbon Monoxide Production

Marijuana smoking has been reported to increase blood levels of carbon monoxide.[32] This is of importance to CBF research. Carbon monoxide production and formation of carboxy hemoglobin interfere with transport and delivery of oxygen to the brain. The brain compensates for cerebral anoxia through a variety of mechanisms, including increased oxygen extraction from the blood and cerebral vasodilation.[33] Thus, increased carbon monoxide production during marijuana smoking can lead to a CBF increase.

Behavioral Effects of Chronic Marijuana Smoking

Long-term effects of drugs on CBF involve different mechanisms (Table 2). Chronic substance abuse is associated with diseases of cerebral blood vessels (embolization, vasculitis, etc.), brain atrophy, and functional psychiatric disorders. There is no reason to suspect that marijuana increases the risk for cerebrovascular disorders. Campbell and associates found brain atrophy with air encephalography in 10 cannabis users;[34] however, subsequent CT scan studies did not confirm this finding.[35,36] Chronic marijuana smoking has been associated with several behavioral syndromes. The precise nature and frequency of these are controversial topics. Of the various behavioral syndromes, marijuana-induced amotivational syndrome characterized by lowering of motivation to engage in ordinary social activity, apathy-reduced drive and ambition, impaired ability to carry out complex tasks, loss of initiative and effectiveness, failure to pursue long-term plans,

TABLE 3
Techniques for the Measurement of CBF and CMR

Technique	Invasiveness	Irradiation	CBF/CMR	Bilateral/ unilateral	Carotid/ vertibrobasilar	Spatial resolution
Nitrous oxide inhalation	Arterial and int. jugular puncture	Absent	CBF and CMR	Ipsilateral hemisphere only	Both	A single value for the entire hemisphere
Intracarotid ^{133}Xe injection	Carotid puncture	0.5 mrad (gonads)	CBF	Ipsilateral hemisphere	Carotid vascular bed only	1 cm
^{133}Xe inhalation	Venous puncture	2.9 mrad (gonads) 102 mrad (lung)	CBF	Both hemispheres	Both carotid and vertibrobasilar beds	3–4 cm
Single photon emission computed tomography (^{133}Xe)	Venous puncture	5 mrad (gonads) 360 mrad (lung)	CBF	Both hemispheres	Both carotid and vertibrobasilar beds	1.7 cm
Position emission tomography (F-18) FDG	Arterial puncture	63 mrad/mCi (gonads) 43 mrad/mCi (total body)	CBF and CMR	Both hemispheres	Both carotid and vertibrobasilar beds	0.7–1.7

difficulty in concentrating, and decline in school work and performance received most attention.[37,38] It needs to be pointed out that reduction in CBF and CMR have been reported in depression, which is also associated with psychomotor slowing with decrease in motivation and performance.[39,40]

TECHNIQUES FOR THE MEASUREMENT OF CBF AND CMR

The Nitrous Oxide Technique

Techniques for the measurement of CBF and CMR are collected in Table 3. The first quantitative technique for CBF and CMR measurement developed by Kety and Schmidt[19] involved inhalation of subanesthetic doses of nitrous oxide and estimation of the concentration of the gas in blood samples drawn from the jugular bulb and femoral artery. CBF is calculated using a modification of the Fick Principle. The arteriovenous (internal jugular) difference in oxygen and glucose multiplied by CBF yielded values for CMR_{oxygen} and $CMR_{glucose}$. The major disadvantages of the technique were its invasive nature (arterial puncture and jugular venous puncture) and inability to provide regional (hemispheric, cortical, and subcortical) values.

The Intracarotid ^{133}Xe Injection Technique

The first isotope clearance technique developed by Ingvar and Lassen represented significant advancement. For the first time it was possible to measure regional flow values.[41] The technique involved intracarotid injection of ^{133}Xe dissolved in saline, followed by recording of progressive decline in radioactivity over the ipse-lateral cerebral cortex by a system of scintillation detectors or a gamma camera applied over the scalp. The clearance curves were analyzed with a bicompartmental model that permitted separation of flow to the grey matter from the slower-clearing white matter. The intracarotid isotope injection resulted in high isotope concentration in the ipse-lateral hemisphere that improved the spatial resolution of the technique to a considerable extent. The major disadvantages of the technique were its invasiveness, inability to measure contralateral hemispheric CBF, inability to measure subcortical CBF, and the look-through effect (low perfusion regions eclipsed by overlapping high perfusion regions). It measured CBF only in the regions supplied by the internal carotid artery (in most instances). Although global CMR could be calculated if arteriovenous (internal jugular vein) differences in glucose or oxygen were known, it was not possible to measure regional CMR with this technique.

The ^{133}Xe Inhalation Technique

Obrist and associates modified the intracarotid injection technique to make it totally noninvasive.[42] ^{133}Xe was administered through a face mask instead of intracarotid injection. The scalp clearance curves were analyzed by a bicompartmental model and the slower-clearing compartment contaminated with extracranial flow was discarded. Scalp clearance curve contamination by radioactivity from air passages was eliminated by commencing the scalp clearance curve analysis until radioactivity from the respiratory passage had dropped substantially. The technique was associated with low radiation exposure and therefore it could be repeated back to back in the same subject several times (similar to the intracarotid injection technique). Test retest reliability was good. It measured CBF to both hemispheres and in both carotid and vertibrobasilar systems. The major disadvantages of the technique include limited spatial resolution, reduced sensitivity to changes in grey matter flow (inferior to the intracarotid injection technique in both respects), inability to measure subcortical CBF, regional CMR, and look-through effect.

Tomographic Techniques

With the advent of tomographic techniques, the measurement of subcortical flow and metabolism became possible. Positron emission tomography (PET) and single photon emission computed tomography (SPECT) involved administration of a radiotracer, followed by reconstruction of tomographic slices of the brain depicting isotope concentrations in various cortical and subcortical brain regions. CBF was computed using appropriate mathematical models.[17,18,43] PET techniques that utilized ^{15}O-labeled water as tracer provided absolute quantitation of CBF. The autoradiographic version of the technique was associated with low levels of radiation exposure and could be repeated in the same subject several times, back to back. However, absolute quantitation still required arterial puncture. PET scans are technologically the most sophisticated CBF measurement techniques, but they also suffer from several shortcomings.[17,18] In spite of its improved spatial resolution over previous techniques, the resolution was still insufficient to image small brain regions satisfactorily. The cerebral cortex is less than 2 mm thick and even the best PET scanners did not have the spatial resolution necessary to image cerebral cortex satisfactorily. This gave rise to some uncertainty as to whether 2-D or 3-D techniques were superior in quantifying cortical flow.[44] PET techniques were developed to measure regional CMR_{oxygen}. However, these techniques are used less frequently because of their complexity (concomitant measurement of CBF and cerebral blood volume) and increased radiation exposure.

PET scans have been used most extensively to quantify cerebral glucose metabolism with the ^{18}F deoxyglucose method.[45,46] The technique is well validated and useful in a wide range of research projects. The major disadvantages of the technique are that it quantifies glucose metabolism over a 90-min period (and therefore is unsuitable to examine rapid changes in CMR after the administration of a drug) and that it cannot be repeated back to back. Long intervals between two $CMR_{glucose}$ measurements is likely to lower the stability of the CMR values across measurements and increase the influence of nonspecific factors. PET scans tend to be very expensive and this is another limiting factor.

SPECT is far less expensive than PET. However, most SPECT CBF techniques do not quantitate CBF satisfactorily. One exception to this, using ^{133}Xe as a tracer, suffers from several limitations, including poor spatial resolution.[47] Although some of the most recent SPECT techniques approach PET in spatial resolution, most SPECT techniques are inferior to PET in this regard. No SPECT techniques are currently available to measure CMR.

MEASUREMENT OF CEREBRAL BLOOD VELOCITY (CBV)

The velocity of blood flow through intracranial arteries is related to CBF, but is not identical to it. CBF denotes cerebral capillary perfusion. The rate at which blood flows through major cerebral arteries that feed cerebral capillaries depends upon two sets of factors, i.e., perfusion pressure and resistance to flow in the cerebral capillary bed. Cerebral blood velocity (CBV) is also sensitive to changes in intracranial pressure, blood viscosity, and diameter of the blood vessel.[48] Factors that increase the perfusion pressure, decrease the resistance through cerebral capillaries (cerebral vasodilation), decrease the intracranial pressure and the blood viscosity, and increase the blood vessel diameter will increase the velocity of flow. Thus, CBV may be used as an indirect index of cerebral capillary perfusion provided changes in perfusion pressure, intracranial pressure, blood viscosity, and blood vessel diameter can be controlled.

Transcranial Doppler (TCD) is a totally noninvasive technique with which blood velocity through major intracranial vessels can be quantified. The shift in frequency of a wave when either the transmitter or receiver is moving in respect to the wave propagating medium was first described by Doppler. In TCD, which is a pulsed ultrasound instrument, the same transducer is used for both transmitting and emitting wave energy. The effect detected by this configuration is that of a frequency shift caused by a moving deflector, namely blood. Bursts of ultrasonic energy are transmitted at regular intervals with a burst of pulse repetition frequency. These bursts travel at velocity C towards the reflector, the two being separated by a distance L. Some of the energy reflected back travels at the same propagation velocity.

Thus, at time T after the emission of the burst, the reflected echoes come back through the transducer (T = 2, 2 L/C). The echoes are transformed into electrical signals and amplified. The principle of range-gating (an electronic gate that opens and samples the signal only for a short period around time T) is used. Signals arriving from reflectors at different distances arrive at different times and their effects are therefore eliminated from the signal sample. The Doppler shift is high-pass filtered to remove wall thumps and probe position movements, and further information processing is usually fast Fourier analysis to display the spectral content of the signal.

With TCD, it is possible to measure CBV through all major intracranial arteries. It quantifies CBV at 30-s intervals and, since it is totally noninvasive, the technique can be used to monitor changes over a prolonged period of time. The major disadvantages are that CBV can be influenced by factors other than CBF. CBV is sensitive to blood vessel diameter and, therefore, it will not be the same in different segments of the same artery. There also is marked variation in CBV between subjects. Difficulties in obtaining good ultrasonic signals in certain subjects, especially older ones, is another drawback. CBV values also vary with the angle between the ultrasonic signal and direction of flow (vessel) and, thus, CBV varies with different ultrasonic probe positions (angles).[49,50]

ANIMAL STUDIES

There have been relatively few investigations conducted regarding animals on the effects of marijuana and/or tetrahydrocannabinol (THC) on cerebral circulation. Beaconsfield and associates examined the effect of Δ^9-THC on cerebral circulation and function. THC was given intravenously in doses of 100 μg/kg in 4 adult male mongrel dogs.[51] CBF was estimated using a venous outflow technique. Arterial, venous (superior vena cava) and cerebrospinal fluid pressures, and end tidal CO_2 ($PECO_2$) were also measured. After THC administration, there was an increase in cerebral venous outflow in each dog (1.0 and 1.8 ml/min), an increase in cerebrospinal fluid pressure (1.8 to 2.5 cm water), an increase in pulse rate (10 to 20%), and a decrease in systolic blood pressure (20 to 30 mmHg). Three out of the four dogs showed no change in central venous pressure and the fourth an increase of only 0.8 cm water. $PECO_2$ remained unchanged in all animals. The findings of this experiment suggest increased cerebral circulation after THC infusions.

Goldman and associates studied the effects of Δ^9-THC on CBF to 10 brain regions in conscious, unrestrained male rats.[52] Regional CBF was estimated based on the distribution pattern (post-mortem) of an indicator (antipyrine) administered in a single i.v. injection. The study was conducted on a total of 43 male rats housed under rigorous environmental conditions. The animals received either Δ^9-THC (1 mg/kg), isotonic saline, or the vehicle alone, given intravenously over a 1-min period. Twenty minutes later, the indicator was given as a bolus injection. Animals were sacrificed 20 s after the indicator and areas of interest (cortical and subcortical) were dissected and tissue indicator concentration determined. Cardiac output was estimated from the arterial concentration of the indicator.

Twenty minutes after THC infusion, when brain blood flow was determined, cataleptoid behavior reached full intensity and blood pressure and heart rate had returned to preinjection levels. During the period of cataleptoid behavior, cardiac output was reduced by about 20% and there was a significant hypercarbia and lowered blood pH. The vehicle placebo did not induce any changes in behavior, blood characteristics, or regional blood flow. However, after THC infusion, blood flow was significantly reduced to the cerebellum, hypothalamus, basal ganglia, and most significantly to dorsal hippocampus. The drug did not have any significant effects on blood flow to the cortical regions.

HUMAN STUDIES

Acute Changes in CBF After Marijuana: Study I

In 1989, we investigated CBF changes after marijuana administration in experienced and inexperienced marijuana smokers with the ^{133}Xe inhalation technique. Experienced smokers (ES) consisted of subjects who had smoked a minimum of 10 joints a week for 3 years and inexperienced smokers (IS) comprised subjects who had not smoked any marijuana for a minimum of 3 years. Several IS subjects had not smoked marijuana for several years and some had never smoked any.

The demographic characteristics of the three groups is given in Table 4. The subjects were contacted through local advertising and they were not paid for their participation. The subjects were free of significant medical and psychiatric illnesses. All participants were predominantly right-handed. They were required to be free of medication for a minimum of 1 month, alcohol and marijuana for 12 h, and coffee and tobacco for 3 h before the experiment.

A control group of 14 subjects was chosen from our CBF database of normals who had 2 resting CBF measurements; they also met all inclusion

TABLE 4
Demographic Characteristics of the Three Groups

		Experienced smokers	Inexperienced smokers	Controls	P
Sex	M	7	10	9	NS
	F	2	7	5	
Age (years)	Mean	25.9	28.3	26.9	NS
	SD	3.9	8.3	7.5	
Trait anxiety	Mean	28.2	37.1	32.1	NS
	SD	7.3	10.5	5.9	
State anxiety	Mean	27.2	31.4	30.0	NS
	SD	4.5	9.6	7.5	
Beck depression	Mean	0.9	2.8	1.2	NS
	SD	1.0	4.8	1.5	

Note: Values obtained during the first CBF measurement. Comparisons were made with analyses of variance.

criteria except that their marijuana smoking history was not known. They were matched to ES and IS on the basis of age and sex as closely as possible. The control subjects had CBF measured twice at an interval of 60 min under resting conditions and they had agreed to refrain from coffee, cola beverages, cigarettes, and any drug or alcohol before being assessed. ES and IS subjects were told that they would be asked to smoke a cigarette that could be either marijuana or a placebo (marijuana with THC removed) between two CBF measurements during the first visit and again a second time, during another visit 1 week later.

The subjects were asked to report to the laboratory in the morning. CBF was measured once under resting conditions and 60 min after smoking a high-potency marijuana cigarette (THC content = 2.2%). The subjects were not told about the identity of the cigarettes. To standardize the smoking process as much as possible, the subjects were instructed to inhale the smoke fully and retain it for 15 s. The procedure was repeated 9 times over a period of 15 min.

Not all the participants returned to the laboratory for the second visit, i.e., only 6 ES and 11 IS came back. During the second visit, 2 additional CBF measurements at 60-min intervals were obtained. After the first resting CBF, they were given a placebo cigarette (THC content = 0.004%). During the CBF measurements, pulse rate, respiratory rate, $PECO_2$, and blood pressure were monitored. None of the subjects became drowsy or sleepy.

Before and after marijuana, the subjects completed Profile of Mood States (POMS)[53] and the state anxiety scale of State Trait Anxiety Inventory (STAI).[54] A blood sample was drawn after each measurement for the determination of hemoglobin and hematocrit.

Differences between groups and change within groups across the repeated assessments were analyzed with multivariate analysis of variance (repeated measures model). This approach was used for regional and hemispheric CBF and for rating scale as well as physiological variables. The model for CBF was: group by period (pre-post), by hemisphere, by region. The difference between the groups was followed up with simple main effect analysis and, as appropriate, Scheffe's test of multiple comparisons among means was used as a post-hoc procedure. Pearson correlations were computed between change scores (post-minus-pre) to relate changes in CBF to other variables, and a multiple regression procedure was used to examine the partial correlations of the different POMS factor scores to the change in hemispheric CBF.

Analysis of demographics and baseline assessments indicated no significant differences between ES, IS, and controls (Table 4). CBF before and after marijuana in ES and IS were compared with the two resting CBF runs in controls in an overall multivariate analysis of variance. For this analysis, the second CBF was corrected for pre- to post-change in $PECO_2$ with a 3% CBF per millimeter $PECO_2$ correction factor. This analysis indicated a group by period interaction ($F = 15.7, p < 0.001$) and a group by region interaction ($F = 2.53, p < 0.02$), but no other main effects or interactions reached statistical significance. These interactions were assessed for each group using a period × hemisphere × region repeated measures model and followed, where appropriate, with post-hoc Scheffe's test on each region (pre-post).

For the IS, these analyses revealed significant CBF decreases in all regions following marijuana; the extent of change in different regions varied from –3.6 to –10.9 CBF units. However, there were no significant differences between regions for this group. For the ES, there was a pre- to post-CBF increase in both hemispheres. However, regional CBF changes reached significance only in the right and left frontal regions and left temporal region. The control group (two resting CBF runs) revealed no change on the full model and none of the regions or the hemisphere means demonstrated a significant change. Differences in hemispheric changes are presented in Figure 1.

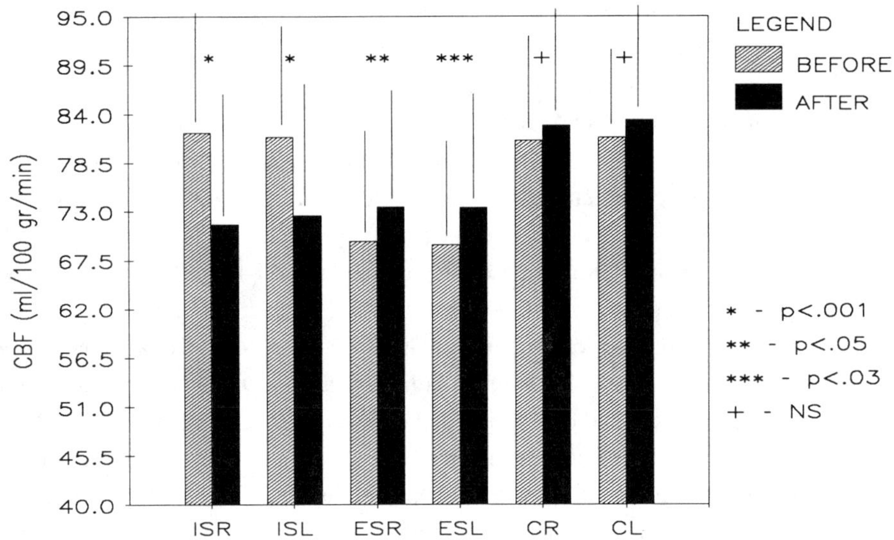

FIGURE 1. Comparisons between the three groups on hemispheric CBF before and after marijuana (analysis of variance with repeated measure). ISR, inexperienced smoker, right hemisphere; ISL, inexperienced smoker, left hemisphere; ESR, experienced smoker, right hemisphere; ESL, experienced smoker, left hemisphere; CR, control group, right hemisphere; and CL, control group, left hemisphere. Diagonal lines, before; and solid blocks, after. (From Mathew et al., *Acta Psychiatria Scand.*, 79, 118, 1989.)

Comparisons between ES and IS on baseline CBF showed lower CBF for the ES. However, even after this difference was taken into account by analysis of covariance on the change scores with the baseline value as the covariate (regressed change scores), the results given above remained unchanged.

A similar set of analyses were employed to examine changes in physiological variables and state anxiety (STAI). The IS had significant increases in blood pressure, pulse rate, respiration, and state anxiety; the ES had significant increases in diastolic blood pressure and pulse rate and the controls had a significant increase in respiratory rate. None of the groups had significant changes in hematocrit or $PECO_2$.

The control group did not complete the POMS; therefore, analysis of the POMS data were conducted only for the IS and ES. After smoking marijuana, IS had an increase in state anxiety while ES had a decrease. Neither group indicated any change in anger or fatigue and both indicated a significant increase in confusion. IS had a significant increase in depression and a decrease in vigor, but neither of these variables changed in the ES. To examine the relationship between change in CBF and change in the physiological and rating scale data, we computed Pearson correlations based on change score. Pulse rate showed significant negative correlation ($p < 0.05$) with all regions except left frontal. Correlations ranged between –0.46 and –0.63. Change in state anxiety correlated negatively ($p < 0.05$) with all regions, the correlations ranging from –0.51 to –0.84. State anxiety showed significant negative correlations with all regional change scores except left frontal region. The correlations ranged from –0.39 to –0.81, with the larger correlations in the right hemisphere. Similarly, POMS factors, anger, fatigue, confusion, and depression had significant inverse correlations with at least some regions with confusion having the greatest number of significant correlations and again the larger correlations tend to be in the right hemisphere (Table 5). By contrast, vigor correlated positively with all regions. Three of the four significant correlations were in the right hemisphere. When relationships between POMS factors were taken into consideration in a multiple regression analysis, only the anxiety factor had a significant partial correlation with right and left hemispheric change.

Results from the marijuana session (visit l) were then compared with the placebo marijuana session (visit 2). As was mentioned before, only 6 ES and 11 IS returned for the second visit (i.e., placebo marijuana session). Analysis to compare those who did not return with those who did failed to indicate any systematic differences. CBF (corrected for changes in $PECO_2$) were analyzed with a group × hemisphere × region repeated measures multivariate analysis of variance.

When compared with the placebo measurements, the IS had a significant decrease in CBF following marijuana (Figure 2). Their global and regional CBF, in fact, decreased both after placebo and marijuana, and the amount of decrease was significantly greater following marijuana than the placebo. There were, however, no significant differences (global or regional) compared with the placebo for the ES whose CBF revealed an increase both after marijuana and placebo. In other words, CBF increase following marijuana was not significantly greater than that seen after placebo.

ES and IS showed different behavioral and CBF response to marijuana in spite of there being no differences between the two groups and controls on such indices relevant to CBF as age, male/female ratio, and trait anxiety. IS showed significant increase in anxiety scores on STAI and POMS, respiratory rate, blood pressure, and pulse rate. ES showed increase in blood

TABLE 5
First-Order (Pearson) and Partial Correlations between Profile of Mood States and CBF Change Scores before and after Marijuana

	POMS factors					
	Anxiety	Anger	Fatigue	Confusion	Depression	Vigor
Right[a]						
F	–0.75[c]	–0.29	–0.47	–0.74[c]	–0.53[d]	0.57[d]
C	–0.53[d]	–0.29	–0.39	–0.53[d]	–0.45	0.30
T	–0.64[d]	–0.06	–0.36	–0.531[d]	–0.27	–0.20
P	–0.81[c]	–0.22	–0.28	–0.72[c]	–0.58[d]	0.73[c]
O	–0.70[c]	–0.51[d]	–0.37	–0.72[c]	–0.51[d]	0.61[d]
Hemispheric mean[b]	–0.56[d]	0.02	–0.11	–0.14	–0.14	0.13
Left[b]						
F	–0.49	0.01	–0.02	–0.49	–0.29	0.30
C	–0.39	–0.30	–0.38	–0.33	–0.38	0.24
T	–0.58[d]	0.00	–0.51[d]	–0.65[d]	–0.28	0.30
P	–0.60[d]	–0.14	–0.22	–0.61[d]	–0.35	0.36
O	–0.79[c]	–0.26	–0.40	–0.69[d]	–0.33	0.49[d]
Hemispheric mean[b]	–0.51[d]	0.07	–0.15	–0.26	–0.04	0.27

Note: F, frontal; C, central; T, temporal; P, parietal; O, occipital.

[a] Pearson's r first-order correlations.
[b] Partial correlations from multiple regression.
[c] $p < 0.01$.
[d] $p < 0.05$.

pressure and pulse rate only. In ES, state anxiety decreased after marijuana. Both IS and ES reported increased confusion, but it was more marked in IS, who also reported increased depression and decreased vigor. This is in keeping with previous reports of dysphoric reactions, including anxiety, being more common among inexperienced smokers. IS showed a highly significant bilateral CBF decrease after marijuana, both in comparison to the control group (2 resting CBF runs) and with their own CBF runs before and after placebo/marijuana. ES, on the other hand, showed a significant bilateral CBF increase most marked in the frontal lobes in comparison with the controls. This CBF difference following marijuana between IS and ES could not be explained on the basis of $PECO_2$. Since CBF is insulated from peripheral circulatory changes via autoregulatory mechanisms, differences

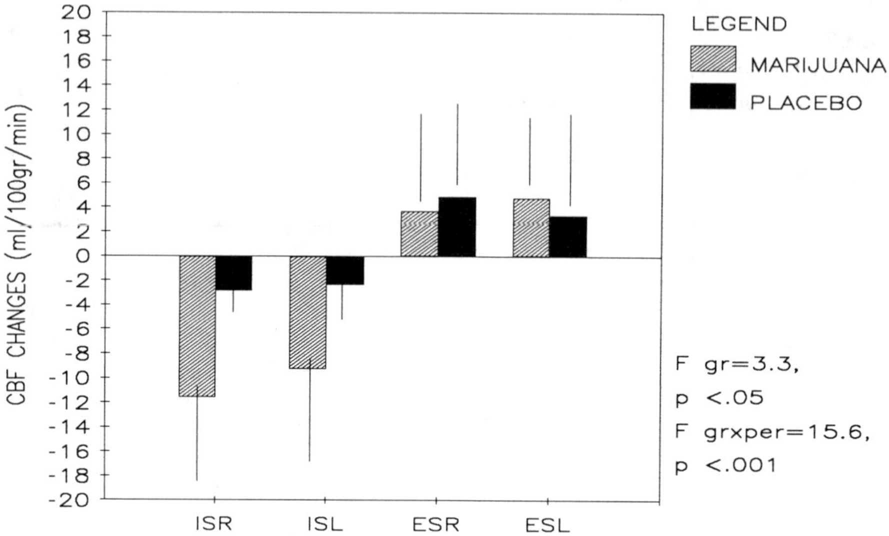

FIGURE 2. CBF changes after marijuana (diagonal lines) and placebo (solid blocks) in inexperienced and experienced marijuana smokers (analysis of variance with repeated measure). ISR, inexperienced smoker, left hemisphere; ISL, inexperienced smoker, left hemisphere; ESR, experienced smoker, right hemisphere; and ESL, experienced smoker, left hemisphere. (From Mathew et al., *Acta Psychiatria Scand.*, 79, 118, 1989.)

between the two groups on post-marijuana pulse rates are, therefore, unlikely to be of relevance. Correlational analysis indicated an inverse relationship between anxiety, which was a dominant mood change in IS, and CBF. The finding of an inverse correlation between anxiety and CBF is in keeping with several other research reports that also indicate decreased CBF in accompaniment of severe anxiety.[56] In other words, the finding of reduced CBF after marijuana in IS may not be specific to marijuana smoking, i.e., it may be a function of severe anxiety regardless of how the anxiety is induced. The marked increase in CBF in ES is more likely to be related to the behavioral effects of the drug.

Acute Changes in CBF After Marijuana: Study II

The second study utilized a more thorough study design to evaluate acute changes in regional CBF induced by marijuana smoking. Male volunteers who had previous exposure to marijuana were recruited through local advertising. None reported any present or past complications (physical or mental) of marijuana. Although some gave a history of abusing other drugs

TABLE 6
Marijuana and Cerebral Blood Flow Participants — All Right-Handed Males

	Mean (range)	SD
Age (years)	25.3	6.4
MAST score	3.1	3.3
BECK score	1.0	2.3
Trait anxiety score	29.1	6.0
Number of marijuana cigarettes per year	71 (2–350)	103.0
	Yes	No
Number who have tried other drugs	12	8

in the past, none had abused any drugs other than marijuana during the previous year. Special attention was paid to alcohol consumption and subjects who drank heavily or showed symptoms suggestive of alcohol dependence or abuse (DSM-III-R) were excluded from the project. In addition, the subjects completed the Michigan Alcoholism Screening Test.[57] None of the participants had significant psychiatric disorders at present or in the past. Trait anxiety and depression were quantified with the State Trait Anxiety Inventory and Beck Depression Inventory,[58] respectively. A physical examination was performed and all were found to be in good physical health.

All participants were questioned about their use of medicines for 3 months before the experiment and none had used any during this period. The subjects were instructed not to consume marijuana or any other drug for 2 weeks before the start and until the end of the study. They were told to avoid alcohol for 24 h and nicotine and caffeine for 4 h before each visit to the laboratory. The subjects were paid for participation in this study on an hourly basis. Clinical and demographic characteristics of the subjects are given in Table 6.

Subjects came to the laboratory on 3 d, separated by a minimum of 1 week so that the effects of marijuana administered during one visit would not affect the measurements made during the next one. In experienced marijuana smokers, elimination half-life for THC is less than 56 h; thus, a 7-d interval was more than 3 half-lives.[59,60] A urine drug screen was performed during each visit.

CBF was measured with the ^{133}Xe inhalation technique,[42] four times during each visit; once before and thrice after marijuana/placebo. The first CBF measurement was used for baseline values after which marijuana/

placebo was administered over a 10-min period. The second CBF was performed 30 min after the drug/placebo, the third measurement after 60 min, and a fourth after 120 min. Blood pressure and pulse were taken before and after each CBF measurement. $PECO_2$ was monitored during the entire procedure. End tidal carbon monoxide concentration was estimated by having the subjects exhale fully into a carbon monoxide analyzer before each CBF measurement. Forehead skin perfusion was quantified with a laser Doppler flow meter[61,62] applied to the skin approximately 1/2 in. above the bridge of the nose.

Mood states before and after marijuana smoking were quantified with POMS, STAI, and a somatic symptom scale.[63] Two rating scales were utilized to quantify altered time sense[64-66] and depersonalization.[67] An analogue scale was used to quantify the degree of "high". All rating scales were administered after each CBF measurement.

After the first CBF measurement, the subjects were taken to a room specially equipped with exhaust fans for smoking marijuana/placebo cigarettes. During the three visits, the subjects received a high-potency marijuana cigarette (THC = 3.55%), low-potency (THC = 1.75%), and a marijuana/placebo cigarette, on a random basis. Neither the subjects nor the laboratory staff knew about the identity of the cigarettes. Attempts at standardizing the smoking routine in study I resulted in a high incidence of dysphoric reactions and dizziness without any reduction in the intersubject variations in plasma levels of THC. Thus, subjects were allowed to smoke ad lib; but they were encouraged to finish the entire cigarette in 10 min which all of them did. Two members of the research team remained with the subjects at all times. After 10 min of smoking, the subjects were taken back to the CBF laboratory in a wheel chair (because of possible postural hypotension and dizziness).

After the subjects were brought to the room for marijuana/placebo smoking, an i.v. line was started for blood sample withdrawal. A single 10-ml blood sample was initially taken for hemoglobin (necessary for computing CBF). Another 10-ml blood sample was taken before the administration of cigarettes for baseline plasma THC assay and this was repeated at 5, 10, 20, 40, and 120 min after the commencement of smoking. Δ^9-THC was assayed by radioimmunoassay with a kit supplied by NIDA.

Procedures of analysis included multivariate analysis of variance for all repeated measures variables, followed where appropriate by *t*-tests (paired). To examine the relationships among regional CBF, rating scales and physi-

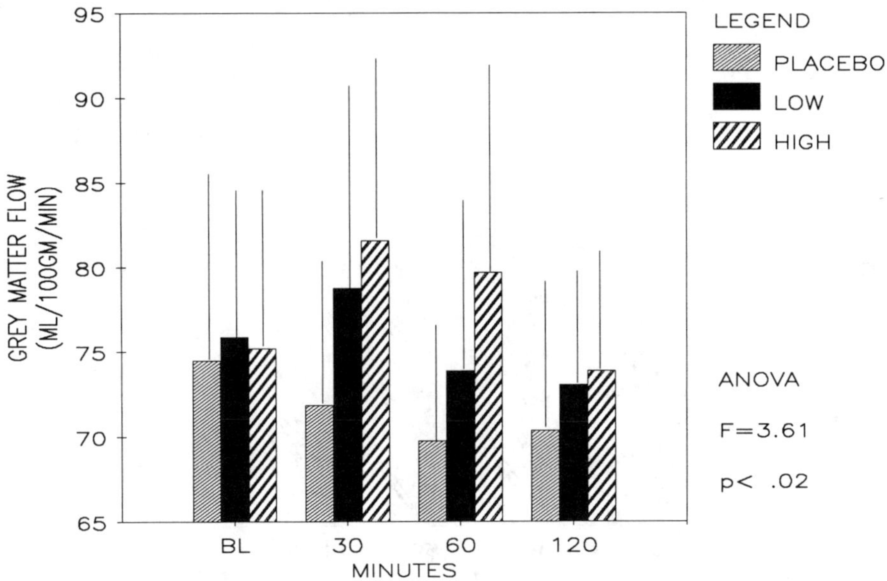

FIGURE 3. Right hemispheric CBF before and after marijuana and placebo.

ological measures, first-order Pearson correlations and multiple regression analyses were employed.

An analysis of the hemispheric mean flows was performed using a drug/placebo × hemisphere × time model. In this model, drug/placebo represented the two doses of marijuana and placebo. The results of this analysis (Figures 3 and 4) indicated a significant drug/placebo by time interaction ($F = 3.61$; $p < 0.02$) with significant global increases in CBF following low and high dose marijuana.

To examine possible regional CBF differences, this analysis was followed by an analysis of the regional flow values. Anterior and posterior flows in each hemisphere were determined by computing weighted means (anterior = frontal + central; posterior = temporal + parietal + occipital). This MANOVA analysis of the anterior and posterior flows used a drug/placebo × time × hemisphere × anterior-posterior model. This analysis indicated a significant drug/placebo by time interaction ($F = 3.24$; $df = 6,14$; $p < 0.033$) and a four-way interaction that approached statistical significance ($F = 2.75$; $df = 6,14$; $p < 0.056$). CBF changes were, in general, more marked in anterior (frontal) as compared to posterior (occipital) regions (Tables 7

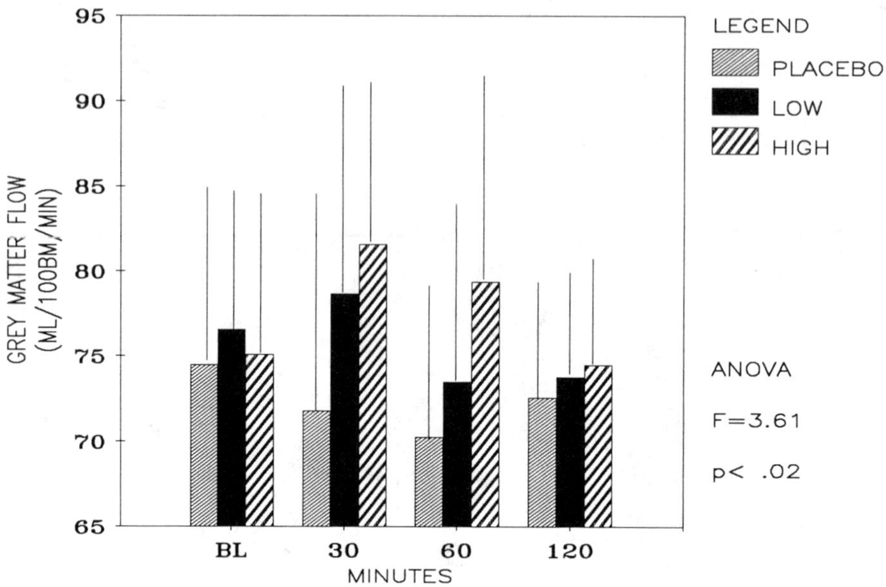

FIGURE 4. Left hemispheric CBF before and after marijuana and placebo.

and 8). To explore this four-way interaction, multiple paired *t*-tests were conducted on the individual regions (frontal, central, temporal, parietal, and occipital). In these analyses, placebo values were compared with low dose and separately compared with high dose. There were no differences at baseline between placebo and low dose or placebo and high dose for either the left or right hemisphere. At 30 min, there were significant differences between placebo and both low and high dose in both the left and the right hemisphere. Consistent with the analysis of anterior and posterior, the magnitude of effect (as evidenced by the *t*-values) was greater in the frontal regions and was greater in the right hemisphere than in the left hemisphere. Also, both in the left and right hemispheres, significant increases in CBF persisted 60 min following smoking high-dose marijuana across almost all regions. These effects were also present in the frontal region for low dose in both left and right hemisphere. These same findings were present after correcting CBF for $PECO_2$ using a predetermined factor of 3% CBF per millimeter of $PECO_2$. Thus, it would appear that marijuana causes a dose-related increase in global CBF, but it also appears to have regional effects, with a greater increase in the frontal and in particular, in the right hemisphere.

TABLE 7
Gray Matter Flow (ml/100 g/min) to Lobes of the Right Hemisphere before and after Marijuana and Placebo (Not Corrected for $PECO_2$)

	Drug/placebo			Contrasts			
	Means			PL vs. Low		PL vs. High	
Baseline	PL	Low	High	*t*	*p* <	*t*	*p* <
F	77.4	81.0	79.5	1.29	NS	<1.0	NS
C	77.9	80.0	77.8	<1.0	NS	<1.0	NS
T	72.1	72.0	71.2	<1.0	NS	<1.0	NS
P	72.4	73.4	71.8	<1.0	NS	<1.0	NS
O	72.0	73.4	70.8	<1.0	NS	<1.0	NS
30 min							
F	75.3	86.5	88.4	3.74	0.001	4.52	0.001
C	73.8	83.2	83.0	3.08	0.006	3.44	0.002
T	67.0	78.0	76.8	3.21	0.005	4.19	0.001
P	69.6	76.6	76.6	3.05	0.007	2.85	0.010
O	69.3	79.3	77.1	2.88	0.010	2.67	0.014
60 min							
F	72.4	80.1	84.7	3.26	0.004	3.98	0.001
C	74.5	79.6	81.7	1.69	NS	2.18	0.041
T	65.7	71.9	75.4	2.56	0.019	2.85	0.010
P	68.0	70.6	75.5	<1.0	NS	2.38	0.026
O	70.6	70.3	76.0	<1.0	NS	1.47	NS
120 min							
F	75.3	79.1	77.9	1.82	NS	1.05	NS
C	74.0	78.6	75.7	1.80	NS	<1.0	NS
T	65.8	72.0	69.8	4.16	0.001	1.94	NS
P	67.7	71.9	70.6	2.29	0.034	1.36	NS
O	68.5	72.1	71.2	1.43	NS	<1.0	NS

Note: F, frontal; C, central; T, temporal; P, parietal; O, occipital; PL, placebo; High = 3.55% THC; Low = 1.75% THC.

Plasma levels of THC (nanograms per milliliter) were analyzed across time comparing the low- and high-dose marijuana and placebo sessions. Levels were compared for baseline and at 5, 10, 20, 40, and 120 min in the model (Figure 5). This analysis indicated a significant drug/placebo by time interaction ($F = 6.81$; $p < 0.001$). Peak plasma levels were at 5 min. For both the low- and high-dose conditions, a great deal of variability was present in the plasma levels, with the standard deviation being almost as large as the mean.

TABLE 8
Gray Matter Flow (ml/100 g/min) to Lobes of the Left Hemisphere before and after Marijuana and Placebo (Not Corrected for $PECO_2$)

	Drug/placebo			Contrasts			
	Means			PL vs. Low		PL vs. High	
Baseline	PL	Low	High	*t*	*p* <	*t*	*p* <
F	77.8	80.3	78.0	1.17	NS	<1.0	NS
C	79.8	80.8	78.2	<1.0	NS	<1.0	NS
T	72.7	75.2	71.5	1.10	NS	<1.0	NS
P	71.9	74.9	72.3	1.27	NS	<1.0	NS
O	73.2	73.7	71.6	<1.0	NS	<1.0	NS
30 min							
F	74.8	85.2	85.8	3.28	0.004	3.50	0.002
C	74.8	83.9	83.8	3.60	0.002	2.98	0.007
T	69.6	78.6	78.3	2.91	0.009	2.98	0.011
P	68.6	75.1	77.5	1.84	NS	4.14	0.001
O	69.3	78.2	77.3	3.07	0.006	2.79	0.011
60 min							
F	74.2	78.9	82.7	2.19	0.041	2.54	0.019
C	73.8	67.8	82.3	1.02	NS	2.28	0.046
T	68.2	70.6	77.0	1.07	NS	2.37	0.027
P	69.0	72.1	75.3	1.08	NS	2.14	0.045
O	67.7	73.1	76.6	2.28	0.034	2.37	0.028
120 min							
F	76.3	78.8	77.7	<1.0	NS	<1.0	NS
C	75.7	78.4	76.5	<1.0	NS	<1.0	NS
T	70.7	73.8	71.2	1.59	NS	<1.0	NS
P	71.0	72.0	72.1	<1.0	NS	<1.0	NS
O	69.0	73.4	70.0	2.29	0.035	<1.0	NS

Note: F, frontal; C, central; T, temporal; P, parietal; O, occipital; PL, placebo; High = 3.55% THC; Low = 1.75% THC.

Analyses were conducted for the systolic and diastolic blood pressures. Neither systolic nor diastolic blood pressure indicated significant drug/placebo or drug/placebo by time effects. There was, for all three conditions, a slight, but statistically nonsignificant, decrease in systolic blood pressure over the first 40 min. A similar pattern was observed for the diastolic blood pressure, again with no statistically significant effects.

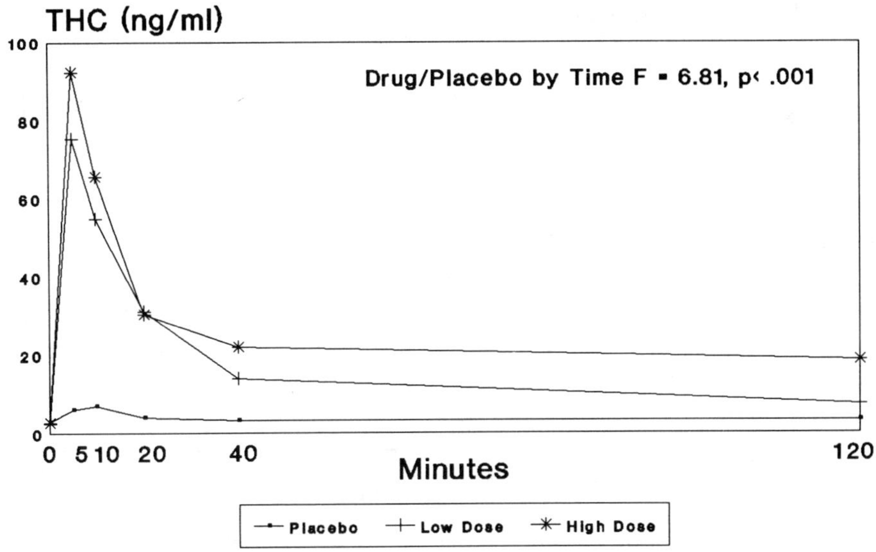

FIGURE 5. Plasma levels of Δ^9-THC.

As presented in Figure 6, pulse rate increased significantly and in a fashion consistent with dose, leading to a highly significant drug/placebo by time interaction (F = 9.48; p <0.001). Peak pulse rate was at 10 min.

Analysis of respiration rate indicated a significant time effect for marijuana and placebo sessions, with a slight but statistically significant increase from baseline with the greatest increase being at 60 min leading to a significant time effect (F = 6.78; p <0.004). However, there was no significant drug/placebo effect and no significant drug/placebo by time interaction.

Results of the analysis of forehead skin flow indicated no significant effects of time or drug/placebo by time. The results of analysis of $PECO_2$ indicated no significant change across time and no drug/placebo by time effect. It should be noted that CBF corrections at 30, 60, and 120 min for differences in $PECO_2$ using standard correction values (3% CBF per torr $PECO_2$) did not change the results. Analysis of end tidal carbon monoxide indicated a highly significant increase across time (F = 30.95; p <0.001). However, there was no drug/placebo difference and no drug/placebo by time interactions (Figure 7). Thus, it would appear from the lack of change in the placebo condition that this level of CO did not alter CBF.

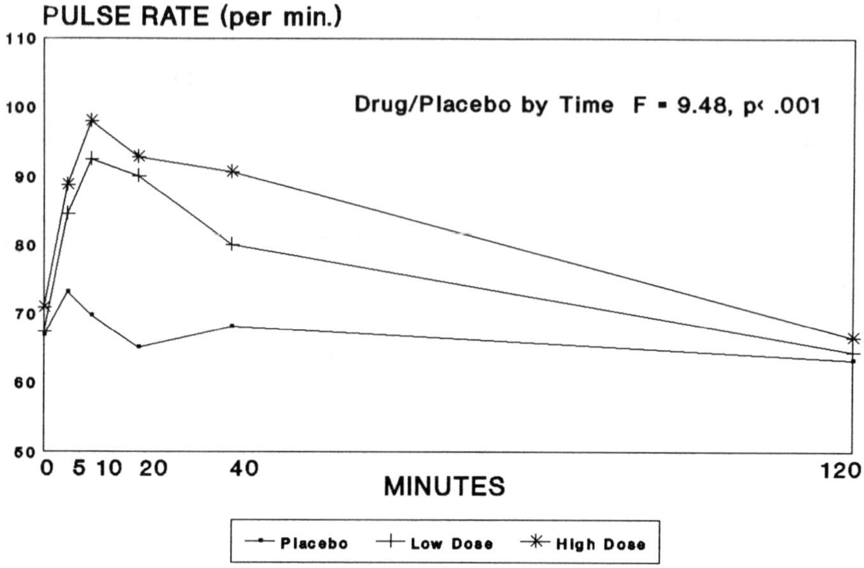

FIGURE 6. Pulse rate following marijuana or placebo.

The six self-rating scales (State Anxiety, Somatic Symptom Scale, Temporal Disintegration Inventory, Depersonalization Inventory, POMS, and Analogue Rating of Intoxication) were each analyzed using a drug/placebo by time multivariate model.

The State Anxiety Scale indicated a trend toward a significant drug/placebo by time interaction (F = 1.96; $p < 0.08$) with somewhat greater effects for the high dose. The Somatic Symptom Scale indicated a significant drug/placebo by time interaction (F = 3.06; $p < 0.008$). There were no significant changes related to placebo across time, but both low- and high-dose marijuana produced change across time.

The Temporal Disintegration Inventory yields scores ranging from –35 to +35, with the lower values (–35) indicating better integration. Analysis of this scale indicated a highly significant drug/placebo by time interaction (F = 4.10; $p < 0.001$) with marked changes in the high-dose condition (Table 9). Analysis of the Depersonalization Inventory also indicated a statistically significant drug/placebo by time interaction (F = 3.33, $p < 0.004$). However, the effects were approximately equal for both the low and high dose. It should be noted that for both of these scales there was essentially no change across the placebo condition.

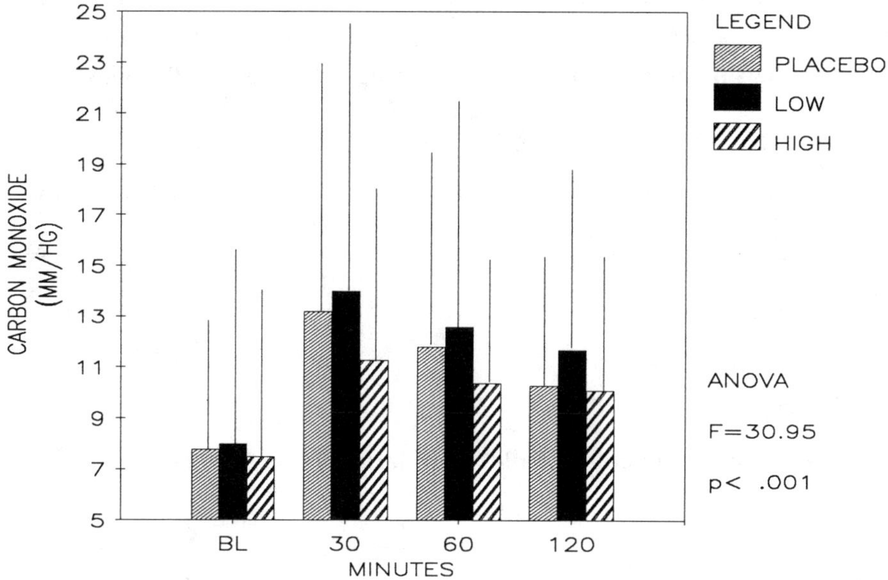

FIGURE 7. End-tidal carbon monoxide before and after marijuana and placebo.

Analysis of POMS consisted of separate analyses for each of the six factors. Factors 1 (tension) and 3 (anger) indicated significant effects (Table 10). Factor 1 indicated a slight increase with time ($F = 2.38$; $p < 0.05$) with the effects being in the low- and high-dose conditions. Factor 3, although also indicating a significant drug/placebo by time interaction, revealed no consistent pattern in relationship to dose or time. Figure 8 presents the results of analysis of ratings of intoxication. A significant drug/placebo by time effect was found ($F = 18.68$; $p < 0.001$). The ratings changed in a pattern consistent with dose of THC, but there was little difference between the two doses.

CBF increase in both hemispheres correlated (Pearson correlation) significantly with degrees of "high", plasma levels of THC, and pulse rate. Loss of time sense, depersonalization, and State Anxiety and Somatic Symptoms scales also correlated significantly with hemispheric CBF. Separate sets of correlations were determined between regional CBF and physiological measures and the rating scales. These correlations were based on means determined across the 20 subjects, providing 12 data points representing the three drug/placebo sessions and 4 time points for each session. For the physiological measures, time points were selected to correspond to

TABLE 9
Temporal Disintegration Inventory Following Marijuana or Placebo

	Placebo		Low		High	
Time	**M**	**SD**	**M**	**SD**	**M**	**SD**
Baseline	–20.92	9.42	–21.72	8.94	–20.60	7.96
30 min	–20.80	9.31	–15.84	11.10	–12.96	12.13
60 min	–20.48	10.25	–17.40	9.39	–16.84	10.01
120 min	–20.28	10.95	–20.20	10.26	–19.32	11.28

Note: Significant Drug/placebo by time interaction (F = 4.10; *p* <0.001).

Depersonalization Inventory Following Marijuana or Placebo

	Placebo		Low		High	
Time	**M**	**SD**	**M**	**SD**	**M**	**SD**
Baseline	13.68	2.98	13.84	3.01	14.20	4.34
30 min	13.64	2.90	19.08	9.63	19.60	10.90
60 min	12.96	2.28	17.12	10.05	15.84	5.81
120 min	13.16	2.62	14.52	4.87	14.08	4.45

Note: Significant Drug/placebo by Time interaction (F = 3.33; *p* <0.004). PL, placebo; High = 3.55% THC; Low = 1.75% THC.

the time of the CBF measurement. All rating scales indicated statistically significant (Table 11) correlations with both left and right regional values (with the exception of POMS factor 3). For the physiological measures, only plasma level and pulse rate indicated statistically significant correlations with CBF values.

Because of the multitude of measures and to further explore these relationships, a three-step, multiple regression analysis was performed for each of the regional values using all variables that had significant first-order correlations. Regional CBF values were expressed as percent of hemispheric mean to remove global CBF change. In step 1, the physiological measures (plasma level and pulse rate), and in step 2, the rating scales were employed in stepwise additive regression models. In step 3, the physiological and rating scale measures that had entered significantly into a predictive model at step 1 (or 2) were included in a combined model. Presented in Table 12 is a summary of findings from the regression analyses. Only

TABLE 10
Profile of Mood State Factors Following Marijuana or Placebo

Factor I

	PL		Low		High	
Tension	**M**	**SD**	**M**	**SD**	**M**	**SD**
Baseline	4.52	4.27	4.00	3.98	4.68	5.09
30 min	3.72	3.26	6.72	7.32	5.96	6.73
70 min	3.40	3.27	5.28	6.36	3.92	3.55
120 min	3.48	3.73	3.72	3.21	3.56	3.54

Note: Significant Drug/placebo by time interaction ($F = 2.38$, $p < 0.032$).

Factor 3

	PL		Low		High	
Anger	**M**	**SD**	**M**	**SD**	**M**	**SD**
Baseline	4.44	7.01	3.52	5.67	3.56	5.91
30 min	3.32	5.19	4.68	6.82	3.00	4.74
60 min	2.68	4.49	4.08	6.45	2.76	5.02
120 min	2.84	4.81	2.80	4.19	3.20	5.89

Note: Significant Drug/placebo by time interaction ($F = 2.42$; $p < 0.029$). PL, placebo; High = 3.55% THC; Low = 1.75% THC.

plasma level entered significantly as a physiological predictor of CBF. In step 2, different rating scales proved predictive for the frontal and parietal regions. Frontal right and left were predicted by the Depersonalization Inventory, right parietal by the Depersonalization Inventory, the State Anxiety and the Self-Rating of Intoxication, parietal left by the somatic sensation scale. No other regions were significantly predicted by either physiological or rating scale variables. In step 3, the rating scales proved to be significant predictors of CBF change, while plasma level did not enter the model. Owing to the limited number of degrees of freedom, we did not further analyze these regressions.

Subjects who participated in the study had previous exposure to marijuana; several gave a history of marijuana smoking, on and off. Unlike the previous study (Acute Changes in CBF After Marijuana—Study I), none reported anxiety or panic during the study and none experienced any other

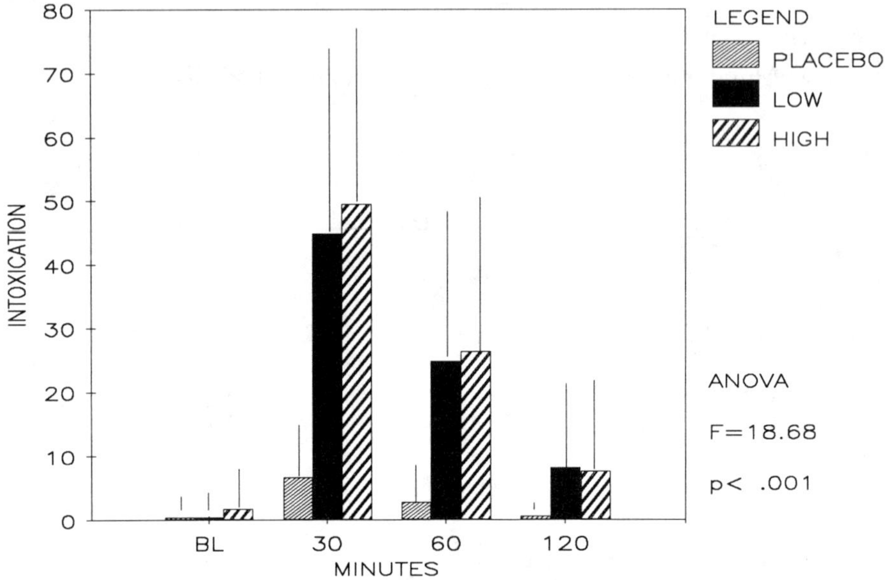

FIGURE 8. Analogue intoxication before and after marijuana and placebo.

adverse effects. Analogue rating scale indicated significant increase in "high" after marijuana. The other rating scale indices which showed significant increases after marijuana were altered time sense and depersonalization. Attention and anger as quantified by POMS increased significantly after the drug and so did the somatic symptom score. The state anxiety scale of STAI showed an increase which approached statistical significance. CBF to both hemispheres showed significant increase, with and without carbon dioxide correction which was maximal 30 min after marijuana administration. Analysis of regional flow patterns indicated greater increase in the frontal region and the right hemisphere. Perez-Reyes and associates[34] reported peaking of THC plasma levels during the 10 min of marijuana smoking and sharp reduction there after, similar to our finding. Pulse rate increased significantly, as did respiratory rate. Blood pressure, forehead skin blood flow, and $PECO_2$ levels did not show significant changes following the drug or placebo. End tidal carbon monoxide increased to the same degree after the two doses of marijuana and placebo. CBF increase in both hemispheres correlated significantly with degrees of "high", plasma levels of THC, and pulse rate. Loss of time sense, depersonalization, State Anxiety, and Somatic Symptoms scales also correlated significantly with hemispheric CBF.

TABLE 11
Correlations Between rCBF and Physiological and Rating Scale Measures

Physiological measures[a]

	Plasma	Pulse	Respiration	BPS	BPD
Right					
F	0.59[b]	0.51[b]	0.22	0.38	–0.27
C	–0.34	–0.39	–0.33	–0.28	–0.25
T	0.23	0.40	–0.19	0.03	–0.39
P	–0.74[b]	–0.63[b]	–0.06	–0.31	0.17
O	–0.12	–0.16	–0.06	–0.09	0.74
Left					
F	0.85[b]	0.60[b]	0.33	0.26	0.23
C	–0.35	–0.10	–0.35	0.01	–0.32
T	0.16	0.34	0.01	0.39	0.04
P	–0.81[b]	–0.75[b]	–0.27	–0.39	–0.09
O	0.28	0.34	0.11	–0.13	0.18

Rating scales[c]

	State	TII	DPI	SSS	POM1	POM3	INTOX
Right							
F	0.47	0.59[b]	0.70[b]	0.60[b]	0.62[b]	0.27	0.63[b]
C	–0.29	–0.20	–0.18	–0.14	–0.10	–0.06	–0.26
T	0.51	0.19	0.33	0.35	0.40	0.51	0.23
P	0.52	0.69[b]	0.85[b]	–0.82[b]	–0.77[b]	–0.35	–0.76
O	–0.32	–0.22	–0.32	–0.26	–0.34	–0.26	–0.20
Left							
F	0.57	0.84[b]	0.85[b]	0.85[b]	0.69[b]	0.06	0.85[b]
C	0.00	–0.28	–0.32	–0.27	–0.13	–0.01	–0.35
T	0.16	0.07	0.03	0.06	0.04	–0.14	0.11
P	–0.65[b]	–0.73[b]	–0.76[b]	–0.83[b]	–0.60[b]	–0.06	–0.79[b]
O	0.42	0.20	0.33	0.43	0.34	0.50	0.30

Note: RCBF expressed as percent of hemispheric mean. F, frontal; C, central; T, temporal; P, parietal; O, occipital.

[a] Plasma, Δ^9–THC levels; Pulse, pulse rate per minute; Respiration, respiration rate per min; BPS, systolic blood pressure mmHg, BPD, diastolic blood pressure mmHg.

[b] To be significant, r must be greater than or equal to 0.58 at p <0.05 (2-tail).

[c] State, state anxiety; TII, temporal disintegration inventory; DPI, depersonalization inventory; SSS, somatic symptom scale; POM1, profile of mood states factors tension (1) and anger (3); Intox, self rating of intoxication.

TABLE 12
Multiple Regression to Predict Each Regional CBF

Right	Predictor variables	Left	Predictor variables
	Step 1 Rating scales[a]		
F	DPI	F	DPI
C	—	C	—
T	—	T	—
P	DPI, State, Intox	P	SSS
O	—	O	—
	Step 2 Physiology measures[b]		
F	Plasma	F	Plasma
C	—	C	—
T	—	T	—
P	Plasma	P	Plasma
O	—	O	—
	Step 3 Combined model		
F	DPI	F	DPI
C	—	C	—
T	—	T	—
P	DPI, State, Intox	P	SSS
O	—	O	—

Note: F, frontal; C, central; T, temporal; P, parietal; O, occipital.

[a] State, State anxiety inventory; DPI, depersonalization inventory; SSS, somatic symptom scale; Intox, self rating of intoxication.

[b] Plasma, Δ^9-THC levels.

When correlations between rating scales and regional flow values were examined, loss of time sense, depersonalization, Somatic Symptom scale, tension, anger, and levels of "high" correlated positively with frontal flow bilaterally, while inverse correlations were seen between these indices and parietal flow.

As was pointed out earlier, marijuana might induce CBF changes through mechanisms other than alteration in brain function. However, the results of the present study suggest that altered brain function was mainly, if not exclusively, responsible for the CBF changes. Forehead skin perfusion did not change appreciably after marijuana smoking. Forehead skin capillaries receive sympathetic innervation from the superior cervical ganglia which also innervate intracranial blood vessels. Thus, marijuana did not appear to increase sympathetic vasomotor tone. Sympathetic stimulation reduces CBF, but CBF increased after marijuana. Tachycardia seen after marijuana is believed to be due to vagal blockade and not sympathetic stimulation.

Although marijuana smoking was associated with increased carbon monoxide production, smoking marijuana cigarettes of two potencies and the placebo produced the same amount of carbon monoxide. Marijuana smoking increased CBF significantly, but placebo smoking did not. Carbon monoxide and CBF did not correlate. Thus, carbon monoxide is unlikely to be responsible for the post-marijuana CBF increase.

CO_2 is a potent cerebrovasodilator and even modest fluctuations in CO_2 cause marked CBF changes. Marijuana, in the present study, did not influence $PECO_2$ levels. Since CO_2 has such a potent effect on CBF, it is conceivable that even nonsignificant changes might influence CBF. However, correction of CBF for changes in $PECO_2$ with a standard correction factor did not alter the results.

The role played by the vascular effects of marijuana on cerebral blood vessels was difficult to evaluate. The drug does have a dilatory effect on conjunctival and muscle blood vessels.[24-26] In the present study, marijuana had no effect on forehead skin perfusion. It may be argued that marijuana may have a direct effect on cerebral vasculature similar to its dilatory effects on conjunctival and muscle vessels. It needs to be noted that marijuana-induced "red eye" lasts for several hours, while in the present study, CBF increase came down significantly within 2 h after the drug.[24,25] In spite of the significant correlation between post-marijuana changes in mood and behavioral indices and regional CBF and finding of a regional difference in CBF response, the possibility of drug-induced vascular (cerebral) change cannot be dismissed.

CBF is autoregulated, i.e., insulated from moderate changes in perfusion pressure.[29] Thus, marijuana-induced changes in peripheral circulation are unlikely to influence brain circulation. In addition, marijuana smoking did not bring about significant changes in blood pressure in spite of the tachycardia.

In the combined model regression analyses, none of the physiological changes after the drug correlated significantly with CBF while several rating scales did. Although plasma levels of THC and CBF correlated (first-order) significantly, the two followed different time courses. Previous research has shown differences in the time courses for plasma THC and intoxication after marijuana smoking. The time course of the CBF changes resembled that of mood changes more closely than plasma THC levels.

Experienced marijuana smokers who participated in the previous study (Acute Changes in CBF After Marijuana—Study I) also showed a similar post-marijuana CBF increase. However, in that study, inexperienced smokers who reported severe anxiety showed CBF decrease after marijuana. Findings from a number of studies indicate cerebral vasoconstriction during moderate to severe anxiety.[56] None of the subjects who participated in the present study reported significant increases in anxiety after marijuana. Absence of an anxiety response in these subjects can be attributed to two factors. The inexperienced marijuana smokers who participated in our previous study (Acute Changes in CBF After Marijuana—Study I) were not currently using marijuana. Participants in this study were current although infrequent marijuana users. Adverse reactions including anxiety are more common among inexperienced individuals.[55] In the previous study, the smoking routine was standardized, which resulted in several subjects receiving more drug than "what they could handle". Many subjects, especially from the inexperienced group, reported dysphoric reactions and severe anxiety. In study II, the subjects were allowed to smoke *ad libitum,* although they were encouraged to finish the cigarette in 10 min.

The association between marijuana-induced changes in regional CBF and behavior are complicated and difficult to explain. Right and left hemispheric flow increased significantly after the drug. Global CBF is closely related to levels of arousal mediated by the reticular activity system.[44,56] Conditions associated with high levels of activation show CBF increase, while low arousal states show CBF decrease.[56] Of the various cortical regions, frontal lobe has the most intimate connections with thalamus which mediates arousal.[68] In the present study, CBF increase after marijuana was most pronounced in the frontal lobe. The right hemisphere has been implicated in the mediation of emotions and most marked CBF changes were seen in the right hemisphere.[69] It needs to be noted that frontal flow and global flow correlated with degrees of "high".

The most marked rating scale change after marijuana was in time sense and depersonalization. Both of these conditions are closely associated with the temporal lobe[70] and, surprisingly, in the present study these two rating scale scores did not correlate with temporal flow. The exact reasons for this are unclear. It is conceivable that these phenomena are related to regions in temporal lobe that are too small to be detected with the ^{133}Xe inhalation technique.[42] The ^{133}Xe inhalation technique does not measure medial temporal lobe blood flow and the region involved in depersonalization and time sense might be located there. Both phenomena correlated inversely with parietal flow.

Parietal lobes are intimately associated with perception and cognition. The effects of marijuana on perception are complex. The drug is known to reduce perceptual acuity, as evidenced by the increase in the critical flicker fusion threshold.[71,72] However, when intoxicated, subjects report increased awareness of tactile, visual (with bright colors), and auditory (more discriminative) stimuli. It is possible that both altered time sense and depersonalization are related to the altered sense of awareness.

Acute Changes in $CMR_{glucose}$ After Marijuana Smoking

Gillespie and associates investigated the acute effects of THC on regional CMR using [^{18}F]-deoxyglucose and PET in a group of eight normal control subjects and three chronic marijuana abusers.[73] CMR glucose was measured under resting conditions and 24 h later, 40 min after i.v. injection of 2 mg of THC. Six of the normal controls and one of the marijuana abusers showed increased regional CMR, while two normal controls and two marijuana abusers showed regional decrease. In 10 out of the 11 subjects, cerebral metabolism showed an increase in the cerebellum.

Acute Changes in Cerebral Blood Velocity after Marijuana

We conducted a study on the effects of marijuana on the right middle cerebral artery blood velocity in experienced marijuana smokers. There were ten healthy, right-handed male volunteers (mean age 25.9 years, S.D. 6) who took part in the project titled "Acute Changes in CBF After Marijuana—Study II". Subjects visited the laboratory two times. A urine drug screen was performed at the beginning of the study during both visits. The volunteers rested on a couch during the entire experiment for 1 h. Blood pressure was taken every minute with an automated blood pressure apparatus from the right arm. The equipment also provided pulse rate measurements. $PECO_2$ was also continuously monitored during this period.

Middle cerebral artery (right side) velocity was measured with TCD through a temporal window with a 2-mHz ultrasonic probe.[49,50] In young adults, it is usually possible to obtain good signals from a relatively large area. The middle cerebral artery runs laterally and slightly anterior as continuation of intracranial intracarotid artery. It has the highest volume flow of all the branches from the circle of Willis, carrying about 80% of the flow to the hemisphere. First, ultrasonic gel was placed between the probe and the patient's skin in the temporal region. The probe was moved by a few millimeters over the surface and placed at different angles until a good signal (CBV more than 50 cm/s) was received; the depth setting used was 50 mm. Once the ideal probe position was found, it was locked in that position by a specially designed headband. A flat transcranial Doppler probe was used in this study. CBV was measured at 1-min intervals.

After the equipment was attached, the subject remained still in the reclining position for 5 min. Afterwards, while still reclining, the subject smoked a cigarette containing marijuana (THC = 3.55%) or a placebo. During the first visit, the subject received marijuana and the second one, a placebo. The laboratory staff knew about the identity of the cigarette, but the subjects did not. After 10 min of smoking, the recording continued for the next 50 min, with the subject in the reclining position.

The data were plotted for every 5 min up to 60 min. In addition, on each measure, paired t-tests were conducted to compare change from baseline for the post-marijuana condition at 5, 10, 15, 20, 30, 40, and 60 min. Presented in Figure 9 is a plotting of the effects of marijuana and placebo on percent change in CBV. CBV is expressed as a percent difference from resting value. During the 10-min smoking period, there was an increase in CBV which was then maintained for the remainder of the 60 min of recording time, whereas the effect of the placebo cigarette was negligible and CBV remained approximately at baseline throughout the 60 min. The t-test indicated statistically significant increases in CBV at all but the 20- and 60-min time points.

Presented in Figure 10 is a summary of the effects of marijuana on pulse rate and blood pressure. Marijuana substantially increased pulse rate during the 10 min of smoking, and it gradually dropped over time during the 60 min, while the placebo had essentially no effect, remaining at approximately

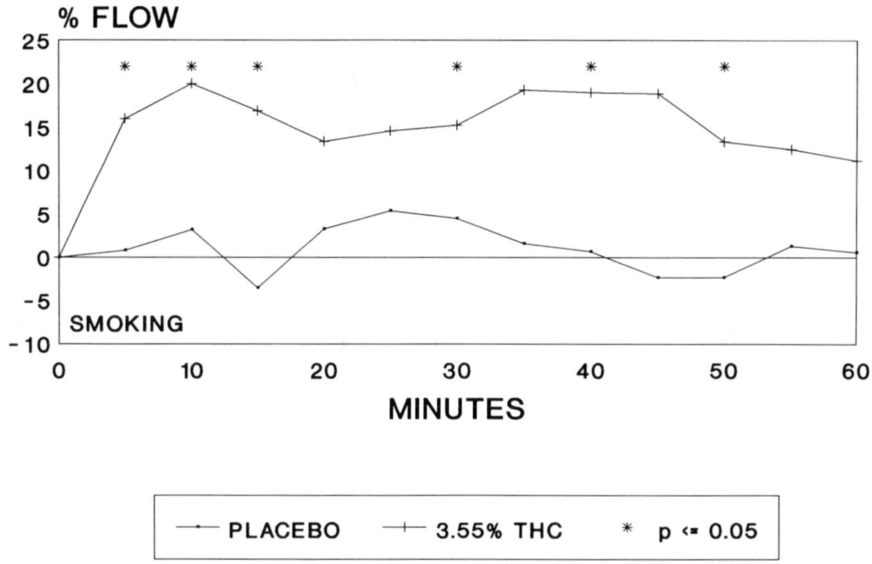

FIGURE 9. Cerebral blood velocity while reclining following placebo and marijuana (percent change from baseline).

the same level throughout the hour of recording. All of the *t*-tests indicated significant differences from baseline during the marijuana session. Comparisons of change in mean arterial blood pressure after marijuana from baseline (*t*-tests) indicated no statistically significant changes. $PECO_2$ did not change significantly after marijuana or placebo.

Marijuana smoking was followed by significant increase in CBV for the entire period of recording. The difference between post-marijuana CBV and baseline (resting) CBV was statistically significant up to 40 min after smoking. Changes in CBV followed an entirely different time course as compared to plasma levels of THC after marijuana smoking in the study titled "Acute Changes in CBF After Marijuana—Study II" (in which the same subjects participated) and other reports. There were no significant changes in $PECO_2$ after marijuana and placebo smoking as compared to the presmoking values.

The reason for the CBV increase seems to be an increase in brain capillary perfusion after marijuana smoking. In the previous two studies

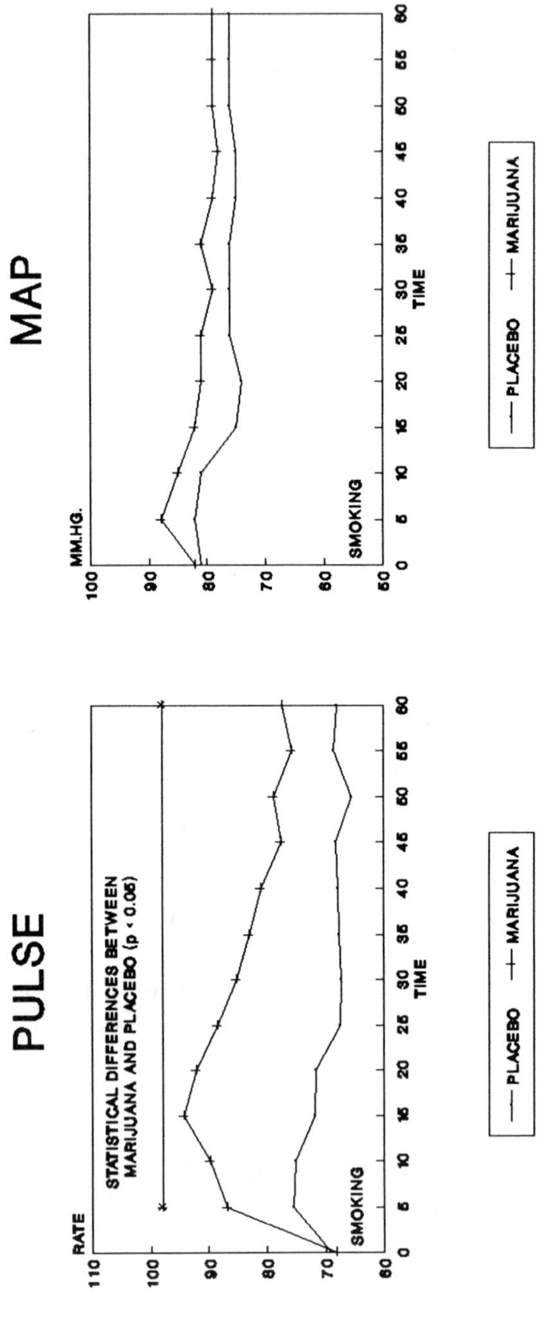

FIGURE 10. Pulse rate and mean arterial pressure following marijuana and placebo. MAP, mean arterial blood pressure.

(Acute Changes in CBF After Marijuana—Study I and Study II), experienced marijuana smokers who did not have panic attacks and severe anxiety showed a post-marijuana CBF increase. This seems to be related to changes in brain function associated with marijuana intoxication. The CBF increase to the hemisphere supplied by the middle cerebral artery should reduce resistance to flow and, thus, increase flow velocity.

CBV is also sensitive to changes in perfusion pressure. Although marijuana smoking was associated with significant increase in pulse rate, mean arterial blood pressure did not change appreciably.[48] Similarly, there were no significant differences between post-marijuana and post-placebo arterial blood pressure. Thus, the CBV increase is unlikely due to changes in perfusion pressure. There is no evidence to suggest that marijuana has an effect on intracranial tension or middle cerebral artery caliber. It does lower intraocular pressure,[74] and therefore it may be argued that it might lower intracranial tension through a similar mechanism. Reduced intracranial tension can increase CBV. We did not measure intracranial tension and there is no literature on this topic. Middle cerebral artery caliber may be sensitive to sympathetic stimulation.[30,31] However, marijuana did not stimulate sympathetic fibers from the superior cervical ganglia (which innervate cranial vessels), as evidenced by the absence of changes in extracranial capillary perfusion after marijuana smoking (Study II). Marijuana can act directly upon vascular beds such as the muscle capillaries and conjunctival vessels.[24-26]

CO_2 is a potent cerebral vasodilator, but marijuana smoking was not associated with significant changes on this index and there were no differences between placebo and marijuana on $PECO_2$.[27,28] Cerebral blood vessels receive sympathetic innervation and it is conceivable that marijuana may have altered sympathetic vasomotor tone.[30,31] Sympathetic stimulation decreases cerebral capillary perfusion and should therefore decrease CBV. Post-marijuana CBV showed an increase and *not* a decrease. Carbon monoxide is produced during marijuana smoking and it is possible that the consequent formation of carboxy hemoglobin will reduce oxygen delivery to the brain.[32] In such an event, cerebral capillary perfusion would show a compensatory increase. In study II, we could not detect any differences between marijuana and placebo smoking on carbon monoxide production, and changes in carbon monoxide production did not correlate with cerebral capillary perfusion (see above). Thus, the CBV increase may be related to increase in CBF related to drug-induced changes in brain function.

Postural Hypotension and Cerebral Blood Velocity After Marijuana

Orthostatic hypotension is an excessive drop in blood pressure while attaining an upright posture. Dizziness and fainting, which are the most common symptoms of orthostatic hypotension, result from cerebral ischemia. Complications of orthostatic hypotension include dizziness, fainting, myocardial infarction, and sudden death. Lacerations and fractures can happen during dizziness and fainting. A number of drugs of abuse including solvents, nitrites, ethyl alcohol, and marijuana are known to cause orthostatic hypotension.[75-78] The precise mechanism responsible for drug-induced postural hypotension and associated cerebral ischemia are unclear.

Normally, upon standing up, there is a caudal shift of 500 to 700 ml of thoracic venous blood and a decrease in venous return by about 20%, stroke volume by about 40%, and cardiac output of between 15 and 20%. As blood pressure begins to fall, splanchnic, renal, and skeletal muscle vascular beds contract, thereby increasing the peripheral vascular resistance by about 40%; this is accompanied by a 20% increase in heart rate. These mechanisms are brought about via baroreceptors which cause sympathetic activation and vagal blockade. Diastolic blood pressure and mean arterial blood pressure increase by about 10%, with systolic pressure showing minimal if any changes. These reflexes which are short lived are followed by a 400% increase in plasma catecholamine levels and a 10% increase in plasma vasopressin. With prolonged standing, renal vasoconstriction leads to increases in plasma renin, angiotensin, and aldosterone which maintain the plasma volume.[79]

CBF is insulated from modest changes in perfusion pressure (blood pressure) through autoregulatory mechanisms.[80,81] When blood pressure goes down, cerebral arterioles dilate to maintain cerebral circulation. However, when posture-related hypotension is profound and sustained (mean arterial blood pressure below 60 mmHg), the cerebral autoregulatory mechanism may fail and such symptoms of cerebral ischemia as dizziness and fainting may occur.

Cerebral ischemia during upright position can result from failure to maintain blood pressure, impaired cerebral autoregulation, or both. Orthostatic hypotension is common after marijuana smoking.[82] We examined changes in blood pressure and CBV (to evaluate cerebral autoregulation) during the upright position after marijuana.

Some seven right-handed male subjects (mean age 25.9 years, S.D. 4.4) from the project titled "Acute Changes in CBF After Marijuana—Study II" with a history of previous exposure to marijuana smoking participated in the project. They were free of medicines for a minimum of 3 months before the experiment.

The subjects came to the laboratory two times. Urine drug screens were performed during each visit. The following measurements were made throughout the experiment. CBV (middle cerebral artery—MCA) was averaged every minute with TCD.[49,50] Systolic and diastolic blood pressures and pulse rate were quantified every minute with an automated blood pressure machine. $PECO_2$ was continuously monitored.

The subjects rested in a reclining position for the first 10 min. During the next 10 min, they were given a cigarette to smoke while still reclining. During the first visit to the laboratory, it was a high-potency marijuana cigarette (THC content = 3.55%); during the second visit, it was a placebo cigarette (marijuana with all THC extracted). The laboratory staff knew about the identity of the cigarettes, but the subjects did not. Measurements continued during marijuana smoking. The subjects were then asked to stand for 5 min. During the last 5 min, they assumed a reclining position again. The subjects were asked to quantify degrees of dizziness on a 4-point rating scale at the end of the experiment.

The data for CBV, pulse rate, and systolic and diastolic blood pressures were plotted across time with 4 5-min intervals; the last 5 min of the baseline resting period, the last 5 min of the smoking period, the 5 min of standing, and the 5 min of the second resting (lying down) period being the time intervals plotted. In addition, mean arterial blood pressure was determined, and this was also plotted. To analyze changes from baseline following marijuana, *t*-tests were conducted to compare last recording on the baseline resting period with the end of smoking, each 1-min recording during standing, and the second period of resting. To evaluate change in $PECO_2$ a repeated measures ANOVA was used.

Presented in Figure 11 is a plot of the orthostatic changes in CBV. This plot contains three separate lines. The line for the placebo and marijuana conditions represent means across the seven subjects across time. The line indicating marijuana is for all subjects across time. In addition, a line depicting the results from five subjects who had rated themselves as either a "2" or a "3" on the dizziness scale was also plotted. There was no

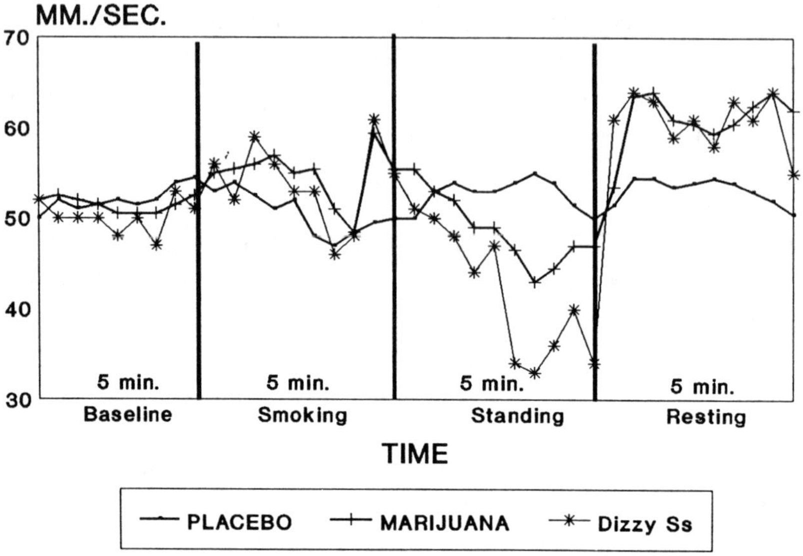

FIGURE 11. Orthostatic changes in blood velocity.

significant difference in the mean CBV following marijuana and placebo conditions at baseline. During smoking, there was a gradual rise in the CBV following marijuana. Upon standing after marijuana, there was a significant drop over time of CBV which began to recover at approximately 3 1/2 min after standing. A similar orthostatic drop in CBV was not seen following placebo. Upon lying down again, there is a significant increase in CBV when compared to baseline after marijuana, which again was not found following placebo. For the five people who rated themselves as very dizzy, the same pattern is present, but the drop in CBV is greatly accentuated. Contrasts (*t*-tests) comparing final baseline value with values at the end of smoking, for each minute during the standing and second resting period, indicated a significant increase over baseline after marijuana. At 3 1/2 min into standing, there was a significant decrease *from baseline;* during the second resting period, all values were significantly greater than baseline.

Presented in Figure 12 is a plot of the orthostatic changes in pulse rate. Consistent with previous reports, pulse rate increased significantly while smoking marijuana and remained significantly elevated over baseline throughout the remainder of the study.[1-5] There was no significant change following placebo.

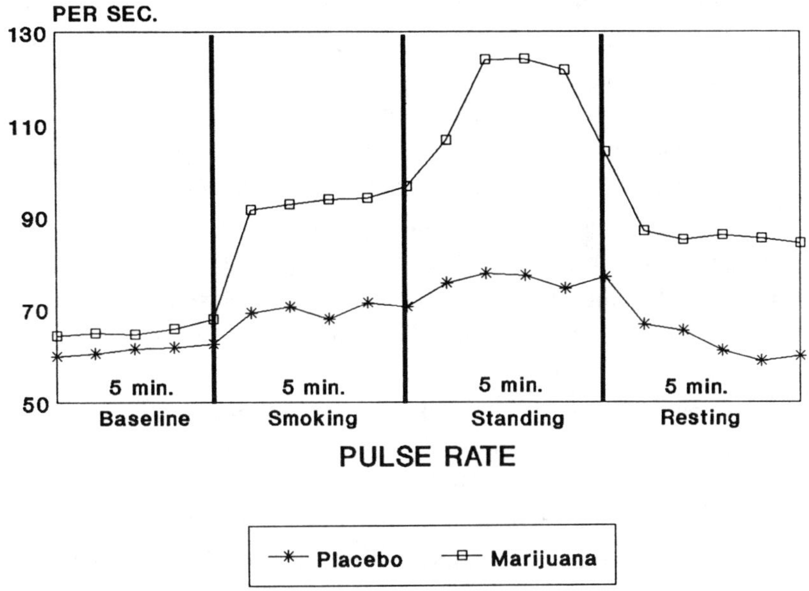

FIGURE 12. Orthostatic changes in pulse rate.

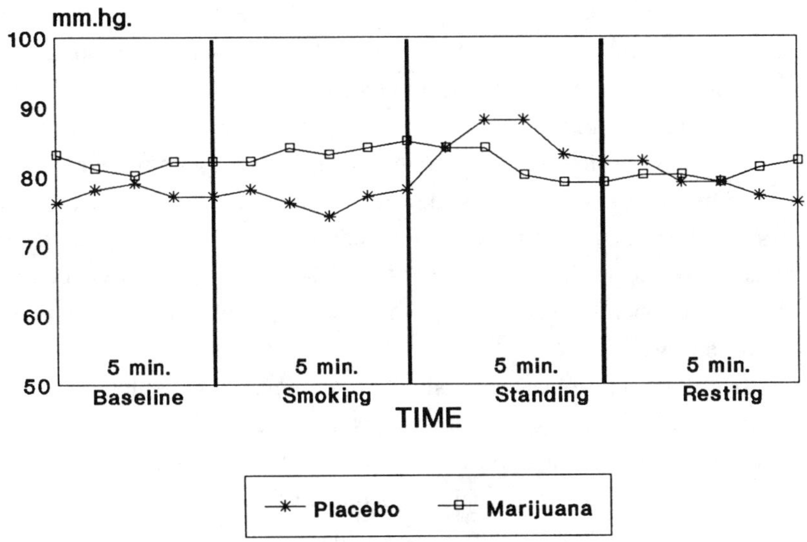

FIGURE 13. Orthostatic changes in mean arterial pressure.

Presented in Figure 13 is a plot of the mean arterial blood pressure (BP). Following marijuana, there were no significant changes from baseline either after smoking or after standing. There was a slight but nonsignificant increase in blood pressure following placebo while standing. Following standing, there was a slight decrease in systolic blood pressure after marijuana, which returned to approximately baseline level. There was no such decrease following placebo. Diastolic blood pressure remained essentially unchanged after marijuana, while there was a slight rise in diastolic blood pressure following placebo during standing, which returned to baseline. $PECO_2$ did not change significantly during standing up after marijuana or placebo.

Five out of seven subjects reported significant dizziness during standing after marijuana smoking. Two of these subjects had to sit down immediately, with severe dizziness. Standing after marijuana was associated with a modest reduction in blood pressure which normalized within 1 min. Standing after placebo was not associated with any changes in blood pressure or pulse rate. Pulse rate increase seen after marijuana smoking increased even further when they stood up. CBV did not change during standing after placebo. However, after marijuana smoking, CBV decreased significantly even after blood pressure returned to normal. Subjects who reported dizziness showed steeper reduction in CBV as compared to others (Figure 11). In this group of subjects, CBV continued to drop until the patients assumed the supine position again.

The decrease in blood pressure is often held responsible for the reports of dizziness during standing after marijuana.[82] Increase in pulse rate during standing after marijuana would seem to suggest reduction in peripheral vascular resistance. Skin circulation shows minimal, if any, changes after marijuana (Acute Changes in CBF After Marijuana—Study II). The reduction in peripheral vascular resistance has been attributed to an increase in muscle flow.[26]

In the present experiment, standing after marijuana smoking was not associated with any significant changes in arterial blood pressure. None of the subjects showed a significant decrease in blood pressure; mean arterial blood pressure should be below 60 mmHg for cerebral ischemia to result.[80,81] Continued reduction in CBV and reports of dizziness in the face of normal blood pressure, would seem to suggest impaired cerebral autoregulation after marijuana. Subjects who complained of severe dizziness showed continued decrease in CBV even after 5 min of standing.

The physiological basis for autoregulation is controversial. A number of explanations — including tissue pressure theory — neurogenic theory, myogenic theory, and metabolic theory have been put forward. The precise mechanism responsible for the effect of marijuana on cerebral autoregulation is even more unclear. Normally, decrease in perfusion pressure will trigger dilation of cerebral arterioles, with a compensatory increase in cerebral circulation.[80,81] In conditions where cerebral blood vessels are maximally dilated (hypercapnia and severe anoxia) and/or where additional vasodilation during reduced perfusion pressure is not possible, cerebral autoregulation is impaired.[81]

Experienced marijuana smokers who do not become anxious after marijuana show a CBF increase after the drug. In the present study, marijuana smoking was associated with a CBV increase while lying down, in the absence of changes in blood pressure or $PECO_2$. Thus, this CBV increase may be related to a parallel increase in cerebral capillary perfusion after the drug. The post-marijuana impairment of cerebral autoregulation may be due to cerebral vasodilation induced by the drug.

Effects of Chronic Marijuana Smoking on Cerebral Blood Flow

Tunving and associates measured CBF with the ^{133}Xe inhalation technique in nine male patients (mean age 24.2 years, S.D. 4.8) with heavy chronic marijuana abuse (mean duration of use 9.7 years, S.D. 4.7) after they were hospitalized for detoxification;[83] eight out of nine patients reported daily use and all except one were using a minimum of 1 g of marijuana daily. The subjects were in good physical health. Four out of the nine patients denied abusing other drugs and the remaining five reported periodic marijuana abuse. None had any alcohol problems. Urine drug screens showed high amounts of THC metabolites, but no other drugs. A group of physically and mentally healthy volunteers matched to the patients on age, sex, and hand preference was used for comparison purposes.

The first CBF measurements were between 1 to 12 d after the last marijuana use and, in four patients, CBF measurements were repeated a second time at 9, 14, 15, and 60 d after the first measurement. Some of the patients received oxazepam before the first CBF measurement, but none received any afterwards.

Cannabis abusers showed reduced hemispheric CBF as compared to the controls during the first measurement. There were no regional flow differences between the two groups of patients. When CBF was repeated in four cases, the hemispheric flow increased by 12%, which was statistically significant. Once again, the CBF increase did not show regional variations. In one patient, CBF was measured four times over a 9-month period. These measurements did not show additional changes in CBF. CBF values were adjusted for differences in $PECO_2$.

The results of Tunving's study indicated reduced CBF in heavy marijuana users immediately after they stopped the habit. CBF increased during abstinence. The findings of this study are complicated by the use of benzodiazepines in some subjects before the first CBF measurements. Benzodiazepines lower CBF and therefore this is a possible explanation for the reduced CBF during the first measurement and higher CBF measurements during the second measurement after abstinence from marijuana and benzodiazepines.[21]

We measured CBF with the same technique in 17 male subjects (age 25.5 years, S.D. 8) who gave a history of smoking a minimum of 6 joints per week (mean 16 joints, S.D. 9) for 6 months or more (mean 83 months, S.D. 67).[84] An equal number of normal volunteers who had never smoked marijuana served as controls. Both groups were matched on age, sex, and hand preference and all participants were in good physical health with no identifiable psychiatric disorders. They were not on any medicines for a minimum of 2 weeks before the study. Four marijuana subjects reported sporadic cocaine use in the past and one subject had taken mushrooms. Nobody reported alcohol abuse or dependence.

The participants were required to avoid coffee and tobacco for 2 h prior to this study and the marijuana smokers did not consume any marijuana during the previous 2 weeks. CBF was measured just once with the ^{133}Xe inhalation technique under resting conditions. There were no differences between the marijuana users and controls on any of the variables, including CBF.

Experienced marijuana smokers (see above) who participated in "Acute Changes in CBF After Marijuana—Study I" showed lower baseline CBF values as compared to inexperienced smokers and normal controls. This finding is at variance with the study mentioned immediately above and the report by Tunving and associates.[83] This discrepancy can be attributed to differences in the characteristics of the subjects who participated in the three projects. Tunving and associates studied patients who were hospitalized for marijuana abuse. Experienced marijuana smokers mentioned above were chronic, heavy marijuana users. These two groups of subjects were found to have reduced CBF. Marijuana smokers whose CBF was found to be normal in the study mentioned above were different from the other two groups. Most of them were staff and students of the university who were not impaired in any identifiable way because of marijuana use. They were not regular marijuana smokers. It is unclear whether the reduced baseline CBF found in heavy marijuana users was specifically related to marijuana. These individuals are likely to differ from the others on several other variables, including tobacco smoking. Chronic tobacco smoking is associated with a CBF decrease.[85] THC has a long half-life.[59,60] It is possible that the reduced baseline CBF in these subjects was influenced by THC left over from previous marijuana smoking.

SUMMARY AND CONCLUSIONS

In the normal brain, CBF is closely coupled to brain function and has been shown to be useful as an index of brain function. Drug-induced changes in brain function are therefore likely to be associated with parallel changes in CBF and CMR. However, pharmacological agents may trigger mechanisms other than brain function such as carbon dioxide levels, autonomic activation, perfusion pressure changes, and direct stimulation of vascular smooth muscle which may independently induce CBF changes.

Acute and long-term use of marijuana are known to be accompanied by a number of behavioral manifestations. They are likely to be accompanied by global and regional CBF changes. Marijuana smoking is also accompanied by autonomic changes and alterations in carbon dioxide and carbon monoxide, and by changes in peripheral vascular beds such as the conjunctiva and muscles. Studies conducted on the acute effects of marijuana on CBF indicate an increase in CBF which tends to peak at 30 min. Greater CBF increases are seen in the frontal region and right hemisphere. Pulse rate, respiratory rate, and end tidal carbon monoxide tend to increase after marijuana (although CBF increases in both hemispheres correlate significantly with degrees of high plasma levels of THC and pulse rate). Behavioral changes seem to be the most significant predictors of the change in CBF. Effects of marijuana on forehead skin circulation seem unrelated to the CBF changes. Marijuana smoking also increases $CMR_{glucose}$.

Inexperienced marijuana smokers differ from experienced ones in behavioral and CBF responses to the drug. They are more likely to experience dysphoric reactions, especially anxiety. Inexperienced smokers also show a decrease in CBF after marijuana. This is in keeping with other studies that point to cerebral vasoconstriction in accompaniment of severe anxiety. Animal studies show increased global CBF after THC administration. However, regional CBF showed a decrease in several subcortical areas.

Marijuana smoking is also accompanied by a significant increase in middle cerebral artery velocity. Increased cerebral artery velocity is unrelated to an increase in pulse rate and changes in blood pressure or end tidal carbon dioxide. In all probability, increased cerebral blood velocity is directly related to increased capillary perfusion, i.e., CBF.

Marijuana is known to cause symptoms suggestive of orthostatic hypotension such as dizziness and fainting during upright posture. Standing after marijuana smoking is associated with a moderate blood pressure decrease that normalizes within 1 min. CBV after marijuana smoking decreased significantly during standing, even after the blood pressure returned to normal. Subjects who reported dizziness tend to show sharper reduction in CBV as compared to the others. These findings suggest that marijuana might interfere with cerebral autoregulatory mechanisms in addition to decreasing perfusion pressure.

The few available studies on the chronic effects indicate a CBF decrease in heavy users with mild to moderate users showing no changes. The CBF decrease associated with chronic use tends to normalize with abstinence.

The number of studies available on the effect of marijuana on CBF is relatively small. The findings mentioned above need independent confirmation by other laboratories. Most of the available studies utilized the ^{133}Xe inhalation technique which does not quantify subcortical flow. Use of techniques with better spatial resolution and the ability to quantify subcortical flow, such as positron emission tomography, are likely to yield more useful findings.

ACKNOWLEDGMENT

The research reported was supported by RO1 DA04985.

REFERENCES

1. **Brill, H. and Nahas, G. G.**, Cannabis intoxication and mental illness, in *Marijuana in Science and Medicine*, Nahas, G. G., Ed., Raven Press, New York, 1984, 263.
2. **Institute of Medicine,** *Marijuana and Health. Report of a Study by a Committee of the Institute of Medicine*, Division of Health Sciences Policy, National Academy Press, Washington, D.C., 1982, 112.
3. **Mendelson, J. H., Babor, T. F., Kuehnle, J. C., Rossi, A. M., Bernstein, J. G., Mello, N. K., and Greenberg, I.**, Behavioral and biologic aspects of marijuana use, *Ann. N.Y. Acad. Sci.*, 282, 186, 1976.
4. **Maykut, M. O.,** Health consequences of acute and chronic marijuana use, *Prog. Neuro-Psychopharmacol. Biol. Psychiat.*, 9, 209, 1985.
5. **Hollister, L. E.,** Health aspects of cannabis, *Pharmacol. Rev.*, 38, 1, 1986.
6. **Benedikt, R. A., Cristofaro, P., Mendelson, J. H., and Mello, N. K.,** Effects of acute marijuana smoking in post-menopausal women, *Psychopharmacology,* 90, 14, 1986.
7. **Meyer, R. E., Pillard, R. C., Shapiro, L. M., and Mirin, S. M.,** Administration of marijuana to heavy and casual marijuana users, *Am. J. Psychiatry*, 128, 198, 1971.
8. **Babor, T. F., Mendelson, J. H., Greenberg, L., and Keuhnle, J. C.,** Marijuana consumption and tolerance to physiological and subjective effects, *Arch. Gen. Psychiatry*, 32, 1548, 1975.
9. **Mendelson, J. H. and Mello, N. K.,** Re-enforcing properties of oral delta-9-tetrahydrocannabinol, smoked marijuana and nabilone influence of previous marijuana use, *Psychopharmacology,* 83, 351, 1980.
10. **Lex, B. W., Mendelson, J. H., Bavli, S., Harvey, K., and Mello, N. K.,** Effects of acute marijuana smoking on pulse rate and mood states in women, *Psychopharmacology,* 84, 178, 1984.
11. **Nahas, G. G.,** Toxicology and pharmacology, in *Marijuana in Science and Medicine*, Nehas, G. G., Ed., Raven Press, New York, 1974, 131.
12. **Harris, L. S.,** General and behavioral pharmacology of delta-9-THC, *Pharmacol. Rev.*, 23, 285, 1971.
13. **Schurr, A.,** Marijuana: much ado about THC, *Comp. Biochem. Physiol.*, 80C, 1, 1985.
14. **Moreau, J. J.,** *Du Hachisch et de l'Alienation Mentale*, Librari de Fortin, Masson, Paris, 1945 (English Edition: Raven Press, New York, 1972).
15. **Mathew, R. J., Margolin, R. A., and Kessler, R. M.,** Cerebral function, blood flow and metabolism: a new vista in psychiatric research, *Integr. Psychiatry*, 3, 214, 1985.
16. **Risberg, J.,** Regional cerebral blood flow measurements by ^{133}Xenon inhalations: methodology and applications in neuropsychology and psychiatry, *Brain Lang.*, 9, 9, 1980.
17. **Phelps, M. E., Mazziotta, J. C., and Huang, S. C.,** Study of cerebral function with positron emission tomography, *J. Cereb. Blood Flow Metab.,* 2, 13, 1982.
18. **Mazziotta, J. C.,** PET scanning principles and applications, *Disc. Neuro. Sci.*, 11, 9, 1985.
19. **Kety, S. S. and Schmidt, C. F.,** The nitrous oxide method for the quantitative determination of cerebral blood flow in man: theory, procedure and normal values, *J. Clin. Invest.*, 27, 476, 1948.
20. **Sokoloff, L.,** The action of drugs on the cerebral circulation, *Pharmacol. Rev.*, 11, 1, 1959.
21. **Mathew, R. J. and Wilson, W. H.,** Substance abuse and cerebral blood flow, *Am. J. Psychiatry*, 148, 292, 1991.

22. **Mathew, R. J. and Wilson, W. H.,** Caffeine-induced changes in cerebral circulation, *Stroke*, 16, 814, 1985.
23. **Mathew, R. J., Wilson, W. H., and Tant, S. R.,** Regional cerebral blood flow changes associated with amyl nitrite inhalation, *Br. J. Addict.*, 84, 293, 1989.
24. **Hollister, L. E., Gillespie, H. K., Ohlsson, A., Lindgren, J.-E., Wahlen, A., and Agurell, S.,** Does plasma concentrations of Δ^9-tetrahydrocannabinol reflect the degree of intoxication?, *J. Clin. Pharmacol.*, 21, 1715, 1981.
25. **Ohlsson, A., Lindgren, J.-E., Wahlen, A., Agurell, S., Hollister, L. E., and Gillespie, H. K.,** Plasma Δ^9-tetrahydrocannabinol concentrations and clinical effects after oral and intravenous administration and smoking, *Clin. Pharmacol. Ther.*, 28, 409, 1980.
26. **Weiss, J. L., Watanabe, A. M., Lemberger, L., Tamarkin, N. R., and Cardon, P. V.,** Cardiovascular effects of delta-9-tetrahydrocannabinol in man, *Clin. Pharmacol. Ther.*, 13, 671, 1972.
27. **Maximilian, V. A., Prohovnik, I., and Risberg, J.,** Cerebral hemodynamic response to mental activation during normo- and hyper-carbia, *Stroke*, 11, 342, 1980.
28. **Mathew, R. J. and Wilson, W. H.,** Cerebral blood flow changes induced by CO_2 in anxiety, *Psychiatry Res.*, 23, 285, 1988.
29. **Strandgaard, S. and Paulson, O. B.,** Cerebral autoregulation, *Stroke*, 15, 413, 1984.
30. **Edvinsson, L.,** Sympathetic control of cerebral circulation, *Trends Neurosci.*, 5, 425, 1982.
31. **Busija, D. W. and Heistead, D. D.,** Factors involved in the physiological regulation of the cerebral circulation, *Rev. Physiol. Biochem. Pharmacol.*, 101, 161, 1984.
32. **Aranow, W. S. and Cassidy, J.,** Effect of marijuana and placebo marijuana smoking on angina pectoris, *N. Engl. J. Med.*, 291, 65, 1979.
33. **Frackowiak, R. S. J.,** *Pathophysiology of Human Cerebral Ischemia: Studies with Positron Tomography and ^{15}Oxygen in Brain Imaging and Brain Functions*, Sokoloff, Ed., Raven Press, New York, 1985, 139.
34. **Campbell, A. M. G., Evans, M., Thomson, J. L. G., and Williams, J. M.,** Cerebral atrophy in young cannabis smokers, *Lancet*, 2, 1219, 1971.
35. **Co, B. T., Goodwin, W. N., Gado, M., Mikhoul, M., and Hills, S. Y.,** Absence of cerebral atrophy in chronic cannabis users, *JAMA*, 237, 1229, 1977.
36. **Kuehnle, J., Mendelson, J. H., Davis, K. R., and New, P. F. J.,** Computed tomographic examination of heavy marijuana smokers, *JAMA*, 237,1231,1977.
37. **Phil, R. O. and Sigal, H.,** Motivation levels and marijuana high, *Can. J. Abnormal Psychol.*, 87, 280, 1970.
38. **Georgotas, A. and Zeidenberg, P.,** Observations on the effects of 4 weeks of heavy marijuana smoking on group interaction and individual behavior, *Compr. Psychiatry*, 20, 427, 1979.
39. **Mathew, R. J., Meyer, J. S., Frances, D. J., Semchuk, K. M., Mortel, K., and Claghorn, J. L.,** Cerebral blood flow in depression, *Am. J. Psychiatry*, 137, 1449, 1980.
40. **Baxter, L. R., Phelps, M. E., Mazziotta, J. C., Schwartz, J. M., Gerner, R. H., Selin, C. E., and Sumida, R. M.,** Cerebral metabolic rate for glucose in mood disorders, *Arch. Gen. Psychiatry*, 42, 441, 1985.
41. **Ingvar, D. H. and Lassen, N. A.,** Quantitative determination of cerebral blood flow in man, *Lancet*, 2, 806, 1961.
42. **Obrist, W. D., Thompson, H. K., Jr., Wang, H. S., and Wilkinson, W. E.,** Regional cerebral blood flow estimated by ^{133}-Xenon inhalation, *Stroke*, 6, 245, 1975.

43. **Price, R. R., Patton, J. A., and Rollo, F. D.,** Single photon emission computed tomography: an overview, in *Single Photon Emission Computed Tomography and Other Selected Computer Topics*, Sorenson, J. A., Ed., Society of Nuclear Medicine, New York, 1980.
44. **Mathew, R. J.,** Hyperfrontality of regional cerebral blood flow distribution in normals during resting wakefulness: fact or artifact?, *Biol. Psychiatry*, 26,717,1989.
45. **Sokoloff, L. and Smith, C.,** Biochemical principles for the measurement of metabolic rate *in vivo,* in *Positron Emission Tomography of the Brain*, Heiss, W.-D. and Phelps, M. E., Eds., Springer-Verlag, New York, 1983, 2.
46. **Alavi, A., Dann, R., Chawluk, J., Alavi, J., Kushner, M., and Reivich, M.,** Positron emission tomography of regional cerebral glucose metabolism, *Sem. Nuclear Med.*, 16, 2, 1986.
47. **Lassen, N. A., Henriksen, L., and Paulson, O.,** Regional cerebral blood flow in stroke by ^{133}xenon inhalation and emission tomography, *Stroke*, 12, 284, 1981.
48. **Aaslid, R. and Lindergaard, K.-F.,** Cerebral hemodynamics, in *Transcranial Doppler Sonography*, Aaslid, R., Ed., Springer-Verlag, New York, 1986, 60.
49. **Aaslid, R.,** The Doppler principle applied to measurement of blood velocity in cerebral arteries, in *Transcranial Doppler Sonography,* Aaslid, R., Ed., Springer-Verlag, New York, 1986.
50. **Arnolds, B. J. and Von Ruetern, G.-M.,** Transcranial Doppler sonography examination technique and normal reference values, *Ultrasound Med. Biol.,* 12, 115, 1986.
51. **Beaconsfield, P., Carpi, A., Cartoni, C., De Basso, P., and Rainsbury, R.,** *Lancet*, 2, 1146, 1972.
52. **Goldman, H., Dagirmanjian, R., Drew, W. G., and Murphy, S.,** Δ^9-Tetrahydrocannabinol alters flow of blood to subcortical areas of the conscious rat brain, *Life Sci.*, 17, 477, 1975.
53. **McNair, D. M., Lorr, M., and Doppleman, L. F.,** *Manual for the Profile of Mood States*, Educational and Industrial Testing Service, San Diego, 1971.
54. **Spielberger, C. D., Gorsuch, R. L., and Lushene, R. D.,** *STAI Manual*, Consulting Psychologists Press, Palo Alto, CA, 1970.
55. **Weil, A. T.,** Adverse reactions to marijuana, *New Engl. J. Med.,* 282, 997, 1970.
56. **Mathew, R. J. and Wilson, W. H.,** Anxiety and cerebral blood flow, *Am. J. Psychiatry,* 147, 838, 1990.
57. **Selzer, M. L.,** The Michigan Alcoholism Screening Test: the quest for a new diagnostic instrument, *Am. J. Psychiatry,* 127, 1653, 1971.
58. **Beck, A. T. and Beck, R. W.,** Screening depressed patients in family practice, *Post Grad. Med.,* 52, 81, 1972.
59. **Lemberger, L., Silberstein, S. D., Axelrod, J., and Kopin, J. J.,** Marijuana: studies of the disposition and metabolism of Δ^9-tetrahydrocannabinol in man, *Science,* 170, 1320, 1970.
60. **Lemberger, L., Tamarkin, N. R., Axelrod, J., and Kopin, J. J.,** Δ^9-Tetrahydrocannabinol and disposition in long-term marijuana smokers, *Science,* 173, 72, 1971.
61. **Fagrell, B.,** Dynamics of skin microcirculation in humans, *J. Cardiovasc. Pharm.,* 7 (Suppl. 3), S53, 1985.
62. **Holloway, J. R. and Watkins, D. W.,** Laser Doppler measurement of cutaneous blood flow, *J. Invest. Dermatol.,* 69, 306, 1977.
63. **Tyrer, P.,** *The Role of Bodily Feelings in Anxiety*, Oxford University Press, London, 1976.

64. **Melges, F. D., Tinklenberg, J. R., and Hollister, L. E.,** Marijuana and temporal span of awareness, *Arch. Gen. Psychiatry,* 24, 567, 1971.
65. **Melges, F. D., Tinklenberg, J. R., Hollister, L. E., and Gillespie, H. K.,** Marijuana and temporal disintegration, *Science,* 168, 1118, 1970.
66. **Melges, F. D., Tinklenberg, J. R., Hollister, L. E., and Gillespie, H. K.,** Temporal disintegration and depersonalization during marijuana intoxication, *Arch. Gen. Psychiatry,* 23, 204, 1970.
67. **Dixon, J. C.,** Depersonalization phenomenon in a sample population of college students, *Br. J. Psychiatry,* 109, 371, 1963.
68. **Nauta, W. J. H.,** The problem of frontal lobe: a reinterpretation, *J. Psychiatr. Res.,* 8, 167, 1971.
69. **Ross, E. D.,** Right hemisphere's role in language, affective behavior and emotion, *Trends Neurosci.,* 7, 342, 1984.
70. **Lishman, W. A.,** Cardinal psychological features of cerebral disorder, in *Organic Psychiatry: The Psychological Consequences of Cerebral Disorder*, Lishman, W. A., Ed., Blackwell Scientific, Oxford, 1987, 3.
71. **Casswell, S. and Marks, D.,** Cannabis-induced impairment of performance of a divided attention task, *Nature (London),* 241, 60, 1973.
72. **Sharma, S. and Moskowitz, H.,** Effects of two levels of attention demand on vigilance performance under marijuana, *Percept. Mot. Skills,* 38, 967, 1974.
73. **Gillespie, H., Volkow, N. D., Tancredi, L. R., Mullani, N., Grant, C., and Hollister, L. E.,** Actions of marijuana on brain metabolism, presented at the 144th Annual Meeting American Psychiatric Association, May 11–16, 1991, New Orleans, LA, Abstr. #47-C.
74. **Hepler, R. S. and Frank, I. M.,** Marijuana smoking and intraocular pressure, *JAMA,* 217, 1392, 1971.
75. **Cohen, S.,** The volatile nitrites, *JAMA,* 241, 2077, 1979.
76. **Shafer, N.,** Hypotension due to nitroglycerine combined with alcohol, *N. Engl. J. Med.,* 273, 1169, 1965.
77. **Stimmel, B.,** Narcotic analgesics: morphine, synthetic opiates and narcotic antagonists, in *Cardiovascular Effects of Mood-Altering Drugs*, Raven Press, New York, 1979.
78. **Merritt, J. C., Cook, C. E., and Davis, K. H.,** Orthostatic hypotension after Δ^{9-} tetrahydrocannabinol marijuana inhalation, *Ophthalmic Res.,* 14, 124, 1982.
79. **Smith, J. J. and Campine, J. P.,** Circulatory response to non-exercise stress, in *Circulatory Physiology—The Essentials*, Smith, J. J. and Campine, J. P., Eds., Williams and Wilkins, Baltimore, MD, 1980.
80. **Harper, A. M.,** Autoregulation of cerebral blood flow: influence of the arterial blood pressure on blood flow through the cerebral cortex, *J. Neurol. Neurosurg. Psychiatr.,* 29, 398, 1966.
81. **McCulloch, C. H. J.,** The physiology and regulation of cerebral blood flow, in *Handbook of Regional Cerebral Blood Flow*, Knezevic, S., Maximilian, V. A., Mubrin, Z., Prohovnik, I., and Wade, J., Eds., Lawrence Erlbaum Associates, New Jersey, 1988, 1.
82. **Merrit J. C., Cook, C. E., and Davis, K. H.,** Orthostatic hypotension after Δ^{9-} tetrahydrocannabinol marijuana inhalation, *Ophthalmic Res.,* 14, 124, 1982.
83. **Tunving, K., Thulin, O., Risberg, J., and Warkentin, S.,** Regional cerebral blood flow in long-term heavy cannabis use, *Psychiatry Res.,* 17, 15, 1986.
84. **Mathew, R. J., Tant, S., and Berger, C.,** Regional cerebral blood flow in marijuana smokers, *Br. J. Addict.,* 81, 567, 1986.
85. **Kubota, K., Yamaguchi, T., Abe, Y., Fujiwara, T., Hatazawa, J., and Matsuzawa, T.,** Effects of smoking on regional cerebral blood flow in neurologically normal subjects, *Stroke,* 14, 720, 1983.

10 Effects of Smoked Marijuana on Human Performance: A Critical Review

INTRODUCTION

Marijuana is the most widely used illicit drug in this country, with an estimated 12 million current users.[1] Marijuana (and other dosage forms of cannabis) have been used as intoxicants throughout the world for thousands of years.[2] One of the greatest concerns to society regarding marijuana is the behavioral toxicity produced by the drug. As is the case with alcohol, marijuana intoxication is often associated with impairments in sensory, psychomotor, and cognitive functions. The costs to society of marijuana-induced behavioral impairment are difficult to estimate, but are likely to be high, given the number of users of the drug and the activities in which some users engage while under the influence of the drug (e.g., attending school, driving an automobile, and working in a factory).

While studying the published literature on the effects of marijuana on psychomotor/cognitive performance, we noted a substantial number of inconsistencies as to which types of behavior are most affected by marijuana. Previous reviews of the subject have not characterized these inconsistencies or critically examined factors that might account for them.[3-7] The present review was designed with several goals in mind: (1) to provide an up-to-date summary of and reference source for the effects of smoked marijuana on human performance; (2) to determine the relative sensitivity of different behavioral tasks to marijuana. Besides being of practical use to researchers, such an analysis might provide insights into the neurochemical or neurological mechanisms by which the drug acts;[8,9] (3) to identify factors that might predict the nature and magnitude of the behavioral effects of marijuana, particularly those methodological factors that might account for discrepant findings in the published literature; and (4) to identify neglected areas worthy of further study and to recommend methods to avoid the shortcomings of past studies.

METHOD

We attempted to identify as many published studies as possible that met the following criteria:

1. Acute administration of marijuana, hashish, or Δ^9-tetrahydrocannabinol (THC) to humans by smoking
2. Post-smoking measurement of simple motor, psychomotor, or cognitive function
3. Use of a placebo control condition
4. Statistical analysis of results

The survey method included searching computer databases (*Medline, Psychological Abstracts*) and annotated bibliographies, and cross-checking references from published studies and review articles. Studies designed to assess possible long-term behavioral effects of chronic marijuana use were not included. Studies were limited to those that involved marijuana administered by smoking since this is the route of administration most commonly used in this and most other countries. Studies that exclusively examined the effects of marijuana smoking on simple sensory or oculomotor functions, psychophysiological parameters, or social behavior were not included. The sample is biased towards studies published in peer-reviewed journals; excluded were studies not published in English, studies published as technical reports or conference proceedings, and abstracts.

We organized the studies in tabular form according to the nature of the behavioral task employed, or the type of basic skill required to perform the task (Tables 1 through 23). In some cases, classification was difficult (and somewhat arbitrary) due to the fact that certain tasks require more than one type of skill. The variables chosen for inclusion in the tables were those that were available from most of the studies and that could reasonably be expected to influence the probability of detecting or the magnitude of behavioral effects of marijuana. These variables are described here in more detail.

Sample Size (n)

For within-subject designs, this number represents the number of subjects who completed the study and for whom data are presented. For group designs, the value is the number of subjects comprising both the placebo group and the high-dose marijuana group.

Marijuana Use (Use)

These codes represent the best estimate that could be obtained of the average frequency of marijuana use of the sample. The classification scheme was as follows: N, never users; O, occasional users (less than once a month); R, regular users (more than once a month but less than 5 d/week); and H, heavy users (5 or more d/week). Studies that stated only that subjects were experienced marijuana users were classified as ">N". Studies that compared two groups with different marijuana use patterns were designated with both codes separated with a slash (e.g., N/R). If a group average could not be determined from the information given, the range of the group is shown (e.g., O-R).

Time Course (Time)

This value represents the longest period of time [in minutes (min) or hours (h)] after smoking at which behavioral effects were determined. In many cases, an exact figure was not available, in which case the value is simply listed as being less than the session duration. Studies that measured effects at more than one time point after smoking are designated with a capital M (for multiple).

Heart Rate (HR)

The tabulated value is the peak heart rate increase (in beats per min) after smoking active, relative to placebo marijuana. In most cases, this was calculated by taking (or estimating from values in figures) the post-smoking minus pre-smoking HR for active marijuana and subtracting from this the post-smoking minus pre-smoking HR for placebo. When no baseline HR was provided, it was estimated to be 70 bpm.

Subjective "High" Ratings (High)

Methods used to measure the magnitude of subjective intoxication from marijuana varied across studies, but the most common were ratings of "high" or "potency" on a 0 to 10 or 0 to 100 scale. When feasible, all scales were converted to a 0 to 100 scale. The tabulated value is the peak high rating after smoking active marijuana minus the rating obtained at the same time point after smoking placebo. As with heart rate, high values were corrected for baseline (pre-smoking) values when these were provided.

Magnitude of Effect (Effect)

For the sake of simplicity, only effects of the highest dose of marijuana tested in each study are presented. (Most reports studied the effects of only a single dose of active marijuana.) Effects that were not significant at the 0.05 level were coded as 0. Significant effects were rated from + to ++++ as follows: + ≤10% effect; ++ = 11 to 50% effect; +++ = 51 to 100% effect; and ++++ >100% effect. Significant effects of unstated magnitude were coded with a question mark (?). Magnitude of effect was calculated from values in tables or text or estimated from data presented in figures. Effects were calculated as percent change from baseline after active marijuana minus percent change from baseline after placebo. For example, if group mean reaction times before and after smoking placebo marijuana were 350 and 345 ms, respectively, and the values before and after active marijuana were 355 and 390 ms, then the effect would be calculated as $100(390 - 355)/355 - 100(345 - 350)/350 = 9.9 - (-1.4) = 11.3\%$, which would be coded as ++. In cases when baseline values were not measured or reported, effect was simply calculated as the value after active marijuana expressed as a percent of the placebo value. For example, if the mean number of words recalled was 9.0 after placebo and 7.5 after active marijuana, then the effect was calculated as $100(7.5 - 9.0)/9.0 = -16.7\%$, coded as ++. The direction of change is not specified in the tables; unless otherwise stated, all effects were in the direction indicative of performance deficits. We recognize that this method of calculating magnitude of drug effect is somewhat arbitrary and that alternative methods might be equally valid. One should also bear in mind that the magnitude of a drug effect as reported here can be highly dependent upon the nature of the dependent variable. Behaviors that occurred with a low frequency under baseline conditions (e.g., intrusions during free recall tasks or false alarm responses during vigilance tasks) often showed large percentage increases after marijuana by this method of calculation. In contrast, variables like reaction time often showed only small percentage increases after active marijuana. One should thus be cautious in interpreting the tabulated effect sizes in terms of practical or clinical significance.

Comments

This column provides additional details about the nature of the tasks used and the particular dependent measures analyzed.

Notably absent as a factor in the tables is any direct indicator of the dose of marijuana administered. This is because it is virtually impossible to estimate the amount of Δ^9-THC actually absorbed when marijuana is smoked. Therefore, rather than list such factors as the Δ^9-THC content of the cigarettes or the number of puffs taken or cigarettes smoked, we decided to include the two most commonly used indirect measures of Δ^9-THC absorption, heart rate and subjective high ratings.

RESULTS

Characteristics of Selected Studies

We found 73 publications that met our selection criteria. A few of these included more than one study, so data were available for a total of 81 studies. General characteristics of these studies are described in this section.

Year of Publication

As Figure 1 indicates, the bulk of the studies (58%) were published within a fairly narrow period of time, from 1972 to 1977.

Sample Size and Experimental Design

The number of subjects studied ranged from 4 to 80, with the most common sample size being 12 (18 studies). Most of the studies used within-subject (crossover) designs, in which each subject was exposed to both a placebo and active drug condition on different occasions. A quarter of the studies, primarily those with the larger sample sizes, used between-subject (group) designs.

Subject Characteristics

Most studies employed exclusively young males. In fact, only 12 of the 81 studies used any female subjects. In terms of drug use history, most studies (56%) used subjects who were regular marijuana users, as defined above. Five studies used only subjects with no reported history of marijuana use, and three used heavy marijuana smokers. Six studies were designed specifically to compare non-users with regular users. Most of the remaining

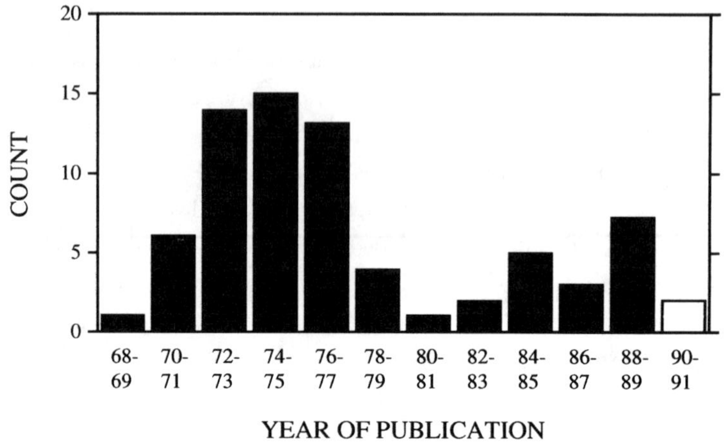

FIGURE 1. Number of selected studies by year of publication. The open bar at 90-91 reflects the fact that data for that time period are incomplete.

studies did not clearly specify the marijuana use histories of their subjects, with 13 stating only that subjects had some history of marijuana use.

Time Course of Drug Effect

Most of the studies did not obtain measurements beyond 90 min after marijuana smoking. Only 12 studies measured behavior beyond 2 h after smoking. Only 13 studies reported results at more than 1 time period post-smoking. For many studies, it was not possible to determine from the described methods exactly when behavioral observations were obtained after smoking.

Heart Rate

Forty-nine studies (60%) presented data on the heart rate-increasing effect of marijuana. The median heart rate increase, relative to placebo, was 25 bpm (range 8 to 43).

Subjective High Ratings

Forty-seven studies (58%) presented data on subjective high that could reasonably be converted to a 0 to 100 scale. The median high rating, relative to placebo, was 42 (range 4 to 71).

Relationship between Heart Rate and Subjective High Rating

Since Δ^9-THC is assumed to be the component of marijuana responsible for both the cardiovascular and subjective effects of the drug, one would predict that there would be a positive correlation between post-smoking high ratings and heart rate increase. Demonstration of such a correlation across the studies reviewed here would further support our decision to use these two measures as indicators of marijuana dose. Figure 2 shows a scatterplot of these two variables from the 36 studies for which data for both measures were available. There was a positive correlation between heart rate increase and increase in high ratings after marijuana smoking, but the correlation was weak and nonsignificant (Pearson r = 0.28). Additional analysis indicated that three studies substantially reduced the strength of the correlation (closed points in Figure 2). Exclusion of these three studies increased the correlation substantially ($r = 0.46$, $p < 0.01$). Fitting a regression line through these 33 points yielded a y-intercept of 15 and a slope of 0.26. This finding suggests that at low levels of smoke exposure heart rate is a more sensitive indicator of Δ^9-THC absorption than subjective high.

This analysis probably underestimates the true correlation between heart rate and subjective response to marijuana. Reported heart rate measurements and ratings of subjective high were not necessarily obtained at the time of peak effect for each measure. In addition, methods of defining and measuring subjective high varied considerably across studies, and some imprecision was necessarily introduced in attempting to convert some of the rating scales to a 0 to 100 standard scale. Finally, most of the studies administered doses of marijuana designed to produce a moderate level of intoxication, which resulted in a clumping of studies around the intermediate level of high ratings (Figure 2). Such a limited range of values tends to lead to underestimation of correlation coefficients.[10]

Summary of Behavioral Effects

Gross and Simple Motor Effects (Tables 1 to 3)

Body Sway. Marijuana consistently produced moderate increases in body sway (standing steadiness) as measured on a "wobble board" (Table 1). The generality of this finding is somewhat limited, however, due to the fact that each of the studies listed in Table 1 was conducted by the same group of investigators.

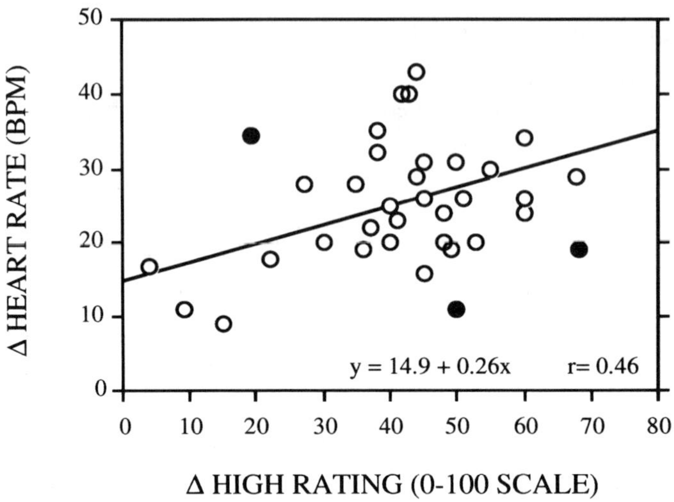

FIGURE 2. Scatterplot of change in heart rate after active marijuana vs. change in ratings of high after active marijuana, for the studies that presented results for both variables. Values were calculated as described in the Methods section. Filled points show outliers that were excluded from calculation of the plotted regression line.

Hand Steadiness. Marijuana smoking also consistently resulted in increased hand tremor (Table 2).

Tapping Speed. Marijuana did not generally affect finger or toe tapping speed (Table 3); only one study reported an effect of unspecified magnitude on finger tapping speed.

Psychomotor Behavior (Tables 4 to 11)

Simple Reaction Time. Marijuana appears to produce, at most, weak and inconsistent effects on simple reaction time (Table 4).

Rotary Pursuit. Rotary pursuit (also referred to as pursuit rotor, pursuit meter) measures the ability of subjects to track a rotating target with a handheld stylus. Some eight of ten studies reported rotary pursuit performance to be significantly impaired by marijuana (Table 5). The two studies that found no effect used regular smokers, whereas the other studies tended to employ subjects with less frequent marijuana use; this suggests that regular use may induce tolerance to the effect of marijuana on this psychomotor task.

Other Hand-Eye Coordination Tasks. A wide variety of other tasks have been employed to test the effects of marijuana on hand-eye coordina-

TABLE 1
Body Sway

n	Use	Time	HR	High	Effect	Ref.
15	>N	<90 min-M	24	48	++	52
15	N->N	<90 min	23	—	+++	53
11	>N	70 min	19	49	++	54
8	>N	25 min-M	—	—	++	55
12	>N	50 min	21	—	++	56

TABLE 2
Hand Steadiness

n	Use	Time	HR	High	Effect	Ref.
11	>N	70 min	19	49	+++	54
32	N/R	40 min	—	—	++	25
40	>N	<2 h	—	—	?	57

TABLE 3
Tapping Speed

n	Use	Time	HR	High	Effect	Comments	Ref.
42	O-R	<2 h	—	49	0	Finger	22
11	>N	70 min	19	49	0	?	54
32	N/R	40 min	—	—	0	Finger, toe	25
12	>N	50 min	21	—	0	Finger	56
40	>N	<2 h	—	—	0	Toe	57
					?	Finger	

TABLE 4
Simple Reaction Time

n	Use	Time	HR	High	Effect	Comments	Ref.
5	R	<90 min	31	—	+	Visual	32
10	>N	<90 min	25	—	+	Visual, auditory	58
12	R	30 min	—	—		Auditory	59
					0	Mean	
					?	Variability	
11	>N	70 min	19	49	0	?	54

TABLE 5
Rotary Pursuit

n	Use	Time	HR	High	Effect	Ref.
15	>N	<90 min-M	24	48	++	52
11	>N	70 min	19	49	+++	54
12	N	1 h	25	—	+++	60
8	N-O	<1 h	32	—	++++	61
15	N->N	<90 min	23	—	++	53
12	>N	50 min	21	—	++	56
8	>N	25 min-M	—	—	++	55
9	N	2 h-M	8	—	?	62
20	R	50 min	—	—	0	63
40	R	?	—	39	0	64

TABLE 6
Other Eye-Hand Coordination Tasks

n	Use	Time	HR	High	Effect	Comments	Ref.
15	>N	<90 min-M	24	48	+	Pegboard	52
11	>N	70 min	19	49	++	Pegboard	54
12	>N	50 min	21	—	++	Pegboard	56
20	R	50 min	—	—	0	Pegboard	63
13	H	<1 h	29	44	+	Card sorting	65
9	R	<4 h	—	—	0	Bingo-chip or word sorting	66
32	N/R	40 min	—	—	++	Vert groove	25
					+++	Horiz groove	
40	>N	<2 h	—	—	?	Vert groove, grooved peg-board, trail-making	57
12	N-O	30 min	—	—	0	Trail-making	67
22	R	<4 h-M	—	—	++	Tracking	68
10	R	35 min	23	41	0	Tracking	69
6	R	<5 h-M	19	36	0	Tracking, circular lights	39
4	>N	<4 h-M	—	42	?	Circular lights	70

tion. As Table 6 indicates, results have been inconsistent, and we were not able to identify any obvious factors that might account for the discrepant findings for those studies that used similar tasks.

Digit Symbol Substitution Test (DSST). The DSST, a component of the Wechsler Adult Intelligence Scale, is a speed-based task of associative

TABLE 7
Digit Symbol Substitution Test (DSST)

n	Use	Time	HR	High	Effect	Ref.
80	R	<3 h	28	27	++	11
25	O	<3 h	31	45	++	11
40	N	<2 h	—	32	++	71
8	R	<1 h	—	65	++	72
40	R	?	—	39	++	64
9	R	<4 h	—	—	++	66
12	R	65 min	35	38	+	17
6	R	<5 h-M	19	36	+	39
5	R	<90 min	31	—	+	32
48	N/R	<1 h	—	—	?	26
9	N	2 h-M	8	—	?	62
13	H	<1 h	29	44	0	65
10	H	?	11	9	0	34
25	H	<3 h	17	4	0	11

ability which requires subjects to copy symbols which correspond to particular digits. All but three studies showed significant effects of marijuana on the DSST, whether the task was administered in the standard pencil-and-paper form or in modified versions designed for microcomputer administration (Table 7). Interestingly, the three studies reporting no effects on this task all used heavy marijuana smokers, suggesting that these subjects were tolerant to the effects of the drug on this task.

Complex or Choice Reaction Time. These tasks (Table 8) involve a decision component in that subjects are required to respond differentially, depending upon which of two or more stimuli is presented. It is difficult to summarize the effects of marijuana on these tasks due to substantial differences between studies in the nature of the tasks and the method of data analysis and presentation. For example, some studies report effects only on mean reaction time, whereas others report effects on reaction time variability and error rates. It does appear that marijuana affects mean reaction time on a number of these tasks without influencing accuracy of response. Most of these studies used regular marijuana users, but the study by Jones showed effects on reaction time in occasional users but not heavy users.[11]

Divided Attention. As was the case for choice reaction time tasks, the methods used to study subjects' ability to perform more than one type of task at a time varied widely from study to study (Table 9). These tasks yield several possible dependent variables, and signal detection analysis was

TABLE 8
Complex or Choice Reaction Time

n	Use	Time	HR	High	Effect	Comments[a]	Ref.
5	R	<90 min	31	—	+	Visual	32
12	O-R	<1 h	—	—	++	2nd stimulus (visual)	73
					+	1st stimulus (auditory)	
48	N/R	<1 h	—	—		Visual	26
					++	Variability	
					+	Mean	
24	R	2.5 h	—	—	+	Words or faces (tachist)	74
24	R	<1 h	—	—		Letter matching	75
					++	RT	
					0	Errors	
7	R	40 min	24	60		Visual, form and color matching (no practice w/o drug)	76
					++	RT	
					0	Errors	
7	R	40 min	34	60		Visual, form and color matching (much practice w/o drug)	76
					0	RT, errors	
12	R	<90 min	40	43	+	Visual, oddity discrim.	77
57	>N	45 min	35	—	++++	Auditory discrim., errors	78
36	R	>2 h	—	14		Word categories	79
					++	RT	
					0	Errors	
25	O	<3 h	31	45	?	?	11
25	H	<3 h	17	4	0	?	11
12	R	<1 h	40	42	0	Visual	16
40	R	>2 h	—	13		Word categories	79
					0	RT, errors	

[a] RT, reaction time.

TABLE 9
Divided Attention

n	Use	Time	HR	High	Effect	Comments	Ref.
8	R	<20 min	18	22		Visual/visual	38
					+++	FAs	
					0	RT	
8	N/R	<2 h	—	—		Visual/visual	27
						Main signal	
					++++	Misses	
					0	FAs	
						Secondary signal	
					++++	Misses, FAs	
12	N/R	45 min	—	—		Visual/visual	28
						Main signal	
					?	Misses	
					0	FAs	
					0	RT	
						Secondary signal	
					?	Misses	
					0	FAs	
20	N/R	70 min	—	—		Visual/visual	29
					++++	Misses	
6	R	8 h-M	43	44		Visual/visual	80
					+	Accuracy, RT	
11	>N	70 min	19	49		Visual/rotary pursuit	54
					++	RT	
12	>N	50 min	21	—		Visual/rotary pursuit	56
					0	RT	
23	R	50 min	26	60		Auditory/digit recall	33
					++++	FAs	
					++	Accuracy	
					0	Digit recall	
12	>N	<1 h	15	—		Visual/visual	81
					++++	Misses (secondary stimulus)	
					+++	Hits (primary stimulus)	
					0	RT	

TABLE 9 (continued)
Divided Attention

n	Use	Time	HR	High	Effect	Comments	Ref.
23	O-R	<70 min	21	—		Visual/simulated driving	82
					++	RT	
					0	Within-subject variability	
12	R	65 min-M	35	38	0	Visual/tracking	17

[a] The two types of tasks used are shown in the first line of Comments, separated by a slash (/). FAs, false alarms.

frequently used to characterize the effects of the drug.[12] The particular dependent variables affected varied across studies, but marijuana frequently produced substantial increases in measures such as misses (failure to respond to a target stimulus) and false alarms (responding to a non-target stimulus). Effects of marijuana on mean reaction time were not common with most of these divided attention tasks.

Sustained Attention. Sustained attention (also referred to as vigilance) refers to the ability to maintain concentration over an extended period of time, particularly on a task that is relatively simple and boring. The studies presented in Table 10 used such tasks with a minimum duration of 5 min. Three of the studies showed significant effects of marijuana. Three studies that showed no drug effects used versions of the continuous performance test (CPT), a widely-used diagnostic test for children with attention deficit disorder.[13,14] One methodological factor that might be relevant here is task duration (time-on-task) — the studies that showed drug effects used task durations of 50 min or more, whereas the studies using the CPT employed much shorter durations (<10 min). Therefore, it cannot be determined whether the lack of marijuana effect on the CPT is due to the nature of the task itself or to the short time periods for which it was used.

Driving, Flying. It is not surprising that marijuana can impair the ability to drive a car or fly an airplane, since the skills required for these tasks are a composite of many of the psychomotor skills shown to be affected by marijuana in the previous tables. Most studies listed in Table 11 showed some impairment of performance after marijuana smoking, although impairment was not observed in all individuals, as noted by Klonoff.[15]

TABLE 10
Sustained Attention

n	Use	Time	HR	High	Effect	Comments[a]	Ref.
12	O-R	<1 h-M	16	—		Visual SD (60 min)	83
					++++	Errors (high demand)	
					+++	Errors (low demand)	
					0	FAs	
23	R	50 min	26	60		Auditory SD (50 min)	33
					++++	FAs	
					++	Accuracy	
12	>N	<1 h	15	—		Visual SD (60 min)	81
					++++	Misses	
					0	RT	
9	R	<4 h	—	—	0	Visual SD (variable)	66
9	N	2 h-M	8	—	0	CPT (5 min)	62
8	R	<1 h	—	65	0	CPT (7 min)	72
48	N/R	<1 h	—	—	0	CPT (? min)	26

[a] SD, signal detection; CPT, continuous performance test. Time in parentheses indicates duration of task.

Memory and Learning (Tables 12 to 17)

Remote Memory. Remote memory refers to the ability to retrieve already-learned material. It was most commonly measured by having subjects generate as many words as possible within a specified time period that met some established criterion (e.g., nouns beginning with the letter "B"). As shown in Table 12, studies are consistent in showing that marijuana has no significant effect on this memory task.

Digit Span. The digit span task is a measure of short-term memory in which subjects are required to reproduce increasingly longer strings of digits, either in the order presented (forward) or in the reverse order presented (backward). As Table 13 indicates, the effects of marijuana on this task have been inconsistent. There are no clear differences between the positive and negative studies that might explain the inconsistency, although it is interesting to note that the two studies reporting the strongest effects are also the two that reported strong effects of marijuana on both heart rate and

TABLE 11
Driving, Flying

n	Use	Time	HR	High	Effect	Comments	Ref.
10	R	6 h-M	34	—		Simulated flying	84
					++++	Major, minor errors; course, altitude deviations	
					+++	Heading deviations	
18	R	48 h-M	—	—		Simulated flying	85
					+	Composite measure	
16	R	3 h-M	31	50		Real driving	68
					++	Errors	
					+	Time	
43	>N	<1 h	22	37	++	Real driving	15
9	R	<1 h	29	68	0	Real driving	86
23	O-R	<70 min	21	—	0	Simulated driving	82

TABLE 12
Remote Memory

n	Use	Time	HR	High	Effect	Ref.
48	R	<2 h	—	23	0	87
12	R	<1 h	9	15	0	35
42	O-R	<2 h	—	49	0	22
40	R	1 h	20	53	0	88
10	>N	<2 h	—	—	0	89

TABLE 13
Digit Span

n	Use	Time	HR	High	Effect	Comments	Ref.
8	R	<20 min	18	22	+	Forward	38
12	R	<1 h	40	42	++	Forward, backward	16
12	R	65 min	35	38	++	Backward	17
					0	Forward	
10	R	<1 h	20	30	0	Backward	90
12	R	<1 h	9	15	0	Forward, backward	35
18	N/R	<1 h	—	—	0	Forward, backward	30
10	R	35 min	23	41	0	Backward	69

subjective high,[16,17] suggesting that dose might have played an important role.

Free Recall. Most free recall tasks involved presenting a list of words to subjects after smoking and having them write down as many words as they could remember, either immediately after presentation or after a period of delay. A few studies used pictorial stimuli rather than words. Nearly all studies showed a consistent decrement (on the order of 10 to 50%) in the number of words or pictures recalled (Table 14). Most studies reporting data on memory intrusions (stimuli listed by subjects that were not actually presented) found that marijuana produced substantial increases in intrusions.

Recognition Memory. Tasks assessing recognition memory require subjects to discriminate between stimuli previously presented to them and similar stimuli not previously presented. As shown in Table 15, effects of marijuana on recognition memory performance have been inconsistent. With the exception of the Klonoff study,[5] all used signal detection analysis to evaluate the effects of the drug; none of these studies found effects of marijuana on hits, but two found significant increases in false alarm rates.

Prose Recall. All studies found some effect of marijuana on memory for prose (narrative) material (Table 16), although the study by Hooker and Jones[35] found significant effects only after a period of delay after smoking. Marijuana seemed to have a greater effect on certain specific types of recall errors (such as out-of-sequence distortions) than on simpler measures such as the number of factual elements recalled.

Paired-Associate Learning. For this task, subjects are first presented with a list of pairs of words. They are subsequently presented with one member of the pair and are asked to recall the associated word. We found only three studies which examined the effects of marijuana on paired-associate learning (Table 17), and none found significant drug effects.

Other Behavioral Effects (Tables 18 to 23)

Time Perception. Most studies on the effects of marijuana on time perception have used one of two methods — time estimation or time production.[18] In time estimation tasks, the experimenter generates a time interval of a certain duration, and the subject is then asked to estimate the duration of the interval. In time production tasks, the subject is asked to generate an interval of a duration specified by the experimenter. All but two of the studies shown in Table 18 found significant effects of marijuana on time perception. Significant and stronger effects tended to be associated

TABLE 14
Free Recall

n	Use	Time	HR	High	Effect	Comments[a]	Ref.
10	>N	<90 min	25	—	++	4	58
13	H	<1 h	29	44	++	12	65
12	R	1 h	20	40	++	30	91
16	R	90 min	28	35	++	40, immed, delayed	92
					++++	Intrusions	
40	R	1 h	20	53	++	14	88
12	R	<2 h	—	71	++	24	93
10	R	<1 h	20	30	++	80	90
					?	Intrusions	90
34	R	100 min	32	38	++	15	94
					?	Intrusions	
28	R	<90 min	26	45	++	40, words and pictures	95
8	R	30 min	11	50	++	20	96
22	R	<1 h	26	51	++	16, geometric figures	97
					++++	Intrusions	
40	R	45 min	25	40	++	35	98
					0	Intrusions	
16	R	30 min	16	45	++	16	99
12	R	<90 min	34	19	++	15, immed, delayed	100
					+++	Intrusions	
10	>N	<2 h	—	—	++	10	89
					0	80, delayed	
					++++	Intrusions, delayed	
40	>N	<2 h	—	—	?	6, delayed	57
12	R	<1 h	9	15	0	5	35
10	R	35 min	23	41	0	20	69

[a] The number in the "Comments" column shows the number of words presented to subjects in each trial.

with longer intervals and multiple trials. The nature of the drug effect was consistent across studies, with subjects overestimating the amount of elapsed time with the estimation method, and producing shorter than requested intervals with the production method (underproduction). Both of these results are consistent with subjects experiencing time as passing more

TABLE 15
Recognition Memory

n	Use	Time	HR	High	Effect	Comments	Ref.
40	>N	<2 h	—	—	?	Pictures	57
34	R	100 min	32	38	+++	FAs	94
					0	Hits	
10	>N	<2 h	—	—	++	FAs, correct rejections	89
					0	Hits, misses	
12	R	<90 min	34	19	0		100
16	R	<1 h	20	48	0		101
16	R	90 min	28	35	0		92

TABLE 16
Prose Recall

n	Use	Time	HR	High	Effect	Comments[a]	Ref.
12	R	<1 h	9	15	++++	Delayed, intrusions	35
					+++	Delayed, omissions	
					0	Immediate	
40	R	<1 h	22	—	++	Immediate and delayed	102
12	N-O	30 min	—	—	++++	Immediate, O-O-S distortions	67
					0	Immediate, other measures	
12	N-O	<1 h	—	—	++++	Immediate, major distortions, O-O-S distortions	103
					++	Immediate, memory units	
					0	Immediate, minor distortions, fragmentary memories	
40	>N	<2 h	—	—	?	Immediate	57

[a] O-O-S, out-of-sequence distortions.

TABLE 17
Paired-Associate Learning

n	Use	Time	HR	High	Effect	Ref.
48	R	<90 min	—	23	0	104
12	R	<1 h	9	15	0	35
40	N	<2 h	—	32	0	71

TABLE 18
Time Perception

n	Use	Time	HR	High	Effect	Comments	Ref.
12	N	<20 min	30	55	++	Production (60 s)	105
13	N	<20 min	19	68	++	Production (60 s)	105
4	R	120 min-M	—	—	++	Production (5–45 s)	19
6	>N	80 min-M	—	—	++	Production (30–120 s)	19
8	R	<1 h	—	65	+	Production (10 s)	72
13	H	<1 h	29	44	+	Production (30 s)	65
					0	Production (10 s)	
10	H	?	11	9	0	Production (15 s)	34
					+	Estimation (15 s)	
5	R	<90 min	31	—	?	Production (10, 30 s)	32
					?	Estimation (15, 30, 45 s)	
10	R	<1 h	20	30	+	Estimation (50–150 s)	90
12	>N	30 min	25	40	++	DRL operant responding	106
40	R	?	—	39	0	Production (15 s)	64
					0	Estimation (15 s)	
10	>N	<90 min	25	—	0	Reproduction (1–5 s)	58

quickly relative to clock time while under the influence of marijuana; that is, marijuana increases the subjective time rate.[19]

Stroop Color and Word Test. This task measures the ability to inhibit an overlearned response in order to maintain a new conceptual set.[20] The studies conducted to date (Table 19) have produced mixed results. A number of the studies (all from the same laboratory) measured Stroop performance under conditions of delayed auditory feedback, a stress-inducing

TABLE 19
Stroop Test

n	Use	Time	HR	High	Effect	Comments[a]	Ref.
15	N->N	<90 min	23	—	+++	DAF	53
40	N	<2 h	—	32	++		71
40	R	?	—	39	++		64
12	R	<1 h	9	15	+		35
12	N-O	<1 h	—	—	0		103
12	N-O	30 min	—	—	0		67
11	>N	70 min	19	49	0	DAF	54
12	N	1 h	25	—	0	DAF	60
8	>N	25 min-M	—	—	0	DAF	55
12	>N	50 min	21	—	0	DAF	56
8	N-O	<1 h	32	—	0	DAF	61

[a] DAF, delayed auditory feedback.

TABLE 20
Embedded Figures

n	Use	Time	HR	High	Effect	Ref.
12	R	<2 h	—	71	+++	93
40	R	?	—	39	+++	64
40	N	<2 h	—	32	++	71

procedure. All but one of these studies found no effect of marijuana. Of the other three studies that did report significant effects of marijuana on Stroop performance, two employed substantially larger sample sizes than the other studies. No other factors that might explain the discrepant findings were apparent. Different studies used different versions of this task, and the studies also varied in the specific measures of performance presented.

Embedded Figures. This task measures the ability of subjects to find a simple geometric figure embedded within a more complex design. It is considered to be a measure of field dependence.[21] All three studies using this task showed marijuana to exert significant effects (Table 20).

Mental Arithmetic. As Table 21 indicates, a wide variety of tests of mental arithmetic have been conducted with marijuana, and the findings have been inconsistent. Most of the studies from the laboratory that used these tasks under conditions of delayed auditory feedback found significant

TABLE 21
Mental Arithmetic

n	Use	Time	HR	High	Effect	Comments	Ref.
18	N/R	<1 h	—	—	+++	Complex addition and subtraction (GDSA)	30
					++	Serial sevens	
48	N/R	<1 h	—	—	+++	GDSA	26
					++	Serial sevens	
15	>N	<90 min-M	24	48	+++	Back count (DAF)	52
12	>N	50 min	21	—	+++	Back count (DAF)	56
					0	Addition, subtraction (DAF)	
12	N	1 h	25	—	+++	Back count, progressive count, addition, subtraction (DAF)	60
					++	Addition+7 (DAF)	
					0	Subtraction+7 (DAF)	
8	N-O	<1 h	32	—	+++	Progressive count, subtraction (DAF)	61
					++	Addition+7 (DAF)	
					0	Back count, addition, subtraction+7 (DAF)	
15	N->N	<90 min	23	—	+++	Back count, progressive count (DAF)	53
					0	Addition, subtraction, addition+7, subtraction+7 (DAF)	
11	>N	70 min	19	49	0	Back count, addition, subtraction (DAF)	54
8	>N	25 min-M	—	—	0	Addition, progressive count (DAF)	55
12	R	<1 h	9	15	0	Serial addition	35
10	R	<1 h	20	30	0	Arithmetic problems	90

TABLE 22
Verbal Output

n	Use	Time	HR	High	Effect	Comments	Ref.
12	N	1 h	25	—	++	Verbal, reverse verbal (DAF)	60
8	N-O	<1 h	32	—	++	Verbal, reverse verbal (DAF)	61
12	>N	50 min	21	—	++	Verbal (DAF)	56
15	N->N	<90 min	23	—	++	Verbal (DAF)	53
					0	Reverse verbal (DAF)	
11	>N	70 min	19	49	0	Verbal (DAF)	54

drug effects, but the particular task affected varied from study to study, making interpretation difficult.

Verbal Output. All the studies listed in Table 22 had subjects read aloud an unfamiliar and incomprehensible passage of text either normally or backwards (reverse), under delayed auditory feedback conditions. All but one of these studies found that marijuana produced moderate impairment of subjects' performance.

Tests of "Creativity". Several studies have attempted to confirm in the laboratory anecdotal reports that marijuana can promote creative thought processes and enhance artistic creativity. Various tests have been developed purporting to measure processes such as "associational fluency", "conceptual flexibility", and "original thinking", which might be related to creativity.[22,23] The studies conducted to date have not found consistent effects of marijuana on these types of tasks (Table 23).

CONCLUSIONS

Several general conclusions can be drawn from the foregoing analysis.

1. At doses that produce typical, moderate levels of intoxication, marijuana can affect a wide range of learned and unlearned behaviors, including simple motor tasks, more complex psychomotor and cognitive tasks, and the ability to accurately estimate the passage of time.

TABLE 23
Tests of Creativity

n	Use	Time	HR	High	Effect	Comments	Ref.
5	R	<90 min	31	—		Free association	32
					+	Speed	
					?	Uniqueness	
40	N	<2 h	—	32	++	Associations	71
					0	Alternate uses	
40	R	?	—	39	++	Associations	64
					++	Alternate uses	
48	R	<2 h	—	23	++	Word generation (uniqueness)	87
12	R	<1 h	40	42	0	Free association (uniqueness)	16
42	O-R	<2 h	—	49	0	Alternate uses, other creativity tasks	22

2. Some behavioral tasks are more sensitive to disruption by marijuana than others. For example, within the memory domain, free recall tasks are more consistently affected than recognition or remote memory tasks.
3. For many behavioral tasks, the effects of marijuana varied considerably from study to study.

With regard to the last point, our attempt to identify factors that might account for discrepant results met with rather limited success. The most consistent finding was evidence that heavy marijuana smokers were relatively insensitive to marijuana-induced impairment on certain psychomotor tasks (rotary pursuit, DSST, or choice reaction time). This suggests the development of tolerance to the behavioral effects of marijuana, a phenomenon that has frequently been observed in both laboratory animals and humans.[24] On the other hand, those studies that specifically compared experienced with naive marijuana users (i.e., never users) found no consistent evidence that naive users were more sensitive to behavioral disruption after smoking marijuana.[25-30]

Dose undoubtedly plays an important role in determining the nature and magnitude of behavioral impairment caused by marijuana, as demonstrated by several studies that varied dose over a wide enough range.[31-33] At a high enough dose, CNS function would probably be affected to the extent that performance on virtually any laboratory behavioral task would be disrupted. (Such a dose might well be so aversive, however, that it would rarely be self-administered intentionally by users in the natural environment.) The importance of dose was not clear in the present analysis, however, probably because we examined only the highest dose tested in each study and because, in many cases, there was a limited range of values of our indirect measures of dose (heart rate and high ratings) across studies for a particular behavioral task. Nevertheless, it is probably not a coincidence that the three studies that showed the least evidence of effective dosing — the three points lying furthest to the left in Figure 2 — also showed either no or unusually weak behavioral effects.[11,34,35]

The most likely explanation for the discrepant findings across studies is that they are the result of factors other than those we chose to examine. The factors we chose were those for which data were most commonly presented in the studies we reviewed. Unfortunately, these constituted only a small fraction of the nonpharmacological factors that can influence the observed effects of drugs on behavior (Table 24). The most notable and frustrating aspect we encountered in conducting our analysis was the failure of many of the reviewed studies to provide the most basic methodological information that would be required to replicate the research. Many studies provided virtually no information as to the marijuana use history of the subject population, the instructions or incentives given to subjects regarding their performance, or pertinent facts concerning the exact nature of the behavioral tasks, the extent of practice subjects received, when the tasks were administered relative to drug treatment, or how long the tasks took to complete. On the other hand, nearly all studies provided information on the amount of marijuana burned and the marijuana Δ^9-THC content (often to two decimal places), despite the fact that this information is largely useless for estimating the amount of THC actually delivered to the subject, given the uncertainties and variability inherent in the smoking process.[36] Given the lack of relevant methodological information, as well as the lack of standardization in the application of the behavioral tasks, it is not surprising that we were unable to account for most of the incongruent findings.

TABLE 24
Nonpharmacological Factors That May Influence Behavioral Responses to Drugs

Subject factors
- History of drug use
- Current level of drug use
- Attitude toward drug
- Motivation level for performing behavioral tasks
- Expectations of effects of drug treatment
- Personality

Situational factors
- Physical environment during testing (noise, lighting level)
- Social environment (presence of other subjects or experimenter during testing)
- Time of day

Other methodological factors
- Information provided in consent form or instructions
- Extent of practice of task before study (under or not under influence of drug)
- Time interval between drug administration and testing
- Task duration, number of trials
- Task difficulty
- Monetary or other reinforcement of task performance (feedback)
- Selection of dependent variables
- Method of data analysis

RECOMMENDATIONS

METHODOLOGICAL ISSUES

Provide Sufficient Methodological Detail to Allow Cross-Study Comparisons and Replications

In terms of characterizing the marijuana use history of the subject population, information (range and median) should be provided for both frequency of *current* marijuana use as well as an approximation of number of *lifetime* uses. In this way, total experience with marijuana can be distinguished from physiological tolerance. With regard to the other factors listed in Table 24, as much information should be provided as journal editors will allow.

Collect and Present Standard Measures of Marijuana Effect

In order to document that pharmacologically effective doses of marijuana were administered, as well as to allow cross-study comparisons, all studies of the behavioral effects of marijuana should measure and report heart rate and subjective high. Both of these measures can be obtained quickly and easily, with no special equipment and minimal interference with other measures or activities. Another measure that investigators might wish to consider is expired air carbon monoxide (CO) levels, a noninvasive and immediate indicator of alveolar smoke absorption. Several laboratories have measured CO change after marijuana smoking, and have shown reliable changes as a function of number of puffs administered or marijuana THC content.[37-39]

Studies examining the effects of marijuana on a new task or type of behavior might also consider including a behavioral task that has been shown to be consistently disrupted by marijuana. Based on our analysis, a good candidate would be a simple free recall task. This task can be conducted quickly, requires no special apparatus (flash cards can be used), and is relatively unconfounded by practice effects. Use of such a task as a positive control would provide a basis for assessing the relative sensitivity of the new task to disruption by marijuana.

Fully and Properly Analyze Data Collected

Given all the time, effort, expense, and risks involved in conducting human drug research, it is lamentable for drug effects to be missed, obscured, or misinterpreted because of faulty or inadequate data collection, presentation, and analysis. There were many examples of such deficiencies among the studies reviewed here. In general, there was an overreliance on measures of central tendency as opposed to measures of variability (both within- and between-subject) and a failure to display or discuss individual differences in responses to the drug. Another common shortcoming was insufficient examination of relationships among different outcome measures. Perhaps the most common problem was failure to identify all relevant dependent variables that could be derived from the behavioral task employed. As an example, consider simple reaction time tasks. As discussed by Crabtree and Antrim (in 1988) and Dinges and Kribbs (in 1990),[40,41] such tasks are anything but "simple". Dinges and Kribbs point out that at least six conceptually distinct performance parameters can be derived from simple reaction time tasks: (1) optimum response capability (fastest reaction times), (2) response slowing (middle reaction times), (3) lapsing (slowest reaction times), (4) false responding (responses without stimuli), (5) response blocking or gaps (reaction times beyond a predetermined limit), and (6) habituation (regression of reaction times across time during the trial). It is possible that marijuana could affect one or more of these measures without affecting the overall mean. Nearly all of the studies reviewed here that used reaction time tasks presented and analyzed only the mean of raw reaction times. From a public safety standpoint, it may be more appropriate to be concerned with the occurrence of unusually long reaction times than with the overall mean.

The same principle applies to many of the other behavioral tasks that were used. Studies using memory tasks often reported only the number of items correctly recalled and not the number of intrusions. Studies using choice reaction time or divided attention tasks often reported only the number or percentage of correct responses and not the number of false alarm responses. Ideally, when one chooses a behavioral task, one should consider all possible performance measures that could be differentially affected by the drug, and analyze each in an appropriate manner.[42] Techniques such as signal detection analysis can also be useful in providing insights into the mechanism by which the drug influences behavior.[12]

Areas Worthy of Additional Study

Sites and Mechanisms of Action

The recent identification and cloning of a cannabinoid receptor in brain tissue could greatly facilitate research on possible sites and mechanisms underlying specific behavioral effects of marijuana, and also raises the possibility of the development of specific antagonists of THC effects.[43]

Duration of Action

Although there was little evidence from the studies reviewed here that behavioral effects of marijuana smoking last longer than about 4 h, few of the studies measured performance at longer time periods, and some studies that did not meet our criteria for inclusion have presented evidence that behavior might be disrupted for considerably longer periods, perhaps up to 24 h after smoking.[44,45] Such findings need to be replicated, since behavioral impairment that long after drug use could be of considerable public health significance, particularly if users do not believe that their performance is impaired after the subjective high has worn off.

Other Types of Behaviors

There has been little systematic study of the effects of marijuana on important but more complex and subtle behavioral processes such as judgement and decision-making, motivation,[46] and risk-taking.[47] Also useful would be more study of the effects of marijuana on behavior in the natural environment or on "real-life" tasks, since drug effects in the laboratory do not always generalize to the "real" world.[48,49] Finally, more attention should be devoted to the relationship between actual marijuana-induced performance deficits and users' subjective assessment of their ability to perform while under the influence of the drug — an issue likely to be relevant to risk-taking behaviors. Studies have shown that this relationship can vary across drugs from different pharmacological classes.[50,51]

Individual Differences

Few studies have systematically examined individual differences in the effects of marijuana on behavior. It would be useful to identify individual difference characteristics that are associated with marijuana-induced psychomotor and cognitive impairment. It would be particularly interesting to study between-subject differences in the relationship between behavioral impairment and subjective effects. For example, are there people who show little or no performance deficit after marijuana smoking while still experiencing the subjective high, and, if so, how do these individuals differ from others?

ACKNOWLEDGMENTS

Work on this review was supported by National Institute on Drug Abuse grant DA-03517. The authors would like to thank Dr. Harriet de Wit for reading a preliminary version of the manuscript, and John Perry for research assistance.

APPENDIX

Tables in Which Cited Studies Appear

Reference	Citation No.	Tables
Belmore and Miller, 1980	99	14
Block and Wittenborn, 1984a	104	17
Block and Wittenborn, 1984b	79	8
Block and Wittenborn, 1985	87	12, 23
Block and Wittenborn, 1986	75	8
Borg et al., 1975	32	4, 7, 8, 18, 23
Braden et al., 1974	59	4
Cappell and Pliner, 1973	90	13, 14, 18, 21
Cappell et al., 1972	106	18
Carlin et al., 1972	64	5, 7, 18, 19, 20, 23
Carlin et al., 1974	71	7, 17, 19, 20, 23
Carlini et al., 1974	105	18
Casswell, 1975	26	7, 8, 10, 21
Casswell and Marks, 1973a	29	9
Casswell and Marks, 1973b	30	13, 21
Chait et al., 1985	65	6, 7, 14, 18
Chait et al., 1988	38	9, 13
Cone et al., 1986	70	6
Dalton et al., 1975	56	1, 3, 5, 6, 9, 19, 21, 22
Dalton et al., 1976	52	1, 5, 6, 21
Dornbush, 1974	89	12, 14, 15
Dornbush et al., 1971	58	4, 14, 18
Drew et al., 1972	67	6, 16, 19
Evans et al., 1973	55	1, 5, 19, 21
Evans et al., 1976	54	1, 2, 3, 4, 5, 6, 9, 19, 21, 22
Galanter et al., 1973	16	8, 13, 23
Hansteen et al., 1976	68	6, 11
Heishman et al., 1989a	39	6, 7
Heishman et al., 1989b	17	7, 9, 13
Hicks et al., 1984	19	18
Hooker and Jones, 1987	35	12, 13, 14, 16, 17, 19, 21
Janowsky et al., 1976	84	11

Reference	Citation No.	Tables
Jones, 1971	11	7, 8
Jones and Stone, 1970	34	7, 18
Kelly et al., 1990	66	6, 7, 10
Kiplinger et al., 1971	53	1, 5, 19, 21, 22
Klonoff, 1974	15	11
Klonoff et al., 1973	57	2, 3, 6, 14, 15, 16
Leirer et al., 1989	85	11
Low et al., 1973	78	8
MacAvoy and Marks, 1975	27	9
Manno et al., 1970	61	5, 19, 21, 22
Manno et al., 1971	60	5, 19, 21, 22
Marks and MacAvoy, 1989	28	9
Miller and Cornett, 1978	92	14, 15
Miller et al., 1972	103	16, 19
Miller et al., 1976	98	14
Miller et al., 1977a	102	16
Miller et al., 1977b	94	14, 15
Miller et al., 1977c	101	15
Miller et al., 1977d	95	14
Miller et al., 1978a	91	14
Miller et al., 1978b	97	14
Miller et al., 1979	100	14, 15
Milstein et al., 1975	25	2, 3, 6
Moskowitz and McGlothlin, 1974	33	9, 10
Moskowitz et al., 1972	81	9, 10
Moskowitz et al., 1974	73	8
Moskowitz et al., 1976	82	9, 11
Pearl et al., 1973	93	14, 20
Peeke et al., 1976	76	8
Perez-Reyes et al., 1988	80	9
Salvendy and McCabe, 1975	63	5, 6
Schaefer et al., 1977	77	8
Sharma and Moskowitz, 1974	83	10
Stillman et al., 1977	74	8
Sutton, 1983	86	11
Vachon et al., 1974	72	7, 10, 18
Weckowicz et al., 1975	22	3, 12, 23
Weil et al., 1968	62	5, 7, 10
Wetzel et al., 1982	88	12, 14
Zacny and Chait, 1991	69	6, 13, 14
Zacny and de Wit, 1989	96	14

REFERENCES

1. **Goldstein, A. and Kalant, H.,** Drug policy: striking the right balance, *Science*, 249, 1513, 1990.
2. **Abel, E. L.,** *Marijuana, The First Twelve Thousand Years*, Plenum Press, New York, 1980.
3. **Jones, R.,** Human effects, in *Marijuana Research Findings: 1976*, NIDA Research Monograph 14, Petersen, R. C., Ed., U.S. Department of Health and Human Services, Rockville, MD, 1977, 128.
4. **Ferraro, D. P.,** Acute effects of marijuana on human memory and cognition, in *Marijuana Research Findings: 1980*, NIDA Research Monograph 31, Petersen, R. C., Ed., U.S. Department of Health and Human Services, Rockville, MD, 1980, 98.
5. **Klonoff, H.,** Acute psychological effects of marihuana in man, including acute cognitive, psychomotor, and perceptual effects on driving, in *Cannabis and Health Hazards,* Fehr, K. O. and Kalant, H., Eds., Addiction Research Foundation, Toronto, Canada, 1983, 433.
6. **Murray, J. B.,** Marijuana's effects on human cognitive functions, psychomotor functions, and personality, *J. Gen. Psychol.*, 113, 23, 1986.
7. **Moskowitz, H.,** Marihuana and driving, *Accid. Anal. Prev.*, 17, 323, 1985.
8. **Mirsky, A. F. and Kornetsky, C.,** On the dissimilar effects of drugs on the digit symbol substitution and continuous performance tests, *Psychopharmacologia*, 5, 161, 1964.
9. **Miller, L. L. and Branconnier, R. J.,** Cannabis: effects on memory and the cholinergic limbic system, *Psychol. Bull.*, 93, 441, 1983.
10. **Halperin, S.,** Spurious correlations—causes and cures, *Psychoneuroendocrinology,* 11, 3, 1986.
11. **Jones, R. T.,** Marihuana-induced "high": influence of expectation, setting and previous drug experience, *Pharmacol. Rev.*, 23, 359, 1971.
12. **Egan, J. P.,** *Signal Detection Theory and ROC Analysis*, Academic Press, New York, 1975.
13. **Rosvold, H. E., Mirsky, A. F., Sarason, I., Bransome, E. D., and Beck, L. H.,** A continuous performance test of brain damage, *J. Consult. Psychol.*, 20, 343, 1956.
14. **Rapport, M. D., DuPaul, G. J., Stoner, G., and Jones, J. T.,** Comparing classroom and clinic measures of attention deficit disorder: differential, idiosyncratic, and dose-response effects of methylphenidate, *J. Consult. Clin. Psychol.*, 54, 334, 1986.
15. **Klonoff, H.,** Effects of marijuana on driving in a restricted area and on city streets: driving performance and physiological changes, in *Marijuana: Effects on Human Behavior*, Miller, L., Ed., Academic Press, New York, 1974, 359.
16. **Galanter, M., Weingartner, H., Vaughan, T. B., Roth, W. T., and Wyatt, R. J.,** Delta-9-transtetrahydrocannabinol and natural marihuana, *Arch. Gen. Psychiatry*, 28, 278, 1973.
17. **Heishman, S. J., Stitzer, M. L., and Yingling, J. E.,** Effects of tetrahydrocannabinol content on marijuana smoking behavior, subjective reports, and performance, *Pharmacol. Biochem. Behav.*, 34, 173, 1989.
18. **Fraisse, P.,** Perception and estimation of time, *Annu. Rev. Psychol.*, 35, 1, 1984.
19. **Hicks, R. E., Gualtieri, C. T., Mayo, J. P., and Perez-Reyes, M.,** Cannabis, atropine, and temporal information processing, *Neuropsychobiology*, 12, 229, 1984.

20. **Golden, C. J.,** *Stroop Color and Word Test. A Manual for Clinical and Experimental Users*, Stoelting Company, Chicago, 1978.
21. **Witkin, H. A.,** Individual differences in ease of perception of embedded figures, *J. Personality*, 19, 1, 1950.
22. **Weckowicz, T. E., Fedora, O., Mason, J., Radstaak, D., Bay, K. S., and Yonge, K. A.,** Effect of marijuana on divergent and convergent production cognitive tests, *J. Abnorm. Psychol.*, 84, 386, 1975.
23. **Guilford, J. P.,** *The Nature of Human Intelligence,* McGraw-Hill, New York, 1967.
24. **Dewey, W. L.,** Cannabinoid pharmacology, *Pharmacol. Rev.*, 38, 151, 1986.
25. **Milstein, S. L., MacCannell, K., Karr, G., and Clark, S.,** Marijuana-produced impairments in coordination, *J. Nerv. Ment. Dis.*, 161, 26, 1975.
26. **Casswell, S.,** Cannabis intoxication: effects of monetary incentive on performance, a controlled investigation of behavioural tolerance in moderate users of cannabis, *Percep. Motor Skills*, 41, 423, 1975.
27. **MacAvoy, M. G. and Marks, D. F.,** Divided attention performance of cannabis users and non-users following cannabis and alcohol, *Psychopharmacologia*, 44, 147, 1975.
28. **Marks, D. F. and MacAvoy, M. G.,** Divided attention performance in cannabis users and non-users following alcohol and cannabis separately and in combination, *Psychopharmacology*, 99, 397, 1989.
29. **Casswell, S. and Marks, D. F.,** Cannabis induced impairment of performance of a divided attention task, *Nature (London)*, 241, 60, 1973.
30. **Casswell, S. and Marks, D. F.,** Cannabis and temporal disintegration in experienced and naive subjects, *Science*, 179, 803, 1973.
31. **Kiplinger, G. F. and Manno, J. E.,** Dose-response relationships to cannabis in human subjects, *Pharmacol. Rev.*, 23, 339, 1971.
32. **Borg, J., Gershon, S., and Alpert, M.,** Dose effects of smoked marihuana on human cognitive and motor functions, *Psychopharmacologia*, 42, 211, 1975.
33. **Moskowitz, H. and McGlothlin, W.,** Effects of marihuana on auditory signal detection, *Psychopharmacologia*, 40, 137, 1974.
34. **Jones, R. T. and Stone, G. C.,** Psychological studies of marijuana and alcohol in man, *Psychopharmacologia*, 18, 108, 1970.
35. **Hooker, W. D. and Jones, R. T.,** Increased susceptibility to memory intrusions and the Stroop interference effect during acute marijuana intoxication, *Psychopharmacology*, 91, 20, 1987.
36. **Petersen, R. C.,** Importance of inhalation patterns in determining effects of marihuana use, *Lancet*, i, 727, 1979.
37. **Herning, R. I., Hooker, W. D., and Jones, R. T.,** Tetrahydrocannabinol content and differences in marijuana smoking behavior, *Psychopharmacology*, 90, 160, 1986.
38. **Chait, L. D., Corwin, R. L., and Johanson, C. E.,** A cumulative dosing procedure for administering marijuana smoke to humans, *Pharmacol. Biochem. Behav.*, 29, 553, 1988.
39. **Heishman, S. J., Stitzer, M. L., and Bigelow, G. E.,** Alcohol and marijuana: comparative dose effect profiles in humans, *Pharmacol. Biochem. Behav.*, 31, 649, 1989.
40. **Crabtree, D. A. and Antrim, L. R.,** Guidelines for measuring reaction time, *Percep. Motor Skills*, 66, 363, 1988.
41. **Dinges, D.F. and Kribbs, N. B.,** Comparison of the effects of alcohol and sleepiness on simple reaction time performance: enhanced habituation as a common process, *Alcohol Drugs Driving,* 5/6, 329, 1990.

42. **Wesnes, K., Simpson, P., and Christmas, L.,** The assessment of human information-processing abilities in psychopharmacology, in *Human Psychopharmacology. Measures and Methods,* Vol 1, Hindmarch, I. and Stonier, P. D., Eds., John Wiley & Sons, Chichester, 1987, 79.
43. **Matsuda, L. A., Lolait, S. J., Brownstein, M. J., Young, A. C., and Bonner, T. I.,** Structure of a cannabinoid receptor and functional expression of the cloned cDNA, *Nature (London)*, 346, 561, 1990.
44. **Barnett, G., Licko, V., and Thompson, T.,** Behavioral pharmacokinetics of marijuana, *Psychopharmacology*, 85, 51, 1985.
45. **Yesavage, J. A., Leirer, V. O., Denari, M., and Hollister, L. E.,** Carry-over effects of marijuana intoxication on aircraft pilot performance: a preliminary report, *Am. J. Psychiatry,* 142, 1325, 1985.
46. **Foltin, R. W., Fischman, M. W., Brady, J. V., Bernstein, D. J., Nellis, M. J., and Kelly, T. H.,** Marijuana and behavioral contingencies, *Drug Dev. Res.*, 20, 67, 1990.
47. **McMillen, D. L. and Wells-Parker, E.,** The effect of alcohol consumption on risk-taking while driving, *Addict. Behav.*, 12, 241, 1987.
48. **Clayton, A. B.,** The effects of psychotropic drugs upon driving-related skills, *Human Factors*, 18, 241, 1976.
49. **Mann, A. H.,** The clinical relevance of psychometric testing, *Eur. J. Clin. Pharmacol.*, 28(Suppl.), 31, 1985.
50. **Roache, J. D. and Griffiths, R. R.,** Comparison of triazolam and pentobarbital: performance impairment, subjective effects and abuse liability, *J. Pharmacol. Exp. Ther.*, 234, 120, 1985.
51. **Roache, J. D. and Griffiths, R. R.,** Lorazepam and meprobamate dose effects in humans: behavioral effects and abuse liability, *J. Pharmacol. Exp. Ther.*, 243, 978, 1987.
52. **Dalton, W. S., Martz, R., Lemberger, L., Rodda, B. E., and Forney, R. B.,** Influence of cannabidiol on delta-9-tetrahydrocannabinol effects, *Clin. Pharmacol. Ther.*, 19, 300, 1976.
53. **Kiplinger, G. F., Manno, J. E., Rodda, B. E., and Forney, R. B.,** Dose-response analysis of the effects of tetrahydrocannabinol in man, *Clin. Pharmacol. Ther.*, 12, 650, 1971.
54. **Evans, M. A., Martz, R., Rodda, B. E., Lemberger, L., and Forney, R. B.,** Effects of marihuana-dextroamphetamine combination, *Clin. Pharmacol. Ther.*, 20, 350, 1976.
55. **Evans, M. A., Martz, R., Brown, D. J., Rodda, B. E., Kiplinger, G. F., Lemberger, L., and Forney, R. B.,** Impairment of performance with low doses of marihuana, *Clin. Pharmacol. Ther.*, 14, 936, 1973.
56. **Dalton, W. S., Mart, R., Lemberger, L., Rodda, B. E., and Forney, R. B.,** Effects of marihuana combined with secobarbital, *Clin. Pharmacol. Ther.*, 18, 298, 1975.
57. **Klonoff, H., Low, M., and Marcus, A.,** Neuropsychological effects of marijuana, *Can. Med. Assoc. J.*, 108, 150, 1973.
58. **Dornbush, R. L., Fink, M., and Freedman, A. M.,** Marijuana, memory, and perception, *Am. J. Psychiatry*, 128, 194, 1971.
59. **Braden, W., Stillman, R. C., and Wyatt, R. J.,** Effects of marihuana on contingent negative variation and reaction time, *Arch. Gen. Psychiatry,* 31, 537, 1974.
60. **Manno, J. E., Kiplinger, G. F., Scholz, N., and Forney, R. B.,** The influence of alcohol and marihuana on motor and mental performance, *Clin. Pharmacol. Ther.*, 12, 202, 1971.

61. **Manno, J. E., Kiplinger, G. F., Haine, S. E., Bennett, I. F., and Forney, R. B.,** Comparative effects of smoking marihuana or placebo on human motor and mental performance, *Clin. Pharmacol. Ther.*, 11, 808, 1970.
62. **Weil, A. T., Zinberg, N. E., and Nelsen, J. M.,** Clinical and psychological effects of marihuana in man, *Science*, 162, 1234, 1968.
63. **Salvendy, G. and McCabe, G. P.,** Marijuana and human performance, *Human Factors*, 17, 229, 1975.
64. **Carlin, A. S., Bakker, C. B., Halpern, L., and Post, R. D.,** Social facilitation of marijuana intoxication: impact of social set and pharmacological activity, *J. Abnorm. Psychol.*, 80, 132, 1972.
65. **Chait, L. D., Fischman, M. W., and Schuster, C. R.,** Hangover effects the morning after marijuana smoking, *Drug Alcohol Depend.*, 15, 229, 1985.
66. **Kelly, T. H., Foltin, R. W., Emurian, C. S., and Fischman, M. W.,** Multidimensional behavioral effects of marijuana, *Prog. Neuro-Psychopharmacol. Biol. Psychiatry*, 14, 885, 1990.
67. **Drew, W. G., Kiplinger, G. F., Miller, L. L., and Marx, M.,** Effects of propranolol on marihuana-induced cognitive dysfunctioning, *Clin. Pharmacol. Ther.*, 13, 526, 1972.
68. **Hansteen, R. W., Miller, R. D., Lonero, L., Reid, L. D., and Jones, B.,** Effects of cannabis and alcohol on automobile driving and psychomotor tracking, *Ann. N.Y. Acad. Sci.*, 282, 240, 1976.
69. **Zacny, J. P. and Chait, L. D.,** Response to marijuana as a function of potency and breathhold duration, *Psychopharmacology*, 103, 223, 1991.
70. **Cone, E. J., Johnson, R. E., Moore, J. D., and Roache, J. D.,** Acute effects of smoking marijuana on hormones, subjective effects and performance in male human subjects, *Pharmacol. Biochem. Behav.*, 24, 1749, 1986.
71. **Carlin, A. S., Post, R. D., Bakker, C. B., and Halpern, L. M.,** The role of modeling and previous experience in the facilitation of marijuana intoxication, *J. Nerv. Ment. Dis.*, 159, 275, 1974.
72. **Vachon, L., Sulkowski, A., and Rich, E.,** Marihuana effects on learning, attention and time estimation, *Psychopharmacologia*, 39, 1, 1974.
73. **Moskowitz, H., Shea, R., and Burns, M.,** Effect of marihuana on the psychological refractory period, *Percep. Motor Skills*, 38, 959, 1974.
74. **Stillman, R. C., Wolkowitz, O., Weingartner, H., Waldman, I., DeRenzo, E. V., and Wyatt, R. J.,** Marijuana: differential effects on the right and left hemisphere functions in man, *Life Sci.*, 21, 1793, 1977.
75. **Block, R. I. and Wittenborn, J. R.,** Marijuana effects on the speed of memory retrieval in the letter-matching task, *Int. J. Addict.*, 21, 281, 1986.
76. **Peeke, S. C., Jones, R. T., and Stone, G. C.,** Effects of practice on marijuana-induced changes in reaction time, *Psychopharmacology*, 48, 159, 1976.
77. **Schaefer, C. F., Gunn, C. G., and Dubowski, K. M.,** Dose-related heart-rate, perceptual, and decisional changes in man following marihuana smoking, *Percep. Motor Skills*, 44, 3, 1977.
78. **Low, M. D., Klonoff, H., and Marcus, A.,** The neurophysiological basis of the marijuana experience, *Can. Med. Assoc. J.*, 108, 157, 1973.
79. **Block, R. I. and Wittenborn, J. R.,** Marijuana effects on semantic memory: verification of common and uncommon category members, *Psychol. Rep.*, 55, 503, 1984.
80. **Perez-Reyes, M., Hicks, R. E., Bumberry, J., Jeffcoat, A. R., and Cook, C. E.,**

Interaction between marihuana and ethanol: effects on psychomotor performance, *Alcoholism: Clin. Exp. Res.*, 12, 268, 1988.

81. **Moskowitz, H., Sharma, S., and McGlothlin, W.,** Effect of marihuana upon peripheral vision as a function of the information processing demands in central vision, *Percep. Motor Skills*, 35, 875, 1972.
82. **Moskowitz, H., Hulbert, S., and McGlothlin, W. H.,** Marihuana: effects on simulated driving performance, *Accid. Anal. Prev.*, 8, 45, 1976.
83. **Sharma, S. and Moskowitz, H.,** Effects of two levels of attention demand on vigilance performance under marihuana, *Percep. Motor Skills*, 38, 967, 1974.
84. **Janowsky, D. S., Meacham, M. P., Blaine, J. D., Schoor, M., and Bozzetti, L. P.,** Marijuana effects on simulated flying ability, *Am. J. Psychiatry*, 133, 384, 1976.
85. **Leirer, V. O., Yesavage, J. A., and Morrow, D. G.,** Marijuana, aging, and task difficulty effects on pilot performance, *Aviat. Space Environ. Med.*, 60, 1145, 1989.
86. **Sutton, L. R.,** The effects of alcohol, marihuana and their combination on driving ability, *J. Stud. Alcohol*, 44, 438, 1983.
87. **Block, R. I. and Wittenborn, J. R.,** Marijuana effects on associative processes, *Psychopharmacology*, 85, 426, 1985.
88. **Wetzel, C. D., Janowsky, D. S., and Clopton, P. L.,** Remote memory during marijuana intoxication, *Psychopharmacology*, 76, 278, 1982.
89. **Dornbush, R. L.,** Marijuana and memory: effects of smoking on storage, *Trans. N.Y. Acad. Sci.*, 36, 94, 1974.
90. **Cappell, H. D. and Pliner, P. L.,** Volitional control of marijuana intoxication: a study of the ability to "come down" on command, *J. Abnorm. Psychol.*, 82, 428, 1973.
91. **Miller, L. L., Cornett, T., and McFarland, D.,** Marijuana: an analysis of storage and retrieval deficits in memory with the technique of restricted reminding, *Pharmacol. Biochem. Behav.*, 8, 327, 1978.
92. **Miller, L. L. and Cornett, T. L.,** Marijuana: dose effects on pulse rate, subjective estimates of intoxication, free recall and recognition memory, *Pharmacol. Biochem. Behav.*, 9, 573, 1978.
93. **Pearl, J. H., Domino, E. F., and Rennick, P.,** Short-term effects of marijuana smoking on cognitive behavior in experienced male users, *Psychopharmacologia*, 31, 13, 1973.
94. **Miller, L. L., McFarland, D., Cornett, T. L., and Brightwell, D.,** Marijuana and memory impairment: effect on free recall and recognition memory, *Pharmacol. Biochem. Behav.*, 7, 99, 1977.
95. **Miller, L. L., McFarland, D. J., Cornett, T. L., Brightwell, D. R., and Wikler, A.,** Marijuana: effects on free recall and subjective organization of pictures and words, *Psychopharmacology*, 55, 257, 1977.
96. **Zacny, J. P. and de Wit, H.,** Effects of food deprivation on responses to marijuana in humans, *Behav. Pharmacol.*, 1, 177, 1989.
97. **Miller, L., Cornett, T., and Nallan, G.,** Marijuana: effect on nonverbal free recall as a function of field dependence, *Psychopharmacology*, 58, 297, 1978.
98. **Miller, L., Cornett, T., Brightwell, D., McFarland, D., Drew, W. G., and Wikler, A.,** Marijuana and memory impairment: the effect of retrieval cues on free recall, *Pharmacol. Biochem. Behav.*, 5, 639, 1976.
99. **Belmore, S. M. and Miller, L. L.,** Levels of processing and acute effects of marijuana on memory, *Pharmacol. Biochem. Behav.*, 13, 199, 1980.
100. **Miller, L. L., Cornett, T. L., and Wikler, A.,** Marijuana: effects on pulse rate, subjective estimates of intoxication and multiple measures of memory, *Life Sci.*, 25, 1325, 1979.

101. **Miller, L., Cornett, T., Drew, W., McFarland, D., Brightwell, D., and Wikler, A.,** Marijuana: dose-response effects on pulse rate, subjective estimates of potency, pleasantness, and recognition memory, *Pharmacology*, 15, 268, 1977.
102. **Miller, L. L., Cornett, T. L., Brightwell, D. R., McFarland, D. J., Drew, W. G., and Wikler, A.,** Marijuana: effects on storage and retrieval of prose material, *Psychopharmacology*, 51, 311, 1977.
103. **Miller, L., Drew, W. G., and Kiplinger, G. F.,** Effects of marijuana on recall of narrative material and Stroop colour-word performance, *Nature (London)*, 237, 172, 1972.
104. **Block, R. I. and Wittenborn, J. R.,** Marijuana effects on visual imagery in a paired-associate task, *Percep. Motor Skills*, 58, 759, 1984.
105. **Carlini, E. A., Karniol, I. G., Renault, P. F., and Schuster, C. R.,** Effects of marihuana in laboratory animals and in man, *Br. J. Pharmacol.*, 50, 299, 1974.
106. **Cappell, H., Webster, C. D., Herring, B. S., and Ginsberg, R.,** Alcohol and marihuana: a comparison of effects on a temporally controlled operant in humans, *J. Pharmacol. Exp. Ther.*, 182, 195, 1972.

11 Dissociative and Discriminative Stimulus Functions of Cannabinoids/ Cannabimimetics

INTRODUCTION

Cannabis has been used for many centuries. It has even been found in Chinese herbal compendiums from the 28th century B.C.[1] Though it has been mainly used for recreational purposes, it also has medicinal properties. Recreationally, cannabis induces euphoria in low doses. Medicinally, it reduces intraoccular pressure and nausea.[2] Ironically, the psychoactive effects that are responsible for its abuse are the same properties that limit its medical use. Thus, it is important not only to understand the psychoactive properties of cannabis as they relate to dependence, but also to dissociate those properties from its medicinal effects in order to develop medically useful compounds.

The subjective effects of cannabis primarily result from the activity of tetrahydrocannabinol (THC). The most psychoactive of these compounds are the levorotated, i.e., (*l*)-isomers, Δ^9- and Δ^8-THC. However, because Δ^8-THC is less potent[3-6] and less commonly found in the various species of cannabis,[7,8] most of the initial studies have focused on the effects of the Δ^9-THC isomer. Thus, for investigations of other compounds with cannabimimetic properties, Δ^9-THC is used as the standard or reference compound.

In order to isolate the chemical components that are responsible for the psychoactive property of cannabis, experimental methods for detecting and quantifying cannabimimetic activity in animals have been developed. These studies are important for identifying commonalities in mechanisms between cannabinoids and cannabimimetics, investigating factors that contribute to cannabis dependence, and assessing the therapeutic potential of newly synthesized cannabinoids (for previous reviews, see References 9 to 13).

Most commonly, investigations of the subjective effects of THC have utilized drug discrimination (DDL) and state-dependent (SDL) learning paradigms. Though both procedures focus on the psychoactive properties of drugs, their purposes differ. The DDL studies are designed to examine drugs as cues for instrumental behavior. Performance is then used as a measure for qualitative and quantitative similarities and differences among drugs. On the other hand, SDL studies examine drugs as part of the process of conditioning, or memory formation and retrieval. Expression of the response, or recall of the memory, is then measured to determine the effects of that drug on neural and psychological processes.

STATE-DEPENDENT LEARNING

SDL is also termed dissociated learning or state-dependent retrieval.[14] It is observed when an association formed in one particular condition is more easily retrieved in that same condition than in a different one. Frequently, these conditions refer to drug or nondrug states (for historical background and reviews of this phenomenon, see References 14 to 18).

During the 20th century, the phenomenon of SDL has been studied under rigorous experimental conditions. Generally, these studies include four groups: two no-change groups that are trained and tested in the same drug or no-drug condition; and two change groups that are trained in one condition, and tested for retention in the other state.[19]

The experimental analysis of SDL began with the report that a leg-flexion conditioned response in dogs was elicited only when the state (curare or no drug) induced during the initial learning was reinstated.[23] Though this finding was generally not confirmed in subsequent studies,[24-26] it did provide an impetus for examining the SDL phenomena. From the late 1950s to the early 1960s, several reports were published regarding drug-induced change of state effects, some of which indicated very marked SDL (e.g., see References 27 and 28). For example, these studies indicated that a response learned while being drugged with high doses of pentobarbital was completely dissociated from a no-drug state (see Reference 29 for a comprehensive theoretical review of SDL investigations).

For cannabinoids, investigations using the SDL paradigm have been useful for delineating some of its disruptive effects on memory and performance,[30] and for exploring the development of tolerance to those effects.[15] These deficits are an important aspect of cannabinoid intoxication and comprise a major unwanted side effect of cannabinoid therapy. Thus, a selective review of literature regarding the psychoactive properties of Δ^9-THC (animals) and marijuana preparations (humans), and its influence on memory will follow.

Animal Studies

An early demonstration of the SDL effects of Δ^9-THC was offered by Barry and Kubena.[31] For a study originally designed to investigate the influence of Δ^9-THC on avoidance learning, rats were trained on a free-operant, conditioned avoidance task, and injected before a training session with either Δ^9-THC (8 mg/kg) or vehicle. Tests were then conducted with all subjects after injections of Δ^9-THC and vehicle. Rats given Δ^9-THC before each training session had significantly enhanced avoidance performance during acquisition, but showed a significant deficit in performance when tested in the vehicle condition (see also Reference 32).

Avoidance Learning

The first demonstration of SDL with Δ^9-THC in a study specifically designed to investigate this phenomenon was reported by Henriksson and Järbe[33] using a signaled conditioned avoidance procedure with a shuttle-box task. The result of that initial study suggested that the subjective effects of Δ^9- and Δ^8-THC were unlike those of the minor tranquilizers such as diazepam. For example, when conditioned avoidance was acquired during a nondrugged state, then tested after administration of diazepam, retention between training and test conditions was observed.[34] However, when rats were in a nondrug state, then tested in a THC state, retention between training and test conditions was lost. Thus, the SDL effect for THC was found even when behavior was learned under a nondrugged state.[33]

One explanation for this difference between the effects of these compounds might be that THC, unlike diazepam, disrupts established conditioned avoidance responding.[35-37] Thus, for rats tested in the diazepam condition, performance deficits induced by THC itself may have resulted in an apparent retrieval deficit.

The difference between benzodiazepine and cannabinoid SDL effects on signaled avoidance behavior is also demonstrated in a study of change between two drug states.[36] Mice were injected with either chlordiazepoxide (CDP), Δ^9-THC, or vehicle, trained on a conditioned avoidance task, then tested for retention in the same or other-drug state. For mice trained under either CDP or Δ^9-THC and tested in the vehicle condition, a significant reduction in performance was obtained. In addition, mice trained under Δ^9-THC, then tested with CDP (5 mg/kg and higher), showed no SDL effects. However, when mice were trained under CDP and then tested under Δ^9-THC, a performance deficit was observed.

A study by Robichaud et al.[38] examined the effects of Δ^9-THC on unsignaled avoidance. Mice were trained to avoid shock and administered Δ^9-THC 30 min before each session throughout training. Conditioned avoidance acquisition was enhanced by drug exposure when performance by these subjects was compared to the no-drug controls. In addition, when the drug-exposed group was tested nondrugged, there was a decrease in the conditioned avoidance response, suggesting state dependency of the avoidance learning.

Observation of SDL for Δ^9-THC states in animals has not been obtained in all studies. Herring[39] reported an absence of SDL for rats trained under a free-operant, signaled avoidance procedure. Rats were trained to bar press to avoid shock under a drug or no-drug state. Each response delayed shock which was signaled by a light-buzzer combination. When tests were conducted under the same or different state, no significant SDL effects were obtained. In addition, on a one-trial passive avoidance task, administration of Δ^9-THC in doses up to 20 mg/kg was ineffective in disrupting avoidance behavior; higher doses caused only a minimal impairment of learning.[38]

Conditioned Suppression

Another aversively motivated task that is apparently sensitive to SDL effects is conditioned suppression. For example, rats were trained on a one-trial conditioned suppression task. Animals were injected with either Δ^9-THC (2.5 mg/kg) or vehicle, then shocked in a cage previously associated with milk drinking. Some 5 d later, rats were administered either the same or other injection, then tested for suppression of milk drinking. SDL effects were obtained. For animals shocked after administration of Δ^9-THC and then tested without drug, little retention of suppression was obtained. For rats trained and tested in the drug state, suppression of drinking was substantially enhanced compared to rats shocked and tested in the no-drug state. However, these experiments also demonstrated that Δ^9-THC suppressed drinking when given to rats that had not been shocked, and thus had unconditioned effects on drinking behavior.[38]

Factors Influencing State Dependent Learning

Elimination Rate

Some of the negative results mentioned above may be explained by the long half-life of Δ^9-THC. Glick and Milloy[40] examined the elimination rate and dose effectiveness of Δ^9-THC on SDL effects using one-trial passive avoidance in mice. Mice were injected with Δ^9-THC (0, 2, 4, or 8 mg/kg) before training, then tested in the no-drug condition for deficits in avoidance learning. At 1-d post-administration, only the group that was injected with 2 mg/kg Δ^9-THC showed any learning deficit. However, 7 d later, deficits were obtained for animals administered the 4-mg/kg dose; at 4 weeks post-administration, they were obtained for the 8-mg/kg dose. The authors suggest that because Δ^9-THC has a rather long half-life, a longer period of time is required for Δ^9-THC and its metabolites to clear the organism and for a no-drug state to be obtained. In a second experiment using only the 2-mg/kg dose, mice were injected with either drug or vehicle prior to training, then tested in either the same or different state. Results indicated that SDL effects were limited to those animals trained in the drug condition and tested in the no-drug state.

Tolerance

Another important factor influencing the SDL effects of Δ^9-THC is the development of tolerance to Δ^9-THC intoxication. An early demonstration of tolerance to Δ^9-THC was suggested by the Barry and Kubena[31] report. When rats were trained on conditioned avoidance task, and injected after each training session with either Δ^9-THC (8 mg/kg) or vehicle, tests conducted in the drug condition resulted in no performance deficit.

Tolerance to the disruptive effects of Δ^9-THC on performance was investigated with a reversal learning procedure.[41] Rats were trained in a nondrugged condition to escape water immersion via one arm of a maze, then tested under a Δ^9-THC condition for escape via another arm. Those rats that were given daily post-session injections of Δ^9-THC showed facilitation of this reversal task in comparison to rats that were given no preexposure. In addition, rats given no preexposure performed even worse than rats trained initially to traverse the maze under the Δ^9-THC condition.

Similar results were obtained using a signaled, conditioned avoidance task. Rats were administered either Δ^8-THC or vehicle 2 h after each training session, then tested in the Δ^8-THC condition. There was a significant decrement in performance for the vehicle-exposed rats, but not for the Δ^8-THC-exposed group. Recovery from the performance decrement for the vehicle-treated group was observed when the animals were retested on the following day. In addition, performance decrements were again observed after the effects of preexposure to Δ^8-THC had dissipated. When both groups were given neither training nor injections for 12 d, then retrained in the no-drug condition, testing under the Δ^8-THC state caused significant decreases in the number of conditioned avoidance responses by both groups, suggesting that tolerance to THC had dissipated.[42]

Studies by Järbe[43] also demonstrated that tolerance developed to not only the disruptive effects of Δ^9-THC, but also to its subjective drug-state effects. Rats were pretreated with Δ^9-THC daily for 2 weeks, then trained on a conditioned avoidance task under the same drug state for 2 daily sessions of 100 trials each. Then, drug administration was halted and animals tested on day 3. No significant deterioration in performance occurred due to this change in state. The performance of the THC-pretreated animals was similar to that of the vehicle controls throughout training and test periods. For nontolerant animals (i.e., not preexposed to Δ^9-THC), drug administration during training had marked suppressive actions. When tested nondrugged, this group evidenced a decrement in conditioned avoidance responses compared to their training performance under THC the previous day. This change-of-state effect was especially evident during the first block of daily training sessions.

Human Studies

The first indication of a possible marijuana-induced state dependency was reported by Abel[44] using a memory task. Subjects were required to learn narrative material under the influence of marijuana, then recall the material in either a sober or marijuana-intoxicated condition. A greater deficit in recall was reported for the subjects tested in the sober state than for those tested while marijuana intoxicated. Deficits in recall were also reported for changes to either sober or drugged states. For example, light marijuana users who were exposed to a paired-associate list in a sober condition and then tested for retention after smoking marijuana (THC content approximately 7 mg/cigarette) showed significant memory deficits.[45]

Sequential Memory

Some reports have suggested that the SDL effects for marijuana may be most apparent for tasks using sequential memory. Hill et al.[46] selected subjects who had smoked marijuana (at least 1/2 of the subjects smoked at least once a week for 4 years) and were drug free for the previous 2 d. They were administered either marijuana or placebo, then asked to memorize word strings, word associations, and the order of objects. When tested for recall under same or different conditions, significant SDL effects were obtained, but only for recall of the sequential order of objects. Similar results were obtained by Stillman et al.[47] using a lower drug concentration (7.5 mg/cigarette). When subjects were tested for retention of the sequential arrangement of pictures, rather strong state dependency effects were obtained.

Active Recall

Other investigations of SDL and marijuana suggest that more difficult tasks utilizing active recall are most affected. After administration of fairly strong drug concentrations (20 mg/cigarette) to marijuana-experienced subjects, tests for word recall or word recognition in either a change or no-change state yielded SDL effects, but only for the word recall task.[48] In a similar study, THC (10 mg/cigarette) was administered to marijuana-experienced subjects who were then exposed to lists of categorized words. When asked to recall the words in either a sober or drugged state, and with or without retrieval cues, an SDL effect was found only for subjects who were tested with no retrieval cues available.[49]

Psychomotor and Avoidance Tasks

Some evidence suggests that the SDL effect for marijuana is most apparent for memory-related skills; SDL is not a robust phenomenon with psychomotor or aversively motivated tasks. For example, both marijuana users and nonusers were administered THC (7 mg/cigarette), and tested for performance on tests of card sorting, block turning, and pursuit rotor, during same- or changed-state conditions.[50] Though THC impaired performance on these tests, no change-of-state effects were obtained. In addition, when subjects were trained under a signaled avoidance procedure to avoid the onset of a noxious tone, tests conducted during same- or changed-state conditions yielded no SDL effects.[46]

Tolerance

One problem with the above studies is the lack of control for tolerance to the effects of marijuana. Though most of the subjects in the prior studies were defined as light marijuana users, the frequency of THC exposure was varied. For example, some studies limited participants to those who only smoked one to two times per week;[48] others used participants who smoked cannabis *at least* once a week,[46,49] and frequency of smoking could range up to six times per week.[47]

In order to examine the influence of frequency of use on marijuana's effects on memory, Cohen and Rickles[51] differentiated between two groups of marijuana-experienced subjects: heavy or social users. Subjects in the heavy-user group had smoked marijuana four or more times per week over the past year (most smoked five to six times per week), went to school/work intoxicated, and smoked both socially and alone. The social-user group included those who smoked only on the weekends, and rarely went to school/work intoxicated. They found that frequency of use had profound effects on the SDL effects of marijuana. When tested for recall of a paired associate list, social users exhibited state-dependent effects; heavy users did not. In fact, the heavy-user group performed equally well whether intoxicated or not, and their performance during acquisition or recall was better than intoxicated social users.

Summary

Thus, cannabinoid SDL effects have been demonstrated with a number of studies. More importantly, both animals and humans exhibit SDL for cannabinoids; and both animals and humans are affected by the nature of the task, elimination rate of the compound, and tolerance induced by prior exposure of the drug.

DRUG DISCRIMINATION LEARNING

In DDL procedures, a drug state is used as a discriminative stimulus for a choice between different response alternatives. The training methods that are used to establish a drug discrimination have been substantially reviewed elsewhere.[52-56] The most common procedure uses a food- or water-motivated operant task (lever pressing), and a fixed ratio (FR) schedule of reinforcement (see References 57 to 60). Then, tests for generalization to the subjective effects of other drugs is used to identify similarities or differences between the training drug state and the test compound.

Δ^9-THC

The DDL paradigm has been quite useful for examining the subjective effects of THC. One outcome of these experiments was the demonstration that the cannabimimetic discriminative stimulus[61] and the analgesic, antiemetic, and anticonvulsant activity of cannabinoids can be dissociated.[2] Though initial studies were limited to naturally occurring compounds (e.g., marijuana, hashish smoke, or derived tinctures), more recently the active component(s) of cannabinoids and their chemical structure have been elucidated,[62] and synthetic compounds have been made available to DDL research. Thus, an interdisciplinary approach using organic chemistry and choice behavior of laboratory animals has been used to determine some chemical requirements for inducing cannabis intoxication. This chapter will review the DDL literature that focused on the Δ^9-THC stimulus, as well as other cannabinoids, in order to delineate the THC state and identify those compounds that do not mimic it.

Specificity

The psychoactive effect of cannabis is unique (see References 4 and 63 to 66). Administration of agents affecting various neurotransmitter systems does not elicit or block the THC discriminative response. For example, several compounds that act on adrenergic, dopaminergic, cholinergic, serotonergic, opiate, and prostaglandin systems do not antagonize the THC cue.[4,64,66,67] Though the cannabinoids, cannabinol and cannabidiol, have been reported to attenuate the effects of THC on other measures (see References 68 to 70), they do not block the THC stimulus (see References 5, 64, 68, and 69). In addition, a dose of 3 mg/kg of the (+)-isomer of the very potent cannabimimetic dimethylheptyl homolog of the 11-hydroxylated Δ^8-THC (11-OH-Δ^8-THC-DMH) does not prevent the response generalization from (–)-11-OH-Δ^8-THC-DMH to the Δ^9-THC stimulus in either rats or pigeons (unpublished).

Thus, the subjective properties of Δ^9-THC are generally pharmacologically specific, with the possible exception of some overlap with the benzodiazepine, diazepam.[64,71] A possible neurochemical explanation for this overlap may be the reports that THC shows some affinity for benzodiazepine receptor sites.[72,73] However, for THC-discriminating pigeons and gerbils, very little generalization of the THC response to benzodiazepines such as diazepam and Ro 11-3128 has been found.[74,75]

Biochemical and electrophysiological data have shown that cannabinoids affect a wide range of neurotransmitter systems and brain areas (see Reference 76 for review). However, the lack of generalization to and blocking of the THC stimulus by these agents, as well as stereochemical considerations, suggest that the subjective effects of Δ^9-THC are mediated through undefined, specific neurochemical systems. Järbe et al.,[77] Browne and Weissman,[64] and others have speculated that these subjective effects may be mediated through a specific recognition site or receptor system. Thus, recent reports[78-80] concerning a cannabinoid binding site in the brain are intriguing and should assist in clarifying the mode(s) and site(s) of action of THC. An initial attempt has been made to characterize specific brain sites responsible for the stimulus effects of THC. Mokler and Rosecrans[81] used intracranial administration of Δ^9-THC to the prefrontal cortex, dorsal hippocampus, and retricular formation. However, this attempt was met with limited success.

Comparison with Hashish Smoke

The subjective effects of Δ^9-THC on animals is apparently directly comparable to the effect of hashish smoke. Rats[4,67] and gerbils[82] perceive the effects induced by hashish inhalation as being similar to the effects induced by i.p. administered Δ^9-THC. When these animals were trained to discriminate Δ^9-THC, and were then exposed to hashish smoke, they performed the THC-appropriate response. Similarly, when rats were trained to discriminate between active smoke inhalation (hashish smoke) and placebo (chevril smoke), they performed the hashish-appropriate response after i.p. injections of THC. This effect was stronger for the Δ^9-THC isomer than it was for the Δ^8-THC isomer.[4] In addition, other effects of Δ^9-THC (e.g., depression of water intake) are similar to the effects of hashish smoke exposure.[83] Thus, most DDL investigations of cannabinoid activity have focused on training animals to discriminate the Δ^9-THC stimulus.

Species Differences

When animals are trained to detect the presence or absence of THC, speed of acquisition is a function of both training dose and species of subject. Animals may differ in their inherent initial sensitivity to THC.[84]

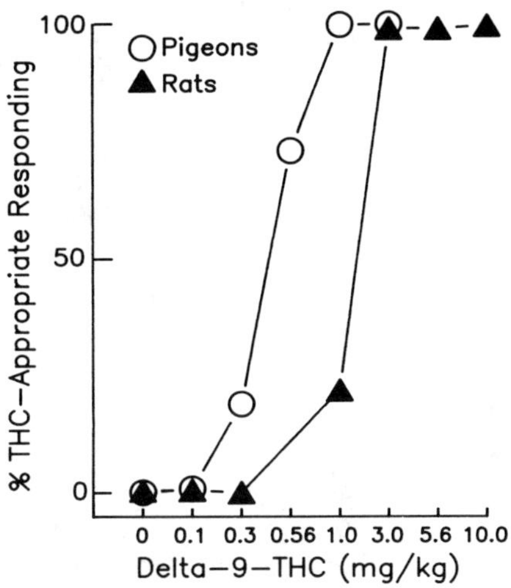

FIGURE 1. Dose-generalization gradients of Δ^9-THC for pigeons and rats trained to discriminate between Δ^9-THC and vehicle. Pigeons were trained with 1 mg/kg, injected i.m. 90 min prior to session onset. Rats were trained with 3 mg/kg, injected i.p. 30 min prior to session onset. n = 8 at all points. (From Järbe, T. U. C. and Mathis, D. A., The discriminative stimulus functions of cannabinoids and cannabimimetics, in *Drug Discrimination: Applications to Drug Abuse Research*, NIDA Res. Monogr., 116, Glennon, R. A., Järbe, T. U. C., and Frankenheim, J., Eds., U.S. Department of Health and Human Services, Alcohol, Drug Abuse and Mental Health Admin., Rockville, MD, 1991, 75.)

Since pigeons tend to be more sensitive than rats to its stimulus effects (e.g., see Reference 85), many investigations of THC utilize pigeons as subjects. They acquire the discrimination fairly rapidly (approximately 20 to 25 training trials with each stimulus condition), and tests for stimulus generalization of the THC cue result in a characteristic dose-effect relationship. Generalization from the Δ^9-THC training stimulus to different doses of the training drug is illustrated in Figure 1. From these graphs, it can be seen that the animals emit progressively fewer drug responses when tested with lowered amounts of the training drug. In addition, similar to generalization tests for exteroceptive stimulus intensity,[86-88] for doses higher than the training dose there is no change in the curvature of the gradients (see also Reference 64). The data in this figure are obtained from pigeons (trained with 1 mg/kg) and rats (trained with 3 mg/kg), and demonstrate the good stimulus control acquired with the Δ^9-THC stimulus. The correlation coefficients (r) and ED_{50} values for the generalization gradients were estimated by logarithmic regression analysis: for pigeons, r = 0.935 (ED_{50} = 0.43 mg/kg); for rats, r = 0.929 (ED_{50} = 1.17 mg/kg).

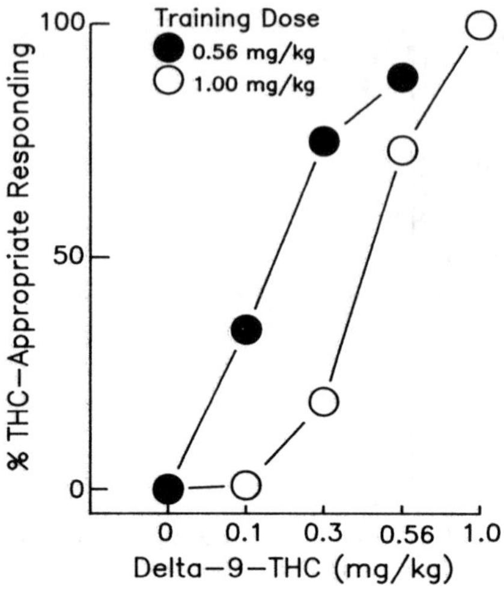

FIGURE 2. Dose-generalization gradients of Δ^9-THC for pigeons trained to discriminate between Δ^9-THC (0.56 mg/kg, n = 7; 1 mg/kg, n = 8) and vehicle. Training and test sessions were conducted 90 min post-i.m. injection. Test sessions utilized a repeated test procedure (see References 68 and 77). Data represent the average of at least two determinations for each animal for the 0.56-mg/kg training condition, and the average of one determination for the 1.0-mg/kg training condition. (From Järbe, T. U. C. and Mathis, D. A., The discriminative stimulus functions of cannabinoids and cannabimimetics, in *Drug Discrimination: Applications to Drug Abuse Research*, NIDA Res. Monogr., 116, Glennon, R. A., Järbe, T. U. C., and Frankenheim, J., Eds., U.S. Department of Health and Human Services, Alcohol, Drug Abuse and Mental Health Admin., Rockville, MD, 1991, 75.)

Effect of Training Dose

Similar to the ED_{50} values obtained with other drugs used as discriminative stimuli, the magnitude of the ED_{50} for Δ^9-THC is affected by the training dose. The affect of training dose within one species is illustrated in Figure 2. Pigeons were trained to discriminate between Δ^9-THC (0.56 or 1 mg/kg) and vehicle, then tested for generalization to different doses of the training drug. The obtained ED_{50} value for the high-dose training condition was approximately threefold higher than for the low-dose condition. In addition, the magnitude of an ED_{50} for Δ^9-THC is affected by the route of and time since drug administration.

Elimination Rate

When rats are trained to discriminate between i.p. injected Δ^9-THC (2 mg/kg) and no drug, tests conducted with i.v. administration of lower doses of THC elicit more THC-appropriate responding than similar tests using i.p.

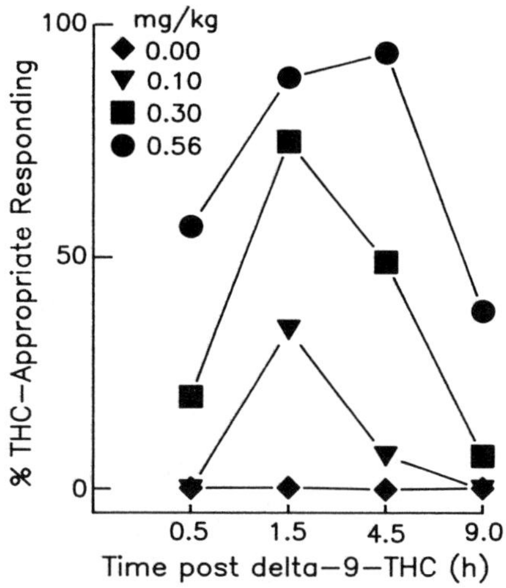

FIGURE 3. Time course for dose-generalization of Δ^9-THC for pigeons trained to discriminate between the presence and absence of 0.56 mg/kg of Δ^9-THC. Training sessions were conducted 90 min after i.m. injections. Test sessions utilized a repeated test procedure (see References 68 and 77), and data represent the average of at least two determinations. Data for the 90-min test interval are the same as those shown for the lower training dose in Figure 2. n = 7 at all points. (From Järbe, T. U. C. and Mathis, D. A., The discriminative stimulus functions of cannabinoids and cannabimimetics, in *Drug Discrimination: Applications to Drug Abuse Research*, NIDA Res. Monogr., 116, Glennon, R. A., Järbe, T. U. C., and Frankenheim, J., Eds., U.S. Department of Health and Human Services, Alcohol, Drug Abuse and Mental Health Admin., Rockville, MD, 1991, 75.)

administration. Conversely, tests conducted with p.o. administration of low THC doses elicit less THC-appropriate responding.[89] The time course for the generalization of the Δ^9-THC discriminative stimulus is illustrated in Figure 3.

Pigeons were trained to discriminate between the presence and absence of 0.56 mg/kg of Δ^9-THC, 90 min after intramuscular (i.m.) administration of THC or vehicle. Using a repeated test procedure, a generalization curve for doses of THC was determined at 30, 90, 270, and 540 min after drug administration. The peak of generalization to Δ^9-THC generally occurred at 90 min after THC administration. These data for generalization to the stimulus effects of the training dose are also similar to data obtained using a single test procedure.[68]

However, just as sensitivity to the THC stimulus varies among species, there are also species differences in the time course of this stimulus. For

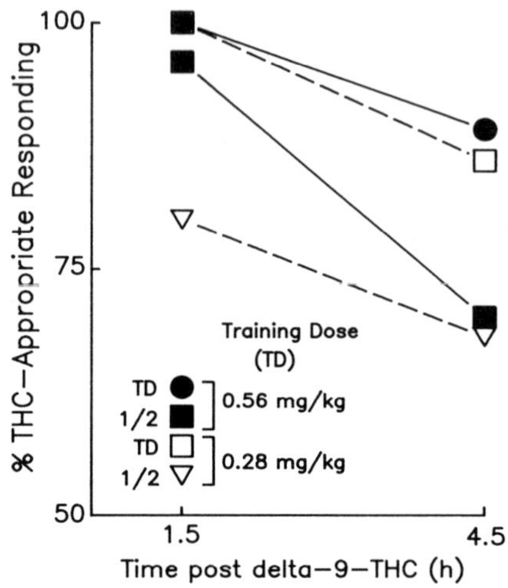

FIGURE 4. Time course for dose-generalization of Δ^9-THC for pigeons trained to discriminate the presence or absence of low or high doses of Δ^9-THC. Subjects were first trained to discriminate 0.28 mg/kg (Low), then tested with 0.28 mg/kg (total dose, TD) and 0.14 mg/kg (1/2 TD). Then, subjects were retrained with 0.56 mg/kg, and retested with 0.56 mg/kg (TD) and 0.28 mg/kg (1/2 TD). Training sessions were conducted 90 min after i.m. injections. Test sessions were conducted using a repeated test procedure,[68,77] and data represent the average of one determination for each animal. n = 5 at all points. (From Järbe, T. U. C. and Mathis, D. A., The discriminative stimulus functions of cannabinoids and cannabimimetics, in *Drug Discrimination: Applications to Drug Abuse Research*, NIDA Res. Monogr., 116, Glennon, R. A., Järbe, T. U. C., and Frankenheim, J., Eds., U.S. Department of Health and Human Services, Alcohol, Drug Abuse and Mental Health Admin., Rockville, MD, 1991, 75.)

example, in rats trained to discriminate THC (3 mg/kg) 30 min post-administration, the peak of generalization to different THC doses occurs at 30 min,[77,90,91] rather than at 90 min post-injection. Thus, the time of peak generalization of THC varies among species; however, within a species, it does not vary as a function of training dose.

Pigeons were trained to discriminate Δ^9-THC (0.28 mg/kg or 0.56 mg/kg) and vehicle, then tested for generalization to the drug (full or 1/2 training dose) at 1.5 and 4.5 h post-administration (Figure 4). The time course of generalization did not differ, regardless of the original THC training dose. Similar effects have been found with rats[67] and gerbils.[82]

Tolerance

In general, retention of the THC discrimination is robust and tolerance to the stimulus effects of Δ^9-THC under normal training and testing procedures has not been reported. However, administration of high or noncontingent doses of this compound does alter the subjective effects of the THC stimulus. When animals are trained to discriminate Δ^9-THC from nondrug and then administered high doses of THC, tests conducted 24 to 48 h after high-dose administration result in an attenuation of the Δ^9-THC cue.[92,93] However, noncontingent exposure apparently is not sufficient to disconnect the future association between the drug stimulus and the behavioral consequences.

Bueno and Carlini[94] observed tolerance to behavior patterns initially disrupted by the administration of a marijuana extract. Despite the fact that seemingly complete tolerance developed to rope climbing, subsequent employment of a DDL task showed that the drug still induced measurable effects. Unfortunately, this study did not include animals without pretreatment. Therefore, it is not possible to evaluate effects on the speed of acquisition of the DDL task by preexposure to Δ^9-THC. Järbe and Henriksson[95] included animals that were not pretreated with drug and concluded that prior exposure to THC did have an effect on the subsequent establishment of the Δ^9-THC discrimination. However, subsequent data by Järbe (unpublished) that utilized several pretreatment doses of Δ^9-THC indicate that tolerance effects to the THC stimulus are small in magnitude.

Other Naturally Occurring Cannabinoids

A variety of other substances with potential psychoactive properties have been detected in cannabis preparations (for an overview of these cannabinoids, see Reference 70), including cannabidiol (CBD) and cannabinol (CBN).[96] As discussed below, though the cannabinoid activity of CBD may be dissociated from its cannabimimetic activity, CBN may share at least some of the cannabimimetic effects of THC.

Cannabidiol

Investigations with CBD have shown that this compound has other cannabinoid properties, such as effects on sleeping-time,[97] anticonvulsive activity,[98] and operant rate-depressant properties.[99] However, generalization of CBD to the subjective, discriminative stimulus effects of THC has consistently not been found (e.g., see Reference 68). Thus, though this compound apparently does contribute to other effects of cannabis preparations, it lacks cannabimimetic properties. In addition, abnormal CBD,[100] the CBD monomethyl ether, and cannabichromene are not cannabimimetic.[91]

These findings parallel reports that CBD lacks cannabimimetic activity in humans.[3,101-103]

Cannabinol

On the other hand, the cannabimimetic activity of CBN is more certain, at least for animals. Most authors report that the CBN stimulus is generalized from thc THC discriminative stimulus.[5,6,69,104]

For humans, research results are more mixed. Some authors report that CBN induces THC-like effects;[101] others report that it does not.[3] However, the negative findings are reported for oral application of CBN,[3] while the positive findings are reported for i.v. administration.[101]

Nevertheless, these investigations of the THC-like effects of CBD and CBN indicate that only CBN has cannabimimetic activity. However, because cannabis preparations contain all three cannabinoids (i.e., THC, CBN, and CBD), these compounds may have different effects when administered in combination. For example, administration of THC has been reported to induce anxiety in humans, but coadministration of CBD with THC reduces that anxiety.[103]

Interactions with Δ^9-THC

The results obtained from studies that assessed the interaction of the stimulus effects of Δ^9-THC, CBN and CBD are summarized in Tables 1A and B. These data indicate that THC does not appear to be solely responsible for the cannabimimetic effects of cannabis preparations. The main interest of studies on cannabinoid combinations has focused on THC administered with one of the other compounds. The results of these studies indicate that the stimulus effects of THC are generally enhanced or prolonged by coadministration of either CBD or CBN. For example, when THC was given in combination with CBD, the discriminative stimulus effects of THC were prolonged.[68] However, this interaction effect may differ among species; for example, for pigeons tested with CBD and THC, no prolongation of THC-appropriate responding was obtained.[68] On the other hand, for studies in which THC was coadministered with CBN to rats, discrimination of the THC stimulus was enhanced, but not prolonged.[69] Again, species differences are apparent with these compounds; for pigeons, CBN only slightly increased discrimination of the THC. However, it should be borne in mind that these results may not be directly comparable because of differing routes of administration. Rats are usually administered THC intraperitoneally; pigeons are given THC via the intramuscular route. For hu-

TABLE 1A
Generalization[a] of Δ^9-THC (3 mg/kg) to Δ^9-THC Administered in Combination with Cannabinol (CBN) or Cannabidiol (CBD)[68,69]

Drug combination (mg/kg)				Time post-administration (h)			
				0.5	1.5	4.5	6.5
Δ^9-THC	0.0	CBD	0.0	0	0	0	0
			30.0	0	0	2	0
	0.3		0.0	27	8	0	0
			30.0	0	22	0	0
	1.0		0.0	66	41	6	0
			30.0	63	83	69	46
Δ^9-THC	0.0	CBN	0.0	0	0	0	0
			3.0	22	9	0	0
	0.1		0.0	1	0	0	2
			3.0	32	25	0	0
	0.3		0.0	28	8	0	0
			3.0	83	53	5	0
	1.0		0.0	67	41	9	1
			3.0	89	73	10	0

[a] See Table 1B.

From Järbe, T. U. C. and Mathis, D. A., The discriminative stimulus functions of cannabinoids and cannabimimetics, in *Drug Discrimination: Applications to Drug Abuse Research*, NIDA Res. Monogr., 116, Glennon, R. A., Järbe, T. U. C., and Frankenheim, J., Eds., U.S. Department of Health and Human Services, Alcohol, Drug Abuse and Mental Health Admin., Rockville, MD, 1991, 75.)

mans, Hollister[3] reported that coadministration of CBN and THC tended to prolong the cannabimimetic action. In addition, Karniol *et al.*[105] observed additive effects on some measures with humans when CBN was given together with THC (for additional references on interactions between THC, CBD, and CBN see References 68 and 69).

Interactions of Cannabidiol and Cannabinol

Relatively fewer studies have focused on the combined effects of CBN and CBD.[90,99] The data obtained thus far indicate that CBD may reduce the stimulus effects of CBN. As mentioned above, some studies have suggested that when animals are trained to discriminate THC, this discrimination is generalized to the CBN stimulus,[5,6,69,104] but not to the CBD stimulus.[5,64,68,106] When THC-discriminating rats were administered CBN and CBD in com-

TABLE 1B
Generalization[a] of Δ^9-THC (3 mg/kg) to Combinations of CBN and CBD

Drug combination (mg/kg)				Mean time post-administration[b] (h)	
				1.0	5.5
CBN	0.0	CBD	0.0	0.0	0.0
			10.0	0.0	0.0
			30.0	0.0	0.0
	10.0		0.0	31.0	22.0
			10.0	25.0	11.0
			30.0	24.0	5.0
	17.5		0.0	75.0	51.0
			10.0	39.0	30.0
			30.0	30.0	12.0

[a] The percentage of responding to the THC-appropriate position (%RDP) for 8 to 12 rats at each injection-test interval.
[b] Average across two time blocks (0.5 and 1.5; 4.5 and 6.5 h).

From Järbe, T. U. C. and Mathis, D. A., The discriminative stimulus functions of cannabinoids and cannibimimetics, in *Drug Discrimination: Applications to Drug Abuse Research*, NIDA Res. Monogr., 116, Glennon, R. A., Järbe, T. U. C., and Frankenheim, J., Eds., U.S. Department of Health and Human Services, Alcohol, Drug Abuse and Mental Health Admin., Rockville, MD, 1991, 75.)

bination, this generalization of THC to CBN was attenuated.[90] Thus, CBD may antagonize or modulate the THC-like, psychoactive effect of CBN. However, cannabis intoxication is not limited to the direct action of these compounds. As discussed below, cannabimimetic effects may also result partly from some metabolites of THC.

Cannabinoid Metabolites

The metabolism of THC is complex and varies between species.[107] Metabolites that have been isolated and examined with THC-trained animals include 11-hydroxy-Δ-THC (11-OH-THC). Collectively, the data indicate that 11-OH-THC is a major psychotropic metabolite, exerting THC-like activity of biological significance. Investigations with animals trained to discriminate between THC and the no-drug state indicate that the 11-hydroxylated forms of THC are similar to the Δ^9-THC stimulus. The cannabimimetic activity of 11-hydroxy metabolites of THC (11-OH-Δ^9-THC and 11-OH-Δ^8-THC) has been investigated using rats[6,85,100] and pi-

TABLE 2
ED_{50} Values for Generalization of Drug-Appropriate Responding to the Training Stimulus and THC Metabolites by Animals Trained to Discriminate Δ^9-THC

THC	Species	n	Time[a] (h)	ED_{50} (r)[b]
Δ^9	Rat[c]	8	0.5	1.15 (0.93)
			1.0	1.07 (0.98)
	Pigeon[d]	8	1.5	0.43 (0.94)
11-OH-Δ^9	Rat	7–8	0.5	0.65 (0.88)
	Pigeon	8	1.5	0.10 (0.99)
11-OH-Δ^8	Rat	8	0.5	1.12 (0.97)
	Pigeon	8	1.5	0.38 (0.97)
8β,11-di-OH-Δ^9	Pigeon	4–5	1.5	19.00 (0.99)

[a] Hours from injection until testing.
[b] The ED_{50} (mg/kg) values were determined from the correlation coefficient (r) by logarithmic regression analysis of gradients for percentage of drug-appropriate (%RDP) gradients.
[c] Rats were trained to discriminate 3 mg/kg administered i.p. 30 min prior to session onset.
[d] Pigeons were trained to discriminate 1 mg/kg administered i.m. 90 min prior to session onset.

From Järbe, T. U. C. and Mathis, D. A., The discriminative stimulus functions of cannabinoids and cannibimimetics, in *Drug Discrimination: Applications to Drug Abuse Research*, NIDA Res. Monogr., 116, Glennon, R. A., Järbe, T. U. C., and Frankenheim, J., Eds., U.S. Department of Health and Human Services, Alcohol, Drug Abuse and Mental Health Admin., Rockville, MD, 1991, 75.)

geons.[85] Both of these species generalized from the Δ^9-THC stimulus to the test compounds in a dose-related manner. However, 11-OH-Δ^9-THC was more potent than 11-OH-Δ^8-THC. The 11-OH-Δ^8-THC was at least as potent as Δ^9-THC. These potency estimates are similar to those reported for humans.[3,108,109]

Other THC metabolites that have been studied for their cannabimimetic effect are 8-α-OH-Δ^9-THC, 8-β-OH-Δ^9-THC, 8-α,11 di-OH-Δ^9-THC, and 8-β,11 di-OH-Δ^9-THC.[85,100] Only the latter compound, 8-β,11 di-OH-Δ^9-THC, produced complete generalization for the Δ^9-THC stimulus by pigeons. The metabolite, 8-β-OH-Δ^9-THC did produce partial generalization for the THC stimulus.[85] Other investigations utilized rats and these compounds did not find generalization for the THC stimulus with 8-β,11 di-OH-Δ^9-THC; some generalization occurred with 8-β-OH-Δ^9-THC.[100] However, because lower doses of the test compounds were used for the tests with rats and pigeons tend to be more sensitive than rats to the THC cue, further tests using higher doses of these compounds with rats must be conducted before a conclusion about species differences can be drawn. In addition, the two epimers of the 8-OH metabolite of Δ^9-THC have been examined in humans. Perez-Reyes et al.[110] concluded that only 8-β-OH-Δ^9-THC was weakly THC-like; whereas Hollister[3] reported that both epimers exert some cannabimimetic activity. Thus, apart from the 11-OH-THC metabolites, the remaining metabolites may possess THC-like action, but fairly high doses are needed to produce a cannabimimetic effect (see ED_{50} values, estimated according to a logarithmic regression analysis, presented in Table 2). Thus, these data suggest that the contribution of most of these metabolites to the psychoactive properties of cannabis in natural settings is very minor. However, examinations of the chemical structure of cannabinoids have helped to elucidate the compounds contributing to cannabis intoxication.

Summary

Studies using the DDL procedure with naturally occurring compounds have shown that cannabis intoxication is primarily induced by, but not limited to, the direct action of Δ^9-THC. The stimulus effects of THC may be modulated by coadministration of other cannabinoids such as CBD or CBN. In fact, CBN may have direct cannabimimetic properties that are attenuated by coadministration of CBD. In addition, cannibimimetic activity can be induced by some THC metabolites, specifically the 11-OH- and 8-β,11 di-OH-Δ^9-THC metabolites. However, in order to determine the chemical requirements for inducing cannabis intoxication, studies of the structure activity relationship (SAR) of cannabinoids using organic chemistry and DDL are necessary.

Though development of tolerance to the subjective effects of THC on the SDL phenomenon is apparent, its development to the THC discriminative stimulus is less likely. In the DDL paradigm, some perception of the induced effects is necessary in order to solve the discrimination problem. It must be borne in mind that in a task like conditioned avoidance, it is beneficial for the animal to overcome the unconditioned disruptive effects of THC on responding; in the DDL paradigm, it is beneficial to recognize and use the drug effect. Thus, the response contingency involved in a particular task may be critical for the development of tolerance to be observed. Thus, in many instances, tolerance may be best understood as behavioral adaptation to those drug effects that disrupt rewarded behavior,[15] rather than a result of mere exposure to the drug.

TABLE 3
Generalization[a] of Synthetic Compounds[b] to THC-Appropriate Responding

Compound	Species	n	Dose range[b]	ED_{50}	(r)[c]
Δ^8-THC					
(+)-Δ^8-THC	Rat[d]	4	5.6–10	None	(None)
(–)-Δ^8-THC-DMH	Rat	14–16	0.1–0.56	0.21	(0.94)
	Pigeon[e]	6–7	0.0175–0.175	0.05	(0.95)
(–)-11-OH-Δ^8-THC-DMH	Rat	11	0.003–0.03	0.01	(0.99)
	Pigeon	5–7	0.0001–0.0056	0.002	(0.99)
(+)-11-OH-Δ^8-THC-DMH	Rat	4–9	3–10	None	(None)
	Pigeon	4	3–10	None	(None)
Benzofuran 7	Rat	9	Up to 10	None	(None)
	Pigeon	5	1–5.6	3.57	(0.78)
Benzofuran 8	Rat	9	0.3–3.0	0.72	(0.98)
	Pigeon	5	0.1–0.3	0.17	(0.99)
Δ^{10}-THC					
(9*S*,6a*R*)-$\Delta^{10,10a}$-THC	Pigeon	2–6	1–17.5	None	(None)
(9*R*,6a*R*)-$\Delta^{10,10a}$-THC	Pigeon	3–4	3–17.5	4.78	(0.92)
(9*S*)-$\Delta^{10a,6a}$-THC	Pigeon	3–4	1–10	6.63	(0.81)
(9*R*)-$\Delta^{10a,6a}$-THC	Pigeon	4	3–30	11.90	(0.98)
Hexahydrocannabinols (HHC)					
(–)-11-OH-HHC (equatorial)	Rat	14	0.1–1	0.24	(0.95)
	Pigeon	7	0.01–0.1	0.02	(0.87)
(–)-11-OH-HHC (axial)	Rat	9	1–3	1.58	(0.98)
	Pigeon	7	1–3	1.72	(0.97)

[a] The percentage of responding to the drug-appropriate position (%RDP) at the time post-administration when highest potency was disclosed.

[b] Chemical structures were found in original reports cited in text.

[c] The ED_{50} (mg/kg) values were determined from the correlation coefficient (r) by logarithmic regression analysis of gradients.

[d] Rats were trained to discriminate 3 mg/kg administered i.p. 30 min prior to session onset.

[e] Pigeons were trained to discriminate 0.56 mg/kg administered i.m. 90 min prior to session onset.

From Järbe, T. U. C. and Mathis, D. A., The discriminative stimulus functions of cannabinoids and cannibimimetics, in *Drug Discrimination: Applications to Drug Abuse Research*, NIDA Res. Monogr., 116, Glennon, R. A., Järbe, T. U. C., and Frankenheim, J., Eds., U.S. Department of Health and Human Services, Alcohol, Drug Abuse and Mental Health Admin., Rockville, MD, 1991, 75.

STRUCTURE ACTIVITY RELATIONSHIP (SAR)

The SAR between chemical structure and drug action reveals the structural requirements for the stereoselectivity or stereospecificity of a given drug effect or action. Stereoselectivity is evidenced when a potency difference for a given effect is obtained with a particular structure. Stereospecificity refers to a complete absence of a drug effect for a particular structure. A major purpose of the SAR investigations with synthetic cannabinoids and cannabimimetics is to eliminate the enantiomer that induces the undesirable psychoactive properties of these compounds (stereospecificity for a THC-like effect), and retain the structures that induce medically useful effects. As reviewed below, these studies have shown that the THC-like properties of these compounds are highly dependent on particular structural configurations (see Table 3 for compounds tested in our laboratory; chemical structures are indicated in the original reports cited below).

Δ^9-THC

The relationship between the structure of cannabinoids and their stimulus effects has been studied using Δ^9-(11)-THC and (3″S)-3″-OH-Δ^9-THC. The results of these studies indicate that the stimulus effects of Δ^9-(11)-THC are less potent than those of Δ^9-THC[92] (see also References 109 and 111). However, the effect of this compound on the development of rapid tolerance to the stimulus effects of Δ^9-THC is similar to the effect of the parent compound. For animals pretreated with high doses of either Δ^9-THC or Δ^9-(11)-THC, and then tested for discrimination of the THC cue 24 to 48 h after these high-dose administrations, discrimination of the THC stimulus is attenuated. Thus, though Δ^9-(11)-THC may be less potent than the parent compound, it apparently does have some similar psychoactive effects.

On the other hand, (3″S)-3″-OH-Δ^9-THC may be more active than Δ^9-THC and the *R* epimer of this compound, and may be an important stereoselective determinant of the THC cue. For example, for rats trained to discriminate Δ^9-THC, (3″S)-3″-OH-Δ^9-THC was approximately five times more potent than the training compound, and approximately seven times more potent than the *R* epimer.[112] This difference in the effect of epimers demonstrates the stereoselectivity of cannabimimetic effects. Selectivity of the relationship between *R* and *S* epimers and cannabimimetic activity is further demonstrated by studies with Δ^{10}-THC structures.

Δ^{10}-THC

Racemic $\Delta^{10a,6a}$-THC and the related compound Synhexyl were previously reported to be cannabimimetic in humans.[3] Recent synthesis of stereoisomers for both $\Delta^{10,10a}$- and $\Delta^{10a,6a}$-THC[113] has permitted further investigations of the relationship of *R* and *S* epimers to cannabimimetic activity. Investigations of the SAR of $\Delta^{10a,6a}$-THC have focused on the (9*S*) and (9*R*) enantiomers. In general, these studies indicate that, for this structure, though the (9*S*) enantiomer is less potent than Δ^9-THC, the (9*R*) enantiomer may not be active. The psychoactive properties of the $\Delta^{10a,6a}$ enantiomers have recently been tested in humans[114] and pigeons.[115] For humans, the (9*S*) enantiomer had psychoactive properties that were qualitatively similar to the effects of Δ^9-THC, but the effects of this enantiomer were less potent (i.e., References 1 and 3 to 6). For pigeons, discrimination of the Δ^9-THC stimulus was generalized to this enantiomer. When the psychoactive properties of the (9*R*) enantiomer were tested with humans, it was found not to be active.[114] However, for the tests with humans, this compound could not be tested in high doses. In contrast, for pigeons, this (9*R*) enantiomer did generalize for the THC stimulus, but it was two to three times less potent. In addition, these animal tests utilized the acetate form of this compound, rather than the parent phenol that was tested in humans. Thus, further tests with this enantiomer must be conducted to determine its efficacy as a cannabimimetic.

Examinations of the structural activity of $\Delta^{10,10a}$-THC has focused on the (9*R*,6a *R*) and (9*S*,6a *R*) epimers in their acetate form. The (9*R*,6a *R*) epimer disclosed THC-like activity (ED_{50} = 4.78 mg/kg), whereas the (9*S*,6a *R*) epimer did not, even at the highest dose tested (17.5 mg/kg) with pigeons.[115] Thus, these SAR investigations of Δ^{10}-THC compounds suggest a stereoselectivity for induction of cannabimimetic action. The effect of *S* and *R* epimers of $\Delta^{10,10a}$-THC on cannabimimetic activity parallels findings obtained with the epimers of 11-hydroxy-hexahydrocannabinol (see below). Other SAR studies of THC effects have focused on the (+)- and (–)-enantiomers, particularly with regard to Δ^8-THC compounds.

Δ^8-THC

The stereoselectivity of Δ^8-THC may be a function of (–)- and (+)-enantiomers. For example, (+)-Δ^8-THC is considerably less potent[77] than (–)-Δ^8-THC[4,5,67] in eliciting the Δ^9-THC discriminative response. In addition, tests with THC-trained rats and pigeons[77,116] have shown that Δ^8-THC-

DMH in its *levo* rotation [(–)-Δ^8-THC-DMH] is a potent cannabimimetic, having a slow onset and long duration.

In order to examine a possible dissociation between the effects of Δ^8-THC enantiomers, monohydroxylated dimethylheptyl homologues of a major metabolite of Δ^8-THC (11-OH-Δ^8-THC) have been studied [i.e., (–)-11-OH-Δ^8-THC-DMH and (+)-11-OH-Δ^8-THC-DMH]. These studies also indicated that the (–)-enantiomer may elicit the cannabimimetic effects of Δ^8-THC. The compound (–)-11-OH-Δ^8-THC-DMH was more potent than naturally occurring Δ^9-THC by approximately 87 times in rats (ED_{50} = 0.01 mg/kg), and 73 times in pigeons (ED_{50} = 0.002 mg/kg).[116] However, the (+)-enantiomer [(+)-11-OH–Δ^8-THC-DMH] was much less potent than the (–)-enantiomer in generalization for the THC cue. It was inactive at doses higher than the ED_{50} value by approximately 1000 times for rats and 4500 times for pigeons. Furthermore, similar effects of these enantiomers on cannabinoid activity have been obtained for performance on the ring test with mice and the rotorod neurotoxicity test with rats.[2,117] Recently obtained data (unpublished) also indicate that this stereoisomeric differentiation is important for cannabimimetic activity by naturally occurring compounds. Stereospecificity of the (+)-isomer was obtained with pigeons trained to discriminate between Δ^9-THC (0.56 mg/kg) and vehicle. Tests with the (+)- and (–)-isomers of the major psychoactive metabolite of Δ^8-THC (11-OH-Δ^8-THC) indicated that the (+)-isomer did not induce THC-like responding after administration of doses up to 10 mg/kg. Studies with hexahydrocannabinols have further demonstrated the structural requirements for cannabimimetic effects, including the importance of the position of the methyl group for cannabimimetic activity.

Hexahydrocannabinols

Tests with hexahydrocannabinols indicate that conversion of THC to 11-OH-THC is *not* a necessary prerequisite to achieve a THC effect,[6,64,100] and that hexahydrocannabinols may be highly stereoselective and stereospecific. For example, examinations of the relative potencies of the hexahydrocannabinols — levonantradol and dextronantradol — have demonstrated that levonantradol was considerably more potent than dextronantradol for eliciting the THC-appropriate response. In addition, the dextro form did not substitute for Δ^9-THC at the dose of 3.2 mg/kg, whereas the levo form disclosed an ED_{50} value of 0.02 mg/kg, making it one of the more potent cannabimimetics examined thus far with drug discrimination of Δ^9-THC.[64] Levonantradol also has been found to exhibit THC-like activity in humans.[118] Accordance between humans and animals with other cannabimimetics has been demonstrated.[6,64,100,119,120]

Two epimers of 11-OH hexahydrocannabinol elicited cannabimimetic activity in both rats and pigeons.[91] The difference in the potency of these isomers was determined by the position of the methyl group in relation to the plane of the cyclohexane ring. Generalization to the THC stimulus was more potent for the isomer in which the methyl group is equatorial (i.e., lies approximately in the plane of the cyclohexane ring) than for the isomer in which the methyl group was axial (i.e., protrudes from the plane of the cyclohexane ring). However, species differences in response to these compounds make elucidation of their SAR less clear. For example, the difference in the relationship between the structure of isomers and cannabimimetic activity is less for rats than for pigeons (for chemical structures, see Reference 91).

Noncannabinoid Structures

Though most of the studies on the SAR of cannabinoids support the contention that a dihydrobenzopyran structure is a necessary component for cannabimimetic activity,[2,8] some studies suggest that this structure is not a central requirement for THC-like activity. For example, an examination of a derivative of 3-phenylcyclohexanol (CP47497), which lacks this structure, indicates that this compound has potent cannabimimetic activity.[121] In addition, other tests conducted with rats and pigeons trained to discriminate Δ^9-THC indicated that there was profound generalization to a DMH homolog of CP47497 (unpublished data).

Other compounds that lack this structure but have cannabimimetic properties are benzofuran cannabimimetics. These compounds have a benzofuran rather than the Δ^9-THC benzopyran moiety and possess marked cannabimimetic activity in both rats and pigeons.[122]

Analgesia and Cannabimimetic Activity

It is not yet known why compounds that lack a dihydrobenzopyran structure have cannabimimetic properties. One suggestion might be that generalization of the THC response to the effects of these compounds may result from analgesic actions. CP47497, levonantradol, and Δ^9-THC all have analgesic properties. Thus, if analgesia is part of the stimulus effects of THC, then generalization to these compounds may reflect generalization of THC analgesia. However, the analgesic properties of THC have been shown to be dissociated from its discriminative stimulus effects. Other compounds having a standard cannabinoid configuration and strong analgesic properties [e.g., the dextro form of THC such as (+)-11-OH-Δ^8-THC-DMH] do not have cannabimimetic properties.[116,117] Therefore, the possibility that the cannabimimetic properties of compounds that lack a dihydrobenzopyran structure reflect their analgesic action seems unlikely, and firm confirmation of the chemical structure needed to induce THC intoxication remains to be elucidated.

Potency

The relative activity of the cannabimimetic compounds examined with DDL procedures disclose a potency range of several thousand. At the lower end of the scale we find 9-normethyl-Δ^8-THC, which has a cannabimimetic activity approximately 50 times less than that of Δ^9-THC.[100] On the upper end of the scale, we find the very potent cannabimimetics (–)-11-OH-Δ^8-THC-DMH and levonatradol, which display potencies of around 80 to 40 times, respectively, more than that of Δ^9-THC.[64,116] Where direct comparisons are possible, a strong concordance between the potency of cannabimimetic compounds in humans and animals has been obtained. For example, comparisons of the clinical potency of Δ^8-, Δ^9-, and the 11-OH-metabolites of THC with data obtained from animals tested for the intensity of a THC-like stimulus induced by these compounds indicate that there is a strong agreement between the human and animal results. This concordance has also been shown with other cannabimimetics.[6]

Summary

Determining the necessary chemical components for inducing the discriminative stimulus properties of Δ^9-THC is important for understanding the factors that contribute to cannabis abuse and the unwanted side effects of cannabis therapy. Using DDL procedures, the cannabimimetic and medicinal properties of cannabinoids (e.g., analgesia, antiemesis, and anticonvulsive) have been dissociated, and some requirements for inducing THC intoxication have been delineated.

As mentioned above, examinations of cannabinoid SAR have shown that the structures determining THC intoxication vary among compounds and are highly stereoselective, and may even be stereospecific. Some of the relevant structural features that have been identified are epimeric and enantiomeric preferences [e.g., *R* and *S* epimers, (+) and (–) enantiomers], and the position of the C-11 substituent on the cyclohexane ring. For example, the (3″S)-3″-OH-Δ^9-THC is more potent than the *R* epimer. For $\Delta^{10,10a}$-

THC, the (9*R*,6a *R*) epimer disclosed THC-like activity, whereas the (9*S*,6a *R*) epimer did not. In addition, the (9*S*) enantiomer for $\Delta^{10a,6a}$-THC is more potent than the (9*R*) enantiomer.

Stereoselectivity of THC intoxication is also demonstrated with Δ^8-THC enantiomers and hexahydrocannabinol isomers. For example, generalization of Δ^8-THC for the THC cue may be induced by the (–)- but not the (+)-enantiomer. In addition, SAR studies of hexahydrocannabinols have demonstrated that cannabimimetic activity also shows isomeric selectivity. For the compounds examined with both rats and pigeons, the equatorial isomer is more potent than the axial isomer. Furthermore, some SAR studies suggest that a dihydrobenzopyran structure is not a necessary requirement for cannabimimetic activity. A derivative of 3-phenylcyclohexanol that lacks this structure (CP47497), its DMH homolog, a hexahydrocannabinol (levonantradol), and a benzofuran cannabinoid, have potent cannabimimetic activity.

Though the complexity of the SAR of cannabinoids and cannabimimetics is apparent, the utility of DDL and synthesis of new compounds as research tools for understanding this SAR is also apparent. It is only with further use of these research tools that a firm understanding of cannabimimetic activity may be gained.

ACKNOWLEDGMENTS

This work was supported by the Swedish Medical Research Council (MFR 09321-02B). The manuscript was completed while T. U. C. Järbe was Visiting Scientist at Hahnemann University, Philadelphia, PA.

REFERENCES

1. **Culliton, B. J.,** Pot facing stringent scientific examination, *Sci. News*, 97, 102, 1970.
2. **Mechoulam, R. and Feigenbaum, J. J.,** Towards cannabinoid drugs, in *Progress in Medicinal Chemistry,* Vol. 24, Ellis, G. P. and West, G. B., Eds., Elsevier/North Holland Biomedical Press, Amsterdam, 1987, 159.
3. **Hollister, L. E.,** Structure-activity relationships in man of cannabis constituents, and homologs and metabolites of delta-9-tetrahydrocannabinol, *Pharmacology*, 11, 3, 1974.
4. **Järbe, T. U. C. and Henriksson, B. G.,** Discriminative response control produced with hashish, tetrahydrocannabinols (delta-8-THC and delta-9-THC), and other drugs, *Psychopharmacologia (Berlin),* 40, 1, 1974.

5. **Järbe, T. U. C., Henriksson, B. G., and Ohlin, G. C.,** Delta-9-THC as a discriminative cue in pigeons: effects of delta-8-THC, CBD, and CBN, *Arch. Int. Pharmacodyn. Ther.*, 228, 68, 1977.
6. **Weissman, A.,** Generalization of the discriminative stimulus properties of delta-9-tetrahydrocannabinol to cannabinoids with therapeutic potential, in *Stimulus Properties of Drugs: Ten Years of Progress*, Colpaert, F. C. and Rosecrans, J. A., Eds., Elsevier/North Holland Biomedical Press, Amsterdam, 1978, 99.
7. **Mechoulam, R. and Edery, H.,** Structure-activity relationships in the cannabinoid series, in *Marijuana*, Mechoulam, R., Ed., Academic Press, New York, 1973, 101.
8. **Bhargava, H. N.,** Potential therapeutic applications of naturally occurring and synthetic cannabinoids, *Gen. Pharmacol.*, 9, 195, 1978.
9. **Barry, H., III and Kubena, R. K.,** Discriminative stimulus characteristics of alcohol, marihuana and atropine, in *Drug Addiction: Experimental Pharmacology*, Singh, J. M., Miller, L., and Lal, H., Eds., Futura Publishing, New York, 1972, 3.
10. **Krimmer, E. C. and Barry, H., II I.,** Discriminable stimuli produced by marihuana constituents, in *Discriminative Stimulus Properties of Drugs*, Lal, H., Ed., Plenum Press, New York, 1977, 121.
11. **Balster, R. L. and Ford, R. D.,** The discriminative stimulus properties of cannabinoids: a review, in *Drug Discrimination and State Dependent Learning*, Ho, B. T., Richards, D. W., III, and Chute, D. L., Eds., Academic Press, New York, 1978, 131.
12. **Järbe, T. U. C., Hiltunen, A. J., and Mechoulam, R.,** Subjectively experienced cannabis effects in animals, *Drug Dev. Res.*, 16, 385, 1989.
13. **Järbe, T. U. C. and Mathis, D. A.,** The discriminative stimulus functions of cannabinoids and cannabimimetics, in *Drug Discrimination: Applications to Drug Abuse Research*, NIDA Res. Monogr., 116, Glennon, R. A., Järbe, T. U. C., and Frankenheim, J., Eds., U.S. Department of Health and Human Services, Alcohol, Drug Abuse and Mental Health Admin., Rockville, MD, 1991, 75.)
14. **Overton, D. A.,** Contextual stimulus effects of drugs and internal states, in *Context and Learning*, Balsam, P. D., and Tomie A. M., Eds., L. Erlbaum Associates, Hillsdale, NJ, 1985, 357.
15. **Järbe, T. U. C.,** State-dependent learning and drug discriminative control of behaviour: an overview. *Acta Neurol. Scand.*, 74(Suppl. 109), 37, 1986.
16. **Siegel, S.,** Drug dissociation in the nineteenth century, in *Drug Discrimination: Applications in CNS Pharmacology*, Colpaert, F. C. and Slangen, J. L., Eds., Elsevier/North Holland Biomedical Press, Amsterdam, 1982, 257.
17. **Siegel, S.,** Wilkie Collins: Victorian novelist as psychopharmacologist, *J. Hist. Med. Allied Sci.*, 38, 161, 1983.
18. **Overton, D. A.,** State dependent learning and drug discrimination, in *Handbook of Psychopharmacology*, Vol. 18, Iversen L. L., Iversen, S. D., and Snyder, S. H., Eds., Plenum Press, New York, 1984, 59.
19. Most experiments in SDL utilized the 2 × 2 design. Although this design was important for evaluating and understanding SDL phenomena, there are many problems inherent in using a single response performed in a particular state (see References 18 and 20). It is beyond the scope of this chapter to detail the confounding factors examined by Overton[18,20] and statistical considerations outlined by Swanson and Kinsbourne.[21] However, many more groups of experimental subjects than those envisaged in the 2 × 2 design may be needed to evaluate SDL effects properly. For example, Bignami and Rosic[22] used an elaborate 2 × 2 × 2 × 2 design and demonstrated that for the anticholinergic agent, scopolamine, an SDL effect was not the main causative factor for the results of that experiment.

20. **Overton, D. A.,** Experimental methods for the study of state-dependent learning, *Fed. Proc.*, 33, 1800, 1974.
21. **Swanson, J. M. and Kinsbourne, M.,** The 2 × 2 design reconsidered: limitations imposed by the statistical models, in *Stimulus Properties of Drugs: Ten Years of Progress*, Colpaert, F. C. and Rosecrans, J. A., Eds., Elsevier/North Holland Biomedical Press, Amsterdam, 1978, 467.
22. **Bignami, G. and Rosic, N.,** Acquisition and performance effects of scopolamine and of treatment withdrawal in avoidance situations, *Physiol. Behav.*, 8, 1127, 1972.
23. **Girden, E. and Culler, E. A.,** Conditioned responses in curarized striate muscle in dogs, *J. Comp. Psychol.*, 23, 261, 1937.
24. **Case, T. J. and Funderbank, W. H.,** An effect of curare on the central nervous system, *Trans. Am. Neurol. Assoc.*, 72, 195, 1947.
25. **Gardner, L. A.,** An Experimental Re-Examination of the Dissociative Effects of Curareform Drugs, Masters thesis, Department of Psychology, Oberlin College, Oberlin, OH, 1961.
26. **Gardner, L. A. and McCollough, C.,** A reinvestigation of the dissociative effect of curareform drugs, *Am. Psychol.*, 17, 398, 1962.
27. **Overton, D. A.,** State-dependent or "dissociated" learning produced with pentobarbital, *J. Comp. Physiol. Psychol.*, 57, 3, 1964.
28. **Holmgren, B.,** Conditioned avoidance reflex under pentobarbital, *Biol. Inst. Estud. Med. Biol. (Mexico)*, 22, 21, 1964.
29. **Overton, D. A.,** Major theories of state dependent learning, in *Drug Discrimination and State Dependent Learning*, Ho, B. T., Richards, D. W., III, and Chute, D. L., Eds., Academic Press, New York, 1978, 283.
30. **Eich, J. E., Weingartner, H., Stillman, R. C., and Gillin, J. C.,** State-dependent accessibility of retrieval cues in the retention of a categorized list, *J. Verb. Learn. Verb. Behav.*, 14, 408, 1975.
31. **Barry, H., III and Kubena, R. K.,** Repeated high doses of delta-1-tetrahydrocannabinol enhance acquisition of shock avoidance by rats, *Proc. 79th Ann. Conv., APA*, 6, 747, 1971.
32. **Waser, P. G., Martin, A., and Heer-Carcano, L.,** The effect of delta-9-tetrahydrocannabinol and LSD on the acquisition of an active avoidance response in the rat, *Psychopharmacologia (Berlin)*, 46, 249, 1976.
33. **Henriksson, B. G. and Järbe, T.,** The effect of two tetrahydrocannabinols (delta-9-THC and delta-8-THC) on conditioned avoidance learning in rats and its transfer to normal state conditions, *Psychopharmacologia (Berlin)*, 22, 23, 1971.
34. **Henriksson, B. G. and Järbe, T.,** Effects of diazepam on conditioned avoidance learning in rats and its transfer to normal state conditions, *Psychopharmacologia (Berlin)*, 20, 186, 1971.
35. **Delini-Stula, A.,** Suppression of conditioned avoidance behavior and the development of tolerance to delta-8-tetrahydrocannabinol and dimethylpyran in rats, *Pharmacopsychiat. Neuropsychopharmacol.*, 6, 189, 1973.
36. **Goldberg, M. E., Hefner, M. A., Robichaud, R. C., and Dubinsky, B.,** Effects of delta-9-tetrahydrocannabinol (THC) and chlordiazepoxide (CDP) on state-dependent learning: evidence for asymmetrical dissociation, *Psychopharmacologia (Berlin)*, 30, 173, 1973.
37. **Grunfeld, Y. and Edery, H.,** Psychopharmacological activity of the active constituents of hashish and some related cannabinoids, *Psychopharmacologia (Berlin)*, 14, 200, 1968.

38. **Robichaud, R. C., Hefner, M. A., Anderson, J. E., and Goldberg, M. E.,** Effects of delta-9-tetrahydrocannabinol (THC) on several rodent learning paradigms, *Pharmacology*, 10, 1, 1973.
39. **Herring, B.,** The effect of l-delta-1-*trans*-tetrahydrocannabinol on learning and retention of avoidance performance in rats, *Psychopharmacologia (Berlin)*, 26, 401, 1972.
40. **Glick, S. G. and Milloy, S.,** Tolerance, state-dependency and long-term behavioral effects of delta-9-THC, in *Current Research in Marijuana*, Lewis, M. F., Ed., Academic Press, New York, 1972.
41. **Järbe, T. U. C. and Henriksson, B. G.,** Effects of delta-8-THC an delta-9-THC on the acquisition of a discriminative positional habit in rats. The transitions between normal and tetrahydrocannabinol-induced states on reversal learning, *Psychopharmacologia (Berlin)*, 31, 321, 1973.
42. **Johansson, J. O., Henriksson, B. G., and Järbe, T. U. C.,** Effects of delta-8-THC on dissociation of conditioned avoidance responding in tolerant and nontolerant rats, *Physiol. Psychol.*, 2, 431, 1974.
43. **Järbe, T. U. C.,** Delta-9-tetrahydrocannabinol: tolerance after noncontingent exposure in rats, *Arch. Int. Pharmacodyn. Ther.*, 231, 49, 1978.
44. **Abel, E. L.,** Marijuana and memory, *Nature (London)*, 227, 1151, 1970.
45. **Rickles, W. H., Jr., Cohen, M. J., Whitaker, C. A., and McIntyre, K. E.,** Marijuana induced state-dependent verbal learning, *Psychopharmacologia (Berlin)*, 30, 349, 1973.
46. **Hill, S. H., Schwin, R., Powell, B., and Goodwin, D. W.,** State-dependent effects of marihuana on human memory, *Nature (London)*, 243, 241, 1973.
47. **Stillman, R. C., Weingartner, H., Wyatt, R. J., Gillin, C., and Eich, J.,** State-dependent (dissociative) effects of marihuana on human memory, *Arch. Gen. Psychiatry*, 31, 81, 1974.
48. **Darley, F. C., Tinklenberg, J. R., Roth, W. T., and Atkinson, R. C.,** The nature of storage deficits and state-dependent retrieval under marihuana, *Psychopharmacologia (Berlin)*, 37, 139, 1974.
49. **Eich, J. E., Weingartner, H., Stillman, R. C., and Gillin, J. C.,** State-dependent accessibility of retrieval cues in the retention of a categorized list, *J. Verb. Learn. Verb. Behav.*, 14, 408, 1975.
50. **Beautrais, A. L. and Marks, D. F.,** A test of state dependency effects in marihuana intoxication for the learning of psychomotor tasks, *Psychopharmacologia (Berlin)*, 46, 37, 1976.
51. **Cohen, M. J. and Rickles, W. H., Jr.,** Performance on a verbal learning task by subjects of heavy past marihuana usage, *Psychopharmacologia (Berlin)*, 37, 323, 1974.
52. **Overton, D. A., Leonard, W. R., and Merkle, D. A.,** Methods for measuring the strength of discriminable drug effects, *Neurosci. Biobehav. Rev.*, 10, 251, 1986.
53. **Colpaert, F. C.,** Drug discrimination: methods of manipulation, measurement and analysis, in *Methods of Assessing the Reinforcing Properties of Abused Drugs*, Bozarth, M. A., Ed., Springer-Verlag, New York, 1987, 341.
54. **Järbe, T. U. C.,** Drug discrimination learning — cue properties of drugs, in *Experimental Psychopharmacology*, Greenshaw, A. J. and Dourish C. T., Eds., Humana Press, Clifton, NJ, 1987, 433.
55. **Järbe, T. U. C.,** Discrimination learning with drug stimuli, in *Neuromethods, Vol 13: Psychopharmacology.*, Boulton, A. A., Baker, G. B., and Greenshaw, A. J., Eds., Humana Press, Clifton, NJ, 1989, 513.
56. **Overton, D. A.,** Applications and limitations of the drug discrimination method for the study of drug abuse, in *Methods of Assessing the Reinforcing Properties of Abused Drugs*, Bozarth, M. A., Ed., Springer-Verlag, New York, 1987, 340.

57. **Lal, H. and Emmett-Oglesby, M. W.,** Behavioral analogues of anxiety: animal models, *Neuropharmacology*, 22, 1423, 1983.
58. **Spealman, R. D.,** Discriminative-stimulus effects of midazolam in squirrel monkeys: comparison with other drugs and antagonism by Ro 15-1788, *J. Pharmacol. Exp. Ther.*, 235, 456, 1985.
59. **Woolverton, W. L., Kamien, J. B., and Goldberg, L. I.,** Pharmacological analysis of the apomorphine discriminative stimulus in Rhesus monkeys, *J. Pharmacol. Exp. Ther.*, 241, 213, 1987.
60. **Mathis, D. A. and Emmett-Oglesby, M. W.,** Quantal vs. graded generalization in drug discrimination: measuring a graded response, *J. Neurosci. Meth.*, 31, 23, 1990.
61. **Weissman, A.,** On the definition of cannabinoids; Botanical? Chemical? Pharmacological?, *J. Clin. Pharmacol.*, 21, 159S, 1981.
62. **Mechoulam, R. and Gaoni, Y.,** A total synthesis of delta-9-tetrahydrocannabinol, the active constituent of hashish, *J. Am. Chem. Soc.*, 87, 3273, 1965.
63. **Kubena, R. K. and Barry, H., III**, Stimulus characteristics of marihuana components, *Nature (London)*, 235, 397, 1972.
64. **Browne, R. G. and Weissman, A.,** Discriminative stimulus properties of delta-9-tetrahydrocannabinol: mechanistic studies, *J. Clin. Pharmacol.*, 21, 227S, 1981.
65. **Calil, H. M.,** Screening hallucinogenic drugs. II. Systematic study of two behavioral tests, *Psychopharmacology*, 56, 87, 1978.
66. **Järbe, T. U. C. and Ohlin, G. C.,** Stimulus effects of delta-9-THC and its interaction with naltrexone and catecholamine blockers in rats, *Psychopharmacology*, 54, 193, 1977.
67. **Järbe, T. U. C., Johansson, J. O., and Henriksson, B. G.,** Characteristics of tetrahydrocannabinol (THC)-produced discrimination in rats, *Psychopharmacology*, 48, 181, 1976.
68. **Hiltunen, A. J. and Järbe, T. U. C.,** Interactions between delta-9-tetrahydrocannabinol and cannabidiol as evaluated by drug discrimination procedures in rats and pigeons, *Neuropharmacology*, 25, 133, 1986.
69. **Järbe, T. U. C. and Hiltunen, A. J.,** Cannabimimetic activity of cannabinol in rats and pigeons, *Neuropharmacology*, 26, 219, 1987.
70. **Sofia, R. D.,** Cannabis: structure-activity relationships, in *Handbook of Psychopharmacology*, Iversen, L. L., Iversen, S. D., and Snyder, S. H., Eds., Plenum Press, New York, 1978, 319.
71. **Mokler, D. J., Nelson, B. D., Harris, L. S., and Rosecrans, J. A.,** The role of benzodiazepine receptors in the discriminative stimulus properties of delta-9-tetrahydrocannabinol, *Life Sci.*, 38, 1581, 1986.
72. **Sethi, B. B., Trivedi, J. K., Kumar, P., Gulati, A., Agarwal, A. K., and Sethi, N.,** Antianxiety effects of cannabis: involvement of central benzodiazepine receptors, *Biol. Psychiatry,* 21, 3, 1986.
73. **Sung, S.-L. and Jacubovic, A.,** Interaction of a water-soluble derivative of delta-9-tetrahydrocannabinol with [^{3}H]diazepam and [^{3}H] flunitrazepam binding to rat membranes, *Prog. Neuropsychopharmacol. Biol. Psychiatry*, 11, 335, 1987.
74. **Järbe, T. U. C. and Hiltunen, A. J.,** Limited stimulus generalization between delta-9-THC and diazepam in pigeons and gerbils, *Psychopharmacology*, 94, 328, 1988.
75. **Järbe, T. U. C., Ostlund, A., and Hiltunen, A. J.,** Cueing effects of anxiolytic benzodiazepines (DZP and Ro 11-3128): stereospecificity and antagonism by the convulsive benzodiazepine Ro 5-3663, *Psychopharmacology*, 94, 501, 1988.
76. **Dewey, W. L.,** Cannabinoid pharmacology, *Pharmacol. Rev.*, 38, 151, 1986.

77. **Järbe, T. U. C., Swedberg, M. D. B., and Mechoulam, R.,** A repeated tests procedure to assess onset and duration of the cue properties of (-)-delta-9-THC, (-)-delta-8-THC-DMH and (+)-delta-8-THC, *Psychopharmacology*, 75, 152, 1981.
78. **DeVane, W. A., Dysarz, F. A., Johnson, M. R., Melvin, L. S., and Howlett, A. C.,** Determination and characterization of a cannabinoid receptor in rat brain, *Mol. Pharmacol.*, 34, 605, 1989.
79. **Herkenham, M., Lynn, A. B., Little, M. D., Johnson, M. R., Melvin, L. S., DeCosta, B. R., and Rice, K. C.,** Cannabinoid receptor localization in brain, *Proc. Natl. Acad. Sci. U.S.A.*, 87, 1932, 1990.
80. **Makriyannis, A. and Rapaka, R. S.,** The molecular basis of cannabinoid activity, *Life Sci.*, 47, 2173, 1990.
81. **Mokler, D. J. and Rosecrans, J. A.,** Discriminative stimulus properties of intracranial administration of delta-9-tetrahydrocannabinol, *Drug Dev. Res.*, 16, 395, 1989.
82. **Järbe, T. U. C., Johansson, J. O., and Henriksson, B. G.,** Delta-9-tetrahydrocannabinol and pentobarbital as discriminative cues in the mongolian gerbil *(Meriones unguiculatus), Pharmacol. Biochem. Behav.*, 3, 403, 1975.
83. **Johansson, J., Järbe, T. U. C., and Henriksson, B. G.,** Acute and subchronic influences of tetrahydrocannabinols on water and food intake, body weight, and temperature in male rats, *T.-I.-T. J. Life Sci.*, 5, 17, 1975.
84. **O'Neal, M. F., Means, L. W., Porter, J. H., Rosecrans, J. A., and Mokler, D. J.,** Rats that acquire a THC discrimination more rapidly are more sensitive to THC and faster in reaching operant criteria, *Pharmacol. Biochem. Behav.*, 29, 67, 1988.
85. **Järbe, T. U. C. and McMillan, D. E.,** Delta-9-THC as a discriminative stimulus in rats and pigeons: generalization to THC-metabolites and SP-111, *Psychopharmacology*, 71, 281, 1980.
86. **Mackintosh, N. J.,** *The Psychology of Animal Learning*, Academic Press, New York, 1974.
87. **Heinemann, E. G. and Chase, S.,** Stimulus generalization, in *Conditioning and Behavior Theory*, Estes, W. K., Ed., Lawrence Erlbaum Associates, Hillsdale, NJ, 1975, 305.
88. **DeWitte, P.,** Hierarchy on the perceptual dimensions of hypothalamic stimulation: the psychophysics of rewarding cues, and effects of sensory stimulation on self-stimulation, in *Stimulus Properties of Drugs: Ten Years of Progress*, Colpaert, F. C. and Rosecrans, J. A., Eds., Elsevier/North Holland Biomedical Press, Amsterdam, 1978, 483.
89. **Barry, H., III and Krimmer, F. C.,** Discriminative delta-9-tetrahydrocannabinol stimulus tested with several doses, routes, intervals, and a marihuana extract, in *Pharmacology of Marihuana*, Braude, M. C. and Szara, S., Eds., Academic Press, New York, 1976, 535.
90. **Hiltunen, A. J. and Järbe, T. U. C.,** Cannabidiol attenuates delta-9-tetrahydrocannabinol like discriminative stimulus effects of cannabinol, *Eur. J. Pharmacol.*, 125, 301, 1986.
91. **Järbe, T. U. C., Hiltunen, A. J., Lander, N., and Mechoulam, R.,** Cannabimimetic activity (delta-1-THC cue) of cannabidiol monomethyl ether and two stereoisomeric hexahydrocannabinols in rats and pigeons, *Pharmacol. Biochem. Behav.*, 25, 393, 1986.
92. **Semjonow, A. and Binder, M.,** Generalization of the discriminative stimulus properties of delta-9-THC to delta-9(11)-THC in rats, *Psychopharmacology*, 85, 178, 1985.
93. **Hirschhorn, E. D. and Rosecrans, J. A.,** Morphine and delta-9-tetrahydrocannabinol: tolerance to the stimulus effects, *Psychopharmacologia (Berlin)*, 36, 243, 1974.
94. **Bueno, O. F. and Carlini, E. A.,** Dissociation of learning in marihuana tolerant rats, *Psychopharmacologia (Berlin)*, 25, 49, 1972.

95. **Järbe, T. U. C. and Henriksson, B. G.,** Open-field behavior and acquisition of discriminative response control in delta-9-THC tolerant rats, *Experientia*, 29, 1251, 1973.
96. **Kettenes-Van Den Bosch, J. J., Salemink, C. A., Van Nordwijk, J., and Khan, I.,** Biological activity of the tetrahydrocannabinols, *J. Ethnopharmacol.*, 2, 197, 1980.
97. **Monti, J. M.,** Hypnotic-like effects of cannabidiol in rats, *Psychopharmacology*, 55, 263, 1977.
98. **Consroe, P., Martin, A., and Singh, V.,** Antiepileptic potential of cannabidiol analogs, *J. Clin. Pharmacol.*, 21, 428S, 1981.
99. **Hiltunen, A. J., Järbe, T. U. C., and Wängdahl, K.,** Cannabinol and cannabidiol in combination: temperature, open-field activity, and vocalization, *Pharmacol. Biochem. Behav.*, 30, 675, 1988.
100. **Ford, R. D., Balster, R. L., Dewey, W. L., Rosecrans, J. A., and Harris, L. S.,** The discriminative stimulus properties of delta-9-tetrahydrocannabinol: generalization to some metabolites and congeners, in *The Cannabinoids: Chemical, Pharmacologic and Therapeutic Aspects*, Agurell S., Dewey W. L., and Willette R. E., Eds., Academic Press, New York, 1984, 545.
101. **Perez-Reyes, M., Timmons, M. C., Davis, K. H., and Wall, E. M.,** A comparison of the pharmacological activity in man of intraveneously administered delta-9-tetrahydrocannabinol, cannabinol and cannabidiol, *Experientia*, 29, 1368, 1973.
102. **Zuardi, A. W., Shirakawa, I., Finkelfarb, E., and Karniol, I. G.,** Action of cannabidiol on the anxiety and other effects produced by delta-9-THC in normal subjects, *Psychopharmacology*, 76, 245, 1982.
103. **Karniol, I. G., Shirakawa, I., Kasinski, N., Pfeferman, A., and Carlini, E. A.,** Cannabidiol interferes with the effects of delta-9-tetrahydrocannabinol in man, *Eur. J. Pharmacol.*, 28, 172, 1974.
104. **Bueno, O. F. A., Carlini, E. A., Finkelfarb, E., and Suzuki, J. S.,** Delta-9-tetrahydrocannabinol, ethanol and amphetamine as discriminative stimuli — generalization tests with other drugs, *Psychopharmacologia (Berlin)*, 46, 235, 1976.
105. **Karniol, I. G., Shirakawa, I., Takahashi, R. N., Knobel, E., and Musty, R. E.,** Effects of delta-9-tetrahydrocannabinol and cannabinol in man, *Pharmacology*, 13, 502, 1975.
106. **Henriksson, B. G., Johansson, J. O., and Järbe, T. U. C.,** Delta-9-tetrahydrocannabinol produced discrimination in pigeons, *Pharmacol. Biochem. Behav.*, 3, 771, 1975.
107. **Mechoulam, R., McCallum, N. K., and Burstein, S.,** Recent advances in the chemistry and biochemistry of cannabis, *Chem. Rev.*, 76, 75, 1976.
108. **Lemberger, L., Crabtree, R. E., and Rowe, H. M.,** 11-Hydroxy-delta-9-tetrahydrocannabinol: pharmacology, disposition, and metabolism of a major metabolite of marihuana in man, *Science*, 177, 62, 1972.
109. **Perez-Reyes, M., Timmons, M. C., Lipton, M. A., Davis, K. D., and Wall, M. E.,** Intravenous injection in man of delta-9-tetrahydrocannabinol and 11-OH-delta-9-tetrahydrocannabinol, *Science*, 177, 633, 1972.
110. **Perez-Reyes, M., Timmons, M. C., Lipton, M. A., Christensen, H., Davis, K. D., and Wall, M. E.,** A comparison of the pharmacological activity of delta-9-tetrahydrocannabinol and its monohydroxylated metabolites in man, *Experientia*, 29, 1009, 1973.
111. **Franke, I., Grothey, A., Koch, S., Witteler, F.-J., Binder, M., and Sandermann, H.,** *N*-ethyl-17(*R*,*S*)-methyl-(6a*R*,10a*R*)-delta-8-tetrahydrocannabinol-18-oic amide, *Biochem. Pharmacol.*, 34, 3983, 1985.

112. **Martin, B. R., Kallman, M. J., Kaempf, G. F., Harris, L. S., Dewey, W. L., and Razdan, R. K.,** Pharmacological potency of *R*- and *S*-3′-hydroxy-delta-9-tetrahydrocannabinol: additional structural requirement for cannabinoid activity, *Pharmacol. Biochem. Behav.*, 21, 61, 1984.
113. **Srebnik, M., Lander, N., Breuer, A., and Mechoulam, R.,** Base-catalysed double bond isomerizations of cannabinoids: structural and stereochemical aspects, *J. Chem. Soc., Perkins Trans. I,* 2881, 1984.
114. **Hollister, L. E., Gillespie, H. K., Mechoulam, R., and Srebnik, M.,** Human pharmacology of 1*S* and 1*R* enantiomers of delta-3-THC, *Psychopharmacology*, 92, 505, 1987.
115. **Järbe, T. U. C., Hiltunen, A. J., Mechoulam, R., Srebnik, M., and Breuer, A.,** Separation of the discriminative stimulus effects of stereoisomers of delta-2-and delta-3-tetrahydrocannabinols in pigeons, *Eur. J. Pharmacol.*, 156, 361, 1988.
116. **Järbe, T. U. C., Hiltunen, A. J., and Mechoulam, R.,** Stereospecificity of the discriminative stimulus functions of the dimethylheptyl homologs of 11-OH-delta-8-tetrahydrocannabinol in rats and pigeons, *J. Pharmacol. Exp. Ther.*, 250, 1000, 1989.
117. **Mechoulam, R., Lander, N., Feigenbaum, J. J., Segal, M., Järbe, T. U. C., Hiltunen, A. J., and Consroe, P.,** Enantiomeric cannabinoids: stereospecificity of psychotropic activity, *Experientia*, 44, 762, 1988.
118. **Lazlo, J., Lucas, V. S., Hanson, D. C., Cronin, C. M., and Sallan, S. E.,** Levonantradol for chemotherapy-induced emises: phase I-II oral administration, *J. Clin. Pharmacol.*, 21, 51S, 1981.
119. **Lemberger, L. and Rowe, H. M.,** Clinical pharmacology of nabilone, a cannabinol derivative, *Clin. Pharmacol. Ther.*, 18, 720, 1975.
120. **Weber, P. A., Bianchine, J. R., and Howes, J. F.,** Nabitan hydrochloride: ocular hypotensive effect in normal human volunteers, *Glaucoma*, 3, 163, 1981.
121. **Weissman, A., Milne, G. M., and Melvin, L. S., Jr.**, Cannabimimetic activity from CP-47,497, a derivative of 3-phenylcyclohexanol, *J. Pharmacol. Exp. Ther.*, 223, 516, 1982.
122. **Mechoulam, R., Breuer, A., Järbe, T. U. C., Hiltunen, A. J., and Glaser, R.,** Cannabimimetic activity of novel enantiomeric, benzofuran cannabinoids, *J. Med. Chem.*, 33, 1037, 1990.

12 Potential Role of Cannabinoids for Therapy of Neurological Disorders

INTRODUCTION

The notion that cannabinoids might have value in the treatment of certain neurological diseases is grounded in folkloric use and official medical use of marijuana (*Cannabis sativa* L.) through the 19th century. In more recent times, this notion has lingered because of additional promising reports from laboratory animal studies and preliminary clinical studies of marijuana or its related compounds. In a comprehensive review in 1986[1] and a conference review in 1988,[2] it was concluded that certain cannabinoids may have potential clinical utility for seizures disorders, spasticity, pain, and dyskinesias. However, it was also stressed that further data, especially from rigorously controlled, well-designed clinical trials, were required to confirm these predictions. From 1987 to the present, there have been some more clinical reports that continue to portray an optimistic prospect for cannabinoid use in neurological disorders. However, only marijuana, (–)-Δ^9-tetrahydrocannabinol (THC) or cannabidiol (CBD) has been investigated and, for the most part, only in a preliminary way. Further, the controlled trials that have been carried out with CBD have failed to demonstrate therapeutic efficacy. These facts depict a rather pessimistic outlook. Most recently, there have been revolutionary developments in basic cannabinoid research concerning the possible sites and mechanisms of action of these agents. These findings appear to pertain to the psychotropic actions and to neurologically relevant therapeutic actions of the cannabinoids. Although much more research is needed, one could speculate that these early leads could have a positive impact on the future development of new cannabinoids for the treatment of various neurological conditions.

It is clear from the older and newer findings that the area of therapeutic cannabinoids is in need of critical appraisal and probable redirection of efforts. The purpose of this chapter is to review and update the research of marijuana and cannabinoids in neurological disorders. Some historical and general information on marijuana and cannabinoids will be given. Newer findings from both basic and clinical areas will be presented, and the emphasis will be on research published since the time of the last reviews on this subject. Further, overviews will be given of the neurological conditions that have been, or may become, targets for treatment with cannabinoids. Lastly, an attempt will be made to provide an assessment of the research and the outlook for future uses of cannabinoids in neurological disorders.

HISTORICAL REFLECTIONS OF MEDICINAL MARIJUANA

Marijuana has been used by numerous cultures throughout recorded history, and an engrossing and extensive review of its many uses has been published by Abel.[3] While the folkloric use of marijuana as medicine has been around since ancient times, it was largely during the 19th century that this substance was assimilated into the standard armamentaria of official medical practice. The fascinating, and sometimes foretelling, accounts of the medical uses of marijuana have been skillfully and insightfully reviewed by Mechoulam.[4] Both of these reviews[3,4] cite many original publications on the historical aspects of medicinal marijuana; thus, no attempt will be made to repeat all of these titles here. However, there will be some overlap and, hopefully, some new interpretation of this earlier material which is deemed relevant to the present topic.

From about 1850 to 1900, several published testimonials and case histories indicated that marijuana could ameliorate some major neurologic symptoms and disorders such as spasticity, pain, convulsions, and chorea. For example, the painful, tonic muscular spasms of tetanus and rabies were attenuated.[5-7] The tetanic convulsions of (presumably) idiopathic and symptomatic epilepsy were diminished.[5,8] The intense pains of migraine,[9,10] neuralgias,[8,11] chronic headache,[12] and spasmodic or inflammatory conditions[13] were alleviated. Further, the choreic movements of (most likely)

Sydenham's (rheumatic) chorea were reduced.[14-16] These effects were frequently accompanied by the conspicuous sedative and hypnotic effects of marijuana. For the most part, the latter were welcomed by the patient and physician alike, since adequate rest and sleep improved the chances of success of the treatment.[13]

Despite its reputation as an useful drug, marijuana's actions were often uncertain, and sometimes injurious to the patient. It was not always effective in the relief of symptoms or in the mitigation of the course of the diseases. Further, it occasionally caused unpleasant or alarming effects such as memory disturbances, dizziness, dysphoria, and hallucinations.[17] During this time, dosage forms of marijuana were prepared as alcoholic tinctures or fluidextracts for oral administration. There was no standardization of the marijuana plant or of the methods of its preparation. Moreover, the marijuana was sometimes admixed with other active substances of the day. The variability of the resultant pharmaceutical preparations could have led to errors in dosing and to the resultant contradictory clinical effects that were observed.

A rather explicit clarification of the therapeutic problems and unique benefits of this drug was made by an important pioneer of clinical marijuana experimentation, Sir John Russell Reynolds. Reynolds was a prominent English neurologist, a president of the Royal College of Physicians, and an appointed physician to Queen Victoria.[18] One of his important contributions to the practice of neurology was the publication of a standard textbook on diagnosis of diseases of the brain, spinal cord, and nerves.[19] An analysis of this work indicates that he had considerable skill and insight into the diagnosis and differential diagnosis of the neurological conditions of the day. One of his important contributions to the science of neurology was the quantitative study of the symptoms in epilepsy. Interestingly, his methods, results, and conclusions are still considered valid today.[18] Additionally, Reynolds had over 30 years experience with the use of marijuana in his neurology practice,[20] and this longevity of clinical (pharmacology) experimentation is remarkable even to the present time. In 1859,[8] Reynolds succinctly stated "there were three stumbling-blocks to the therapeutic use of marijuana: viz., the exhibition of an impure drug, the administration of an improper dose, or selection of the wrong case." Moreover, in 1859[8] and again in 1890,[20] he defined which conditions were, and were not, amenable to treatment with marijuana. Most germane to the present discussion are his comments on the effects of marijuana on sensorial and muscular conditions.

"In almost all painful maladies I have found it is by far the most useful of drugs . . . It is especially so in cases of . . . neuralgia of the fifth nerve . . . tumor of brain . . . thickening of spinal meninges . . . the lightning pains of the ataxia patient. . . and, migraine . . . It is almost useless in sciatica and in myodynia, whether in the neck, the thorax or the back . . . Notwithstanding its notable usefulness in the alleviation of pain, its value is, I think, still greater in the treatment of spasm of some kinds. It is of no use in certain forms of convulsion, in other forms it appears almost marvelous in its action . . . in clonic spasms, whether of epileptoid or choreoid type . . . in attacks of violent tetanic convulsions . . . in recurrent convulsions . . . it is the most useful agent with which I am acquainted . . . In cases of genuine epilepsy, where attacks of petit mal are exclusively present and very frequent . . . I have never found it to reduce either the frequency or severity of the seizures, or to materially affect their character . . . In tonic spasms such as torticollis and writer's cramp, in general chorea, in paralysis agitans, in the jerking movements of spinal sclerosis, in trismus and in tetanus, it has proved in my experience absolutely useless."

Obviously, these early accounts should be viewed with the same scientific skepticism that is shown for any case report or uncontrolled trial. Still, the historical findings have often provided the impetus for modern investigations on the effects and potential therapeutic uses of marijuana or cannabinoids.

Furthermore, as a prefatory remark, several of these early marijuana findings, of both benefit and no effect, have received at least some confirmation from modern experiments in laboratory animals or humans.

GENERAL CONSIDERATIONS OF CONTEMPORARY MARIJUANA AND CANNABINOIDS

Marijuana: Pharmacological Aspects and Issues of Safety

Marijuana is widely used to achieve euphoria or other pleasurable effects. In the U.S., it is most commonly smoked, and it is used mainly by young persons who are generally in good health.[21] Patterns of use vary widely, although most persons probably use it intermittently rather than daily.[21] A rather unique subgroup of marijuana users consists of patients with spinal cord injury or other neurological disorders such as epilepsy and some movement disorders who smoke marijuana to self-treat their symptoms. While demographic data are limited to a few small surveys or anecdotal reports,[1] the information suggests that perhaps anywhere from 1 to 56% of these neurologic patients may be involved, and that their pattern of self-medication with marijuana may be more often one of regular use.

Primarily from the 1940s to the 1970s, a multitude of reports was published on the psychological and physiological effects of marijuana smoking, mostly in healthy adult volunteers. The key question that was, and still is, asked of this research is: what is the health hazard of marijuana smoking, especially to the chronic social user? The answer to this question is neither simple nor free of controversy, but it is germane to issues of social use and to any potential therapeutic use of the drug.

On acute administration, marijuana elicits a typical spectrum of psychoactive effects (e.g., the constellation of perceptual and cognitive alterations associated with the subjective "high") and objective effects such as tachycardia, dry mouth, and conjectival injection.[21] With specific regard to neurological function,[1] ataxia, tremulousness, and subjective muscle weakness are reported most frequently. However, these effects are variable in occurrence and are weak in relation to the more obvious psychoactive, cardiovascular, and autonomic effects of marijuana.

Most of the clinical information on chronic effects has been obtained from field studies or short-term trials of marijuana administration in chronic users of the drug.[1,21] Prospective studies of chronic administration are few, with the longest being, to our knowledge, a 94-d study of 64 continuous days of marijuana smoking.[22] Nevertheless, certain findings and inferences from the available studies are, more or less, generally accepted. With repeated, frequent, and adequate dose administration, tolerance develops to the cardiovascular effects and most subjective effects, and a mild withdrawal syndrome occurs on cessation of the drug.[22] Use of marijuana on a regular basis might be expected to be detrimental to psychosocial maturation of young persons.[21] Marijuana has potentially negative effects on the cardiovascular system, and this may be harmful to those with preexisting heart disease.[21] The endocrine effects of the drug might be expected to delay puberty in prepubertal boys, and to adversely influence hormonal and reproductive function in women.[21] The most potentially damaging consequence of marijuana is on the pulmonary system, and the newest research suggests that habitual smoking of only a few marijuana cigarettes a day can have adverse long-term effects on the lung.[23]

In the review of the health aspects of cannabis,[21] Hollister pointed out the factors contributing to the ambiguity surrounding the health hazard of marijuana. Further, and with due acknowledgment of certain caveats such as prevailing patterns and doses of use, he concluded that marijuana does not pose greater risks than licit social drugs such as alcohol, tobacco, and caffeine. While this conclusion will likely continue to be debated for some time, the equating of the risk of marijuana to that of tobacco would appear to have very negative impact on the issue of marijuana use for neurological disorders. Most of the neurological disorders for which marijuana is being used currently or is contemplated for the future are chronic conditions necessitating often life-long daily treatment. Before marijuana could be considered as an established treatment for these conditions, substantial proof of efficacy and safety from prospective trials would be required. Moreover, the most critical trials would necessitate chronic smoking of the drug. Whether it would be possible to obtain the necessary authorization for these studies is uncertain. However, assuming it was granted, we think it would be difficult to prove the safety of chronic daily marijuana smoking to the extent required for official approval and marketing as a new therapeutic drug. Nevertheless, marijuana continues to be used (illegally) as self-medication, and used (legally) as an investigational treatment of various disorders. The reports concerning its effects in epilepsy, movement disorders, conditions associated with spasticity, migraine, and neuropathic pain will be described.

Cannabinoids: General Chemical and Pharmacological Aspects

The First Era: Cannabimimetic Structure-Activity Relationships

A new era of marijuana research began in 1964 when Gaoni and Mechoulam[24] identified (–)-Δ^9-THC as the major psychoactive constituent of marijuana, and devised a chemical synthesis for its production. Since that time, about 60 other cannabinoids and some 260 other noncannabinoid compounds have been identified in marijuana,[25] and many hundreds of analogs of (–)-Δ^9-THC and of other natural and unnatural cannabinoids have been synthesized in the laboratory.[26] Despite the availability of large numbers of cannabinoid and noncannabinoid compounds, the vast majority of pharmacological research has been confined to the study of relatively few cannabinoids. Also, this research has focused mainly on their psychoactive properties. For example, only about 20 or so cannabinoids have been evaluated clinically, and most of them have been tested solely for their ability to produce the marijuana-like "high".[26] Further, only about 10 of the compounds have been evaluated clinically for their therapeutic utility, and most of these have been examined only cursorily.[26]

Pharmacological research in laboratory animals has been more extensive, but has also been somewhat selective. (–)-Δ^9-THC has received much experimental attention, as evidenced by at least 2500 papers on the compound since 1964.[27] Yet only a few of the other natural cannabinoids of marijuana have been evaluated extensively or systematically, and little comparative data are available on the noncannabinoids of the plant or on the plant as a whole.[25,26] The realization that many of the pharmacological activities of marijuana, including potential therapeutic properties, may not be due entirely to the few naturally occurring cannabinoids studied thus far has been underscored previously.[25-29] On the other hand, a plethora of synthetic cannabinoid analogs has been evaluated in laboratory animals, and virtually all of the detailed information on the structure-activity relationships (SAR) has come from these studies. For some cannabinoid prototypes (see next section), extensive SAR data has been accumulated on analgesic or anticonvulsant activities. However, most of the available SAR data reflect the psychoactive properties of THC cannabinoids or other cannabinoids related to THC in structure or cannabimimetic activity.[26] The term "cannabimimetic" has been coined to refer to cannabinoids that produce human subjective effects in common with (–)-Δ^9-THC, or that produce (–)-Δ^9-THC-like activity in laboratory animals that reflects (or parallels) the human activity.[30] Some of the commonly used animal behavioral measures for assessing cannabimimetic effect are: rodent catalepsy and hypoactivity, dog ataxia, monkey behavioral alterations, and, rodent, pigeon, or monkey drug discrimination behavior.[26,30,31] In general, each of these measures appears to provide the same relative information with regard to cannabimimetic potencies of cannabinoids.[26] Additionally, the rank order potencies of cannabinoids in the animal models generally correlate well with the potency rankings in man.[26] Based on the combined data from animal and human studies, tentative rules for cannabimimetic SAR have been formulated.[6,32] Some of the essential structural features include: the presence of a phenolic hydroxyl; the presence of a terpenoid methyl, hydroxyl or ketone group; the appropriate length, substitutions, and orientation of the side chain; and the correct absolute stereochemistry (*trans*, levorotatory for most compounds). Depending on the specific structural modifications, potency differences among compounds are substantial, i.e., ranging from hundreds of times that of (–)-Δ^9-THC to being completely devoid of cannabimimetic activity. Some pertinent examples of the latter are CBD, most (+)-isomers of cannabimimetic cannabinoids, and certain metabolites of the THCs such as (–)-Δ^9-THC-11-oic acid. The elimination or at least reduction of cannabimimetic potency is important for therapeutic concerns, and this has been a goal of some of the drug design in this area. Unfortunately, however, virtually all of the currently available cannabinoids that have medicinal properties or potential also have cannabimimetic activity, with little separation between these activities.[26,33]

The Second Era: Cannabinoid Mechanisms and Sites of Action

The newest era of marijuana research began in 1987 when Howlett, Devane, and colleagues announced the discovery of the cannabinoid receptor from rat brain.[34,35] While the existence of this receptor had been hypothesized for many years, the breakthrough came with the synthesis of a novel tritiated cannabinoid of very high specific activity (see section on "Medicinal Cannabinoid Prototypes"), and the development of specific tissue binding assays that were suitable for the characterization of this receptor. Information on the biochemical, pharmacological, and other aspects of this receptor has been published,[34-37] and will be summarized briefly here. The membrane receptor is coupled to adenylate cyclase via the G_i protein, is saturable, and has high affinity and enantioselectivity for agonist ligands of the cannabinoid class. For example, compounds such as (–)-CP 55940, (–)-HU-210, levonantradol, nabilone, and (–)-Δ^9-THC have substantial affinity for the receptor, whereas the (+)-enantiomers of these cannabinoids, CBD, and (–)-Δ^6-THC-11-oic acid have virtually no affinity for the receptor (see section on "Medicinal Cannabinoid Prototypes" for a description of the compounds). In general, there are high positive correlations between cannabinoid potencies for receptor binding and cannabinoid potencies for adenylate cyclase inhibition, cannabimimetic (behavioral) activities, and analgesic activities. Further, a wide range of noncannabinoids agents such as neurotransmitters, neuromodulators, and hormones has been studied, and none of them bind to the cannabinoid receptor.

In the short time span since the discovery of the cannabinoid receptor, several other important discoveries have been made. Notably, these include additional findings related to the cannabinoid receptor and to other important mechanisms of action of cannabinoids which, interestingly, also might be linked to the cannabinoid receptor.

Firstly, the regional distribution of cannabinoid receptors has been mapped in the brains of rats, guinea pigs, dogs, monkeys, and humans.[38] The location of these receptors in brain sections by autoradiography techniques revealed a heterogenous distribution that was generally similar across the species. In the human, dense binding was found in the following structures: the substantia nigra pars reticulata, globus pallidus, putamen, hippocampal dentate gyrus, cerebral cortex, caudate nucleus, amygdala, and cerebellum. Sparse densities of receptors were found in the thalamus, reticular formation, and spinal cord.

Secondly, the gene (a complementary DNA, SKR6) from rat cerebral cortex that encodes the cannabinoid receptor has been cloned.[39] The primary protein structure of the cannabinoid receptor was defined and the pharmacological attributes of the receptor were characterized.

Thirdly, a possible role for prostaglandins in the analgesic, hyperalgesic, and cannabimimetic effects of cannabinoids has been elucidated, and a potent analgesic effect in mice has been demonstrated for (–)-Δ^8-THC-11-oic acid.[40] As the latter cannabinoid is a noncannabimimetic metabolite of THC, the findings demonstrate that a complete separation of activities can be obtained within the THC series of compounds. Further, some additional preliminary data suggest that the cannabimimetic activity of cannabinoids in mice may be mediated by an interaction between prostaglandins and the G protein.[41] This interesting finding suggests that there might be a functional connection between prostaglandin and cannabinoid receptor mechanisms of action.

Lastly, the noncannabimimetic (+)-enantiomer of HU-210 has been shown in preparations of rat brain cortical membranes to be a functional antagonist of the *N*-methyl-D-aspartate (NMDA) receptor.[42] This finding could have therapeutic ramifications since hyperfunction of the NMDA receptor or of its neurotransmitter agonist, L-glutamate, is believed to be involved in the pathophysiology of epilepsy, Huntington's disease, cerebral ischemia, and other neurodegenerative disorders.[43] Several areas of the brain appear to be implicated, including notably, the cerebral cortex, striatum (putamen and caudate nucleus), dentate gyrus, and cerebellum.[44] Also, NMDA receptors exert a normal tonic excitatory tone within these same brain areas, and by acting specifically in the cerebellum, NMDA receptors exert a normal phasic control over dopaminergic neurotransmission in the nigrostriatum.[45] The fact that NMDA and cannabinoid receptors and/or their activities are colocalized in many of the same brain areas raises an intriguing question about a possible functional connection between these two mechanisms of action.

Therapeutic Implications of Cannabinoid Receptors in Brain

As indicated above, there have been a number of novel findings within the past couple years concerning the mechanisms and sites of action of cannabinoids. These findings provide new information, and new directions for study, of the potential for cannabinoids in the treatment of neurological disorders. For example, prostaglandin and NMDA receptor mechanisms appear to be linked to therapeutic properties of noncannabimimetic cannabinoids, and these mechanisms may be linked in some way to the cannabinoid receptor. Replications and extensions of these preliminary findings are clearly warranted. The most sensational discoveries relate to the pharmacology, location, and identification of the cannabinoid receptor, and these findings have tremendous implications for better understanding of cannabinoid actions in the brain and understanding of normal brain function as well. These findings have also provided a host of important questions about the cannabinoid receptor, including its possible role in the mediation of effects related to therapeutic aspects.

From a functional standpoint, the cannabinoid receptor appears to be coupled to both analgesic and cannabimimetic activities and, thus far, the bond seems to be largely inseparable. Another mysterious aspect pertains to where these receptors predominate in the brain. Many of the brain structures containing cannabinoid receptors are typically associated with motor functions rather than cognition or analgesia, and this is generally contrary to the most conspicuous actions of the psychoactive cannabinoids, at least in man. However, cannabinoids do have effects on motor behaviors in animals and man,[1] including potential side effects such as (cerebellar) ataxia and tremulousness, and potential benefits such as anticonvulsant and antispastic effects. Paradoxically, truncal ataxia and tremor produced by cannabinoids in normal healthy man are also the symptoms that appear to be ameliorated by cannabinoids or cannabis in patients with various spastic conditions (see section on "Marijuana, Cannabinoids, and Spasticity"). Better understanding of these paradoxical responses must await additional breakthroughs, such as the identification of an endogenous ligand, a specific receptor antagonist, and possible identification of subtypes of cannabinoid receptors. The information provided from such findings may facilitate, among other things, the design and synthesis of pharmacologically more selective drugs. In the interim, the regional distribution of cannabinoid receptors in the brain offers some tantalizing clues concerning the potential beneficial effects of cannabinoids in neurological disorders. The following presents more specific information on some of the most important areas.

Substantia Nigra

The nondopaminergic pars reticulata of the substantia nigra contains the greatest abundance of cannabinoid receptors.[38] This structure constitutes one of the major relay-output stations of the basal ganglia that transforms and funnels neuronal activity towards execution of spinal motor activity.[46] It receives major input from γ-aminobutyric acid (GABA)-containing neurons from the basal ganglia, and L-glutamate-containing neurons from the subthalamic nucleus. It sends major GABA-ergic projections to the superior colliculus, reticular formation, and (ventral medial) thalamus. The latter projects to the primary motor cortex and, in turn, the cortical neurons project directly (via the pyramidal system) and indirectly (via the extrapyramidal system) to spinal motor neurons. Since GABA is an inhibitory neurotransmitter and L-glutamate is an excitatory one, both are crucial for normal CNS functions.

While the role of the pars reticulata in controlling normal motor activity has been known for a long time, recent evidence suggests that this structure may also subserve two important pharmacological functions. In genetically spastic rats, the substantia nigra was identified as a site of action for the muscle-relaxing effects of currently used antispasticity drugs.[47] The antispasticity effects of these drugs were topographically restricted to the pars reticulata, and appeared to involve mechanisms of GABA facilitation and NMDA receptor blockade.[47] Additionally, in genetic, electroshock, and chemoshock rat models of epilepsy, the substantia nigra was identified as a site of anticonvulsant activity of GABA-mimetic drugs. [48] Activation of GABA receptors in the pars reticulata and subsequent preferential activation of the GABA-ergic nigrocollicular pathways were responsible for inhibiting the spread of both convulsive (tonic-clonic) and nonconvulsive (absence) generalized seizures.[48,49] Cannabinoids are well-known anticonvulsants in animals.[1] They appear to act primarily by inhibiting seizure spread in the CNS,[1] possibly by a facilitation of GABA mechanisms.[50] Cannabinoids also appear to have antispasticity effects in humans, although there are few pharmacological or other data in animals (or man) that directly relate to this effect. Since there are no related data on cannabinoid effects and their mechanisms in the substantia nigra pars reticulata, further studies of this site and all of its anatomical, biochemical, pharmacological, and receptor connections are clearly warranted.

Corpus Striatum

The corpus striatum (i.e., the globus pallidus, putamen, and caudate nucleus) also has a high density of cannabinoid receptors.[38] The major input to this extrapyramidal area is from the thalamus, cerebral cortex, and substantia nigra. The major output is to the motor cortex through the intermediary of the ventral (anterior and lateral) complex of the thalamus. Projections from the cortex ultimately give rise to the descending motor (pyramidal and extrapyramidal) pathways that affect the activity of lower motor neurons in the spinal cord.[46] The internal operation of the corpus striatum is multifarious, with a complicated interplay of (at least) four major neurotransmitters and their specific receptors: dopamine (DA) (from the pars compacta of the substantia nigra), L-glutamate (from the cerebral cortex), and GABA and acetylcholine (ACh) (both from interneurons within the basal ganglia).[46] The interactions of these neurotransmitters and receptors have great relevance to normal physiology and to the pathophysiology and pharmacological treatment of several movement disorders. From a general functional standpoint,[51,52] DA and ACh produce effects that are mutually opposite. GABA has a dual action on DA-mediated events. Lower levels (or doses) of GABA produce a facilitation of DA activity (via a reduction of ACh activity) and higher levels (or doses) of GABA produce an anti-DA effect (via a greater reduction of DA activity than that of ACh activity). L-glutamate (from the cerebellum) exerts a phasic control over nigrostriatal DA neurotransmission[45] and exerts a modulation of the activities of both ACh and GABA.[43] Paradoxically, and for reasons still unknown, L-glutamate can also cause excitotoxicity and subsequent degeneration of neurons within this and other areas of the CNS.[43]

A more complete discussion of movement disorders is presented in the section on "Overview of Movement Disorders". The following is a brief sketch of the movement disorders in relation to defects of basal ganglia biochemistry. Parkinson's disease (and its triad of hypokinesia, muscular rigidity, and resting tremor), which is caused by degeneration of DA neurons originating from the substantia nigra pars compacta to the corpus striatum, is functionally a basal ganglia disorder of DA deficiency and relative ACh excess. Neuroleptic drug-induced Parkinsonism is caused by blockade of DA receptors in the nigrostriatal system, resulting in a functional DA deficiency and relative ACh excess. The chorea of Huntington's disease results from relative DA excess (i.e., sparing of nigrostriatal DA fibers) associated with L-glutamate-induced neurodegeneration of ACh- and GABA-containing neurons. Tardive dyskinesia is presumably due to a striatal DA receptor supersensitivity caused by prolonged administration of neuroleptic drugs. Further, Tourette's syndrome appears to be a manifestation of DA excess (and possibly ACh deficiency), and dystonia reflects an excess of ACh (and possibly excess of DA). It should be stressed that mechanisms in other brain sites also modulate the expression of these disorders. Nevertheless, drugs that have a predominant effect on one of the neurotransmitters in the basal ganglia can produce different effects on symptoms of these disorders depending on their biochemical abnormalities. One classic example of this is reserpine or neuroleptic drugs that sometimes can ameliorate chorea, dystonia, and the tics of Tourette's syndrome, but can trigger or exacerbate the symptoms of Parkinsonism.[51,52]

Theoretically, cannabinoids should be able to influence the major symptoms of primary basal ganglia disorders since, in animals, they produce hypokinesia,[26,31] augment hypokinesia caused by reserpine or neuroleptic drugs,[53-55] and have a multitude of effects on each of the neurotransmitters mentioned above.[56] However, despite the voluminous information on cannabinoid effects on brain neurotransmitters,[56] results are sometimes conflicting and there is still no clear picture of how or where the cannabinoids act. For example, the basal ganglia activity of DA has been reported to be increased,[56,57] decreased,[56] and not affected[55] by cannabinoids. Also, increased GABA,[58] ACh,[59] and norepinephrine (NE)[60] activities, and decreased DA[56] and serotonin[61] activities have been reported to specifically mediate the cannabinoid-induced hypokinesia. The major site of action for cannabinoid-induced hypokinesia and cannabinoid augmentation of reserpine- or neuroleptic drug-induced hypokinesia has been reported to be the striatum.[53,56] However, other data indicate that it is not the basal ganglia that mediate these effects, but instead, other areas of the brain such as hypothalamus, amygdala, and locus ceruleus.[60,61] Further, different primary mechanisms and sites of action appear to be involved for each of the three drugs (cannabinoids, reserpine, and neuroleptics) used to produce hypokinesia.[60-63] Collectively, these reports illustrate the lack of information and understanding of the specific coupling of this one behavioral effect with sites and mechanisms in the brain. Combined with this, there are few data on cannabinoid effects on other features or models of neurological disorders. It is clear that the multitude of effects and interactions between cannabinoids and the neurotransmitters require scientific clarification and expansion. Moreover, one would suspect that the cannabinoid receptor might have a role in some of these effects of cannabinoids, and possibly a role in the normal activities of the neurotransmitters themselves. Research into these aspects is warranted as well.

Cerebellum

The cerebellum has an abundance of cannabinoid receptors.[38] While the density of these receptors is lower than in other areas mentioned above, we feel that this specific population of receptors could be critically important for the mediation of some potential therapeutic effects of cannabinoids in neurological disorders. The cerebellum regulates the progress of motor activity by signals from the cerebral cortex, hypothalamus, locus ceruleus, raphe nuclei, and various peripheral structures.[46] It influences a wide variety of voluntary and involuntary movements by its efferent connections to the reticular, vestibular and red nuclei, and to the motor cortex via the (ventral) thalamus.[46] Major neurotransmitters that regulate cerebellar functions include NE (from the locus ceruleus), serotonin (from the raphe nuclei), GABA, and L-glutamate (both within the cerebellum).[46] Moreover, the anatomic and functional interactions between the cerebellum, the substantia nigra, the striatum, ventral midbrain regions, and limbic system[45,64] may underlie the modulatory effect of the cerebellum on extrapyramidal motor function and the expression of various neurological disorders. The following are some specific examples.

The cerebellum (via L-glutamate/NMDA receptor activity) exerts a phasic control over nigrostriatal DA neurotransmission.[45] Further, biochemical studies, which show increased forebrain dopamine concentrations following lesions of the cerebellar cortex, indicate inhibitory influences of the cerebellum on DA activity within the striatum.[64] These connections may account for the observations that cerebellar stimulation increases cerebrospinal fluid (CSF) homovanillic acid (HVA) concentrations in patients with dyskinesias,[65] and induces spontaneous orofacial dyskinesias in rats.[66] Moreover, the modulatory effects of the cerebellum on basal ganglia functions may explain the association of cerebellar pathology with spontaneous orofacial dyskinesias in the elderly,[67] and with neuroleptic-induced tardive dyskinesia in schizophrenic patients.[68] Pathological changes in the form of inflated neurons in the dentate nucleus of the cerebellum have been reported in autopsied cases of neuroleptic-treated schizophrenic patients with tardive dyskinesia.[69] These findings are also in accordance with previous studies demonstrating a positive relationship between the degeneration of the dentate nucleus of the cerebellum and the onset of involuntary hyperkinesis in patients on long-term neuroleptic medication.[70]

The dopaminergic pars compacta of the substantia nigra is the area most severely affected in Parkinson's disease. However, neuronal loss in Parkinson's disease occurs also in the locus ceruleus,[71] an area known to give rise to NE projections to various brain regions including the cerebellum.[72] It is further of note that NE mechanisms are specifically involved in the tremor and rigidity of reserpine-treated rats[73] and that, in Parkinson's disease, NE levels are reduced in the cerebellum.[74] These findings suggest that cerebellar dysfunction may substantially contribute to the motor disability of the disease.

Abnormalities in cerebellar NE and decreased Purkinje cell size have been found in the dystonic rat,[75] an autosomal recessive mutant thought to represent an animal model of human dystonia musculorum deformans.[76] Also "dystonia"-like postural asymmetry resembling human dystonia develops in rats following injection of ACTH (adrenocorticotropic hormone) fragments into the locus ceruleus.[77] These findings support the notion that cerulocerebellar pathways are involved in the pathophysiology of some genetic forms of human dystonia.

Cerebellar symptoms such as intention tremor, truncal ataxia, dysarthria, and nystagmus are common clinical features of multiple sclerosis, reflecting foci of demyelination in the brainstem and cerebellar peduncles.[78] Also, improvement has been noted in both spasticity[79] and epilepsy[80] during chronic cerebellar stimulation in humans. It has been suggested[65] that release of striatal dopamine by the cerebellar stimulation may be related to these therapeutic effects.

Lastly, a perplexing dichotomy of cannabinoid action is the causation [1] and the amelioration of the seeming cerebellar symptoms of ataxia and tremor. Recently, the effects of (–)-Δ^9-THC have been studied on regional brain glucose metabolism using positron emission tomography in normal humans.[81] The cerebellum was the only brain site showing a metabolic change, and this effect was correlated with the degree of subjective effects, but not with the severity of ataxia and tremor, produced by the cannabinoid. While further studies are required, one could speculate that cannabinoids might have a selective action on the cerebellum to benefit, rather than cause, ataxia and tremor.

Few studies have targeted the cerebellum with respect to the actions of cannabinoids, and there are no studies coupling therapeutic effects of cannabinoids and cerebellar cannabinoid receptors. Clearly, these issues should be addressed.

Medicinal Cannabinoid Prototypes

Figure 1 illustrates the chemical structures of the principal prototypes of cannabinoids that have therapeutic related activities in humans and/or animals. The prototypes or their analogs also have, or could have, implications for treatment of neurological disorders. The following is a brief synopsis.

(–)-Δ^9-THC

(–)-Δ^9-THC has a broad pharmacological spectrum. With regard to therapeutic aspects, the drug has anticonvulsant properties in animals,[1] and analgesic,[82] ocular hypotensive,[83] bronchodilator,[84] and antiemetic[85] effects in animals and humans. Common side effects of the drug on acute oral administration are dry mouth, tachycardia, orthostatic hypotension, and cannabimimetic effects such as drowsiness, dizziness, ataxia, euphoria, anxiety, and concentration difficulties.[86] In general, there is not a great separation of potencies among its multitude of acute effects and, in particular, between its cannabimimetic and (intended) therapeutic effects.[86] Nevertheless, this cannabinoid has demonstrated efficacy and safety as an oral antiemetic for short-term use, and it is approved and commercially available for this application (Dronabinol; Unimed, Inc.). One study has evaluated the effects of daily oral administration of (–)-Δ^9-THC for 30 d in normal volunteers. Under this schedule and using fairly high doses, tolerance to the cardiovascular and most cannabimimetic effects developed, and a relatively mild withdrawal syndrome was observed after discontinuation of the drug.[87] Additional prospective human studies of similar or longer-term administration apparently are not available; thus, the safety or efficacy of the drug for chronic administration has not been established. Recently, (–)-Δ^9-THC has been investigated as an appetite stimulant in patients with cancer or HIV infection.[88] Additionally, the drug has been evaluated in an *in vitro* model of migraine, and has been evaluated in patients with multiple sclerosis or spinal cord injury. These neurological effects and uses of (–)-Δ^9-THC will be described in later sections.

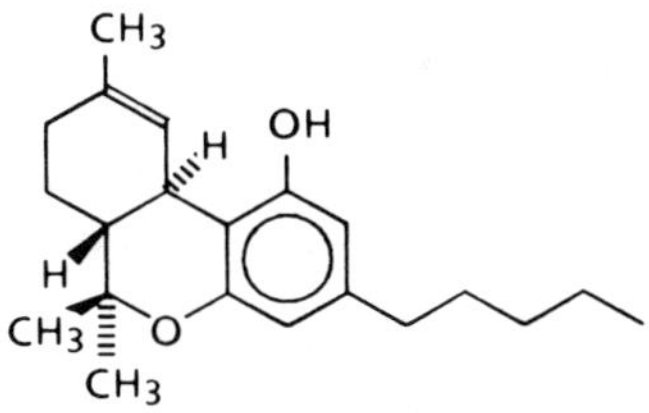

(-)-Δ^9-Tetrahydrocannabinol (Δ^9-THC also named Δ^1-THC)

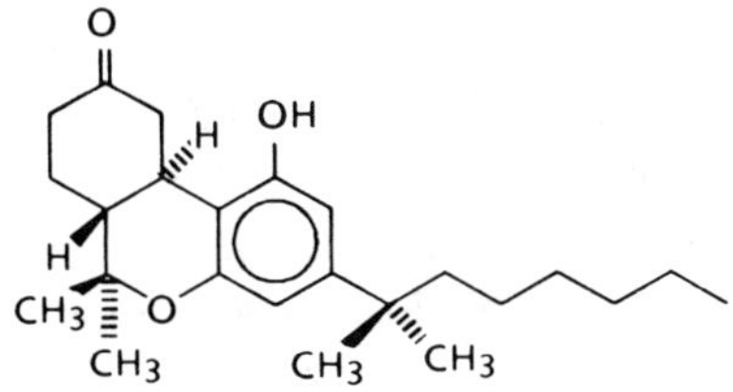

Nabilone

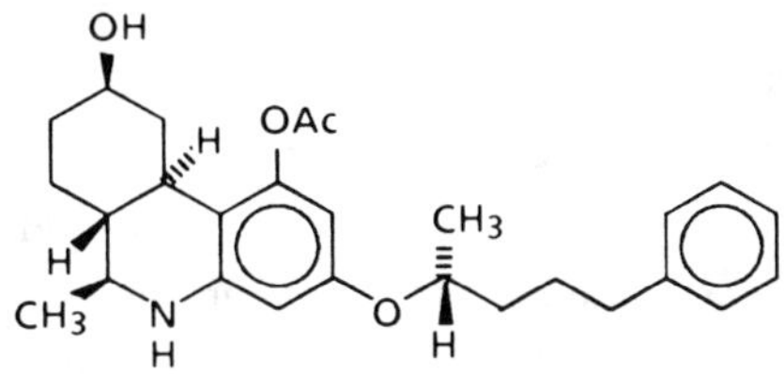

Levonantradol

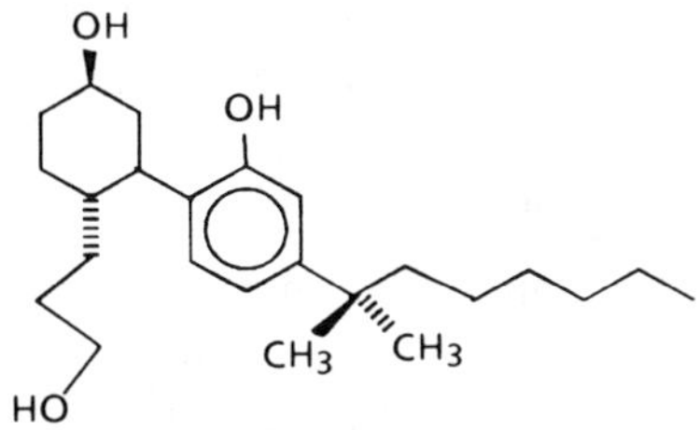

(-)-CP-55940

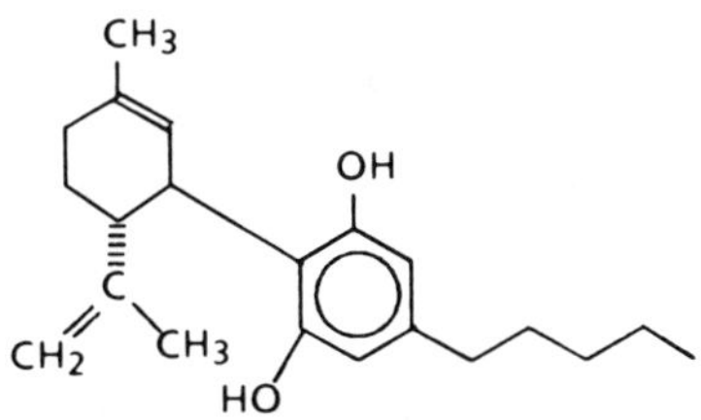

(-)-Cannabidiol (CBD)

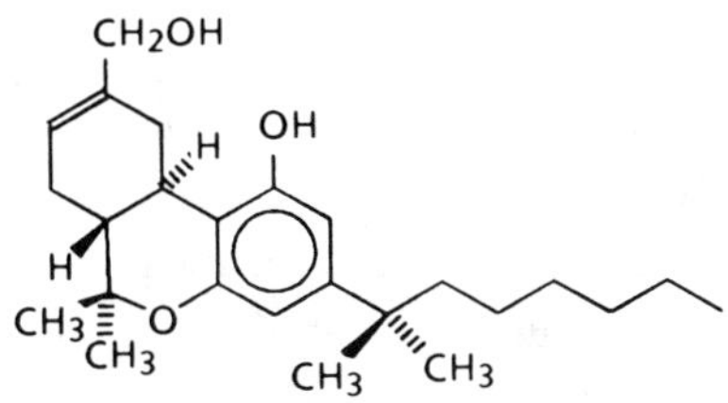

(-)-HU-210

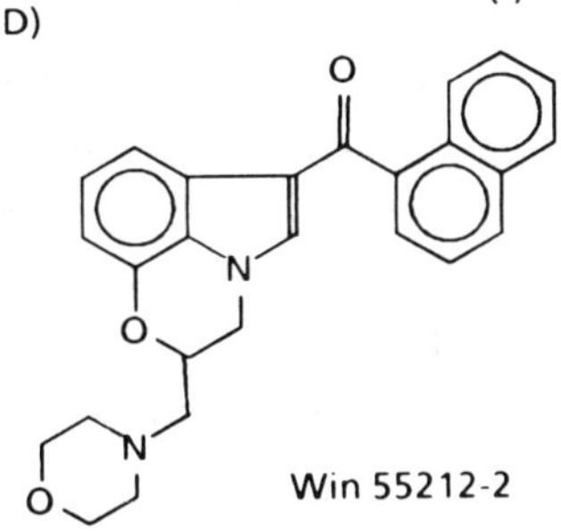

Win 55212-2

FIGURE 1. Structures of prototype cannabinoids that have possible therapeutic implications for neurological disorders.

CBD

(–)-CBD is a naturally occurring and frequently abundant constituent of marijuana. The pertinent pharmacology of this compound has been previously reviewed.[1] Briefly, CBD is completely devoid of cannabimimetic activity in both animals and man. CBD is also devoid of virtually all other major activities in animals or humans that are characteristic of (–)-Δ^9-THC. A major exception is anticonvulsant activity in animals; but even here, the pharmacological profiles of CBD and (–)-Δ^9-THC are different enough to suggest separate mechanisms of action for the two drugs. The lack of cannabimimetic activity, together with the presence of anticonvulsant activity, have made CBD a very attractive candidate for treating patients with epilepsy. Further, these attributes, together with some additional properties gleaned from preliminary animal and/or humans studies, have made CBD an aspirant for treating patients with other neurological conditions such as dystonia, Huntington's chorea, and neuropathic pain. The results and discussion of these endeavors are presented in later sections.

Nabilone

Nabilone is a successful outgrowth of a cannabinoid research program and its spearhead, L. Lemberger, at Lilly Laboratories. Using the typical approach of pharmaceutical industry, the program sought to discover new therapeutic drugs through synthesis and pharmacological evaluation in animals of hundreds of new chemical entities.[89] Nabilone is a non-THC cannabinoid (i.e., a 9-keto analog of (+/–)-hexahydrocannabinol-dimethylheptyl), albeit with a spectrum of activity closely related to that of (–)-Δ^9-THC.[89] In animals[26] and humans,[89] nabilone is about 10 times more potent than (–)-Δ^9-THC. Nabilone also has demonstrated efficacy and safety as an antiemetic and it is the only other cannabinoid currently approved and commercially available (Cesamet; Eli Lilly and Company) for this or any use. Nabilone has cannabimimetic side effects, and data from clinical trials suggest that the drug may have a higher incidence of unacceptable dysphoric effects than (–)-Δ^9-THC.[85] However, more recent findings from post-marketing surveillance studies indicate that the incidence of reported side effects of nabilone in clinical practice is much lower than the incidence of such effects in controlled clinical trials.[90] Further, nabilone appears to have a lower incidence of euphoria, and may have less abuse potential than that of (–)-Δ^9-THC.[85,89] While nabilone might have some degree of clinical advantage over (–)-Δ^9-THC, to our knowledge, nabilone (or any of its analogs) has not been evaluated for potential treatment in any neurological disorders. Furthermore, at the present time, there does not appear to be any obvious programmatic endeavor by the company to develop any new cannabinoids for these or other therapeutic uses.

Levonantradol

Levonantradol is a product of a similar research program at Pfizer Laboratories thath sought to develop new cannabinoids, primarily for their analgesic properties.[91] Levonantradol is one prototype of a large group of non-THC (commonly referred to as "nonclassical") cannabinoids that were synthesized by Johnson and Melvin and subsequently tested for pharmacologic activity. Levonantradol has anticonvulsant properties in animals,[91] and analgesic,[82,91] antiemetic,[85,91] and cannabimimetic[26,33,85] activities in animals and humans. Depending on the animal species, route of administration, and specific test, it is about 4 to 500 times more potent as an analgesic, and about equipotent—30 times more potent as a cannabimimetic, than (–)-Δ^9-THC.[30,92] The compound is also stereospecific in its actions, being about 7 to 80 times more potent than its (+)-enantiomer.[92] Levonantradol is an effective analgesic and antiemetic in patients, although the beneficial effects are accompanied by a high incidence of intolerable dysphoric and other cannabimimetic side effects.[82,85,91] To our knowledge, levonantradol (or any of its analogs) has not been evaluated in any neurological disorder. Further, the corporate sponsored research program on cannabinoids ceased about 6 years ago and, apparently, there has been no additional research of these compounds by the company. However (and mainly through the personal efforts of Johnson and Melvin), other researchers have continued to investigate the nonclassical analogs in animals mainly as probes of the mechanisms of action of the cannabinoids.

(–)-CP 55940

(–)-CP 55940 is another prototype nonclassical cannabinoid developed by the Pfizer group;[91] in animals, its pharmacologic spectrum is similar to, and its potencies are about threefold greater than those of levonantradol.[92] However, this compound (i.e., in a tritiated form) has very special significance since it is the agonist ligand that was used in the recent discovery of the cannabinoid receptor in brain. This discovery and its important ramifications for cannabimimetic and medicinal research were discussed above.

(–)-HU-210

(–)-HU-210 is one of a number of novel crystalline THC cannabinoids of absolute stereochemical purity which recently have been synthesized by Mechoulam and colleagues.[32] (–)-HU-210 is a dimethylheptyl analog of 11-OH-Δ^8-THC, the latter a metabolite of the naturally occurring cannabinoid (–)-Δ^8-THC. In animals, this compound is similar in pharmacological spectrum to levonantradol and (–)-CP 55940, that is, having cannabimimetic,[93] analgesic,[93] anticonvulsant,[94] and cannabinoid receptor binding activities.[95] For these activities, (–)-HU-210 is about 10 to 20 times more potent than (–)-CP 55940,[92,93] and is up to 7500 times more active than the (+)-enantiomer of HU-210.[93,94] The potency of (–)-HU-210 and the degree of separation of activity between the enantiomeric pair are the greatest reported thus far for any cannabinoids. Unfortunately, the extreme potency of (–)-HU-210 as a cannabimimetic certainly would preclude it from any considerations for therapeutic potential. On the other hand, certain noncannabimimetic analogs in this series are reported to have therapeutic activities in animals, but the data are limited, sometimes conflicting, and in need of replication. Examples are the analgesic activity of the noncannabimimetic (–)-Δ^8-THC-11-oic acid, a major metabolite of (–)-Δ^8-THC; the analgesic activity of the noncannabimimetic (+)-HU-210 in one study,[32] but not in another,[93] and the antiemetic activity of (+)-HU-210.[96] Further, (+)-HU-210 has been reported to be a functional antagonist of the NMDA receptor. If confirmed and expanded, this new and surprising discovery could be of great significance for future drug development since hyperactive functions of the NMDA receptor appear to be involved in several neurological disorders.

Win 55212-2

Lastly, Win 55212-2 is a prototype of a novel series of aminoalkylindole analgesics recently synthesized by the research group at Sterling Drug, Inc.[97] This compound and its congeners are structurally different from all other known cannabinoid structures. Nevertheless, studies with tritium-labeled Win 55212-2 indicate that this compound binds with great affinity and stereoselectivity to the cannabinoid receptor.[98] Furthermore, preliminary data suggest that some compounds within this chemical series may have antagonist properties at this receptor.[99] These findings are novel and potentially very important for the development of new probes for study of both cannabimimetic and therapeutic effects of the cannabinoids.

MARIJUANA, CANNABINOIDS, AND EPILEPSY

Overview of Epilepsy and Epileptic Seizures

The following overview is abstracted from general reviews on epilepsy[100,101] and on standard antiepileptic drugs.[102-104] Additional references are cited where appropriate.

Definition and Prevalence

An epileptic seizure is an intermittent derangement of the nervous system presumably due to a sudden, excessive, disorderly discharge of cerebral neurons. The discharge may result in an almost instantaneous disturbance of sensation, loss of consciousness or psychic function, convulsive movements, or some combination thereof. Recurrent seizures are characteristic of idiopathic or primary epilepsy, and may also occur in symptomatic or secondary forms of epilepsy, as well as in a diverse group of chronic or progressive neurological disorders. Epilepsy is one of the most common neurologic disorders, affecting about 1% of the American population.

Classification

Seizures have been classified in several ways: according to their supposed etiology and site of origin, on the basis of their clinical form (generalized or focal), frequency, or physiologic correlates such as the electroencephalogram (EEG), and in terms of their response to therapy. According to the most recent classification by the International League against Epilepsy, seizures are categorized as either partial or generalized. Partial (focal or localization-related) seizures arise in specific loci, and are subdivided into simple partial seizures (consciousness not impaired), complex partial seizures (with impairment of consciousness), and partial seizures (simple and complex) which may become secondarily generalized. On the other hand, primary generalized seizures involve large areas of brain from the onset, are usually bilateral in motor and EEG manifestations, and are associated with impairment of consciousness. Generalized seizures are subdivided into the convulsive myoclonic, tonic, clonic or tonic-clonic (grand mal) seizures, and the nonconvulsive absence (petit mal) seizures. Also, mixed seizure types may be present at the beginning or evolve during the course of the illness.

Clinical Signs

Simple partial seizures develop as a result of a cerebral focal lesion. The specific type and pattern of the seizure vary with the site of the lesion. For example, partial motor seizures originating from the frontal lobe may present with focal clonic or tonic muscular contractions. Migration of the contractions from one muscle group to another may occur (Jacksonian seizures).

Complex partial seizures (temporal lobe epilepsy) differ from primary generalized seizures and absence seizures with respect to two features: (1) the presence of an aura (i.e., the initial event in the seizure) which is often a complex hallucinatory phenomenon or perceptual illusion, suggesting a temporal lobe origin, and (2) patients behave in a confused manner followed by amnesia. The motor components of complex partial seizures usually occur during a period of unresponsiveness where the patient may smack his or her lips and make sucking, chewing, or swallowing movements. About two thirds of patients with complex partial seizures also have generalized seizures or have had them at some earlier time.

The primary generalized convulsive seizures (e.g., grand mal) usually begin with a sudden loss of consciousness and a fall to the ground. The motor signs consist of tonic, and then clonic violent muscular contractions. When fully recovered, the patient has no memory of the attack. About 5 to 8% of patients will at some time have a series of grand mal seizures without completely regaining consciousness between them (status epilepticus).

The generalized absence seizure (petit mal), which is the most characteristic epilepsy of childhood, rarely begins before 4 years of age. These seizures are distinguished by their brevity (2 to 10 s) and paucity of motor activity. The attack consists of interruption of consciousness associated with clonic movements of the eyelids, facial muscles, or fingers. These movements are usually bilateral and symmetrical. A 3-per-second spike-wave discharge characterizes the EEG abnormality of absence seizures. The attacks tend to diminish in frequency in adolescence and may disappear. Often, however, absence seizures give way to grand mal seizures.

Etiology and Pathogenesis

Epileptic seizures have been defined as a sudden alteration of central nervous function, resulting from a paroxysmal high-frequency or synchronous low-frequency, high-voltage electrical discharge. Just why the neurons in or near a focal lesion discharge is not fully understood. Some of the electrical properties of a cortical epileptogenic focus suggest that its neurons have been deafferented. Such neurons are known to be hypersensitive and they may remain chronically in a state of partial depolarization. The cytoplasmic membranes of these cells appear to have an increased permeability which renders them susceptible to activation by hyperthermia, hypoxia, hypoglycemia, hypocalcemia, and hyponatremia as well as repeated sensory stimulation and during certain phases of sleep. Biochemical studies indicate that GABA deficiency and alterations in local regulatory mechanisms of extracellular sodium, potassium, calcium, and magnesium metabolism may partly underlie the heightened excitability of epileptogenic neurons.

In some cases of primary generalized epilepsy and absence seizures, the brain is grossly and microscopically normal. However, many of the cases of focal epilepsy and complex partial seizures have definable pathology. These include zones of neuronal loss and gliosis (scars), vascular malformations, and tumors. In the majority of cases with temporal lobe epilepsy, pathologic studies have revealed neuronal loss and gliosis in the hippocampal and amygdaloid regions. These may be related to the effects of traumatic or hypoxic insults. Further, when sustained seizure activity is induced in animals in specific pathways utilizing L-glutamate as a transmitter, cytopathological changes occur in dendrosomal neural elements postsynaptic to such pathways.[43] This, together with other cytopathology findings from autopsied brains of patients with intractable epilepsy, suggest that NMDA/L-glutamate excitotoxicity is intimately involved in the pathology of chronic epilepsy.[43]

Treatment

The treatment of epilepsy primarily consists of identifying the causative factor(s) that may be identified in patients with partial or complex partial seizures. In approximately 90% of patients with seizure disorders, attacks are controlled completely or reduced in frequency and severity by the appropriate use of anticonvulsant drugs. Certain drugs are more effective in one type of seizure than in another, and it is necessary to use the proper drugs in the optimum dosages for the control of different types of seizures.

Moreover, monotherapy (the use of a single drug) and the judicial use of blood drug level monitoring are also important requirements for optimal seizure control. For generalized (primary or secondary) tonic-clonic seizures, phenytoin, carbamazepine, or valproic acid are the drugs of first choice. In some cases where phenytoin or carbamazepine produce only partial response, phenobarbital may be substituted. Generalized myoclonic seizures are usually managed with valproic acid or clonazepam. Partial (simple and complex) seizures are managed by many of the same drugs, i.e., carbamazepine (usually the initial drug of choice), phenytoin, phenobarbital, or valproic acid. Absence seizures do not respond to agents used for the above generalized or partial seizures. The treatments of choice for absence seizures include ethosuximide, valproic acid, or clonazepam.

All anticonvulsants produce side effects. Phenobarbital may produce drowsiness and mental dullness, nystagmus, and ataxia of gait. Overdose of phenytoin may cause ataxia, stupor, and coma. Its prolonged use may lead to hirsutism, hypertrophy of the gums, and coarsening of the facial features. In addition, chronic phenytoin intoxication may produce peripheral neuropathy and cerebellar degeneration. Carbamazepine causes similar effects as phenytoin, but to a lesser degree. Use of valproic acid is most commonly associated with tremor, weight gain, thinning or loss of hair, and ankle swelling. Hepatotoxicity, which can be fatal, has also been reported with the use of valproic acid, but it is rare. Ethosuximide most commonly produces gastrointestinal distress and psychiatric symptoms (insomnia, lethargy, and psychosis). Clonazepam may produce variable lethargy, or even hyperactivity (especially in children). Additionally, carbamazepine, phenytoin, and phenobarbital may exacerbate absence seizures, and ethosuximide may aggravate tonic-clonic seizures. Finally, some anticonvulsants such as phenytoin, phenobarbital, carbamazepine and valproic acid may have teratogenic effects.

Effects of Marijuana and Cannabinoids

Animal Studies

There is a plethora of data on the effects of CBD, (–)-Δ^9-THC, and their analogs in animal seizure models, and these findings have been detailed in a 1986 review.[1] Since that time there have been few additional studies on the cannabinoids and seizures. Thus, the following discussion is mainly a summary of the previous review, together with supplemental citations as indicated.

There are important similarities and differences between the two cannabinoid prototypes. CBD and its analogs are anticonvulsant and are completely devoid of stimulatory or convulsant activities. Generally, the CBDs are effective against cortical focal (simple partial) seizures produced by topical application of convulsant metals, limbic (complex partial) seizures produced by repetitive electrical stimulation or kindling, and generalized maximal (grand mal) seizures induced in genetically epileptic animals or produced in nonepileptic animals by electroshock or GABA-inhibiting drugs. However, the CBDs are largely inactive in animal models of minor generalized (absence, petit mal) seizures produced by electroshock or chemoshock methods. CBD can increase potencies of antiepileptic drugs used clinically to treat partial and grand mal seizures, but can decrease potencies of clinically used antiabsence drugs. The anticonvulsant effects of CBD are not stereoselective in that the natural (–)- and unnatural (+)-enantiomers are equieffective. CBD, rather than its metabolites, appear to mediate the anticonvulsant effects of the cannabinoid. Further, CBD appears to reduce the spread of the generalized maximal seizures and, in limbic seizure models, may increase the threshold for seizures at the limbic focus.

(–)-Δ^9-THC and other cannabimimetic analogs are also anticonvulsant in most animal models of generalized maximal seizures and limbic seizures. The THCs are stereospecific in that the (–)-enantiomers are anticonvulsant, but the (+)-enantiomers are not.[94] Furthermore, the THCs have excitatory, proconvulsant, or convulsant activity in some genetically epileptic animals, and in some animal models of cortical focal, limbic and generalized absence seizures. Also, (–)-Δ^9-THC can increase seizure susceptibility to electroshock-induced minimal convulsions following withdrawal of the drug.

Collectively, the animal findings predict that CBD could be effective for simple and complex partial seizures and for major (grand mal) seizures, and might potentiate the effects of standard antiepileptic drugs for these conditions. However, CBD could be ineffectual for absence seizures and might aggravate the beneficial effects of standard antiabsence drugs. On the other hand, (–)-Δ^9-THC might exacerbate most of the seizure types and might jeopardize seizure control with standard antiepileptic drugs if the stimulant (proconvulsant or convulsant) properties of the drug predominate over the anticonvulsant properties. Although parallel data are lacking for marijuana itself, one could speculate that marijuana might have anticonvulsant, convulsant, or no effects, depending on seizure type and on which cannabinoid is predominant in the plant.

Human Studies

The reports dealing with marijuana or specific cannabinoids and human seizure disorders have been discussed in the section on "Historical Reflections of Medicinal Marijuana", and in a 1986 review of the topic.[1] There have been only a few reports over the past 150 years and almost all are brief accounts of uncontrolled observations or studies in one or only a few patients. Briefly, marijuana was reported to ameliorate or to exacerbate grand mal seizures and to not affect petit mal seizures. Some dimethylheptyl analogs of (–)-Δ^3-THC, given orally every day for 7 weeks, were reported to alleviate grand mal seizures in institutionalized children. In 1980, a Brazilian group reported the only controlled study of CBD up to that time, i.e., a double-blind, placebo-controlled, randomized, add-on trial of CBD in 15 patients with inadequately controlled complex partial seizures or with secondarily generalized tonic-clonic seizures.[105] The results indicated that CBD, 200 to 300 mg/d for 4.5 months, was effective (in combination with standard antiepileptic drugs) for controlling seizures in these patients. Since the last review and critique of this topic,[1] a few additional reports on marijuana or CBD have appeared.

In 1990, Ellison et al.[106] published a brief case report of an adult man whose alleged regular use of marijuana seemed to suppress his complex partial seizures, but which repeatedly reemerged on sudden withdrawal of the marijuana. The patient also had bipolar affective disorder and a history of alcoholism, and was taking several psychotropic medications (lithium, trifluoperazine, benztropine, disulfiram, lorazepam or clonazepam, and/or carbamazepine) concomitantly with the marijuana. Also in 1990, Ng et al.[107] published the first case control study of illicit drug use and the risk of new onset seizures. The use, frequency, and duration of use before hospitalization of marijuana and other drugs (amphetamines, cocaine, heroin, lysergic acid diethylamide, methadone, methaqualone, and phencyclidine) were ascertained by interview in 308 seizure patients and 294 controls. Marijuana was the drug most commonly reported, with about 25% of each group using it. Marijuana use in both groups was daily or at least weekly and was of long duration (about 8 to 9 years). The seizure types were classed as generalized-onset major motor (63% of seizure patients), partial (29%), or unclassified (8%), and by antecedents as either provoked or unprovoked. These and other variables such as age, sex, history of head trauma, hypertension, stroke, alcohol, and multiple illicit drug use were subjected to multiple logistic regression analyses. The major findings were that heroin use was a risk factor, and marijuana use a protective factor, for new onset seizures. Stratification of the data indicated that marijuana was a protective factor for men, and that the men who had used marijuana within 90 d of admission had the lowest risk for all first seizures. In women, no association was found for marijuana use and seizures. The authors suggested that the low frequency of use by women in their study, a sex difference, or misreporting may be the reason for the failure to detect a relationship in this group.

In 1986, Ames[108] published a brief abstract of a double-blind, placebo-controlled study of oral CBD (200 to 300 mg/d for 4 weeks) in 12 epileptic patients. Details are lacking, but the author states that there was no statistically significant difference in seizure frequency between the CBD and placebo group.

Lastly, Trembly et al.[109] reported in 1990 on a double-blind, placebo-controlled, add-on, crossover study of CBD in 10 (drug refractory) patients with generalized tonic-clonic, cortical focal, or complex partial seizures. All patients were taking one or more standard antiepileptic drugs (phenytoin, carbamazepine, and/or valproic acid) throughout the 2-year study. The sequential design consisted of 3 months of baseline, 6 months of single-blind placebo, 6 months of double-blind CBD or placebo, 6 months of double-blind placebo or CBD (crossover), and 3 months of a final baseline. CBD was administered orally and at 300 mg/d in divided doses. Patients kept a daily log of their seizures. Patient interviews, neuropsychological testing, determinations of serum levels of anticonvulsant drug and of CBD (assayed by GC/MS), and urinary (EMIT) tests of THC cannabinoids were carried out at monthly clinic visits. The results were that CBD (300 mg/d) had no effects on seizure pattern, character or frequency, or on neuropsychological and hematologic tests, and produced no symptomatic side effects. Serum levels of CBD ranged from about 2 to 30 ng/ml while on the CBD. Also, it was reported that seizure frequency was reduced in an additional patient given 900 to 1200 mg of CBD per day for 10 months in an open trial of the drug. Corresponding serum levels of CBD at 1200 mg/d were measurably higher (about 150 ng/ml), and the authors suggested that CBD at the larger dose may deserve further investigation as an anticonvulsant.

Collectively, the limited evidence to date suggests that marijuana (and perhaps THC as well) might have anticonvulsant, rather than proconvulsant, activity in man. These anticonvulsant effects may be specific to partial seizures and major generalized seizures. The activity of CBD in humans, on the other hand, is a puzzle. An earlier report of antiepileptic activity has not been confirmed in one recent report, and it is impossible to tell from these publications which finding is closer to the truth. CBD has also been reported to be ineffective in a third controlled trial, but the short duration of the trial and the lack of details and data in this abstract report preclude any meaningful assessment of this study. Nevertheless, prudent scientific skepticism would lead one to conclude that CBD, at doses of 200 to 300 mg/d, may not possess clinical antiepileptic activity. Whether or not higher doses of CBD might have such activity will have to await findings from rigorous, well designed, controlled clinical trials.

MARIJUANA, CANNABINOIDS, AND MOVEMENT DISORDERS

Overview of Movement Disorders

Movement disorders include a diverse group of idiopathic or drug-induced conditions that have in common a disturbance in extrapyramidal (involuntary) motor function. The following is a brief overview of some predominant features and other aspects of movement disorders that are germane to the present topic.

Symptoms[110,111]

Some symptoms that commonly occur in movement disorders include: akinesia, bradykinesia, or hypokinesia (abrupt stoppage of voluntary movement or slowness in initiating voluntary movement, as seen in Parkinsonism); chorea (arrhythmic, rapid, brief, involuntary muscular twitches as seen in Huntington's disease and Sydenham's chorea); dyskinesia (a generic term for various forms of spontaneous involuntary movements that are usually choreic in nature and often drug induced); dystonia (slow, sustained involuntary muscle contractions resulting in postural abnormality as seen in the dystonic movement disorders); myoclonus (arrhythmic, abrupt, jerky contractions of a single or group of muscles, as may be seen in spinal cord tumor and seen in one type of generalized epilepsy); tardive dyskinesia (neuroleptic-induced rhythmic, stereotyped movements of oral, buccal, and lingual regions); tic (sudden, rapid, brief repetitive stereotyped movements or vocalizations as seen in Tourette's syndrome); and, tremor (rhythmic, oscillatory movement, usually due to alternating contraction of opposing muscles at a joint). The latter may be subdivided into resting tremor (as seen in

Parkinsonism) and intention or action tremor (as may be seen in cerebellar disorders, multiple sclerosis, and spinal cord injury).

Dystonia

The dystonias are a heterogenous group of disorders.[112] The primary (idiopathic; familial and nonfamilial) dystonias include the focal (localized) spasmodic torticollis, writer's cramp, and Meige's syndrome (blepharospasm oromandibular dystonia), and the generalized (and inherited) dystonia musculorum deformans (torsion dystonia). The primary dystonias are chronic and slowly progressive. The secondary (symptomatic) types, such as those caused by neuroleptic or levodopa therapy, are usually acute conditions, although they may become persistent (tardive dystonia).

The pathophysiology of the primary dystonias is an enigma. Based primarily on favorable treatment responses to anticholinergic and antidopaminergic drugs, hyperfunction of both ACh and DA may be involved.[113] However, not all dystonic patients improve with anticholinergic medication, and some dystonic patients may improve with levodopa therapy.[113,114] Further, levodopa can cause dystonic movements in Parkinson's patients and DA blocking drugs can cause dystonic features in schizophrenic patients.[112] As was mentioned in the section on "Therapeutic Implications of Cannabinoid Receptors in the Brain", some experimental data suggests a dysfunction of NE in the cerebellum and its connections in some animal models of dystonia. Additionally, in humans, post-mortem studies have demonstrated decreased concentrations of NE and serotonin in particular brain stem and diencephalic nuclei in patients with dystonia musculorum deformans.[115] Moreover, other studies have revealed decreased CSF tetrahydrobiopterin (BH4) levels in patients with dystonic disorders, including those with inherited dystonia.[116,117] BH4 is a critical cofactor in controlling the rate of tyrosine and tryptophan hydroxylation and is therefore important for the synthesis of DA, NE, and serotonin. Thus, reduced brain BH4 levels in dystonia could reflect a primary metabolic abnormality or a response to neuronal atrophy.

In cases of drug-induced dystonia, the primary approach to management includes the administration of an anticholinergic drug (e.g., trihexyphenidyl or benztropine) or dose reduction (or withdrawal) of the offending drug.[114] Treatment of the primary dystonias is difficult, often unpredictable and frequently accompanied by disabling side effects of the treatments. Anticholinergic drugs, given in high doses orally, have the highest therapeutic benefit, although up to about 50% of children and about 60% of adults may not respond.[113] Some patients may benefit from such diverse agents as diazepam, levodopa, bromocriptine, baclofen, carbamazepine, or tetrabenazine, and others may benefit from combinations of anticholinergics and DA blocking drugs.[113,114] Very recently, botulinum A toxin, a powerful presynaptically active neuromuscular blocking agent, has been approved and marketed in the U.S. as a new (local injectable) treatment for various forms of focal dystonia.[118]

Huntington's Disease or Chorea

Chorea is a major feature of Huntington's disease. It is also a feature of Sydenham's chorea and sometimes a side effect of chronic levodopa therapy. Sydenham's chorea[118] is an acute, often self-limiting disease of childhood caused by a streptococcal infection. It was prevalent during the 19th century and it was probably the form of chorea that was treated with marijuana at that time. However, in modern times, the incidence of Sydenham's chorea has been declining to the point of near disappearance. Huntington's disease[119] is a rare, dominantly (and completely expressed) inherited neurodegenerative disease characterized clinically by the presence of progressive choreiform movements, psychological disturbance, and dementia. In most cases, symptoms of the disease do not emerge until the 4th or 5th decade of life. Save in rare cases, Huntington's disease is progressive and terminates fatally usually within 10 to 15 years.

In Huntington's disease, neurochemical abnormalities in the striatum and associated structures include reductions of GABA, ACh, substance P, enkephalin, and cholecystokinin, and elevations of somatostatin, neuropeptide Y, and DA. Excessive striatal DA activity appears to explain the chorea of Huntington's disease, as well as the chorea-inducing effects of levodopa in patients with Parkinson's disease.[119] Further, this same biochemical abnormality also appears responsible for Sydenham's chorea.[118] The relationships of the biochemical abnormalities to the other clinical symptoms or to the underlying neurodegenerative nature of Huntington's disease are obscure.[119] However, a considerable amount of experimental evidence appears to implicate a hyperfunction of L-glutamate and/or its NMDA receptor as a basic pathophysiological mechanism.[43,44]

The treatment of Huntington's disease is currently purely symptomatic.[119] Pharmacological manipulations have focused almost exclusively on the choreic features of the disease. The most consistent antichoreic effects result from treatment with neuroleptics such as haloperidol, or treatment with other anti-DA drugs (reserpine or tetrabenazine). However, these agents often produce unwanted sedation, lethargy, and depression and are associated with an increased risk of Parkinsonism and tardive dyskinesia. Several attempts to enhance GABA neurotransmission using potent GABA-mimetic agents such as muscimol failed to demonstrate clinical benefits. Cholinergic agonists may temporarily lessen chorea, but the effect is unpredictable and usually attended by increased incoordination and overall functional decline. Since Huntington's disease may be associated with increased glutaminergic neurotransmission, it is possible that agents that diminish L-glutamate activity might retard striatal neuronal degeneration and thereby forestall the functional decline of the patients. However, recent studies with baclofen (a release inhibitor of L-glutamate and a $GABA_B$ agonist)[120] and dextromethorphan (a noncompetitive NMDA receptor antagonist)[121] have failed to demonstrate any therapeutic benefit.

Tourette's Syndrome[122]

Tourette's syndrome is a chronic involuntary movement disorder characterized by both motor and vocal tics with onset in childhood. A consistent feature of the syndrome is the waxing and waning of symptoms over a 6-week to 3-month interval. Also, tics are usually increased by emotional states of anxiety, anger, stress, and fatigue. The exact prevalence is unknown, but a currently accepted figure is 0.5 per 1000. Although Tourette's syndrome is thought to be a lifelong disorder, it has been estimated that tics will disappear or diminish markedly by late adolescence in about 60% of cases.

Current information suggests that Tourette's syndrome is inherited via a sex-influenced autosomal dominant mode with variable expressivity. The exact neuroanatomic localization of the dysfunction is unknown, but circumstantial evidence implicates the basal ganglia and/or its major interconnections in the pathophysiology of the disorder. Similarly, evidence for a defect in the DA system is suggestive but not conclusive. DA has received most support mainly because neuroleptics attenuate symptoms in a large number of patients, while dopaminergic drugs such as pemoline and methylphenidate exacerbate the tics. Other findings have implicated abnormalities of NE, ACh, GABA, serotonin, opioid, and adenylate cyclase systems, but their precise roles have yet to be elucidated.

Symptomatic chemotherapy is, at present, the mainstay of treatment for the motor and vocal symptoms of Tourette's syndrome. Haloperidol is the most widely used drug attenuating symptoms in up to 85% of patients. Side effects include sedation, school and work avoidance, social phobia, and an increased risk of Parkinsonism and tardive dyskinesia. Unfortunately, these side effects of haloperidol (and other neuroleptics such as pimozide) often limit the clinical usefulness of treatment. Other drugs that have been tried with varying success include clonidine, clonazepam, tetrabenazine, imipramine, naltrexone, and calcium channel blockers,[122] as well as nicotine chewing gum.[123]

Parkinsonism[124,125]

Parkinson's disease (primary idiopathic Parkinsonism) is a chronic, progressive neurological disorder and is characterized clinically by the presence of bradykinesia, rigidity, resting tremor, and postural instability. The disease begins most frequently between 40 and 70 years of age. Parkinson's disease is a common disorder affecting about 1% of the U.S. population over the age of 50 years. Secondary (iatrogenic; acute) Parkinsonism is a commonly occurring (about 75 to 90% of patients) acute side effect of therapy with neuroleptics and related DA blocking drugs.

Despite increasing knowledge of the biochemical pathology of Parkinson's disease, the primary cause of the disease is unknown. The recent discovery that the neurotoxin MPTP can induce a Parkinsonian syndrome in humans has refocused attention on the potential role of environmental toxins in the etiology of the disease. It is now accepted that loss of pigmented cells in the substantia nigra pars compacta (and other pigmented nuclei such as the locus ceruleus and dorsal motor nucleus of the vagus) is the most consistent pathologic finding of Parkinson's disease. This results in a reduction of striatal DA (and a "relative" increase of striatal ACh) that underlies the emergence of the motor symptoms of the disease. Parkinson's disease is also associated with reduction in the activity of other neurotransmitters including NE, serotonin, substance P, methionine-enkephalin, and somatostatin. These changes may modulate the severity of major symptoms, or may contribute to the development of other features of the disease such as dementia and mental depression.

At present, levodopa is the most effective agent for Parkinson's disease. The combination of levodopa with a peripherally acting decarboxylase inhibitor (carbidopa) prevents its rapid destruction in the blood and permits the control of symptoms more quickly and with a much lower dose. About two thirds of patients tolerate the drug initially and experience few side effects. The most common side effect of levodopa or levodopa/carbidopa, and the limiting factor in its use, is the induction of involuntary movements such as chorea and dystonia.

Additional side effects may include psychiatric symptoms such as depression, psychotic behavior, and sleep disturbances. Other sometimes useful agents for the management of Parkinson's disease include anticholinergic drugs, amantadine (a neuronal DA releaser), DA receptor agonists such as bromocriptine and pergolide, and monoamine oxidase (MAO)-B inhibitors such as selegiline. New investigational treatment strategies include the neurosurgical transplantation of normal (adrenal or striatal) tissue into the striatum of Parkinson's patients, and the use of selegiline as a neuroprotective agent for retarding the progression of Parkinson's disease. Last, secondary Parkinsonism is usually managed effectively by administration of an anticholinergic drug, or discontinuation of the offending drug.

Tardive Dyskinesia[126,127]

The term tardive dyskinesia refers to a distinct clinical entity — that of a persistent movement disorder developing as a complication of long-term administration of neuroleptic drugs. The most common form is the orobuccal-lingual syndrome of mouth, tongue, and jaw movements, characterized by chewing, lip sucking, and tongue movements. It occurs in 10 to 40% of psychiatric patients given neuroleptics for more than 1 year. The natural history of tardive dyskinesia is highly variable. In general, the onset is insidious, occurring within months to years after the initiation of antipsychotic therapy. Typically, the majority of patients are unaware of their involuntary movements.

Although the exact pathophysiology underlying tardive dyskinesia is unknown, the most accepted theory involves supersensitivity of DA receptors in the basal ganglia following prolonged neuroleptic treatment. Dysfunction of other brain chemicals such as GABA, ACh, serotonin, and melatonin may be involved as well.

Treatment of tardive dyskinesia is difficult. Discontinuation of the neuroleptic drugs can result in a remission of the movement disorder in about 35% of patients, but this leads frequently to reemergence of the underlying psychosis. The tardive dyskinesia can often be suppressed by increasing doses of the antipsychotic medication in some patients; but in others, the symptoms may not change. Another approach is to keep the minimum dose of neuroleptic drug to control psychosis, and add reserpine or tetrabenazine therapy for the control of tardive dyskinesia. While the latter drugs will frequently suppress the movement disorder, side effects of Parkinsonism, mental depression, and orthostatic hypotension are common. Many other types of drugs have been tried, but none has met with great success.

Effects of Marijuana and Cannabinoids

Animal Studies

There is a paucity of data of cannabinoid effects in animal models relevant to human movement disorders. Only a handful of studies have been carried out, and few cannabinoids have been tested. Many of the animal models employed might be considered only general or vague representations of the human disorders, although they appear to have general heuristic value for predicting the potential clinical response to drugs.

In the section on "Therapeutic Implications of Cannabinoid Receptors in Brain", hypokinesia, produced or augmented by (–)-Δ^9-THC, was discussed in reference to potential mediation by neurochemical changes in the basal ganglia and other brain structures. Further, the model and its potential relevance to cannabinoid therapeutics have been recently summarized by Moss and colleagues.[123] While the site(s) and mechanism(s) of action are still debatable, the behavioral model appears to have some predictability for drug responses in movement disorders. In general, drugs that reduce reserpine-, tetrabenazine-, or neuroleptic-induced hypokinesia in animals might be expected to relieve Parkinsonian hypokinesia, but aggravate symptoms of the hyperkinetic movement disorders in man (chorea, tics, dystonia, and tardive dyskinesia). On the other hand, drugs that augment the drug-induced hypokinesia in animals might be expected to exacerbate Parkinsonian hypokinesia but alleviate the symptoms of hyperkinetic movement disorders in humans. With respect to cannabinoids, (–)-Δ^9-THC and other cannabimimetic cannabinoids greatly augment hypokinesia induced in rodents by reserpine or neuroleptic drugs.[53-55] CBD also augments, albeit only modestly, reserpine-[54] or tetrabenazine-induced[128] hypokinesia in rodents. Further, CBD has no effect on tetrabenazine-induced resting tremor or muscle rigidity in these animals.[128]

Another commonly used model is the drug-induced circling (turning or rotational) behavior of rats with 6-hydroxydopamine-induced unilateral lesions of DA nigrostriatal neurons.[129] Generally, drugs that produce or augment circling behavior (contralateral circling with, for example, levodopa or apomorphine, and ipsilateral turning with, for example, anticholinergic drugs, amantadine, or amphetamine) could be predicted to benefit Parkinsonian symptoms, but might worsen some hyperkinetic dyskinesias. Drugs that block apomorphine- or amphetamine-induced circling (e.g., haloperidol) might be expected to exacerbate Parkinson symptoms, but might relieve chorea and other hyperkinetic symptoms. Interestingly, (–)-Δ^9-THC has been reported to produce (ipsilateral) circling in this model,[130] and CBD has been reported to reduce apomorphine-induced (contralateral) circling.[131]

A few other animal models have been used to assess the effects of cannabinoids. (–)-Δ^9-THC was unable to inhibit oxotremorine-induced tremor in mice,[132] the latter being a dubious model of the tremor of Parkinson's disease.[129] CBD has been shown to reduce torticollis and other dyskinesias of rats with inherited dystonia,[128] a putative model of dystonia musculorum deformans.[76] Last, CBD was shown to reduce aggressive behavior in rats following intracaudate injection of the neurotoxin, L-pyroglutamate.[131] The latter is a possible model of the pathological (striatal neuronal degeneration) and behavioral aspects of Huntington's disease.[133]

In summary, the animal data are limited and incomplete. Nevertheless, the general overall impression is that CBD and (–)-Δ^9-THC might be able to alleviate symptoms of several hyperkinetic movement disorders in humans, or augment the marginal efficacy of the standard drugs that are used to treat these symptoms. Additionally, these cannabinoids might exacerbate the hypokinesia of Parkinsonism, but they might not effect the tremor or rigidity of the disorder.

Human Studies

The following presents a synopsis of the reports of marijuana or CBD in dystonia, Huntington's disease, Tourette's syndrome, tardive dyskinesia, and Parkinson's disease. The reports dealing with the effects of marijuana and cannabinoids on the tremor and ataxia associated with conditions with spasticity are presented in the section on "Marijuana, Cannabinoids, and Spasticity".

Dystonia

In 1986, Consroe et al.[134] published the results of a nonblinded, uncontrolled, dose-rising (from 100 to 600 mg/d over 6 weeks) trial of oral CBD in 5 patients with dystonia (Meige's syndrome, spasmodic torticollis, dystonia musculorum deformans, or levodopa-induced dystonia). CBD (given along with standard medication for dystonia, hypertension, or Parkinson's disease) appeared to produce dose-related improvements of dystonia in all patients. In two patients with coexisting Parkinsonian features, CBD (at doses over 300 mg/d) appeared to exacerbate hypokinesia and resting tremor. Side effects of CBD were infrequently reported and consisted of mild lightheadedness, dry mouth, sedation, and hypotension. No further studies have been carried out in dystonic patients.

Huntington's Disease

In 1988, Sandyk et al.[135] published the results of a nonblinded, uncontrolled trial of oral CBD (300 mg/d for 1 week and 600 mg/d for the next 3 weeks) in four patients with Huntington's disease. CBD appeared to produce a mild to moderate improvement of chorea in the three patients with the less advanced stage of the disease. Based upon the encouraging preliminary findings from the human trials and the animal studies, a controlled trial of CBD in patients with Huntington's disease subsequently was undertaken by Consroe and colleagues.[136] The effects of oral CBD (10 mg/kg/d for 6 weeks) and placebo (sesame oil for 6 weeks) were ascertained weekly under a double-blind, randomized crossover design in 15 neuroleptic-free patients. A comparison of the effects of CBD and placebo on chorea severity, many other therapeutic outcome variables, a cannabis side effect inventory, clinical lab tests, and other safety outcome variables indicated no significant or clinically important differences. Correspondingly, weekly plasma levels of CBD (assayed by GC/MS and averaging 6 to 11 ng/ml) did not differ significantly over the 6 weeks of CBD administration. It was concluded that CBD, at an average daily dose of about 700 mg/d for 6 weeks was neither symptomatically effective nor toxic, relative to placebo, in neuroleptic-free patients with Huntington's disease.

Tourette's Syndrome

In 1988, Sandyk and Awerbuch[137] published the case histories of three patients with Tourette's syndrome who reported relief of their motor tics during their alleged marijuana smoking. The patients were also taking, or had taken, anti-tic medications (e.g., haloperidol, clonidine, and/or naltrexone) that were largely ineffective or intolerable. The authors suggested that the beneficial effect of marijuana in these patients may be due primarily to anxiety-reducing properties of the drug rather than to a specific antidyskinetic effect.

Tardive Dyskinesia

A substantial portion of schizophrenic patients abuse marijuana (and other psychoactive drugs),[138] and thus the possible influence of marijuana on the patient's disabilities is of major clinical importance. Many studies have focused on psychotic symptoms, and both detrimental[139] and beneficial[140] effects of marijuana have been reported. Recently, several studies have focused specifically on the occurrence of movement disorders in these patients. In a study of 284 psychiatric inpatients, Olivera et al.[141] reported on the incidence of tardive dyskinesia as related to specific substance abuse. The incidence was not significantly different in patients who chronically abused alcohol alone (25.4%) or who chronically abused alcohol together with marijuana (26.7%). However, these incidences were significantly higher than those in patients who did not use alcohol or who abused other drugs (sedatives, opioids, and stimulants) with or without alcohol. The authors concluded that chronic use of alcohol, but apparently not of the other drugs, enhances the vulnerability of neuroleptic-treated psychiatric patients to tardive dyskinesia.

Additionally, Sandyk and Consroe[142] have studied the relationship of marijuana abuse to both tardive dyskinesia and Parkinsonism in 36 schizophrenic inpatients. Diagnoses of psychiatric illness and substance abuse were made according to DSM-III-R (American Psychiatric Association, 1987) criteria, and diagnoses (presence or absence) of persistent tardive dyskinesia and Parkinsonism were assessed according to standard clinical criteria. Twenty one of the patients had a history of chronic marijuana use for greater than 5 years, and marijuana was the only drug of abuse meeting DSM-III-R criteria for substance abuse in this group. In the other 15 patients, marijuana or other substance abuse was absent based on history and urine screening tests. Also, both groups reported only occasional past use of alcohol or other substances of abuse. Analyses of demographic and treatment (mean) variables of age (31.8 vs. 35.8 years), age of onset of psychiatric illness (22.1 vs. 22.5 years), duration of psychiatric illness (9.7 vs. 13.3 years), and current neuroleptic dosage (in milligrams chlorpromazine equivalents; 1130.5 vs. 1167.8 mg/d) were not significantly different between the two groups (two-tailed t-test, $p > 0.05$ each comparison). Further, the prevalence of persistent tardive dyskinesia (36.3 vs. 53.3%) and drug-induced Parkinsonism (86.3 vs. 86.6%) were not significantly different between the marijuana abusers and the nonmarijuana group (Chi square test, $p > 0.05$ for each comparison). Collectively, these findings suggest that marijuana abuse or dependence does not seem to be a risk factor for neuroleptic-induced tardive dyskinesia or Parkinsonism in chronic schizophrenic patients.

Parkinson's Disease

As described above, marijuana does not appear to enhance the susceptibility of neuroleptic-treated psychiatric patients to the development of Parkinsonism (or tardive dyskinesia). It is hypothetically possible, however, that marijuana might influence the severity of the major signs of the condition. In 1990, Frankel et al.[143] published a short letter of an open study of 5 patients with Parkinson's disease given on consecutive days: a marijuana cigarette (1 g containing 2.9% THC), diazepam (5 mg orally), levodopa/carbidopa (250 mg/25 mg orally), and apomorphine (1 to 5 mg s.c.). The patients were evaluated for Parkinsonian disability (especially tremor) before and after dosing with each treatment. Improvement of tremor was documented with levodopa and apomorphine, whereas no improvement of tremor or other disabilities was documented with diazepam and marijuana.

MARIJUANA, CANNABINOIDS, AND SPASTICITY

Overview of Spasticity

Spasticity is a specific type of hypertonia in which there is an increased velocity-dependent resistance to passive movement of muscles.[144] Spasticity may develop following a lesion in descending cortical, cerebellar and spinal motor pathways involving the pyramidal tract. In patients with spasticity, α-motoneurons in the spinal cord become abnormally excitable.[144] This may be due to the loss of normal supraspinal modulation and/or defects within the spinal cord itself. Impairment of GABA-mediated presynaptic inhibition, glycine-mediated interneuronal inhibition, and L-glutamate (or aspartate) neuronal excitation may be involved to varying degrees in a given patient with spasticity.[145]

Spasticity is not a disease, but a disorder of motor control observed in a number of neurological conditions such as cerebrovascular insult, cerebral palsy, tumors, trauma (e.g., of spinal cord), and multiple sclerosis. The entity of spasticity is characterized by negative signs of weakness and loss of dexterity, and by positive signs of increased muscle tone ("spasticity") and flexor and extensor spasms.[144] Additionally, other features such as pain, tremor, bowel and bladder dysfunctions, may be present. The occurrence, severity, and course of the signs and symptoms vary widely among patients, and this is due to the nature and site of the lesion or abnormality. Examples are the varied clinical presentations of patients with multiple sclerosis and the relapsing and remitting course of this chronic disease.[144] Nevertheless, spasticity is a chronic problem affecting a relatively large number of neurological patients. Moreover, it is a major cause of functional disability and, as such, it is often the target of symptomatic drug treatment. Pertinent to the present topic is spasticity associated with spinal cord injury and multiple sclerosis. The following presents a brief overview of the standard drug therapy of the spasticity and some other common signs and symptoms of patients with these conditions.[144-146]

Spasticity and flexor and extensor spasms of both conditions are often managed with the antispasticity drugs, baclofen, benzodiazepines (diazepam or clonazepam), tizanidine, or dantrolene. Baclofen is a $GABA_B$ agonist that appears to act predominately by suppressing the release of excitatory neurotransmitters (L-glutamate and aspartic acid) from primary afferent fibers and interneurons in the spinal cord. The benzodiazepines, and perhaps tizanidine as well, appear to facilitate $GABA_A$ receptor-mediated presynaptic inhibition in the spinal cord. Moreover, baclofen, tizanidine, and the benzodiazepines also may be acting at supraspinal sites such as the cerebellum and the substantia nigra pars reticulata. Dantrolene, on the other hand, reduces skeletal muscle strength by interfering with calcium-mediated, excitation-contraction coupling directly in the muscle. Although effective for many patients, a major problem with all of these drugs is their tendency to aggravate weakness. Moreover, patients who rely on spasticity or extensor spasms for ambulation and transfers may worsen functionally with treatment of the antispasticity agent. Other side effects of most of these agents include sedation, fatigue, dizziness, and ataxia. In multiple sclerosis, short-term therapy of ACTH or corticosteroids such as prednisone can accelerate recovery from acute attacks (relapses or exacerbations). Concomitantly, patients may experience some relief from their spasticity. It has not been shown definitely, however, that these drugs enhance the degree of recovery.

Acute pain syndromes, most commonly trigeminal neuralgia, occur in about 9% of multiple sclerosis patients. Most patients with acute pain can be managed effectively by carbamazepine, either alone or in combination with phenytoin or baclofen. On the other hand, chronic pain (often occurring at night) is a relatively frequent complication of spasticity in both multiple sclerosis and spinal cord injury. Unfortunately, this type of pain rarely responds fully to carbamazepine, phenytoin, baclofen, tricyclic antidepressants, clonazepam, or nonsteroidal antiinflammatory drugs. Behavioral approaches and careful use of narcotics may be required.

Symptoms of bowel and bladder dysfunction are almost universal in patients with long-standing multiple sclerosis, and they are quite common in spinal cord injured patients as well. Most patients with bowel dysfunction benefit from dietary increases of roughage and appropriate use of laxatives. Problems with bladder control are more difficult to manage. Anticholinergic drugs (e.g., propantheline) are the mainstays of treatment for patients with bladder spasms (detrusor muscle hyperreflexia). Cholinergic medications (e.g., bethanechol) can be helpful in some patients with bladder hypotonia, although these drugs may exacerbate spasticity of the external urethral sphincter. Other treatments for resistant cases of bladder dysfunction include combinations of anticholinergics and α-adrenergic antagonists (e.g., phenoxybenzamine) or antispasticity drugs (e.g., diazepam, baclofen, or tizanidine).

Lastly, chronic fatigue, paroxysmal phenomenon (paroxysmal tonic seizures and paroxysmal dysarthria with ataxia), and persistent cerebellar dysfunction (intention tremor and limb and gait ataxia) may occur in many patients with multiple sclerosis. Also, weakness (e.g., paraparesis), hemiplegia, and paraplegia may be a complication of both multiple sclerosis and spinal cord injury. The fatigue of multiple sclerosis may respond to amantadine, and the paroxysmal phenomena may respond to carbamazepine, phenytoin, or baclofen. However, intention tremor and ataxia are virtually resistant to all known drug treatments. Further, there is no consistently effective treatment for the weakness or paralysis of multiple sclerosis.

Effects of Marijuana and Cannabinoids

Animal Studies

There are no data on the effects of cannabinoids in animal models specifically relevant to spasticity (e.g., the genetically spastic rat[47]) although, presumably, work is in progress on cannabinoid receptor binding in one of these models.[147] There are, however, many data on the neurological effects of cannabinoids in animals that relate to motor behavior and muscle tone. These data have been reviewed.[1] Briefly, the typical gross behavioral effects of (–)-Δ^9-THC and other cannabimimetic cannabinoids include hypomotility, hyperreactivity, catalepsy, abnormal gait, and static ataxia. Electrophysiological studies indicate that the THCs and related cannabinoids can attenuate polysynaptic reflexes, although the (presumably) supraspinal site of action for these effects has not been identified. Also, these cannabinoids can produce both an enhancement (at low doses) and depression (at higher doses) of a spinal monosynaptic reflex, and can produce mixed stimulant and depressant effects in the cortex and other sites within the brain. Furthermore, the THCs can reduce neuronal conduction, block neuromuscular transmission, and depress the propagation of action potentials directly in excitable skeletal muscle. In contrast to the THCs, CBD is devoid of stimulant effects on behavior or neuronal activity. However, CBD produces depression of the spinal monosynaptic reflex and depression of synaptic transmission at other sites in brain.

In spite of the relatively large amount of experimental information, a clear picture of how cannabinoids might act to benefit spasticity, intention tremor, and ataxia, and paradoxically to cause tremors and ataxia, has not emerged. However, the novel findings of cannabinoid receptor localization in mammalian brain provide some clues to this mystery. Additionally, a detailed quantitative localization of cannabinoid receptors in rat brain has recently been published.[147] This study has reaffirmed the paucity of cannabinoid receptors in the spinal cord and the great abundance of receptors in the cerebellum and basal ganglia complex. Moreover, new data indicate an important differential pattern of cannabinoid receptor localization within the corpus striatum (i.e., dense binding in sensorimotor compartments, but sparse binding in the "limbic" domain) and within the cerebral cortex (i.e., greater binding in frontal motor cortex than in other cortical areas). Last, relatively dense binding was found in pyramidal neuronal layers of the cortex that project to spinal cord, brain stem, and striatum. Collectively, it is possible that cannabinoid receptors could mediate motor function via both pyramidal and extrapyramidal systems, and this obviously is germane to neurological disorders with spasticity. Clearly, there is a need for additional studies of these important motor areas of the brain using relevant animal models of spasticity and extrapyramidal (and cerebellar) dysfunction.

Human Studies

The reports of marijuana and cannabinoid effects in conditions associated with spasticity have been reviewed in a 1986 publication.[1] Briefly, there are a number of anecdotal accounts of the antispasticity effects of marijuana smoking in patients with both spinal cord injury and multiple sclerosis. A single, uncontrolled trial of acute oral (–)-Δ^9-THC (5 to 15 mg) in multiple sclerosis patients indicated an objective reduction of intention tremor and ataxia in 2 of 8 patients. A single double-blind, placebo-controlled study indicated that acute oral (–)-Δ^9-THC (10 mg) objectively reduced spasticity in 9 patients with multiple sclerosis. Since the review, there have been a few additional reports.

In 1987, Ungerleider et al.[148] reported on a short-term trial of oral (–)-Δ^9-THC in 13 subjects with multiple sclerosis and spasticity, and a history of intolerable side effects from use of standard antispasticity drugs. There were nine subjects who had used marijuana in the past to treat their disabilities. The study was a double-blind, placebo-controlled, crossover, dose-rising (2.5 to 15 mg) trial of 5 d of (–)-Δ^9-THC and 5 d of placebo, with a 2-d washout period. Subjective ratings of spasticity and systemic side effects were obtained from patient responses, and objective ratings of motor impairment (i.e., ordinal measurements of limb weakness, spasticity and coordination, gait impairment, and reflexes) were recorded by neurologists prior to the study (baseline) and on the last day of each treatment. Results indicated that the subjective level of spasticity decreased significantly as the dose of (–)-Δ^9-THC increased, with 7.5 mg being the minimally effective dose. Further, subjective spasticity scores were significantly lower for (–)-Δ^9-THC (at 7.5 mg) relative to placebo. However, there were no statistically significant differences among baseline, placebo, and (–)-Δ^9-THC conditions for any of the objective measures recorded in the trial. Typical side effects of (–)-Δ^9-THC (e.g., weakness, dry mouth, dizziness, and mental clouding) were reported by virtually all patients while on the cannabinoid (at 7.5 mg), and also were reported by 5 patients while on placebo. With doses above 7.5 mg, side effects were often reported as being intolerable. A perhaps telling comment made by the authors was that most of the patients in the study did not request (–)-Δ^9-THC subsequent to the study.

In 1989, Meinck et al.[149] reported on an open trial of acute marijuana smoking in a single multiple sclerosis patient with severe spastic tetraparesis, intention tremor, and limb and gait ataxia. The patient was bound to a wheelchair and had used marijuana frequently in the past to self-medicate his symptoms. He was hospitalized for the duration of the study and was not taking any drugs. Daily clinical ratings, electromyographic recordings of leg flexor reflexes, and electromagnetic recordings of finger and hand intention tremor were made for 5 d before, 3 d after, and during the smoking of one marijuana cigarette (concentration of cannabinoids not specified). Shortly after marijuana smoking, a marked reduction of spasticity (reduced muscle tone, normalization of deep tendon reflexes, reduction of clonus, attenuation of flexor reflexes, and narrowing of the receptive field of the plantar extensor response, i.e., the Babinski sign) was reported. Also, ataxia (finger-nose testing and evaluation of gait) was conspicuously improved, and intention tremor was completely abolished.

In 1990, Maurer et al.[150] reported the results of a double-blind, placebo/drug-controlled trial of oral (–)-Δ^9-THC in a single patient with spasticity and pain due to spinal cord injury (i.e., following surgical removal of an ependymoma). Prior to the trial, an open dose-finding study (of 2.5, 5, and 10 mg) was carried out with the patient, and 5 mg emerged as an apparently effective and well-tolerated dose. The patient was also taking daily baclofen and clonazepam, and these were maintained throughout the trial. The double-blind trial compared (–)-Δ^9-THC (5 mg), codeine (50 mg), and placebo (each in identically appearing capsules) in a randomized, balanced cross-over design. Each treatment was given 18 times over the course of 5 months. Effects were assessed by the patient (the day after each treatment), who provided subjective ratings of variables on two-pole, visual analog scales of 50-mm length. Results indicated that relative to placebo, (–)-Δ^9-THC was significantly more effective on all variables, i.e., improving spasticity, pain, micturition, sleep, concentration, and mood. Also, relative to placebo, codeine was significantly more effective on all variables except spasticity where there was no difference in scores. Compared with codeine, the cannabinoid was equieffective on all variables except spasticity. For the latter, only (–)-Δ^9-THC had an antispastic effect, and reportedly lasted for more than 12 h after each dose. Following this trial, the patient received a maintenance supply of (–)-Δ^9-THC in addition to his regular medication for about 3 years. Motivated by slow and sustained improvement, the patient gradually reduced dosages of all medications and finally terminated all medications, including (–)-Δ^9-THC.

In summary, spinal cord injury and multiple sclerosis are probably the most common neurological indications for (illicit) marijuana use. Many patients believe that marijuana provides prompt and effective relief of their spasticity, and apparently without major side effects. The data from one clinical trial of marijuana in one patient appears to confirm the anecdotal accounts. Results of clinical trials with (–)-Δ^9-THC are mixed, but the overall impression is that this cannabinoid has antispasticity and analgesic effects, albeit with a greater side effect liability. In addition, both marijuana and (–)-Δ^9-THC appear to have antitremor and antiataxia effects. This would be of major clinical significance since these effects are not ascribed to any medications currently available. Collectively, these encouraging findings warrant additional controlled trials that are of proper design, adequate sample size, and sufficient duration to establish definitively the benefit-risk ratio of marijuana and (–)-Δ^9-THC.

MARIJUANA, CANNABINOIDS, AND NEUROLOGICAL PAIN

Overview of Migraine Headache and Neuropathic Pain

Migraine[151]

Migraine is a familial disorder characterized by periodic, commonly unilateral, throbbing headaches that begin in childhood, adolescence, or early adult life and recur with diminishing frequency during advancing years. Migraine is prevalent, found in an estimated 3 to 5% of the general population, and in a much higher proportion (15%) of women during their reproductive years. The incidence in females is about twice that in males.

Classic (typical) migraine is characterized by a premonitory change in mood and appetite, followed by a visual aura and disturbance of other neurologic function (e.g., paresthesia, dizziness, and weakness of extremities), hemicranial (or occasionally bilateral) headache, and nausea and vomiting. Common (atypical) migraine is characterized by a heralded onset of hemicranial or generalized headache without a typical aura, and with or without nausea and vomiting. Both types of migraine may last for hours or as long as a day or two.

The pathophysiology of migraine is believed to involve a complex interplay of vascular and neuronal factors in genetically susceptible individuals. In one popular hypothetical model,[152] an internal clock, mood changes, or excessive afferent stimuli (e.g., flashing lights, loud sounds, or perfume) influence brain stem structures and neurotransmitters (e.g., the locus ceruleus releasing NE and the nucleus raphe dorsalis releasing serotonin) that project to the cerebral cortex.

This causes constriction of the cerebral microcirculation, ischemia of occipital cortex (associated with the visual aura), and spreading depression (associated with the other neurological symptoms). Peripherally, adrenal NE causes platelet aggregation with the liberation of free serotonin in plasma and in gut (the latter associated with the nausea and vomiting). Subsequently, plasma levels of serotonin fall because of metabolism (to 5-hydroxyindoleacetic acid) and absorption to vessels. The low levels of plasma serotonin lead to release of tonic vasoconstriction and hence to dilation of arteries. The vascular distension of cranial vessels (sensitized by serotonin), coupled with an inflammatory component (involving prostaglandins, histamine, and other substances), causes the pain and throbbing component of migraine headache. The painful aspects of migraine are perpetuated by reverberations between the trigeminal nerve and the intracranial vasculature. Recently, plasma levels of the neuroexcitatory amino acids, L-glutamate, and aspartic acid have been shown to be significantly elevated in patients with migraine.[153] This suggests that the functional activity of the NMDA receptor may be another important factor to consider in the pathogenesis of migraine.

Treatment of migraine includes management of the acute attack and prophylaxis.[154] Aspirin, acetaminophen, and potent nonsteroidal anti-inflammatory agents are effective in a majority of mild to moderate acute attacks. Their mechanism of analgesic and antiinflammatory actions is presumably via prostaglandin inhibition, although it is unclear whether this is the primary basis for their efficacy in migraine. Ergotamine and other ergot preparations are the most effective treatments of an acute attack, and they are commonly used when the attack is severe. Ergotamine abolishes the headache in about 75% of patients, although side effects of nausea and peripheral vasoconstriction (ergotism) may be significant. The ergots are potent vasoconstrictors of cerebral vessels, and this effect may be mediated, in part, by an agonist action on 5-hydroxytryptamine (5-HT_1) receptors. Recently, this mechanism has gained support from other, more potent and selective 5-HT_1 agonists such as sumatriptan. Although still experimental, sumatriptan is reported to be an extremely effective and well-tolerated therapy for acute migraine. Prophylactic treatments include propranolol, tricyclic antidepressants (e.g., amitriptyline), methysergide, and calcium channel blockers (e.g., verapamil or nifedipine). Although some of these drugs show antagonist activity at the 5-HT_2 receptor, no unifying hypothesis explains the effectiveness of all of these prophylactic agents. Moreover, despite these diverse pharmacological interventions, there is a subgroup of patients who fail to respond adequately to any currently available drug.

Neuropathic Pain[155]

Neuropathic pain is associated with a diverse group of chronic disorders, including various causalgias (pain felt in a limb where there has been partial damage to the sympathetic and somatic sensory nerves) and neuralgias (pain often following the course of a nerve). Depending on the condition, the character of the pain may be expressed in various ways. For example, in sciatic nerve causalgia, the pain is felt as an intense and sustained burning pain in the leg, often with a superimposed lancinating (stabbing) pain component. The pain of trigeminal neuralgia is characterized by paroxysms of intense, stabbing pain in the distribution of the mandibular and maxillary divisions of the trigeminal nerve. In this case, the pain is fleeting, lasting a few seconds to a minute. However, the attacks may recur frequently over the course of several weeks at a time. In addition to the paroxysmal pain, patients with trigeminal neuralgia may experience a more or less continuous discomfort and sensitivity of the face.

Patients with neuropathic pain present a great therapeutic problem, since typical analgesic drugs are either ineffective or impractical to use.[155] For nonlancinating neuropathic pain, a tricyclic antidepressant such as amitriptyline may be useful. Antiepileptic drugs (carbamazepine and phenytoin) are presently the drugs of choice in the management of trigeminal neuralgia and other conditions with lancinating pains. Carbamazepine is the most effective drug. Clonazepam and baclofen may be useful in patients who cannot tolerate carbamazepine. However, most of the patients with severe pain may require surgery or other nonpharmacologic methods of pain control.

Effects of Marijuana and Cannabinoids

Animal Studies

As previously discussed above, virtually all cannabimimetic cannabinoids have analgesic and antiinflammatory properties, and many of these are extremely potent. Unfortunately, there is little separation between these potentially beneficial effects and cannabimimetic activity. CBD and virtually all other noncannabimimetic cannabinoids are devoid of these effects. (–)-Δ^8-THC-11-oic acid, and possibly (+)-HU-210, are exceptions, since these cannabinoids have analgesic and antiedema effects and no cannabimimetic activity. Also, cannabinoids influence many of the chemical substances that are believed to be involved in the pathogenesis of migraine. In one recent study, (–)-Δ^8-THC-11-oic acid was reported to inhibit the synthesis of prostaglandins and antagonize the edema and mortality produced by platelet activating factor.[156] In another study, (+)-HU-210 was reported to be a functional antagonist of the NMDA receptor.[42] For the most part, however, these findings are preliminary or incomplete, and in need of further study. There have been many investigations of the effects of cannabimimetic cannabinoids on brain NE and serotonin, yet the results are quite variable.[53] Also, there are no cannabinoid data on 5-HT receptor subtypes (e.g., binding data to these receptors in cranial vessels) or cranial vasculature (e.g., on resistance and blood flow dynamics of the carotid arteries), which are most relevant to migraine. Although more specific data are needed, the fact that the cannabinoids have potent antinociceptive, antiinflammatory, and prostaglandin inhibitory effects suggests that they might have analgesic effects in patients with migraine. It remains to be determined, however, whether the cannabinoids would have any advantage over the multitude of nonsteroidal antiinflammatory agents already on the market for treating migraine.

Animal tests for nociception and inflammation are not applicable models of neuropathic pain, since antiepileptics or antidepressants, rather than typical analgesics, are more effective in treating this kind of pain. Interestingly, CBD has an antiepileptic spectrum of activity in animals similar to that of carbamazepine and phenytoin,[1] the standard drug treatments of these paroxysmal pains. This analogy has prompted a clinical trial of CBD in patients with neuropathic pain and the results of this will be described below.

Human Studies

There are few newer reports that either directly or indirectly relate to the effects of marijuana or cannabinoids in migraine. In 1985, Volfe et al.[156] reported the *in vitro* effects of (–)-Δ^9-THC and CBD on serotonin release from human platelets. The platelets were obtained from patients during and between migraine attacks, and were incubated with low pharmacologically relevant concentrations of the cannabinoids. The results indicated that (–)-Δ^9-THC markedly inhibited the serotonin release during the attack, but not between attacks. CBD had no inhibitory effect at either time period. In 1987, El-Mallakh[157] described three cases in which common migraine headaches appeared shortly after discontinuation of long-term marijuana use. The patients alleged daily marijuana use ranged from 5 to 13 years and, according to their self-reports, there was a total absence of migraine headache during the period of marijuana use. In 1989, Matthew et al.[158] reported the effects of marijuana smoking on cerebral blood flow in experienced and inexperienced marijuana users. Results indicated that, compared with placebo marijuana smoke, marijuana produced a significant decrease of cerebral blood flow across all lobes of the cortex in the inexperienced users. Experienced users, on the other hand, showed no significant differences in cerebral blood flow after marijuana. While this latter study is not directly related to the study of migraine, the finding of cerebral blood flow reduction is rudimentarily similar to that which occurs in patients during migraine headache.[152] Extrapolation of these results could suggest that acute marijuana smoking might precipitate or exacerbate an attack, whereas chronic marijuana smoking might prevent an attack in migraine patients. This is, of course, speculation for which there are no objective data.

Only one study, presented in 1987, has been carried out with cannabinoids in neuropathic pain. Lindstrom et al.[160] reported on the double-blind, placebo-controlled, randomized crossover trial of oral CBD in ten patients with sustained neuropathic pain (e.g., sciatic nerve causalgia, postherpetic neuralgia, etc.). CBD (450 mg/d, divided into three doses) and placebo were given for 1 week, with a 1-week washout between the treatments. The patients estimated their pain daily, using a visual analog scale from 0 (no pain) to 100 (unbearable pain). Subjective side effects were recorded daily, and plasma levels of CBD were measured at the end of the trial. The results indicated that there was no significant difference of pain scores between CBD and placebo. The only reported side effect of CBD was sedation, and this occurred in seven of the patients. Plasma levels of CBD ranged from 50 to 80 ng/ml at the end of CBD administration.

In summary, anecdotal accounts of the benefits of marijuana for migraine have been around for hundreds of years, and the one recent report of three cases is in line with these tales. Further, the serotonin platelet study provides some information that suggests antimigraine activity for (–)-Δ^9-THC. However, there continues to be no hard data to confirm the presumed antimigraine effects of marijuana or cannabinoids. On the other hand, it appears that CBD clearly is not efficacious for treating sustained neuropathic pain. Whether CBD would relieve paroxysmal (e.g., trigeminal neuralgia) neuropathic pain, or whether other cannabinoids would relieve paroxysmal or sustained (e.g., sciatic nerve causalgia) neuropathic pain is unknown. At present, one is left only with the reports from the 19th century indicating that marijuana, at least, "is especially useful for neuralgia of the fifth nerve and is almost useless for sciatica and myodynia".

TABLE 1
Reported Effects of Marijuana Treatment on Features of Neurological Disorders Described in the 19th Century and Possible Contemporary Analogies to These Neurological Disorders

Disorders of the 19th century	Contemporary disorders	Effect of marijuana
Convulsions: recurrent, general, tonic, clonic	Tonic-clonic seizures: generalized epilepsy	Benefit
Convulsions: petit mal	Absence seizures: generalized epilepsy	No effect
Spasms of torticollis and writer's cramp	Dystonic movements: spasmodic torticollis; writer's cramp	No effect
Rheumatic chorea; generalized chorea	Choreic movements: Sydenham's chorea; Huntington's chorea	Benefit; no effect
Tremor of paralysis agitans	Resting tremor: Parkinson's disease	No effect
Tonic painful spasms; jerky movements of spinal sclerosis	Spasticity and ataxia: spinal cord injury; multiple sclerosis	Benefit; no effect
Pain of neuropathy, e.g., sciatica	Sustained pain: neuropathic pain	No effect
Pain of neuropathy, e.g., trigeminal neuralgia	Paroxysmal pain: neuropathic pain	Benefit
Migraine headache	Migraine headache	Benefit

SUMMARY AND CONCLUSIONS

A substantial number of patients with epilepsy, idiopathic and drug-induced dystonias, tardive dyskinesia, Huntington's disease, Tourette's syndrome, advanced Parkinson's disease, multiple sclerosis, spasticity associated with spinal cord injury, severe migraine, and neuropathic pain are recalcitrant to presently available drugs. While there is no doubt that new treatment drugs are needed, the central question of this review is: what is the therapeutic potential of cannabinoids, and of marijuana itself, for these chronic and debilitating neurological conditions? A succinct answer is that the outlook is less than encouraging for the present and near term, but the view is more promising for the future.

Tables 1 to 3 present a synopsis of the reported effects of marijuana and cannabinoids up to the present time. Table 1 presents a summary of the reported effects of marijuana treatment on some of the major features of neurological disorders described in the 19th century. Also listed are the

TABLE 2
Predicted Effects of (–)-Δ^9-Tetrahydrocannabinol (THC) and Cannabidiol (CBD) on Contemporary Neurological Disorders Based on Their Reported Activity in Animal Models

Disorder or symptom	Animal model	Drug	Predicted Effect
Partial (simple and complex) seizures	Focal seizures	THC	Aggravate or benefit
		CBD	Benefit
Tonic-clonic seizures (grand mal)	Maximal generalized seizures	THC	Benefit
		CBD	Benefit
Absence seizures (petit mal)	Minimal generalized seizures	THC	No effect or aggravate
		CBD	No effect
Dystonia, Huntington's chorea, Tourette's syndrome, and tardive dyskinesia	Reserpine-induced hypokinesia	THC	Benefit
		CBD	Benefit
Parkinsonism: hypokinesia	Reserpine-induced hypokinesia	THC	Aggravate
		CBD	Aggravate
Tremor and rigidity	Tremor and rigidity	CBD	No effect
Parkinsonism	Drug-induced circling behavior	THC	Benefit
		CBD	Aggravate
Dystonia	Inherited dystonia	CBD	Benefit
Huntington's chorea	L-Pyroglutamate-induced behavioral disturbance	CBD	Benefit
Spasticity	Polysynaptic and monosynaptic reflexes	THC	Benefit or aggravate
		CBD	Benefit
Cerebellar ataxia	Cannabinoid-induced ataxia	THC	Aggravate
		CBD	No effect
Migraine headache	Nociception, edema, and prostaglandins	THC	Benefit
		CBD	No effect

contemporary neurological disorders that appear to be analogous, and may even be homologous, to their 19th century counterparts. Table 2 displays a summary of the predicted effects of (–)-Δ^9-THC and CBD on contemporary neurological disorders based on their activity in animal studies. Based upon the animal and available human data, it is assumed that other cannabimimetic cannabinoids would have similar neurological effects to that of (–)-Δ^9-THC.

TABLE 3
Reported Effects of Marijuana (MJ), (–)-Δ^9-Tetrahydrocannabinol (THC) and Cannabidiol (CBD) on Contemporary Neurological Disorders

Disorder	Major symptom	Drug	Effect
Focal epilepsy	Partial (simple and complex) seizures	MJ	Benefit
		CBD	Benefit or no effect
Generalized epilepsy	Tonic-clonic seizures (grand mal)	MJ	Benefit
		THC (Analogs)	Benefit
		CBD	Benefit or no effect
Generalized epilepsy	Absence seizures (petit mal)		
Dystonias	Dystonia	CBD	Benefit
Huntington's disease	Chorea	CBD	No effect
Tourette's syndrome	Tics	MJ	Benefit
Tardive dyskinesia	Oro-buccal-lingual dyskinesia	MJ	No effect
Parkinson's disease	Resting tremor; rigidity; hypokinesia	MJ	No effect
		CBD	Aggravate
Multiple sclerosis and spinal cord injury	Spasticity; intention tremor; ataxia	MJ	Benefit
		THC	Benefit
Neuropathic conditions	Sustained pain	CBD	No effect
Migraine	Pain and other symptoms	MJ	Benefit

Table 3 shows a summary of the reported effects of marijuana, (–)-Δ^9-THC and CBD on the contemporary neurological disorders. For both the animal and clinical reports, the strength of the data varies widely. In these tabular presentations, we have used the terse generic descriptions of "benefit, aggravate, and no effect" to provide a tenor of the reports rather than a critique.

Table 4 presents a summary of our appraisal of the collective evidence concerning the therapeutic potential of marijuana, (–)-Δ^9-THC and CBD in these neurological disorders. We have appraised marijuana and (–)-Δ^9-THC as separate entities (which they are) and in conformity with their respective utilization in the reported studies. Concerning (–)-Δ^9-THC, most studies have used this cannabinoid prototype, but it appears that most other currently available cannabimimetic cannabinoids would have a similar qualitative profile of pharmacological activity. Similarly, the qualitative spec-

TABLE 4
Therapeutic Potential of Marijuana (MJ), (–)-Δ^9-Tetrahydrocannabinol (THC) and Cannabidiol (CBD) on Contemporary Neurological Disorders

Disorder	Major symptom	Drug	Therapeutic Potential[a]
Focal epilepsy	Partial (simple and complex) seizures	MJ	+
		THC	–/+
		CBD	+/–
Generalized epilepsy	Tonic-clonic seizures (grand mal)	MJ	+
		THC	+
		CBD	+/–
Generalized epilepsy	Absence seizures (petit mal)	MJ	–/+
		THC	–/+
		CBD	–/+
Dystonias	Dystonia	MJ	+/–
		THC	+/–
		CBD	+
Huntington's disease	Chorea	MJ	+/–
		THC	+/–
		CBD	– –
Tourette's syndrome	Tics	MJ	+
		THC	+/–
		CBD	+/–
Tardive dyskinesia	Oro-buccal-lingual dyskinesia	MJ	–
		THC	+/–
		CBD	+/–
Parkinson's disease	Resting tremor; rigidity; hypokinesia	MJ	–
		THC	–/+
		CBD	–
Multiple sclerosis and spinal cord injury	Spasticity; intention tremor; ataxia	MJ	+
		THC	++
		CBD	+/–
Neuropathic conditions	Sustained pain	MJ	–/+
		THC	
		CBD	– –
Neuropathic conditions	Paroxysmal pain	MJ	+/–
		THC	
		CBD	
Migraine	Pain and other symptoms	MJ	+/–
		THC	+/–
		CBD	–/+

[a] Therapeutic potential is rated from +++ to – – –; ratings indicate excellent (+++), good (++), and fair (+) evidence for efficacy; equivocal or uncertain evidence for (+/–) or against (–/+) efficacy; and fair (–), good (– –), and excellent (– – –) evidence for lack of efficacy. A blank space indicates no data or reports. See text for criteria of rankings.

trum of activity of the frequently studied CBD would appear generally comparable to most analogs of CBD. Also, we have utilized the term "therapeutic potential" in reference to efficacy or lack thereof, and have attempted to provide a measure of the strength of the evidence for these important attributes of the drugs. The benefit-risk (e.g., side effect) ratio would have been a better term to use, but side effects and toxicity of these agents rarely have been evaluated systematically or critically in the reports under review.

Although subjective and arguable, the following ordinal rating scheme and associated criteria were used to describe the therapeutic potential for efficacy (plus designations) and lack of efficacy (minus designations). An "excellent (+++)" potential indicates at least two appropriate (e.g., double-blind, placebo, or active drug-controlled, adequate sample size for the design and measures, adequate duration of treatment, etc.) clinical trials showing efficacy and whose results are not opposed by findings of a lack of efficacy from a similar-type trial(s). A "good (++)" potential indicates at least one appropriate and uncontested clinical trial showing efficacy. A "fair (+)" potential indicates at least one uncontested open trial or case-controlled study suggesting efficacy, or uncontested case reports suggesting efficacy and convincing animal data predicting efficacy. An "equivocal or uncertain potential for (+/–) or against (–/+) efficacy" indicates opposing findings of efficacy and lack of efficacy for similar types of clinical studies or reports, or only isolated anecdotal reports and/or inconclusive or ambivalent animal data that claim efficacy or lack thereof. The "minus" designations refer to fair (–), good (– –), and excellent (– – –) evidence for the lack of efficacy, using the same criteria as for the plus designations regarding efficacy.

Based on these criteria, none of these agents at present would be judged to have an excellent therapeutic potential for any disorder mentioned here. Only (–)-Δ^9-THC was judged to have a good potential, and only for neurological disorders associated with spasticity. Interestingly, some of the more convincing evidence was for a lack of efficacy, i.e., for CBD in Huntington's disease and in sustained neuropathic pain. The other evidence would be judged only fair at best, and most only equivocal. The following describes the spectrum of therapeutic potential for each agent.

Marijuana appears to be efficacious for partial and generalized tonic-clonic seizures, tics of Tourette's syndrome, and spasticity, ataxia, and tremor of multiple sclerosis and spinal cord injury. However, the evidence for clinical efficacy is only fair, since there is not yet an appropriate controlled trial of the substance for any of the conditions. In a similar evidentiary vein, marijuana appears to have a lack of clinical efficacy in tardive dyskinesia and Parkinson's disease. Its therapeutic potential (or lack thereof) in dystonias, chorea, migraine, neuropathic pain, and absence seizures is uncertain since the only evidence consists of few clinical reports, primarily from the previous century.

(–)-Δ^9-THC appears to attenuate spasticity, tremor, and ataxia. Overall, the evidence for clinical efficacy is good. However, no one study has incorporated all of the features of design that would qualify it as completely adequate. The definitive controlled study of daily administration of (–)-Δ^9-THC for at least 6 months in a sufficient number of patients with spasticity, tremor, and ataxia has yet to be done. (–)-Δ^9-THC also appears to have antiepileptic activity for major motor seizures. The evidence for this is only fair and indirect, i.e., data of (–)-Δ^9-THC in various animal models of epilepsy, and a single abstract report of some dimethylheptyl analogs of (–)-Δ^3-THC in children with generalized convulsive seizures. The potential effectiveness or ineffectiveness of (–)-Δ^9-THC in partial seizures, absence seizures, various movement disorders, and migraine is uncertain, as the evidence is based largely on animal data without clinical confirmation.

CBD appears to be ineffective for sustained pain associated with various neuropathic conditions, and ineffective for symptoms of Huntington's disease. In the former case, there is good evidence of lack of clinical efficacy from one adequately controlled trial of the drug. In the latter case, limited data in animal models and precursory clinical data from our open trial in four patients with Huntington's disease suggested that CBD might have beneficial effects in the disorder. Our subsequent controlled trial of CBD (double-blind, placebo-controlled trial of about 700 mg/d CBD for 6 weeks in 15 neuroleptic-free patients with Huntington's disease) failed to demonstrate any beneficial or toxic effects of the cannabinoid. Also, plasma levels of CBD were low (6 to 11 ng/ml) suggesting that the oral bioavailability of CBD is very low. Whether higher plasma levels of CBD would have changed the outcome of the study is not known. At the measured plasma levels of CBD, there was not even a nonsignificant trend in any of the data which would suggest clinical efficacy for the cannabinoid. Thus, within the context of the methods and results of our controlled trial, we must consider the clinical findings to be good evidence for the lack of symptomatic effectiveness of CBD in Huntington's disease.

CBD may have clinical antidystonic properties and may exacerbate the hypokinesia and tremor of Parkinsonism. However, the evidence for these possible therapeutic effects is only fair (i.e., based on findings from our pilot, open trial of CBD in five patients with dystonic movement disorders, and on limited data in animal models). The therapeutic potential of CBD in the remaining disorders is uncertain. Limited or circumstantial data from animal studies suggest that CBD might be effective in Tourette's syndrome, tardive dyskinesia, and perhaps spasticity, and might be ineffective in migraine. Substantial data from well-established animal models predict that CBD is probably ineffective in absence seizures; yet there are no corresponding clinical reports of CBD in humans with any of these conditions.

Extensive data from a number of relevant animal models also predict that CBD may be effective in partial seizures and generalized tonic-clonic seizures. For these seizure disorders, two independent and somewhat similar controlled trials of CBD have been carried out (e.g., double-blind, placebo-controlled, 10 or 15 patients with partial and/or generalized tonic clonic seizures, added on CBD at 300 mg/d for 4.5 or 6 months). Conflicting findings of both clinical benefit and no effect were found. The more recent clinical trial ascertained serum levels of CBD, and the levels were found to be low (averaging about 20 ng/ml). The report also mentioned that much higher serum levels of CBD (averaging about 150 ng/ml) were obtained in one additional epileptic patient given, under nonblinded conditions, 900 to 1200 mg/d CBD for 10 months. Corresponding to this, a reduction of this patient's seizure activity was reported. The reasons for the discrepancies between the two controlled trials, and between the positive animal findings and the negative human findings, are obscure. Since CBD, rather than its metabolites, appears responsible for its anticonvulsant activity in animals at least, the plasma (and of course brain) levels of CBD become a critical issue. In humans, it is now clear that the bioavailability of oral CBD is very low. Massive dose increases or other methods (e.g., formulation changes or chemical modifications of CBD) might result in plasma levels that are commensurate with a reliable antiepileptic effect. However, until verification of these suppositions is obtained, the clinical findings to date do little to instill confidence in CBD as an antiepileptic drug.

In sum, the present reality is that only two cannabinoids, nabilone and (–)-Δ^9-THC, are legend drugs (for antiemetic use), only 20 or so cannabinoids have been clinically evaluated for any purpose, and essentially only marijuana, (–)-Δ^9-THC, and CBD have been clinically evaluated in neurological disorders. The initial promises for the efficacy of CBD in epilepsy and Huntington's disease have not been confirmed in more recent clinical trials, and other presumed therapeutic activities of the drug have yet to be established in appropriate studies. (–)-Δ^9-THC appears to attenuate spasticity, ataxia, and tremor, and may have antiepileptic activity; yet the conclusive clinical studies to prove these properties have not been performed. Further, the patient's tolerance to (–)-Δ^9-THC (as well as to all other cannabimimetic cannabinoids tested clinically) appears to be relatively low at doses that are clinically effective or required for therapeutic activity. The cannabimimetic side effects may prove especially problematic for daily maintenance therapy, which is required for the treatment of spasticity and most of the other neurological disorders. Marijuana appears to have a similar spectrum of clinical activity and similar need for verification of efficacy, as those of (–)-Δ^9-THC. However, there are unique legal, political, and emotional issues that surround marijuana use itself. One has to wonder whether this substance would ever be approved for general clinical use, even if efficacy and safety were conclusively demonstrated in appropriate clinical trials.

There are two additional cannabinoids that, in preliminary animal studies, appear to have an important separation of activities. (–)-Δ^8-THC-11-oic acid and the (+)-enantiomer of HU-210 are devoid of cannabimimetic activity, but appear to have analgesic properties. Also, the (+)-HU-210 has been reported to have functional NMDA receptor blocking activity, and potentially this could be important for therapy of several neurological disorders. Much more research is required to confirm and expand these findings. Virtually all other currently available cannabinoids that have any therapeutically related activity in animals also have significant cannabimimetic activity. Moreover, few if any of these cannabinoids are sufficiently developed (i.e., through evaluation in appropriate animal models of neurologic dysfunction and in prescribed animal toxicity tests) for clinical trials.

Current knowledge pertaining to the potential role of cannabinoids for treatment of neurological disorders is limited for several reasons. Notably among these are the inadequacy of most clinical studies carried out to date, the lack of activity or lack of pharmacological specificity of the few agents which have been used in these clinical studies, and the lack of pharmacological specificity (or brain site specificity) of virtually all of the other cannabinoids presently available.

Within the short span of the past few years, major discoveries have resulted from the basic research of cannabinoids. These are the identification of the cannabinoid receptor in human and animal brains, the mapping of its heterogenous distribution in these brains, and the cloning of this receptor. Collectively, these findings are unprecedented for their potential to lead to the discovery of new cannabinoid drugs for treating neurological disorders.

Cannabinoid receptors have been demonstrated to be preferentially located in brain regions (i.e., substantia nigra pars reticulata, corpus striatum, cerebellum, and hippocampus) which are known to subserve motor, behavioral, and cognitive functions, as seen in both health and disease. It is conceivable that the effects of cannabinoids on motor behavior may be mediated at one brain site or by one receptor subtype, while the nonmotor effects might involve an interaction with different brain sites or receptor subtypes. Further, it is possible that selective agonists or antagonists of the cannabinoid receptor might produce beneficial therapeutic effects in various neurological disorders without producing adverse cannabimimetic side effects.

Several areas of the brain appear to be especially promising for study of the beneficial effects of the cannabinoid receptor drugs. The pars reticulata of the substantia nigra has recently been shown in animal studies to be a site of action for many of the standard drugs used clinically to treat epilepsy and spasticity. Although the relationship of the pars reticulata to the pathophysiology of these disorders is unknown, investigation of cannabinoid action on this brain site and its anatomical connections could provide new insights into the therapeutic mechanism of action of these drugs.

Substantial animal and clinical evidence implicates the corpus striatum in the pathophysiology of various involuntary movement disorders such as Huntington's chorea, Parkinsonism, Tourette's syndrome, and the dystonias. The development of cannabinoids with receptor specificity to this brain area would constitute an important new strategy for treating these disorders.

The cerebellum is another pivotal area for study of cannabinoid actions. The cerebellum has long been associated with the control of balance and motor coordination. New research indicates that the cerebellum is involved in the coordination and integration of a wide variety of processes that are not solely confined to the motor sphere. The cerebellar influences are mediated through widespread anatomical connections with the brain stem motor and sensory nuclei, limbic forebrain, hippocampus, and the cerebral cortex. These extensive neuronal connections provide an anatomical and functional basis for the modulatory effect of the cerebellum on motor activity, behavior, cognition, and affect.

A recent human study has demonstrated that there is an increase of metabolic activity selective to the cerebellum following the administration of $(-)$-Δ^9-THC, and that this change of metabolic activity is correlated with the psychological, but not the neurological (ataxia and tremor), effects produced by the cannabinoid. These results, together with those showing abundant cannabinoid receptors in cerebellum, suggest that the cerebellum is a major target for the cannabimimetic effects of cannabinoids. These findings also suggest the possibility that receptor-selective cannabinoids might be able to influence those neurological deficits that are related to cerebellar dysfunction such as multiple sclerosis, ataxia, tremor, and possibly tardive dyskinesia.

The realization of the potential beneficial effects of cannabinoids in neurological disorders will depend upon new breakthroughs in research, such as the identification of the endogenous ligand, the identification of subtypes of cannabinoid receptors, and the discovery of selective agonists and antagonists for the receptor subtypes. Ultimately, these findings could lead to new cannabinoids with clinically important separation of activities. While the timetable of these events cannot be predicted, the future outlook for the research and development of cannabinoids for therapeutic uses in neurological disorders looks brighter than it has in the past 150 years.

ACKNOWLEDGMENT

This work was supported, in part, by grant Nos. FD-R-000167 and FD-R-000623 from the U.S. Food and Drug Administration.

REFERENCES

1. **Consroe, P. and Snider, S. R.,** Therapeutic potential of cannabinoids in neurological disorders, in *Cannabinoids as Therapeutic Agents*, Mechoulam, R., Ed., CRC Press, Boca Raton, FL, 1986, 21.
2. **Consroe, P. and Lemberger, L.,** Marijuana '87. Summary of the session on therapeutic/clinical aspects, in *Marijuana: An International Research Report*, Chesher, G., Consroe, P., and Musty, R., Eds., National Campaign Against Drug Abuse Monograph Series No. 7, Australian Government Publishing Service, Canberra, 1988, 167.
3. **Abel, E. L.,** *Marijuana, The First Twelve Thousand Years*, Plenum Press, New York, 1980.
4. **Mechoulam, R.,** The pharmacohistory of *Cannabis sativa,* in *Cannabinoids as Therapeutic Agents*, Mechoulam, R., Ed., CRC Press, Boca Raton, FL, 1986, 1.
5. **O'Shaughnessy, W. B.,** On the preparation of the Indian hemp or gunjah (*Cannabis indica*): the effects on the animal system in health and their utility in the treatment of tetanus and other convulsive diseases, *Trans. Med. Phys. Soc. Bombay*, 8, 421, 1839.
6. **Shaw, J.,** On the use of the *Cannabis indica* (or Indian hemp) — 1st in tetanus — 2nd in hydrophobia — 3rd in cholera — with remarks on its effects, *Madras Q. Med. J.*, 5, 74, 1843.

7. **Roemer, B.,** On tetanus and tetanoid affections, with cases, *St. Louis Med. Surg. J.*, 10, 363, 1873.
8. **Reynolds, J. R.,** On some of the therapeutical uses of Indian hemp, *Arch. Med.*, 2, 154, 1859.
9. **Greene, R.,** The treatment of migraine with Indian hemp, *Practitioner*, 41, 35, 1888.
10. **Brookes, W. L.,** A case of recurrent migraine successfully treated with *Cannabis indica*, *Ind. Med. Rec.*, 11, 338, 1896.
11. **Anderson, G. S. D.,** Remarks on the remedial virtues of *Cannabis indica*, or Indian hemp, *West. J. Med. Surg.*, 4, 426, 1855.
12. **Mackenzie, S.,** Remarks on the value of Indian hemp in the treatment of a certain type of headache, *Br. Med. J.*, 1, 97, 1887.
13. **Clendinning, J.,** Observations on the medicinal properties of the *Cannabis sativa* of India, *Medico-Chirurg. Trans.*, 26, 188, 1843.
14. **Corrigan, O.,** Choree (emploi des inhalations de chloroforme dans la traitement de la) — emploi du haschich dans la meme maladie, *J. Pharm. Chim.*, 27, 312, 1855.
15. **Douglas, S.,** On the use of Indian hemp in chorea, *Edinburgh Med. J.*, 14, 777, 1869.
16. **Burton, E. T.,** *Cannabis indica* in chorea and pertussis, *Lancet*, 1, 1859, 1902.
17. **Lewis, H. E.,** *Cannabis indica* — a study of its physiologic action, toxic effects and therapeutic indications, *Merck's Arch.*, 1, 247, 1900.
18. **Buchwald, J. and Devinsky, O.,** J. Russell Reynolds and the study of interictal symptoms in epilepsy, *Arch. Neurol.*, 45, 802, 1988.
19. **Reynolds, J. R.,** *The Diagnosis of Diseases of the Brain, Spinal Cord, Nerves and Their Appendages*, John Churchill, London, 1865.
20. **Reynolds, J. R.,** On the therapeutic uses and toxic effects of *Cannabis indica*, *Lancet*, 1, 637, 1890.
21. **Hollister, L. E.,** Health aspects of cannabis, *Pharmacol. Rev.*, 38, 1, 1986.
22. **Cohen, S.,** The 94-day cannabis study, *Ann. N.Y. Acad. Sci.*, 282, 211, 1976.
23. **Tashkin, D. P., Fligiel, S., Wu, T. C., Gong, H., Jr., Barbers, R. G., Coulson, A. H., Simmons, M. S., and Beals, T. F.,** Effects of habitual use of marijuana and/or cocaine on the lung, in *Research Findings on Smoking of Abused Substances*, Chaing, C. N. and Hawks, R. L., Eds., NIDA Research Monograph 99, National Institute on Drug Abuse, Rockville, MD, 1990, 63.
24. **Gaoni, Y. and Mechoulam, R.,** Isolation, structure elucidation and partial synthesis of an active constituent of hashish, *J. Am. Chem. Soc.*, 86, 1646, 1964.
25. **Turner, C. E., ElSohly, M. A., and Boeren, E. G.,** Constituents of *Cannabis sativa* L. XVII. A review of the natural constituents, *J. Nat. Products*, 43, 169, 1980.
26. **Razdan, R.,** Structure-activity relationships in cannabinoids, *Pharmacol. Rev.*, 38, 75, 1986.
27. **Bhargava, H. N.,** Potential therapeutic applications of naturally occurring and synthetic cannabinoids, *Gen. Pharmacol.*, 9, 195, 1978.
28. **Fairbairn, J. W. and Pickens, J. T.,** Activity of cannabis in relation to its 1-*trans*-tetrahydrocannabinol content, *Br. J. Pharmacol.*, 72, 401, 1981.
29. **Green, K., McDonald, T. F., Cheeks, L., Slagle, T., Forbes, E., Deutsch, H. M., and Zalkow, L. H.,** Comparative morphological and pharmacological effects of water-soluble marihuana-derived material in rabbits and primates, in *Marijuana: An International Research Report*, Chesher, G., Consroe, P., and Musty, R., Eds., National Campaign Against Drug Abuse Monograph No. 7, Australian Government Publishing Service, Canberra, 1988, 125.

30. **Weissman, A.,** On the definition of cannabinoids: botanical?, chemical?, pharmacological?, *J. Clin. Pharmacol.*, 21, 159s, 1981.
31. **Martin, B. R., Compton, D. R., Little, P. J., Martin, T. J., and Beardsley, P. M.,** Pharmacological evaluation of agonist and antagonist activity of cannabinoids, in *Structure-Activity Relationships of the Cannabinoids*, Rapaka, R. S. and Makriyannis, A., Eds., NIDA Research Monograph 79, National Institute on Drug Abuse, Rockville, MD, 1987, 108.
32. **Mechoulam, R., Lander, N., Srebnik, M., Breuer, A., Segal, M., Feigenbaum, J. J., Jarbe, T. U. C., and Consroe, P.,** Stereochemical requirements for cannabimimetic activity, in *Structure-Activity Relationships of the Cannabinoids*, Rapaka, R. S. and Makriyannis, A., Eds., NIDA Research Monograph 79, National Institute on Drug Abuse, Rockville, MD, 1987, 15.
33. **Compton, D. R., Melvin, L. S., Johnson, M. R., and Martin, B. R.,** Nonclassical cannabinoid chemical structures: classification of bicyclic and related nantradol analogs as cannabimimetic agents, *FASEB J.*, 5(4), Part 1, 703, 1991.
34. **Devane, W. A., Howlett, A. C., Johnson, M. R., and Melvin, L. S.,** Nonclassical cannabinoid analgetics inhibit neuronal adenylate cyclase and bind to a receptor site in brain, in *Marijuana: An International Research Report*, Chesher, G., Consroe, P., and Musty, R., Eds., National Campaign Against Drug Abuse Monograph 7, Australian Government Publishing Service, Canberra, 1988, 141.
35. **Devane, W. A., Dysarz, F. A., Johnson, M. R., Melvin, L. S., and Howlett, A. C.,** Determination and characterization of a cannabinoid receptor in rat brain, *Mol. Pharmacol.*, 34, 605, 1988.
36. **Bidaut-Russell, M., Devane W. A., and Howlett, A. C.,** Cannabinoid receptors and modulation of cyclic AMP accumulation in the rat brain, *J. Neurochem.*, 55, 21, 1990.
37. **Howlett, A. C., Bidaut-Russell, M., Devane, W. A., Melvin, L. S., Johnson, M. R., and Herkenham, M.,** The cannabinoid receptor: biochemical, anatomical and behavioral characterization, *Trends Neurosci.*, 13, 420, 1990.
38. **Herkenham, M., Lynn, A. B., Little, M. D., Johnson, M. R., Melvin, L. S., de Costa, B. R., and Rice, K. C.,** Cannabinoid receptor localization in brain, *Proc. Natl. Acad. Sci. U.S.A.*, 87, 1932, 1990.
39. **Matsuda, L. A., Lolait, S. J., Brownstein, M. J., Young, A. C., and Bonner, T. I.,** Structure of a cannabinoid receptor and functional expression of the cloned cDNA, *Nature (London)*, 346, 561, 1990.
40. **Burstein, S. H., Hull, K., Hunter, S. A., and Latham, V.,** Cannabinoid and pain responses: a possible role for prostaglandins, *FASEB J.*, 2, 3022, 1988.
41. **Audette, C. A., Burstein, S. H., Doyle, S. A., and Hunter, S. A.,** G-protein mediation of cannabinoid-induced phospholipase activation, *Pharmacol. Biochem. Behav.*, 40, 559, 1991.
42. **Feigenbaum, J. J., Bergmann, F., Richmound, S. A., Mechoulam, R., Nadler, V., Kloog, Y., and Sokolovsky, M.,** Nonpsychotropic cannabinoid acts as a functional *N*-methyl-D-aspartate receptor blocker, *Proc. Natl. Acad. Sci. U.S.A.*, 86, 9584, 1989.
43. **Olney, J. W.,** Excitatory amino acids and neuropsychiatric disorders, *Biol. Psychiatry.*, 26, 505, 1989.
44. **Young, A. B., Cha, J.-H. J., Greenamyre, T., Maragos, W. F., and Penney, J. B.,** Distribution of PCP and glutamate receptors in normal and pathological mammalian brain, in *Allosteric Modulation of Amino Acid Receptors: Therapeutic Implications*, Barnard, E. A. and Costa, E., Eds., Raven Press, New York, 1989, 357.

45. **Rao, T. S., Cler, J. A., Mick, S. J., Emmett, M. R., Farah, J. M., Contreras, P. C., Iyengar, S., and Wood, P. L.,** Neurochemical interactions of competitive *N*-methyl-D-aspartate antagonists with dopaminergic neurotransmission and the cerebellar cyclic GMP system: functional evidence for a phasic glutamatergic control of the nigrostriatal dopaminergic pathway, *J. Neurochem.*, 56, 907, 1991.
46. **Gilman, S. and Newman, S. W.,** *Manter and Gatz's Essentials of Clinical Neuroanatomy and Neurophysiology*, 7th ed., F. A. Davis Company, Philadelphia, 1989.
47. **Turski, L., Klockgether, T., Schwarz, M., Turski, W. A., and Sontag, K. H.,** Substantia nigra: a site of action of muscle relaxant drugs, *Ann. Neurol.*, 28, 341, 1990.
48. **Iadarola, M. J. and Gale, K.,** Substantia nigra: site of anticonvulsant activity mediated by gamma-aminobutyric acid, *Science*, 218, 1237, 1982.
49. **Depaulis, A., Vergnes, M., Liu, Z., Kempf, E., and Marescaux, C.,** Involvement of the nigral output pathways in the inhibitory control of the substantia nigra over generalized non-convulsive seizures in the rat, *Neuroscience*, 39, 339, 1990.
50. **Consroe, P., Benedito, M. A. C., Leite, J. R., Carlini, E. A., and Mechoulam, R.,** Effect of cannabidiol on behavioral seizures caused by convulsant drugs or current in mice, *Eur. J. Pharmacol.*, 83, 293, 1982.
51. **Menza, M.,** A simple model of the interactions of dopamine, acetylcholine and GABA in movement disorders seen in psychiatry, *Med. Hypothesis*, 27, 285, 1988.
52. **Bartholini, G., Scatton, B., Zivkovic, B., and Lloyd, K. G.,** GABA receptor agonists and extrapyramidal motor function: therapeutic implications for Parkinson's disease, in *Advances in Neurology*, Yahr, M. D. and Bergmann, K. J., Eds., Raven Press, New York, 1986, 79.
53. **Montgomery, S. P., Moss, D. E., and Manderscheid, P. Z.,** Tetrahydrocannabinol and levonantradol effects on extrapyramidal motor behaviors: neuroanatomical location and hypothesis of mechanism, in *Marihuana '84*, Harvey, D. J. and Paton, W. D. M., Eds., IRL Press, Oxford, 1985, 295.
54. **Moss, D. E., Montgomery, S. P., Salo, A., and Steger, R. W.,** Tetrahydrocannabinol effects on extrapyramidal motor behaviors in an animal model of Parkinson's disease, in *The Cannabinoids: Chemical, Pharmacologic, and Therapeutic Aspects*, Agurell, S., Dewey, W. L., and Willette, R. E., Eds., Academic Press, New York, 1984, 815.
55. **Castaneda, E., Moss, D. E., Oddie, S. O., and Whishaw, I. Q.,** Delta-9-tetrahydrocannabinol does not affect levels of extracellular dopamine as measured by in vivo microdialysis in freely moving rats, *Pharmacol. Biochem. Behav.*, 40, 587, 1991.
56. **Pertwee, R. G.,** The central neuropharmacology of psychotropic cannabinoids, *Pharmacol. Ther.*, 36, 189, 1988.
57. **Ton, J. M. N. C., Gerhardt, G. A., Friedemann, M., Etgen, A. M., Rose, G. M., Sharpless, N. S., and Gardner, E. L.,** The effects of delta-9-tetrahydrocannabinol on potassium-evoked release of dopamine in the rat caudate nucleus: an *in vivo* electrochemical and *in vivo* microdialysis study, *Brain Res.*, 451, 59, 1988.
58. **Pertwee, R. G., Greentree, S. G., and Swift, P. A.,** Drugs which stimulate or facilitate central GABAergic transmission interact synergistically with delta-9-tetrahydrocannabinol to produce marked catalepsy in mice, *Neuropharmacology*, 27, 1265, 1988.
59. **Pertwee, R. G. and Ross, T. M.,** Drugs which stimulate or facilitate central cholinergic transmission interact synergistically with delta-9-tetrahydrocannabinol to produce marked catalepsy in mice, *Neuropharmacology*, 30, 67, 1991.
60. **Kataoka, Y., Ohta, H., Fujiwara, M., Oishi, R., and Ueki, S.,** Noradrenergic involvement in catalepsy induced by delta-9-tetrahydrocannabinol, *Neuropharmacology*, 26, 55, 1987.

61. **Ueki, S.,** Abnormal behavior induced by delta-9-tetrahydrocannabinol and its pharmacological characteristics, *Trends Pharmacol. Sci.*, 1, 126, 1980.
62. **Worms, P., Willigens, M. T., Continsouza-Blanc, D., and Lloyd, K. G.,** The effect of different types of cortical lesions on drug-induced catalepsy in rats: a pharmacological analysis, *Eur. J. Pharmacol.*, 113, 53, 1985.
63. **Gao, X.-K. and Song, B.,** Comparison between the catalepsies induced by delta-9-tetrahydrocannabinol and haloperidol in rats, *Acta Pharmacol. Sin.*, 8, 203, 1987.
64. **Snider, R. S., Maiti, A., and Snider, S. R.,** Cerebellar pathways to ventral midbrain and nigra, *Exp. Neurol.*, 53, 714, 1976.
65. **Tabaddor, K., Wolfson, L. I., and Sharpless, N. S.,** Ventricular fluid homovanillic acid and 5-hydroxyindoleacetic acid concentrations in patients with movement disorders, *Neurology*, 28, 1249, 1978.
66. **Snider, S. R., Jackson, V., and Snider, R.,** Increased incidence of drug-induced stereotypes in rats with lesions of the cerebellar "face" area, *Neurology*, 31, 96, 1981.
67. **Appenzeller, O. and Biehl, J.,** Mouthing in the elderly. A cerebellar sign, *J. Neurol. Sci.*, 6, 249, 1968.
68. **Jellinger, K.,** Neuropathologic findings after long-term neuroleptic therapy, in *Neurotoxicity*, Roizin, L., Shiraki, H., and Grcevic, N., Eds., Raven Press, New York, 1977, 25.
69. **Arai, N., Amano, N., Iseki, E., Yokoi, S., Saito, A., Takekawa, Y., and Misugi, K.,** Tardive dyskinesia with inflated neurons of the cerebellar dentate nucleus. Case reports and morphometric study, *Acta Neuropathol.*, 73, 38, 1987.
70. **Okumura, A., Oda, M., Iwasa, S., and Shiraki, H.,** Pathomorphological and clinicopathological study of so-called grumose degeneration in the cerebellar dentate nucleus, *Adv. Neurol. Sci.*, 19, 483, 1975.
71. **Greenfield, J. G. and Bosanquet, F. D.,** The brainstem lesions in Parkinsonism, *J. Neurol. Neurosurg. Psychiatry,* 16, 213, 1953.
72. **Snider, R. S.,** A cerebellar-ceruleus pathway, *Brain Res.*, 88, 59, 1975.
73. **Colpaert, F. C.,** Pharmacological characteristics of tremor, rigidity and hypokinesia induced by reserpine in rat, *Neuropharmacology*, 26, 1431, 1987.
74. **Kish, S. J., Shannak, K. S., Rajput, A. H., Gilbert, J. J., and Hornykiewicz, O.,** Cerebellar norepinephrine in patients with Parkinson's disease and control subjects, *Arch. Neurol.*, 41, 612, 1984.
75. **McKeon, T. W., Lorden, J. F., Beales, M., and Oltmans, G. A.,** Alterations in the noradrenergic projection to the cerebellum of the dystonic (dt) rat, *Brain Res.*, 366, 89, 1986.
76. **Lorden, J. F., McKeon, T. W., Baker, H. J., Cox, N., and Walkely, S. U.,** Characterization of the rat mutant dystonic (dt): a new animal model of dystonia musculorum deformans, *J. Neurosci.*, 4, 1925, 1984.
77. **Jacquet, Y. F.,** "Dystonia"-like postural asymmetry after microinjection of ACTH *N*-terminal fragments but not after ACTH 1-39 in rat brainstem suggests role of neuropeptide mutation in genetic movement disorder, *Brain Res.*, 294, 144, 1984.
78. **Adams, R. D. and Victor, M.,** *Principles of Neurology*, 3rd ed., McGraw-Hill, New York, 1985, 699.
79. **Cooper, I. S., Riklan, M., and Amin, I.,** Chronic cerebellar stimulation in cerebral palsy, *Neurology*, 26, 744, 1976.
80. **Cooper, I. S., Amin, I., and Riklan, M.,** Chronic cerebellar stimulation in epilepsy, *Arch. Neurol.*, 33, 559, 1976.

81. **Volkow, N. D., Gillespie, H., Mullani, N., Tancredi, L., Grant, C., Ivanovic, M., and Hollister, L.,** Cerebellar metabolic activation be delta-9-tetrahydrocannabinol in human brain: a study with positron emission tomography and ^{18}F-2-fluoro-2-deoxyglucose, *Psychiatr. Res. Neuroimaging*, 40, 69, 1991.
82. **Segal, M.,** Cannabinoids and analgesia, in *Cannabinoids as Therapeutic Agents*, Mechoulam, R., Ed., CRC Press, Boca Raton, FL, 1986, 105.
83. **Adler, M. W. and Geller, E. B.,** Ocular effects of cannabinoids, in *Cannabinoids as Therapeutic Agents*, Mechoulam, R., Ed., CRC Press, Boca Raton, FL, 1986.
84. **Graham, J. D. P.,** The bronchodilator action of cannabinoids, in *Cannabinoids as Therapeutic Agents*, Mechoulam, R., Ed., CRC Press, Boca Raton, FL, 1986, 147.
85. **Levitt, M.,** Cannabinoids as antiemetics in cancer chemotherapy, in *Cannabinoids as Therapeutic Agents*, Mechoulam, R., Ed., CRC Press, Inc., 1986, 71.
86. **Devine, M. L., Dow, G. D., Greenberg, B. R., Holsten, D. W., Icaza, L., Jue, P. Y., Meyers, F. H., O'Brien, E., Roberts, C. M., Rocchio, G. L., Stanton, W., and Wesson, D. R.,** Adverse reactions of delta-9-tetrahydrocannabinol given as an antiemetic in a multicenter study, *Clin. Pharmacol.*, 6, 319, 1987.
87. **Jones, R. T., Benowitz, N., and Bachman, J.,** Clinical studies of cannabis tolerance and dependence, *Ann. N.Y. Acad. Sci.*, 282, 221, 1976.
88. **Plasse, T. F., Gorter, R. W., Krasnow, S. H., Lane, M., Shepard, K. V., and Wadleigh, R. G.,** Recent clinical experience with dronabinol, *Pharmacol. Biochem. Behav.*, 40, 695, 1991.
89. **Archer, R. A., Stark, P., and Lemberger, L.,** Nabilone, in *Cannabinoids as Therapeutic Agents*, Mechoulam, R., Ed., CRC Press, Boca Raton, FL, 1986, 85.
90. **Lemberger, L.,** Nabilone, a synthetic, cannabinoid antiemetic, in *Marijuana: An International Research Report*, Chesher, G., Consroe, P., and Musty, R., Eds., National Campaign Against Drug Abuse Monograph No. 7, Australian Government Publishing Service, Canberra, 1988, 131.
91. **Johnson, M. R. and Melvin, L. S.,** The discovery of the nonclassical cannabinoid analgetics, in *Cannabinoids as Therapeutic Agents*, Mechoulam, R., Ed., CRC Press, Boca Raton, FL, 1986, 121.
92. **Little, P. J., Compton, D. R., Johnson, M. R., Melvin, L. S., and Martin, B. R.,** Pharmacology and stereoselectivity of structurally novel cannabinoids in mice, *J. Pharmacol. Exp. Ther.*, 247, 1046, 1988.
93. **Little, P. J., Compton, D. R., Mechoulam, R., and Martin, B. R.,** Stereochemical effects of 11-OH-delta-8-THC-dimethylheptyl in mice and dogs, *Pharmacol. Biochem. Behav.*, 32, 661, 1989.
94. **Consroe, P. and Mechoulam, R.,** Anticonvulsant and neurotoxic effects of tetrahydrocannabinol stereoisomers, in *Structure-Activity Relationships of the Cannabinoids*, Rapaka, R. S. and Makriyannis, A., Eds., NIDA Research Monograph 79, National Institute on Drug Abuse, Rockville, MD, 1987, 59.
95. **Howlett, A. C., Champion, T. M., Wilken, G. H., and Mechoulam, R.,** Stereochemical effects of 11-OH-delta-8-tetrahydrocannabinol-dimethylheptyl to inhibit adenylate cyclase and bind to the cannabinoid receptor, *Neuropharmacology*, 29, 161, 1990.
96. **Feigenbaum, J. J., Richmond, A. A., Weissman, Y., and Mechoulam, R.,** Inhibition of cisplatin-induced emesis in the pigeon by a non-psychotropic synthetic cannabinoid, *Eur. J. Pharmacol.*, 169, 159, 1989.

97. **Bell, M. R., Dambra, T. E., Kumar, V., Eissenstat, M. A., Herrmann, J. L., Wetzel, J. R., Rosi, D., Philion, R. E., Daum, S. J., Hlasta, D. J., Kullnig, R. K., Ackerman, J. H., Haubrich, D. R., Luttinger, D. A., Baizman, E. R., Miller, M. S., and Ward, S. J.,** Antinociceptive (aminoalkyl)indoles, *J. Med. Chem.*, 34, 1099, 1991.
98. **Childers, S. R., Pacheco, M. A., and Ward, S. J.,** Cannabinoids and aminoalkylindoles bind to common receptors to inhibit adenylate cyclase in brain, *Soc. Neurosci. Abstr.*, 16(Part 2), 1302, 1990.
99. **Kuster, J. E., Haycock, D., Howlett, A. C., Subramanium, B., and Ward, S. J.,** Aminoalkylindole binding data suggest a new approach for studying cannabinoid activity, *Soc. Neurosci. Abstr.*, 16 (Part 1) 697, 1990.
100. **Porter, R. J. and Theodore, W. H.,** Epilepsy, *Neurol. Clinics*, 4, 1986.
101. **Dreifuss, F. E.,** Classification of epileptic seizures and the epilepsies, *Ped. Clinics North Am.*, 36, 265, 1989.
102. **Pellock, J. M.,** Efficacy and adverse effects of antiepileptic drugs, *Ped. Clinics North Am.*, 36, 435, 1989.
103. **Bleck, T. P.,** Convulsive disorders: the use of anticonvulsant drugs, *Clin. Neuropharmacol.*, 13, 198, 1990.
104. **Brodie, M. J.,** Established anticonvulsants and treatment of refractory epilepsy, *Lancet*, 336, 350, 1990.
105. **Cunha, J. M., Carlini, E. A., Pereira, A. E., Ramos, O. L., Pimentel, C., Gagliardi, R., Sanvito, W. L., Lander, N., and Mechoulam, R.,** Chronic administration of cannabidiol to healthy volunteers and epileptic patients, *Pharmacology*, 21, 175, 1980.
106. **Ellison, W. R., Gelwan, E., and Ogletree, J.,** Complex partial seizure symptoms affected by marijuana abuse, *J. Clin. Psychiatry*, 51, 439, 1990.
107. **Ng, S. K. C., Brust, J. C. M., Hauser, W. A., and Susser, M.,** Illicit drug use and the risk of new-onset seizures, *Am. J. Epidemiol.*, 132, 47, 1990.
108. **Ames, F. R.,** Anticonvulsant effect of cannabidiol, *S. Afr. Med. J.*, 69, 14, 1986.
109. **Trembly, B. and Sherman, M.,** Double-blind clinical study of cannabidiol as a secondary anticonvulsant, paper presented at Marijuana '90 Int. Conf. on Cannabis and Cannabinoids, Kolympari (Crete), July 8 to 11, 1990.
110. **Yung, C. Y.,** Clinical features of movement disorders, *Brain Res. Bull.*, 11, 167, 1983.
111. **Yung, C. Y.,** Evaluation and differential diagnosis of movement disorders: a flow chart approach, *Brain Res. Bull.*, 11, 149, 1983.
112. **Fahn, S.,** The varied clinical expressions of dystonia, *Neurol. Clinics,* 2, 541, 1984.
113. **Marsden, C. D., Marion, M. H., and Quinn, N.,** The treatment of severe dystonia in children and adults, *J. Neurol. Neurosurg. Psychiatry*, 47, 1166, 1984.
114. **Fahn, S. and Jankovic, J.,** Practical management of dystonia, *Neurol. Clinics*, 2, 555, 1984.
115. **Hornykiewicz, O., Kish, S. J., Becker, L. E., Farley, I., and Shannak, K.,** Brain neurotransmitters in dystonia musculorum deformans, *N. Engl. J. Med.*, 315, 347, 1986.
116. **LeWitt, P. A., Miller, L. P., and Levine, R. A.,** Tetrahydrobiopterin in dystonia: identification of abnormal metabolism and therapeutic trials, *Neurology*, 36, 760, 1986.
117. **Jankovic, J. and Schwartz, K.,** Botulinum toxin injections for cervical dystonia, *Neurology*, 40, 277, 1990.
118. **Nausieda, P. A., Grossman, B. J., Koller, W. C., Weiner, W. J., and Klawans, H. L.,** Sydenham's chorea: an update, *Neurology*, 30, 331, 1980.
119. **Martin, J. B. and Gusella, J. F.,** Huntington's disease pathogenesis and management, *N. Engl. J. Med.*, 315, 1267, 1986.

120. **Shoulson, I., Odoroff, C., Oakes, D., Behr, J., Goldblatt, D., Caine, E., Kennedy, J., Miller, C., Bamford, K., Rubin, A., Plumb, S., and Kurlan, R.,** A controlled clinical trial of baclofen as protective therapy in early Huntington's disease, *Ann. Neurol.*, 25, 252, 1989.
121. **Walker, F. O. and Hunt, V. P.,** An open label trial of dextromethorphan in Huntington's disease, *Clin. Neuropharmacol.*, 12, 322, 1989.
122. **Singer, H. S. and Walkup, J. T.,** Tourette syndrome and other tic disorders diagnosis, pathophysiology and treatment, *Medicine*, 70, 15, 1991.
123. **Moss, D. E., Manderscheid, P. Z., Montgomery, S. P., Norman, A. B., and Sanberg, P. R.,** Nicotine and cannabinoids as adjuncts to neuroleptics in the treatment of Tourette syndrome and other motor disorders, *Life Sci.*, 44, 1521, 1989.
124. **Koller, W. C.,** *Handbook of Parkinson's Disease*, Marcel Dekker, New York, 1987.
125. **Klawans, H. L.,** Emerging strategies in Parkinson's disease, *Neurology*, 40(Suppl. 3), 1, 1990.
126. **Goetz, C. G. and Klawans, H. L.,** Tardive dyskinesia, *Neurol. Clinics*, 2, 605, 1984.
127. **Kang, U. J. and Fahn, S.,** Management of tardive dyskinesia, *Rational Drug Ther.*, 22, 1, 1988.
128. **Consroe, P., Musty, R., and Conti, L.,** Effects of cannabidiol in animal models of neurological dysfunction, in *Marijuana: An International Research Report*, Chesher, G., Consroe, P., and Musty, R., Eds., National Campaign Against Drug Abuse Monograph Series No. 7, Australian Government Publishing Service, Canberra, 1988, 147.
129. **Kaakkola, S. and Teravainen, H.,** Animal models of parkinsonism, *Pharmacol. Toxicol.*, 67, 95, 1990.
130. **Sakurai, Y., Ohta, H., Shimazoe, T., Kataoka, Y., Fujiwara, M., and Ueki, S.,** Delta-9-tetrahydrocannabinol elicited ipsilateral circling behavior in rats with unilateral nigral lesion, *Life Sci.*, 37, 2181, 1985.
131. **Conti, L. H., Johannesen, J., Musty, R., and Consroe, P.,** Anti-dyskinetic effects of cannabidiol, in *Marijuana: An International Research Report*, Chesher, G., Consroe, P., and Musty, R., Eds., National Campaign Against Drug Abuse Monograph Series No. 7, Australian Government Publishing Service, Canberra, 1988, 153.
132. **Sofia, R. D., Kubena, R. K., and Barry, H.,** Inactivity of delta-9-tetrahydrocannabinol in antidepressant screening tests, *Psychopharmacology*, 31, 121, 1973.
133. **Rieke, G. K., Scarfe, A. D., and Hunter, J. F.,** L-Pyroglutamate: an alternate neurotoxin for a rodent model of Huntington's disease, *Brain Res. Bull.*, 13, 443, 1984.
134. **Consroe, P., Sandyk, R., and Snider, S. R.,** Open label evaluation of cannabidiol in dystonic movement disorders, *Int. J. Neurosci.*, 30, 277, 1986.
135. **Sandyk, R., Consroe, P., Stern, L., Snider, S. R., and Bliklen, D.,** Preliminary trial of cannabidiol in Huntington's disease, in *Marijuana: An International Research Report*, Chesher, G., Consroe, P., and Musty, R., Eds., National Campaign Against Drug Abuse Monograph Series No. 7, Australian Government Publishing Service, Canberra, 1988, 157.
136. **Consroe, P., Laguna, J., Allender, J., Snider, S., Stern, L., Sandyk, R., Kennedy, K., and Schram, K.,** Controlled clinical trial of cannabidiol in Huntington's disease, *Pharmacol. Biochem. Behav.*, 40, 701, 1991.
137. **Sandyk, R. and Awerbuch, G.,** Marijuana and Tourette's syndrome, *J. Clin. Psychopharmacol.*, 8, 444, 1988.
138. **Schneier, F. R. and Siris, S. G.,** A review of psychoactive substance use and abuse in schizophrenia, *J. Nerv. Ment. Dis.*, 175, 641, 1987.

139. **Thornicroft, G.,** Cannabis and psychosis is there epidemiological evidence for an association?, *Br. J. Psychiatry*, 157, 25, 1990.
140. **Dixon, L., Haas, G., Weiden, P. J., Sweeney, J., and Frances, A. J.,** Drug abuse in schizophrenic patients: clinical correlates and reasons for use, *Am. J. Psychiatry*, 148, 224, 1991.
141. **Olivera, A. A., Kiefer, M. W., and Manley, N. K.,** Tardive dyskinesia in psychiatric patients with substance use disorders, *Am. J. Drug Alcohol Abuse*, 16, 57, 1990.
142. **Sandyk, R. and Consroe, P.,** unpublished data, 1991.
143. **Frankel, J. P., Hughes, A., Lees, A. J., and Stern, G. M.,** Marijuana for Parkinsonian tremor, *J. Neurol. Neurosurg. Psychiatry*, 53, 436, 1990.
144. **Wuis, E. W.,** Spasticity and drug therapy, *Pharm. Weekbl. (Sci.)*, 9, 249, 1987.
145. **Rice, G. P. A.,** Pharmacotherapy of spasticity: some theoretical and practical considerations, *Can. J. Neurol. Sci.*, 14, 510, 1987.
146. **Noseworthy, J. H.,** Therapeutics of multiple sclerosis, *Clin. Neuropharmacol.*, 14, 49, 1991.
147. **Herkenham, M., Lynn, A. B., Johnson, M. R., Melvin, L. S., de Costa, B. R., and Rice, K. C.,** Characterization and localization of cannabinoid receptors in rat brain: a quantitative *in vitro* autoradiographic study, *J. Neurosci.*, 11, 563, 1991.
148. **Ungerleider, J. T., Andyrsiak, T., Fairbanks, L., Ellison, G. W., and Myers, L. W.,** Delta-9-THC in the treatment of spasticity associated with multiple sclerosis, *Adv. Alcohol Substance Abuse*, 7, 39, 1987.
149. **Meinck, H.-M., Schonle, P. W., and Conrad, B.,** Effect of cannabinoids on spasticity and ataxia in multiple sclerosis, *J. Neurol.*, 236, 120, 1989.
150. **Maurer, M., Henn, V., Dittrich, A., and Hofmann, A.,** Delta-9-tetrahydrocannabinol shows antispastic and analgesic effects in a single case double-blind trial, *Eur. Arch. Psychiatry Clin. Neurosci.*, 240, 1, 1990.
151. **Adams, R. D. and Victor, M.,** *Principles of Neurology*, 3rd ed., McGraw-Hill, New York, 1985, 129.
152. **Lance, J. W.,** A concept of migraine and the search for the ideal headache drug, *Headache*, 30(Suppl. 1), 17, 1990.
153. **Ferrari, M. D., Odink, J., Bos, K. D., Malessy, M. J. A., and Bruyn, G. W.,** Neuroexcitatory plasma amino acids are elevated in migraine, *Neurology*, 40, 1582, 1990.
154. **Peroutka, S. J.,** The pharmacology of current anti-migraine drugs, *Headache*, 30(Suppl. 1), 5, 1990.
155. **Adams, R. D. and Victor, M.,** *Principles of Neurology*, McGraw-Hill, New York, 1985, 144.
156. **Burstein, S. H., Audette, C. A., Doyle, S. A., Hull, K., Hunter, S. A., and Latham, V.,** Antagonism to the actions of platelet activating factor by a nonpsychoactive cannabinoid, *J. Pharmacol. Exp. Ther.*, 251, 1989.
157. **Volfe, Z., Dvilansky, A., and Nathan, I.,** Cannabinoids block release of serotonin from platelets induced by plasma from migraine patients, *Int. J. Clin. Pharm. Res.*, V, 243, 1985.
158. **El-Mallakh, R. S.,** Marijuana and migraine, *Headache*, 27, 442, 1987.
159. **Mathew, R. J., Wilson, W. H., and Tant, S. R.,** Acute changes in cerebral blood flow associated with marijuana smoking, *Acta Psychiatry Scand.*, 79, 118, 1989.
160. **Lindstrom, P., Lindblom, U., and Boreus, L.,** Lack of effect of cannabidiol in sustained neuropathic, paper presented at Marijuana '87 Int. Conf. on Cannabis, Melbourne, September 2 to 4, 1987.

13 Characterization of Mechanisms Mediating the Effects of Δ^9-Tetrahydrocannabinol on Behavior

INTRODUCTION

The main manifestation of the effects of marijuana and its active principle, Δ^9-tetrahydrocannabinol (THC), is change in behavior. These effects are evident in various aspects of behavior and exhibit both species and individual differences. These differences are, in turn, related to the dynamic relationship of the environment and the inner organization of a living system. Thus, distinct functional states of the organism correspond to different types of the goal-directed behavior.

An important characteristic of THC action is its influence on different systems of the organism that differ both in the degree of complexity and in physiological functions. Therefore, meaningful analysis of the influence of THC on the organization of behavior requires consideration of the effects of THC on the relationship of organism with the environment, which is manifested in various components of the behavioral organization, as well as in different functional systems that influence the activity of the CNS. Such a task requires evaluation of a great number of indices that characterize various aspects of the total activity of an organism. We have approached this task by studying the effects of THC on different components of behavior in the course of goal-directed reactions. The relationship of THC to substances that influence the activity of various systems involved in these responses was also studied.

Mechanisms Mediating the Effects of Δ^9-Tetrahydrocannabinol on Animal Behavior

According to several reports, THC in the dosage of 2 mg/kg causes a decline in norepinephrine (NE) and serotonin levels in the thalamic and hypothalamic structures, which is accompanied by a sedative effect and difficulties in coordination of movements.[1,2] The same reports indicate that an increase in the dose of THC results in a more prominent fall of NE levels in the above-mentioned structures as well as in the pons and medulla oblongata that, however, is not accompanied by a fall in serotonin level.[3] Furthermore, injection of THC resulted in a marked decrease in the levels of dopamine (DA) and serotonin in the caudate nucleus, midbrain, and diencephalon, with a concomitant decrease of self-stimulation frequency and motor activity.[4]

Serotonin content in the whole brain was decreased after THC injection in those animals in which THC caused aggressive behavior.[5,6] In contrast, in animals whose aggressive behavior was experimentally stimulated, THC caused an inhibition of aggression that was accompanied by a decrease in DA level in the subcortical structures and a rise of serotonin level in the same region.[7]

When the effects of THC on serotonin turnover were examined using injections of labeled tryptophan, the results indicated that THC does not change the rate of serotonin metabolism.[8] Other studies have, however, shown that THC increases serotonin content in the hypothalamus and midbrain and, at the same time, decreases the rate of serotonin turnover.[9] When injected in small doses, THC reduced body temperature in chronically infused animals and this effect was accompanied by a rise of serotonin concentration with no change in NE or DA content.[8] When labeled tyrosine was injected, there was evidence for a functional relationship between the fall in temperature and the rate of NE and DA synthesis.[10] In an attempt to discover the connection between behavior and neurochemical effects of THC, several authors analyzed behavioral changes induced by THC in conjunction with other substances that alter the activity of monoaminergic systems. It has thus been shown that infusion of 6-hydroxydopamine, which causes degeneration of adrenaline- and DA-containing terminals, prevented hypothermal and hyperreactive effects of THC. The DA receptor blocker pimozide inhibited stereotypic behavior caused by THC.[11] However, aggressive behavior caused by THC was intensified when 6-hydroxydopamine was injected. This aggressive behavior was attenuated when NE was infused,[12] and intensified when dihydroxyphenylalanine was injected.[13] Suppression of serotonin synthesis by reserpine altered the ability of THC to produce a sedative effect[14] and cause hypokinesis.

In addition to studies that provide various kinds of evidence for the dependence of THC action on the activity of the monoaminergic systems, there are reports in which such a connection is denied. Thus, in animals that were treated with an inhibitor of catecholamine synthesis, α-methyl-*p*-tyrosine, there were no additional changes in the rate of catecholamine depletion when THC was injected.[15] Moreover, THC given in a dose capable of causing catalepsy did not change the content of NE, DA, serotonin, or 5-hydroxyindoleacetic acid in the brain regions examined.[16] It has also been discovered that the stimulatory action of THC on behavior does not change when various blockers of catecholaminergic receptors are injected.[17]

The absence of a connection between monoamines and the action of THC has also been shown in studies that demonstrated a decrease in cholinergic activity of the septohippocampal tracts under the influence of THC.[18]

The role of different neurotransmitter systems in mediating the effects of THC has been by studied employing a variety of methods. However, there are contradictions in the results. In addition to these contradictions, there is essentially no information on the time course of the interaction of THC with mediator systems during THC-induced modifications of the various functional systems that are responsible for the organization of the behavioral act. Therefore, a study of this aspect of THC action using pharmacological modification of mediator systems was of great interest.

MATERIALS AND METHODS

The study was performed in cats that were previously trained using a technique of conditional defensive reflexes to distinguish diversely oriented lines presented for 300 to 3000 ms. The training was considered complete when differentiation of stimuli lasting 3000 ms was performed correctly in nearly 100% of the trials. Correct differentiation of shorter stimuli was in the range of 60 to 75%, irrespective of the time of training. Detailed description of the technique was reported previously.[19] The following parameters of the conditioned reflex were taken into consideration: differentiation of a short-term (500 ms) stimulus (DSS), differentiation of a long-term (3000 ms) stimulus (DLS), latent period of a conditioned reflex to differentiated (LPD) and single (LPS) stimuli, and the incidence of the interstimulus reactions. Preliminary experiments indicated that THC, in the dose of 3 mg/kg, aggravates DSS and DLS, lengthens LPD, does not change LPS, and reduces interstimulus reactions.

The following substances were used for the analysis of THC action.

1. Serotonin system: methysergide (a serotonin receptor blocker), 5-hydroxytryptophan (a serotonin precursor), and parachlorophenyl-alanine or PCPA (decreases serotonin synthesis at the stage of tryptophan hydroxylation)
2. Dopaminergic system: α-DOPA (a DA precursor), apomorphine (DA receptor agonist), and α-methylparatyrosine (MPT, catecholamine synthesis blocker)
3. Adrenergic system: phentolamine and yohimbine (α-receptor antagonists), propranolol (β-receptor antagonists), and ephedrine (sympathomimetic drug)

All of these substances were injected in doses that did not change the parameters of conditioned reflexes. The doses of 5-hydroxytryptophan, α-DOPA, PCPA, and methylparatyrosine used in our laboratory have been previously shown to alter the levels of the studied amines in the brain without inducing gross hypo- or hyperfunction of monoaminergic systems.[20-22] All substances were injected intraperitoneally.

RESULTS

In order to identify the mediator(s) of THC action on behavior, we have used pharmacological means to alter specific neurotransmitter systems in the brain without affecting the function of these mediator systems.

Methysergide (1 mg/kg), a blocker of serotonin receptors, weakened the inhibitory action of THC on DSS and DLS without changing the effect of THC on LPD, LPS, and the interstimulus reaction (IR). 5-Hydroxytryptophan (20 mg/kg) intensified the inhibition of DSS and DLS by THC without influencing LPD, LPS, and IR. The serotonin synthesis blocker PCPA (200 mg/kg) injected 48 h before administration of THC weakened the effect of THC on DSS and DLS, while the remainder of the avoidance reaction (AR) parameters were not affected. The administration of α-DOPA (20 mg/kg), 1 h before infusion of THC, attenuated the action of THC on DLS and DSS but intensified its influence on LPD and LPS. The effects of the DA receptor agonist apomorphine were analogous to those of α-DOPA. Treatment with α-MPT (100 mg/kg) 8 h prior to THC did not change THC action on DLS, but intensified the drug's action on DSS and attenuated its effects on LPD and LPS.

The sympathomimetic, ephedrine, reduced the effect of THC on LPD, but did not have an effect on the other behavioral parameters. Blockade of adrenoreceptors also modified the action of THC. Thus, yohimbine (1 mg/kg) and phentolamine (2 mg/kg) intensified the inhibitory action of THC on DSS, and somewhat attenuated the action of THC on LPD and LPS. In contrast, the β-blocker propranolol (3 mg/kg) did not influence the action of THC on any of the AR parameters examined.

DISCUSSION

These results indicate that the actions of THC on different parameters of the conditioned reflex are mediated by different neurotransmitter mechanisms. Suppression of the activity of serotoninergic systems (serotonin receptor blockade or a decrease of the serotonin level) leads to attenuation of THC effects on differentiation. In contrast, stimulation of the activity of the serotoninergic system enhanced the inhibitory action of THC on the ability of the animals to differentiate visual stimuli.

Activation of the dopaminergic system intensified the action of THC on LPD and LPS. However, the action of THC on the ability of animals to differentiate long- and short-term stimuli was weakened. Blockade of α-1 and especially α-2 adrenoreceptors had a limited behavioral effect as compared with the consequences of interfering with the activity of serotoninergic and dopaminergic systems, and led to an intensification of the inhibitory effects of THC on the differentiation of short-term stimuli.

The present results also indicate that the most pronounced changes in THC action were produced by the drugs that altered the activity of the serotoninergic system. The injection of drugs that intensified or weakened the serotoninergic activity produced opposite effects. Thus, drugs that decreased serotonin action eliminated the effect of THC on the ability of the animal to differentiate long- and short-term stimuli, while 5-hydroxytryptophan intensified the effect of THC on differentiation. It should be pointed out that substances influencing serotoninergic transmission did not modify THC effects on LPs of the conditioned reflexes.

The similarity of effects of the substances that are related to the serotoninergic system, but influence its different components, suggests that the effects of THC on conditioned reflex reactions are mediated by an interaction of the cannabinoid at the level of serotonin synthesis and/or serotonin receptors.

Clear-cut results were obtained when substances interfering with the dopaminergic system were injected. Substances that enhance dopaminergic activity attenuated the effects of THC and increased LPs of the conditioned reflexes. Depletion of DA with α-MPT[23,24] produced opposite effects on DSS and reduced LPD and LPS.

Δ^9-THC was previously reported to increase serotonin content in the hypothalamus and midbrain,[25] as well as in the diencephalon.[26,27] Increases in serotonin levels were shown to result in decreased sensitivity to external irritants[28] and difficulty in exhibiting conditioned defense reflexes.[29-32] The content of serotonin in the brain is also known to be reduced in animals after intensive training.[33] Thus, alterations of THC action by drugs that reduced serotonin activity may account for the suppression of the increased activity of the serotoninergic system under the influence of THC. It is not clear why serotoninergic substances fail to alter the LPs of the conditioned reflexes.

It was mentioned above that we have recorded two time parameters of the conditioned reflex. Before the onset of a response reaction, the duration is very sensitive to the changes of the conditioned stimulus parameters. It is lengthened by the transition from a simple reaction to differentiation, as well as by the changes of the meaning of the signal,[34] the increase of the stimulus strength,[35] the change of the emotional state,[36,37] and the difficulty in detecting a stimulus.[38,39] The duration of the movement in these cases either does not change or changes very slightly.[34] Premotor time corresponding to these two parameters and to the time of perception plus the analysis of the information received can thus be distinguished from the motor time that corresponds to the realization of the movement itself. The analysis of the literature indicates that it is the premotor time that is sensitive to the extra receptive influences, and that the motor time is sensitive to such factors as fatigue[36,40] and the complexity of the movement.[41]

Inasmuch as a conditioned irritant and also an acquired reaction[42] are involved in the conditioned reflex, motor time is supposed to be sensitive to pharmacological preparations. In very few studies, premotor and motor time were recorded separately. The LP (latent period) avoidance reaction has been shown to depend on the emotional significance of the stimulus[43] and can be influenced by electrostimulation of the hypothalamic nuclei or periaqueductal substance.[44] The LP can be lengthened by cooling of the caudate nucleus and shortened by cooling the central median hypothalamic nucleus.[45] The LP of the reaction is also lengthened in hyperthermia. In this case, the number of mistakes decreases.[46]

The studies that examined the correlation of regular reactions and a LP of the reaction suggest that the latter generally reflects the excitability level of the animals, whereas durable latent conditioned reflexes indicate perception of a signal denoting the presence of the conditioned irritant.[47] It has also been demonstrated that by strengthening the conditioned defense reflex, the time before the onset of the movement is lengthened, which can be interpreted as perfecting the evaluation of action.[48] Lengthening of the LP accompanied by the increase of regular reactions is also observed in response to many pharmacological preparations.[49] Thus, the correctness of the reaction and the rate of its realization are not positively correlated. Moreover, there is a limit of the reaction rate for any kind of activity. After the onset of the reaction, a further increase in its rate becomes unfavorable.[50] Correlation of the rate and the precision of the choice also depends on the complexity of the task. Harmonious relations of the rate and precision are characteristic of simple tasks, whereas complex tasks are characterized by disharmonious relations.[51]

The use of complex and simple tasks in the same experiment prolongs latency, while the reaction time is similar, possibly reflecting the complexity of preparation to different reactions.[52] A possible explanation of the increase of the latent period, when the complexity of the stimulus rises, is the complexity of the processes that are necessary to organize the reaction. This relationship may indicate that the hypothetical stimulatory force that provides realization of the reaction increases with the increase of time interval between the stimulus and the onset of the reaction.[43] In general, the dependence of the reaction time and the amount of information in the stimuli presented has a regular character and this relationship is referred to as Hyke's law. The general strategy of recognition may be defined as a prediction of the probable appearance of some stimulus.[53] This conclusion agrees with the postulates of Anokhin and Bernstein that the brain contains an apparatus that predicts the results of an action.[41,54] It can be concluded that the reaction rate to an external influence depends on the rate of a number of processes that finally result in the formation of decisions by the character for a future action, the program of action and acceptance of the results of this action. Since many CNS systems that have different levels of excitability participate in realization of this program of actions, an optimum rate is necessary to accomplish it. When this optimal rate is reached, further increase or decrease becomes unfavorable. Therefore, we suspect that when methysergide or PCPA are injected together with THC, some optimum rate of the processes is attained that brings about a response reaction if two durable or short-term stimuli are presented.

With respect to the interaction of THC with the DA system, the injection of THC is known to either decrease or not change DA levels in the hypothalamic caudate nucleus and midbrain.[55] It is also known that THC decreases DA release from the nerve endings and intensifies synaptic DA uptake.[56] There is also evidence that THC slows down DA turnover.[5] Thus, THC appears to slow down metabolic processes in the DA system and to reduce its participation in the CNS activity. It is also known that a decrease in DA decreases attention,[57,58] inhibits orientation reaction,[59] and complicates realization of the tasks associated with space discrimination.[60] Training was shown to increase DA levels in some brain formations,[61] whereas DA agonists increase emotional reactivity to electric shocks.[62]

The effects of a DA agonist are a direct activation of the dopaminergic system. This, in turn, slows down the reaction to the AR, indicating that there is slowing down of processes necessary to organize response reactions. Clearly, effects of a DA agonist differ from the effects of altering the function of the serotoninergic system. Increasing the reaction time for differentiated stimuli corresponds to increasing of the reaction time for a single stimulus. Thus, it can be suggested that dopaminergic substances do not slow down the processes associated with the choice of any action when two stimuli are presented. However, dopaminergic agents do slow down the processes that participate in the realization of both differentiation and simple reflexes, since a simple instrumental reflex requires no new decision for its realization and can be materialized on the basis of a preexisting motor program.[63] Such factors may include the environmental afferentation and the level of dominating motivation, with both elements being closely connected. The dominating motivation is important and involved in all stages of the conditioned reflex act and determines which demand should be satisfied by a behavioral act. This, in turn, influences the general excitement level of the animal. Each type of motivation has been shown to exercise a specific influence on the activity of neuronal networks in the brain. Thus, motivation defines the character of the action[64] by changing the functional tonic state of the nerve centers.

Therefore, the increase in the reaction time under these conditions may be associated with dopaminergic system-mediated changes of motivation level. This possibility is consistent with the known effects of dopaminergic substances on emotional reactivity[65,66] and the activity of limbic structures.[67] Moreover, it has been proposed that mesocortical and mesolimbic dopaminergic tracts play a major role in the regulation of the emotional state.[68]

The effects of α-MPT on THC-induced changes in DSS and in the duration of the reaction time were opposite to the effects of α-DOPA and apomorphine. Modification of the effects of THC by these substances provides evidence for a role of the dopaminergic system in mediating the effects of THC. However, α-MPT interacts with THC with a limited number of effects concerned mainly with the processes that take place in the sensory register. As the action of α-MPT is less selective than that of α-DOPA and apomorphine, it can be concluded that a simultaneous decrease of the DA and NE levels alters the effect of THC on DSS; whereas, modulation of the effects of THC on DLS requires a more local influence on the dopaminergic system. Thus, α-MPT injection produces a complex picture of changes of different AR parameters. Various physiological systems may be involved in this response, including the processes of information storage and its retrieval from long-term repositories.[46]

Injection of an adrenomimetic, ephedrine, markedly intensified both LPS and LPD, while the remainder of the AR parameters were not altered. Evidently, the AR time parameters can change without concomitant alterations of their qualitative character. This could be attributed, among others, to simplification of the search motor activity and a decrease of stereotypic movements when the NE system is stimulated.[69]

The α-adrenergic blocker, phentolamine, and the α_2 blocker, yohimbine, produce similar effects, with the action of yohimbine being more pronounced. THC is known to cause a decrease in NE content in a number of brain regions.[70] Measurements of NE content and the rate of its turnover in THC-treated animals suggest that THC alters NE synthesis,[71] which can lead to reduced NE discharge and consequent excitation of presynaptic α_2-adrenoreceptors.

It is also known that both phentolamine and yohimbine increase NE discharge by blocking α_2-receptors, the effect being more pronounced for yohimbine.[61,72] Moreover, α_2 postsynaptic receptors can cause hyperpolarization.[73] The increase of the discharge of NE can lead to depletion of NE stores and suppression of the activity of the noradrenergic system, i.e. changes equivalent to the effects of α-MPT.

The β-blocker, propranolol, fails to modify the effects of THC and, therefore, a conclusion can be made that β-adrenergic receptors are not involved in the action of THC. This conclusion contradicts previous reports of the effects of THC on β-adrenoreceptors in mouse cerebral cortex.[74]

CONCLUSION

Our experiments have shown that the actions of THC on the complex, multicomponent system of the behavioral organization are mediated by a number of neurotransmitter systems. The present results also suggest that the same component of the organization of behavior will depend, to a certain extent, on several mediator systems. This characteristic of the THC interaction with the mediator systems may make it very difficult, if not impossible, to localize the action of THC at the cellular level.

REFERENCES

1. **Ho Beng, T. and Johnson, K.,** Sites of neurochemical action of Δ^9-tetrahydrocannabinol interaction with reserpine, in *Marijuana: Chemistry, Biochemistry and Cellular Effects*, Springer-Verlag, New York, 1978, 367.
2. **Anlakh, C., Bhattadaryya, A., Hassain, M., and Pradhan, S.,** Behavioral and neurochemical effects of repeated administration of Δ^9- tetrahydrocannabinol in rats, *Neuropharmacology*, 19, 97, 1980.
3. **Miczer, K. and Dexit, B.,** Behavioral and biochemical effects of chronic delta-9-tetrahydrocannabinol in rats, *Psychopharmacology*, 67, 195, 1980.
4. **Bhattacharia, A., Anlakh, C., Pradhan, S., and Ghosh, P.,** Behavioral and neurochemical effects of Δ^9-tetrahydrocannabinol in rats, *Neuropharmacology*, 19, 87, 1980.
5. **Segawa, T.,** Brain serotonin metabolism and Δ^9-THC-induced muricide behavior in rats, *Jpn. J. Pharmacol.*, 24, 581, 1977.
6. **Palermo, N., Carvalho, F., and Melito, I.,** Δ^9-THC blocks the reproduction of the worked out avoiding reaction, *Arch. Int. Pharmacodyn.*, 222, 300, 1973.
7. **Pras, P., Serizet, A., Bayada, J., and Babean, P.,** Neurotransmission cerebral application therapeutiques actuelles, *Can. Med.*, 7, 11, 1981.

8. **Taylor, D. and Fennessy, M.,** The effect of Δ^9-tetrahydrocannabinol (Δ^9-THC) on the turnover rate of brain serotonin in the rat, *Clin. Exp. Pharmacol. Physiol.*, 6, 327, 1979.
9. **Sofia, K., Dixit, B., and Barry, H.,** The effect of Δ^9-tetrahydrocannabinol on serotonin metabolism in the rat brain, *Life Sci.*, Part 1, 10, 425, 1971.
10. **Arrigo, Scoto, G., Spadaro, C., and Mazzone, G.,** Effecti comortamentali nel ratto di un estrato di cannabis: ruolo delle catecolamine, *Boll. Soc. Ital. Biol. Sper.*, 54, 2348, 1978.
11. **Bloom, A., Johnson, K., and Dewey, W.,** The effects of cannabinols on body temperature and brain catecholamine synthesis, Res. Commun., *Chem. Pathol. Pharmacol.*, 20, 51, 1978.
12. **Musty, R., Lindsey, C., and Carlini, E.,** 6-Hydroxydopamine and aggressive behavior induced by marihuana in REM sleep-deprived rats, *Psychopharmacology*, 48, 175, 1976.
13. **Carlini, E., Lindsey, C., and Charles, J.,** Pharmacological manipulations of brain catecholamines and the aggressive behavior induced by marihuana in REM-sleep-deprived rats, *Aggress. Behav.*, 1, 81, 1974.
14. **Adams, P. and Barrat, E.,** Role of biogenic amines of the effects of marijuana on EEG pattern in cats, *Electroencephalogr. Clin. Neurophysiol.*, 39, 621, 1975.
15. **Bertilsson, L.,** Mechanism of action of benzodiazepines—the GABA hypothesis, *Acta Psychiat. Scand.*, 33, 48, 1975.
16. **Leonard, B.,** The effect of Δ^9-tetrahydrocannabinol on biogenic amines and their amino acid precursor in the rat brain, *Pharm. Res. Commun.*, 3, 139, 1972.
17. **Järbe, T. and Ohlin, G.,** Stimulus effects of Δ^9-tetrahydrocannabinol and its interaction with naltrexone and catecholamine blockers in rats, *Psychopharmacology*, 54, 193, 1977.
18. **Lindamood, C. and Colasant, B.,** Involvement of brain monoamines in the delta-9-tetrahydrocannabinol-induced reduction of hippocampal choline uptake, *J. Neurochem.*, 37, 788, 1981.
19. **Machula, A. I. and Barkov, N. K.,** Influence of 9-tetrahydrocannabinol on differentiation of visual stimuli by cats, *Pharmacol. Toxicol.*, 5, 8, 1980.
20. **Gromova, E. A.,** *Emotion Memory*, Moscow, 1980, 181.
21. **Kruglikov, R. I.,** *Neurochemistry Mechanisms of Learning and Memory,* Science, Moscow, 1981, 211.
22. **Fratta, W., Biggio, G., and Mercuro, G.,** The effect of D- and L-*p*-chlorophenylalanine: the metabolism of 5-hydroxytryptamine in brain, *J. Pharm. Pharmacol.,* 25, 908, 1973.
23. **Weiseman, L. and Koe, B.,** Behavioral effects of α-methyltyrosine, an inhibitor of tyrosine hydroxylase, *Life Sci.*, 4, 1037, 1965.
24. **Recht, R., Borys, H., and Moore, K.,** Alterations in behavior and brain catecholamine levels in rats treated with α-methyltyrosine, *J. Pharmacol. Exp. Ther.*, 153, 412, 1966.
25. **Demidovitch, B. P., Maron, I. A., and Shuvalov, E. Z.,** *Numerical Methods of the Analysis*, Science, Moscow, 1968, 387.
26. **Glezer, V. D. J.,** On the principal differences in the functional organization in the visual systems of the representatives of the evolution of the vertebralis, *Evolutionary Biochem. Pharmacol.*, 1, 474, 1975.
27. **Pradhan, S. N., Ghosh, B., Anlakh, C., and Bhatachaki, A.,** Effects of Δ^9-tetrahydrocannabinol on foot shock-induced aggression in rats, *Commun. Psychopharmacol.*, 4, 127, 1980.
28. **Ramyina, L.,** Serotoninergic receptors in the visual system of light-deprived rats, *Int. J. Dev. Neurosci.*, 3, 285, 1985.

29. **Tenen, S. S.,** The effects of *p*-chlorophenylalanine, a serotonin depletor, on avoidance acquisition, pain sensitivity and related behavior in the rat, *Psychopharmacologia*, 10, 204, 1979.
30. **Sorens, S.,** Effect of central analeptics and convulsive drugs on cortical evoked potentials in rats during barbitural sleep, *Activ. Nerv. Super.*, 19(Suppl.), 330, 1976.
31. **Tanaka, C., Yoh, I., and Takaori, S.,** Relationship between brain monoamine levels and Sidman avoidance behavior in rats treated with tyrosine and tryptophan hydroxylase inhibitors, *Brain Res.*, 45, 153, 1972.
32. **Ohmori, K., Kojima, T., and Marumo, H.,** Effects of reduced cerebral serotonin on learning, Nihon Yacurigaku Dzasi, *Folia Pharmacol.*, 72, 23, 1981.
33. **Riego, M. and Morimoto, H.,** Diazepam increases GABA mediated inhibition in the olfactory cortex slice, *Pflugers Arch.*, 397, 312, 1970.
34. **Penteleeva, T.,** About the time structure of sensomotoric reaction of man, *Problems Psychology*, 5, 74, 1981.
35. **Gilyarovsky, V. A.,** *Selected Works*, Medicine, Moscow, 356.
36. **Saburova, G.,** The change of time of reaction to the light and sound signals for the evaluation of functional state of brain structures damaged locally, *Higher Nervous Activity*, 23, 200, 1973.
37. **Clineschmidt, B. V. and Lottri, V. S.,** Indoleamine antagonists relative potencies as inhibitors of tryptamine and 5-hydroxytryptophan-evoked responses, *Br. J. Pharm.*, 50, 311, 1975.
38. **Lines, C. and Milner, A.,** Nasotemporal overlap in the human retina investigated by means of simple reaction time to lateralized light flash, *Exp. Brain Res.*, 50, 166, 1983.
39. **Livanov, M. I. and Shulgina, G. M.,** Sensor deficit and capability of the organism, *Irkutsk*, 5-15, 1986.
40. **Stull, A. and Kearney, J.,** Circadian variation in an amphetamine-induced motor response, *Pharmacol. Biochem. Behav.*, 7, 571, 1977.
41. **Bernstein, N. A.,** The problems of model building in the biology of activity, *Problems Philosophy*, 6, 77, 1961.
42. **Gambaryan, L. S.,** The specifics of the instrumental conditional reflexes, *Higher Nervous Activity*, 23, 331, 1973.
43. **Grice, G., Nullmeyer, R., and Spiker, V.**, Human reaction time: toward a general theory, *Exp. Psychol. Gen.*, 111, 135, 1982.
44. **Milner, J.,** *Magic Figure Seven Plus or Minus Two*, Interesting Psychology, Progress, Moscow, 1964, 192.
45. **Saulskaya, N. B. and Chivilyova, O. G.,** Dopamine's metabolism during working out: the conditional reflexes and the avoidance reflexes of the rats, *Higher Nervous Activity*, 5, 1083, 1981.
46. **Hancock P. D. and Dirkin, G.,** Central and peripheral visual choice-reaction time under conditions of induced cortical hyperthermia, *Percep. Mot. Skills*, 54, 395, 1982.
47. **Zelinsky, K.,** The analysis of the remodelling of the signal significance of the irritators in the avoidance reflex, *Higher Nervous Activity*, 25, 766, 1975.
48. **Bobrovnikov, L. V.,** The dynamics of the formation of two types of instrumental behaviour in rabbits, *Higher Nervous Activity*, 32, 85, 1977.
49. **Talalaenko, A. N. and Krivobok, G. K.,** On neurochemical mechanisms of the blocking influence of hippocampus upon the food finding conditional reaction of the rats, *Higher Nervous Activity*, 35, 1002, 1985.
50. **Golytsyn, G. A.,** *Choice and Dominance in Problems of Making Decisions*, Science, Moscow, 1976, 309.

51. **Kulikov, V. G. and Shechter, M. S.,** The correlation of speed and exactness of identification as the indicator of the way of identification, *N. Stud. Psychol.*, 2/27, 15, 1982.
52. **Kovesdi, J. and Slak, S.,** Simple reaction time as a function of preparation for a subsequent choice reaction time task, *Bull. Psychonom. Soc.*, 20, 45, 1982.
53. **Sentogotan, Ya. and Arbib, M.,** *Conceptual Models of the Nervous System*, Moscow, 1976, 1.
54. **Anokhin, P. K.,** *Biology and Neurophysiology of Conditioned Reflexes*, Science, Moscow, 1968, 627.
55. **Sofia, K., Dixit, B., and Barry, H.,** The effect of repeated administration of Δ^9-tetrahydrocannabinol on serotonin metabolism on the rat brain, *Arch. Int. Pharmacodyn. Ther.*, 229, 52, 1977.
56. **Williams, J.,** The effect of Δ^9-tetrahydrocannabinol on the noradrenaline and dopamine content of the brain and heart of the rat, *Br. J. Pharm.*, 52, 446, 1974.
57. **Oke, A. and Adams, R.,** Selective attention dysfunction in adult rats neonatally treated with 6-hydroxydopamine, *Pharmacol. Biochem. Behav.*, 9, 429, 1978.
58. **Boyer, J., Montaron, M., and Buiser, P.,** Rythmes electrocorticaux parietaux chez le chat: relation avec un comportement d'attention focalisee et argument pour un controle dopaminergique d'origine mesencephalique, *C. R. Acad. Sci.*, 2, 779, 1980.
59. **De Lanerolle, N. and Millan, J.,** Dopamine, chick behavior and states of attention, *J. Comp. Physiol. Psychol.*, 94, 346, 1980.
60. **Brozosky, T. and Brown,** Dopamine in prefrontal cortex of rhesus monkeys: evidence for a role in cognitive function, in *Catecholamines: Basic and Clin. Front.,* Vol. 2, Proc. 4th Int. Catecholamine Symp., Pacific Grove, CA, 1978, New York, 1979, 1681.
61. **Livanov, M. I.,** *Space Organization of Brain Processes*, Science, Moscow, 1972.
62. **Baevsky, R. M.,** *Forecasting of States Margining Norm and Pathology*, Medicine, Moscow, 1979, 294.
63. **Milstein, N. I.,** *Psychotonimetics*, Medicine, Moscow, 1971, 149.
64. **Krylov, S. S.,** The role of the selective blockade of adrenergic brain structures in the removal of drug dependence, in *2nd Soviet-Italian Symp. on Mediators in the Action of Psychotropic Substances, Heads of the Reports,* Moscow, 1974, 1977, 33.
65. **Allikmet, L.,** Neurochemical aspects of behavior, *Psychopharmacologia*, 15, 392, 1974.
66. **Avis, H.,** The neuropharmacology of aggression: a critical review, *Psychol. Bull.*, 81, 47, 1974.
67. **Bernardo, L. and Prince, D.,** Dopamine action on hippocampal pyramidal cells, *J. Neurosci.*, 2, 415, 1982.
68. **Fadda, F., Argiolas, M., Melis, M., et al.,** Stress-induced increase in 3,4-dihydroxyphenylacetic and (DOPAC) levels in the cerebral cortex and *N. accumbeus*: reversal by diazepam, *Life Sci.*, 22, 2219, 1978.
69. **Keyu, B. J.,** Effect of lysergic acid diethylamide on potentials evoked in the specific sensory pathways, *Br. Med. Bull.*, 21, 30, 1965.
70. **Schildkraut, J. and Feron, D.,** The effects of Δ^9-tetrahydrocannabinol on the metabolism of norepinephrine in rat brain, *Psychopharmacologia*, 20, 191, 1971.
71. **Bloom, A.,** Effects of Δ^9-tetrahydrocannabinol on the synthesis of dopamine and norepinephrine in mouse brain synaptosomes, *Pharmacol. Exp. Ther.*, 221, 97, 1982.
72. **Anden, N., Grabowska-Anden, M., and Liljenberg, B.,** On the presence of autoreceptors on dopamine neurons in different brain regions, *J. Neural Transmiss.*, 57, 129, 1982.

73. **Starke, K. and Docherty, J.,** Alpha and alpha$_2$-adrenoreceptors: pharmacology and implications, *J. Cardiovasc. Pharmacol.*, 3(Suppl. 1), 14, 1981.
74. **Hillard, C. J. and Bloom, A. S.,** Delta-9-tetrahydrocannabinol-induced changes in beta-adrenergic receptor binding in mouse cerebral cortex, *Brain Res.*, 235, 370, 1982.

14 Effects of Δ^9-Tetrahydrocannabinol on Pregnancy, Puberty, and the Neuroendocrine System

INTRODUCTION

Cannabis as a drug of abuse was used as far back as 1000 years B.C.[1] It has been widely known in the Middle East since the first millennium.[2] Marijuana and hashish, the best known cannabis preparations, have become the most widely used illicit drugs in the Western world.[3]

There is a considerable amount of evidence that marijuana can exert important effects on all phases of reproduction and ontogenesis.[4-12] After demonstration that the main psychoactive component of marijuana is Δ^9-tetrahydrocannabinol (THC),[13] research on the effects of marijuana and THC on various physiological functions increased rapidly.

The effects of THC on reproduction in general have been extensively reviewed.[14-21] However, only a few papers have addressed the effects of THC and other cannabinoids on neuroendocrinology of pregnancy and puberty.[4,14,22,33] Thus, in the present paper, we will review the effects of cannabinoids on pregnancy and puberty and the neuroendocrine system.

BRIEF REVIEW OF METHODS USED

During the past 2 decades, most of the experimental studies concerned with the effects of THC on reproduction[19,34-41] were performed using rats, rabbits, and monkeys. The results of these studies conducted in laboratory animals[11,18,42] were often contradictory to the effects observed in the human in clinical studies.[12,43-46] It is difficult to explain these differences in terms of species variations. Instead, it appears that discrepancies between the results of animal and clinical studies are due mainly to differences in experimental design. However, interspecies differences in drug tolerance[20,21,47] must also be taken into consideration, particularly in chronic studies.

Table 1 shows the experimental design of some of the published studies: animal species, doses used, route of administration, and period of treatment. We excluded clinical studies from these comparisons because it was difficult to relate doses used in humans with precisely dosed animal experiments. Doses of THC used *in vivo* varied between 1 μg/kg and 100 mg/kg (Table 1). The relationship of the effects of THC with the dose utilized will be discussed later in this chapter. Relatively few chronic experiments were reported. The results of these studies are even more confusing, possibly because of the differences in development of drug tolerance.

EFFECTS OF THC ON HYPOTHALAMO-PITUITARY-GONADAL FUNCTIONS

Before the detailed presentation of the main topics of our paper, we will summarize the general effects of THC on various reproductive parameters.

Effects on Luteinizing Hormone Releasing Hormone (LHRH)

A single injection of THC increased LHRH content in the mediobasal hypothalamus (MBH) in male rats.[70] Kumar and Simpkins[72] found that the MBH LHRH content was increased in ovariectomized rats after systemic administration of a single dose of THC. Doms et al.[8] showed that an accumulation of LHRH containing dense-core vesicles in the hypothalamic median eminence occurred after THC treatment. Ayalon et al.[53] and Tyrey[21] postulated that the action of THC involved primarily central neuroendocrine mechanisms, since its effects could be reversed by administration of exogenous LHRH. In contrast, Kumar and Chen[49] found a decrease in LHRH in the MBH and also in the anterior hypothalamic-preoptic area, while we found no changes in the LHRH content of the anterior hypothalamus[71] after THC administration. Our *in vitro* studies also demonstrated that THC did not alter the release or storage of gonadotropins and did not modify the responsiveness of cultured anterior pituitary cells to LHRH.[71] Moreover, it appears that the effects caused by chronic marijuana administration may be different from those of a single dose.[48]

Results indicate that the inhibitory effects of THC on serum luteinizing hormone (LH) levels might be due to an inhibitory action on MBH LHRH release. The effects of THC on the neuroendocrine synapses of the MBH may be mediated by opioid mechanisms[49,72] or by catecholamines.[74,76] The involvement of neuropeptide Y (NPY) cannot be excluded.[56,77] It is very likely that the central effects of THC are mediated by its direct effects on catecholamines.[21,66,70]

Effects on Pituitary Reproductive Hormones

Since the early studies of Marks[62] on the inhibitory effects of THC on pituitary LH secretion, a number of papers reported similar effects. Studies in ovariectomized monkeys confirmed these observations and defined a dose-response relationship.[20,78,79] In ovariectomized rats, Tyrey showed suppression of episodic LH secretion following THC administration.[80,81] These effects were reversible, as was also reported by other authors.[82,83] THC has also been shown to block ovulation and disrupt the estrous cycle.[9,48,52,84,85]

Results of studies *in vivo* and *in vitro* indicated that THC had no direct effect on pituitary LH secretion. Although THC decreases blood LH levels in several species, it suppresses or has no effects on serum follicle stimulating hormone (FSH).[20,70,80,86]

The primary effect of THC on prolactin (PRL) secretion appears to be inhibitory. It was reported that THC suppresses tonic levels of PRL[38,69,87,88] and blocks the PRL surge in proestrous animals.[85] Studies in monkeys indicated that THC produces a reversible inhibition of PRL secretion.[31,89] The PRL surge induced by cervical stimulation[61] or in suckling animals[64] can also be prevented by acute THC administration. Deafferentation studies[64] demonstrated that the site of action of THC on PRL secretion was rostral to the MBH. It was suggested that hypothalamic neurotransmitters are responsible for this action.[66,72,90]

Effects on the Gonads

Δ^9-THC has an inhibitory effect on serum testosterone levels in male rats.[5,10,49,91] Moreover, administration of THC can decrease testicular weight[92-94] and reduce prostate and seminal vesicle weights.[9,18,65]

TABLE 1
Effects of THC on Reproduction in Laboratory Animals: Experimental Designs

Species	Dose	Route of administration[a]	Timing of administration	Ref.
Rat	50 mg/kg	i.p.	Acute	48
	15 mg/kg	i.p.	Acute	49
	10 mg/kg	i.p.	Acute	10, 40, 50—53
	5 mg/kg	i.p.	Acute	10
	40 µg/kg	i.p.	Acute	54
	2 x 10 mg/kg/d	i.p.	Between 27th day of age and onset of puberty	55
	1 µg/kg/d	i.p.	Throughout pregnancy	32
	1 µg/kg/d	i.p.	During 3rd week of pregnancy	56, 57
	1 µg/kg/d	i.p.	Between 22 d of age and onset of puberty	58
	38 mg/kg/d	s.c.	5 d	59
	19 mg/kg/d	s.c.	5 d	59
	3.8 mg/kg/d	s.c.	5 d	59
	3 mg/kg	s.c.	Acute	60
	8 mg/kg	i.v.	Acute	61
	5 mg/kg	i.v.	Acute	62
	0.5 mg/kg	i.v.	Acute	63, 64
	25 mg/kg	p.o.	Acute	65
	5 mg/kg	p.o.	Acute	65
	1 mg/kg	p.o.	Acute	65
	0.5 mg/kg	p.o.	Acute	66
	10 mg/kg/d	p.o.	2 weeks	41
	5 mg/kg/d	p.o.	2nd and 3rd weeks of pregnancy	67
	10 mg/kg	i.c.v.	Acute	68
	5 mg/kg	i.c.v.	Acute	68
	1 mg/kg	i.c.v.	Acute	68
	1 µg/kg	i.c.v.	Acute	69, 70
	1.6 µg/kg/d	i.c.v.	7 d	71
	3 mg/kg	i.m.	Acute	72

TABLE 1, continued

Mouse	100 mg/kg	p.o.	Acute	73
	50 mg/kg	p.o.	Acute	6, 26, 27, 74
	5 mg/kg	p.o.	Acute	74
	0.5 mg/kg	p.o.	Acute	74
Rabbit	5 mg/kg/d	i.m.	4 d	19
	2.5 mg/kg/d	i.m.	4 d	19
	1.25 mg/kg/d	i.m.	4 d	19
	0.6 mg/kg/d	i.m.	4 d	19
	0.3 mg/kg/d	i.m.	4 d	19
Rhesus monkey	5.0 mg/kg	i.m.	Acute	75
	2.5 mg/kg	i.m.	Acute	34
	2.5 mg/kg/d	i.m.	Throughout pregnancy	23
	2.5 mg/kg/d	i.m.	Second half of pregnancy	35
	0.3 mg/kg	i. placenta	Acute	35

Note: The above table is not a complete review of all studies.

[a] i.c.v., intracerebroventricular; p.o., per os.

Oligospermia has been shown to occur following THC administration in laboratory animals[18] and in marijuana-smoking humans.[31,91]

Studies on the effects of THC on female sex steroids has been primarily concerned with its influence on progesterone production. The inhibition or disruption of ovarian cyclicity in nonhuman primates[19,31] and in rodents[14] was also observed. In marijuana-smoking women, a disruption of the menstrual cycle has been described.[3,43] In spite of its reported estrogenic activity, THC apparently does not interact with estrogen receptors.[79] The action of THC on the growth and function of accessory reproductive organs in both sexes does not seem to reflect a direct action of THC on these structures. It is most probable that the changes in testosterone and estrogen levels[95] by THC are due to THC-induced changes in gonadotropins or hypothalamic regulatory hormones, as has been discussed earlier in this review.

THE EFFECTS OF THC ON PREGNANCY

The use of marijuana by pregnant women is an important health issue concerning both the women and the newborns. The widespread use of marijuana alone or in combination with other drugs is well known.[12] Studies with marijuana and THC in laboratory animals and retrospective studies on women smoking marijuana during pregnancy show that the use of these drugs increases the risks of stillbirth and loss of pregnancy and can produce other adverse effects.[2,4,7,8,16,31,96,97]

The Effects on Dams

It is known that both fetal and maternal endocrine factors are of importance in mechanisms that maintain pregnancy.[98] Administration of THC to pregnant animals was shown to prolong gestation.[35,57] It was suggested that THC may delay fertilization or disrupt hormonal mechanisms that initiate parturition.[23,35] The drug causes an increase in the frequency of stillbirths[15,28,30,47,57] and a decrease in litter size.[32,57] We observed similar results utilizing a very low dose of THC (1 μg/kg) that did not induce any behavioral effects (Table 2). However, no weight loss during the gestation period was recorded.[57]

Changes in serum LH, FSH, and progesterone levels after administration of THC to pregnant rats are shown in Figure 1. THC significantly decreased serum LH levels when administered during the 3rd week of pregnancy. Only a slight decrease in serum FSH levels was observed. Serum progesterone levels exhibited the expected decrease as pregnancy progressed in both THC-treated and control animals, but this decrease was more prominent in THC-treated females.

It has been reported that the placenta is not a barrier to the passage of most drugs, including marijuana or THC, so they may reach the fetus and also act on the development of the placenta itself.[99] THC has a relatively slow clearance in several species;[73,100,101] therefore, it was suggested that accumulation of THC in the placenta and/or the fetuses may play a role in the actions of THC administered to pregnant females. Multiple dosing with THC resulted in a higher concentration of THC in dam's plasma than was observed after a single dose.[11,23,47] Autoradiographic studies demonstrated not only placental transfer of THC, but also concentrations of THC in the corpora lutea of pregnant mice.[102] Other reproductive processes can also be inhibited by cannabinoids,[14] e.g., uterine contractility and ovarian follicular maturation.

Placental steroid secretion can be inhibited by THC administration.[19,83] This effect of THC may be explained by the observations that marijuana

TABLE 2
The Effects of THC on Litter Size

	Control	Treated
Number of deliveries	52	96
Length of pregnancy in days (means ± SEM)	21.8 ± 0.1	23.3 ± 0.1
Average litter size	10.57	6.44
Number of stillbirths/litter	0.39	3.34
Male/female ratio	1.19	0.86

Note: Pregnant rats were treated daily with 1 μg/kg THC i.p. during the 3rd week of pregnancy. Controls received solvent.

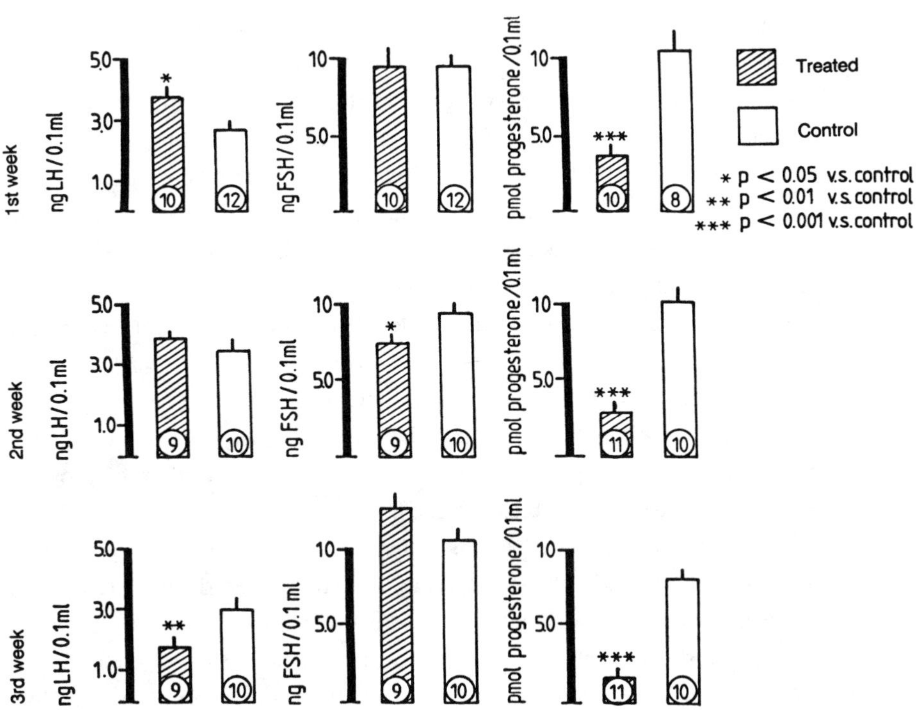

FIGURE 1. The effects of THC on serum LH, FSH, and progesterone levels in pregnant rats. Animals were treated during the 1st, 2nd, and 3rd week of pregnancy with 1 μg THC/kg b.w., i.p., daily. Control animals received solvent. Dams were killed by decapitation on the 22nd or 23rd gestational day. The number of animals is shown within each column.

extract or THC binds substances present in the microsomal fraction of the placenta, interfering with steroid secretion and synthesis in this organ.[19,83]

It has been shown that cannabinoids have an inhibitory action on prostaglandin synthesis in several tissues and organs.[14,103,104] We have demonstrated that the administration of a low dose of THC (1 μg/kg) resulted in a significant reduction of serum prostaglandin levels in pregnant rats (Table 3). If THC does exert an inhibitory effect on prostaglandin synthesis in pregnant rats, its ability to delay the onset of labor may reflect the reduced prostaglandin levels in the placenta and/or uterus.

Effects of Maternal Administration of THC on the Offspring

As was mentioned above, prenatal THC administration caused a significant increase in the incidence of stillbirths and a decrease in litter size. The major question concerning these effects is whether they reflect a direct action of THC on the fetuses or represent consequences of the drug's effect on maternal organs. Owing to the existence of evidence that THC passes through the placental barrier, both effects have to be taken into consideration. Exposure to THC can result in increased resorption of fetuses[16,42] and perinatal deaths.[42] Prenatal administration of THC can also alter the sex ratio of the fetuses (Table 2). Thus, depending on the experimental protocol (i.e., dose administered) either the males (low dose) or the females (higher dose) may be preferentially affected *in utero*.[15,32]

In contrast to other studies,[57] both male and female offspring of THC-treated dams weighed less at birth than the offspring of vehicle-injected controls. Gonadal weights were also significantly decreased in the progeny of THC-treated dams (Tables 4 and 5).

In our experiments, alterations were also recorded in the hypothalamo-pituitary-gonadal axis of the offspring of THC-treated dams. These alterations persisted up to 10 to 15 d after birth (Tables 6, 7, and 8). It would appear that the accumulation of THC or its metabolites during fetal development gradually disappears after birth, and the inhibitory effects of low doses of THC may be temporary.

Of particular interest are the teratological effects of marijuana. Persaud and Ellington[105] found normal offspring after prenatal THC administration in mice. We have obtained similar results with a low dose of THC administered during the entire gestation period.[32,57] A high incidence of mal-

TABLE 3
The Effects of THC on Serum Prostaglandin $F_{1\alpha}$ Levels in Pregnant Rats

Control (17)	765.28 ± 42.35
1st week (11)	718.40 ± 31.50
2nd week (12)	789.60 ± 74.50
3rd week (21)	628.50 ± 37.20[a]

Note: 1 μgTHC/kg b.w./d, i.p. Controls received solvent daily during the 3rd week. Results are expressed in pg/ml ± SEM. THC was injected daily during the respective week; number of animals in parenthesis.

[a] $p < 0.02$ vs. control.

TABLE 4
Effects of Treatment with THC During Pregnancy on Body Weight of Rat Offspring

	Male		Female	
Days of age	**Control**	**Treated**	**Control**	**Treated**
0	6.73 ± 0.10 (21)	6.14 ± 0.10[a] (18)	6 .62 ± 0.07 (17)	6.14 ± 0.15[b] (21)
5	11.75 ± 0.27 (25)	9.70 ± 0.23[a] (24)	11.58 ± 0.49 (22)	12.71 ± 0.23 (24)
10	24.50 ± 0.45 (20)	22.64 ± 0.71[c] (21)	21.49 ± 0.52 (24)	19.50 ± 0.73 (22)
15	31.28 ± 0.44 (15)	24.30 ± 0.73[a] (19)	30.18 ± 0.66 (26)	27.18 ± 1.08 (12)
20	45.26 ± 0.51 (18)	40.21 ± 0.82[a] (21)	41.17 ± 0.51 (18)	40.23 ± 0.52 (24)

Note: Results are expressed in grams, averages ± SEM. Number of animals is in parentheses. THC was administered at 1 μg THC/kg b.w./d, i.p., during the 3rd week of pregnancy. Controls received solvent.

[a] $p < 0.001$.
[b] $p < 0.01$.
[c] $p < 0.05$.

TABLE 5
Effects of Maternal Treatment with THC on Gonadal Weight of the Offspring

	Male		Female	
Days of age	**Control**	**Treated**	**Control**	**Treated**
0	2.70 ± 0.15 (21)	2.00 ± 0.07[a] (18)	1.94 ± 0.17 (17)	1.50 ± 0.06[b] (21)
5	6.30 ± 0.27 (25)	5.17 ± 0.18[b] (24)	2.00 ± 0.10 (22)	2.02 ± 0.04 (24)
10	22.71 ± 0.63 (20)	16.68 ± 0.54[a] (21)	3.94 ± 0.39 (24)	2.60 ± 0.15[b] (22)
15	39.26 ± 1.11 (15)	34.54 ± 2.33[c] (19)	5.85 ± 0.28 (26)	5.00 ± 0.19 (12)
20	99.94 ± 2.24 (18)	89.90 ± 2.29[b] (21)	11.23 ± 0.41 (18)	9.83 ± 0.35 (24)

Note: Results are expressed in milligrams, the average of two gonads/2 ± SEM. Number of animals is in parenthesis. THC was administered at 1 μg/kg b.w./d, i.p., during the 3rd week of pregnancy. Controls received solvent.

[a] $p < 0.001$.
[b] $p < 0.01$.
[c] $p < 0.05$.

formed newborns was observed in rats if a high dose of THC or cannabis extract was injected during the first few days of pregnancy.[36,37] If THC was administered after the period of organogenesis, the teratogenic potency was no longer present. It should also be mentioned that treatment with THC before the end of organogenesis produced teratological effects only at very high doses.[28,105,106] It is possible that THC is, in fact, not teratogenic, at least in the rat and monkey,[19,33] and other substances present in marijuana are responsible for the defects observed.[36,37]

Clinical studies also failed to resolve the possible teratogenicity of marijuana. Although there is evidence that THC crosses the placental barrier in the human,[99] the potent teratogenic effects of marijuana have not been confirmed in clinical studies.[16] However, lower birth weights, a shorter gestation period, and increased incidence of malformations have been found in some marijuana users.[16,19] The interactions of marijuana with other drugs, including alcohol,[4,31,44] and chlorodiazoperoxide (Librium),[96] has also to be taken into consideration when interpreting the results of clinical studies.

TABLE 6
Effects of Maternal Treatment with THC on Hypothalamic LHRH Content in the Progeny

	Mediobasal Hypothalamus		Anterior Hypothalamus	
Days of age	**Control**	**Treated**	**Control**	**Treated**
0	5.00 ± 0.13	1.90 ± 0.25[b]	0.37 ± 0.04	0.67 ± 0.1
	(11)	(12)	(11)	(12)
5	78.37 ± 12.18	32.25 ± 7.50[a]	17.96 ± 5.00	12.40 ± 2.4
	(14)	(13)	(18)	(14)
10	190.23 ± 25.16	118.30 ± 13.70[c]	11.70 ± 2.12	15.65 ± 2.2
	(16)	(17)	(16)	(16)
15	161.11 ± 16.85	131.40 ± 15.20	18.05 ± 2.60	24.60 ± 3.7
	(15)	(12)	(15)	(12)
20	170.48 ± 25.48	145.53 ± 14.54	23.82 ± 4.62	45.24 ± 8.6
	(16)	(23)	(16)	(23)

Note: The results are expressed in fmol tissue ± SEM. Number of animals is in parenthesis. THC was administered at 1 μgTHC/kg b.w./d, i.p., during the 3rd week of pregnancy. Controls received solvent.

[a] $p < 0.0025$.
[b] $p < 0.025$.
[c] $p < 0.01$.

Some clinical data failed to show higher risk in frequent marijuana users[11,12,44] and thus it is difficult to extrapolate the animal data to humans. More studies are needed to substantiate and reconcile experimental data and clinical observations and to explain the differences reported thus far.

The effects of THC on postnatal development also require clarification. Both animal experiments and clinical studies demonstrate THC accumulation in mother's milk and in the plasma of both mother and infant.[46,107] Radioactive THC administered to lactating females could be detected in the urine and feces of their suckling infants in the rat and in the monkey.[108] Moreover, milk production in rodents has been shown to be partially inhibited.[36] Thus, the effects of THC or marijuana on postnatal development depend on whether the animal was given THC (or the mother was smoking marijuana) after birth or prior to delivery.

Dalterio et al. reported that cannabinoids reduced brain dopamine and norepinephrine concentration in adult mice if administered during the perinatal period.[25] The same investigators have demonstrated that postnatal

TABLE 7
Effects of Maternal Treatment with THC on Serum Estrogen Levels in Female Offspring

Days of age	Control	Treated
0	n.d.	n.d.
5	n.d.	n.d.
10	n .d.	n.d.
15	0.10 ± 0.02	0.16 ± 0.01[a]
	(11)	(14)
20	0.13 ± 0.01	0.15 ± 0.01
	(19)	(23)

Note: Results are expressed in pmol 17β-E_2/ml ± SEM. Number of animals is in parentheses. THC (1 μgTHC/kg b.w./d i.p.) was administered during the 3rd week of pregnancy. Controls received solvent. n.d.= non detectable, i.e., less than the sensitivity of our assay.

[a] $p < 0.025$.

TABLE 8
Effects of Maternal Treatment with THC on Serum Progesterone Levels in Female Offspring

Days of age	Control	Treated
0	n.d.	n.d.
5	4.03 ± 0.62	2.53 ± 0.25a
	(19)	(16)
10	2.95 ± 0.66	4.59 ± 0.64
	(15)	(7)
15	11.18 ± 1.14	11.78 ± 1.17
	(18)	(11)
20	18.32 ± 0.90	27.61 ± 2.19
	(19)	(23)

Note: Results are expressed in pmol/ml ± SEM. Number of animals is in parentheses. THC was administered (1 μgTHC/kg b.w./d i.p.) during the 3rd week of pregnancy. The controls received solvent. n.d.= non detectable, i.e., less than the sensitivity of our assay.

[a] $p < 0.025$.

THC exposure resulted in long-term alterations in body weight and regulation of pituitary and gonadal function.[27] Luthra (cited by Nahas in Reference 16) described that THC administered to lactating rats decreased RNA, DNA, and protein concentration in the brain of the developing offspring.

EFFECTS OF THC ON PUBERTY

There is a considerable amount of experimental and clinical evidence that the developing neuroendocrine system during adolescence is particularly vulnerable.[109] Therefore, one of the main questions regarding the risks of marijuana use concerns its possible effects during childhood and adolescence. Little experimental work has been done to address this question and there are practically no clinical studies on the effects of marijuana use on adolescent development (except for clinical epidemiological studies).

The interesting studies of Tyrey[21] and Smith et al.[18] provided evidence that the initiation of the reproductive cycles and first ovulation were delayed after chronic prepubertal administration of THC. We have obtained identical results (Figure 2.). Field and Tyrey demonstrated that the timing of prepubertal THC exposure is an important factor in determining whether or not maturation is delayed.[110]

THC exposure during the first days after birth resulted in long-lasting inhibitory effects on postpuberal reproductive function if the administered dose was large enough.[59] These effects occurred when prolonged doses were given and did not occur in the acute studies.[20,27,76] Exposure to THC during the critical period of sexual differentiation resulted in abnormal postpuberal reproductive hormone levels.[58,71] The onset of puberty depends upon factors affecting both the basal levels of gonadotropin secretion and gonadotropin surges.[109] After prepubertal THC treatment, not only the onset of puberty is delayed, but the number of ova in the first ovulation is also decreased (Figure 3), suggesting an effect of THC on the ovulatory surge of LH during this period of development.[113,114]

Delayed puberty may result from either gonadal or hypothalamo-pituitary disorders.[111] Urbansky and Ojeda[112] demonstrated that prepubertal changes in ovarian function in the rat are due to changes in the pattern of LH release, primarily LH pulse amplitude rather than frequency. These changes presumably correlate with the pattern of pulsatile LHRH release.

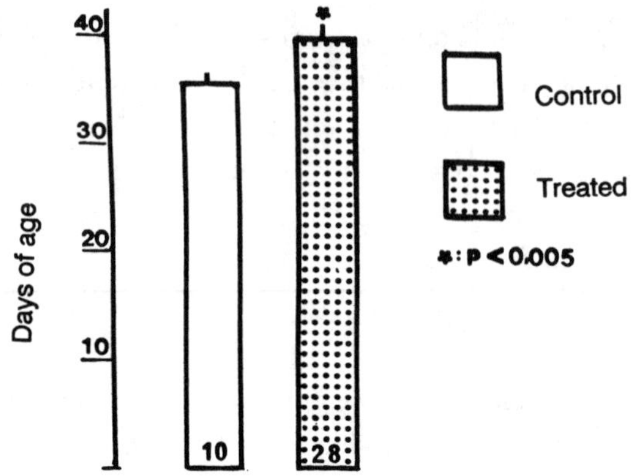

FIGURE 2. Age at first estrus (days). The rats were treated with 1 μgTHC/kg b.w./d, i.p., from the 22nd day of age up to the day of vaginal opening. The controls received solvent. The number in each bar represents the number of animals

As was already mentioned above, it has been demonstrated that THC acts on hypothalamic LHRH secretion[14,53,62] and may affect LHRH pulses.[80] Consequently, the pulsatile LH secretory pattern may be altered by THC, affecting follicular development and the onset of puberty. It appears that THC administered during adolescence may also have an irreversible effect on the reproductive cycle.[27,58,59] Basal LH levels are reduced in rats given THC during this period and cyclicity becomes irregular.[57]

Cannabinoid exposure has also been shown to alter the activity of catecholamines and other monoamines in the hypothalamus.[66,74] Moreover, THC exposure can increase the met-enkephalin immunoreactivity levels in the brain.[49] Hypothalamic levels of NPY can also be affected after THC treatment.[56] Since all these factors are involved in the regulation of LHRH secretion and/or synthesis, it can be suspected that the effect of THC on puberty is mediated by its effects on hypothalamic neuroendocrine systems. The presence of cannabinoid receptors in the brain (including the hypothalamus)[115] supports this assumption.

It is unclear to what extent the responses to THC are affected by the development of tolerance. It was suggested that although tolerance developed in pubertal female rats after chronic THC treatment,[55] the inhibitory effects of THC persisted and resulted in reduced ovarian function and in lower basal LH levels which, however, were still sufficient for partial ovulation.[57]

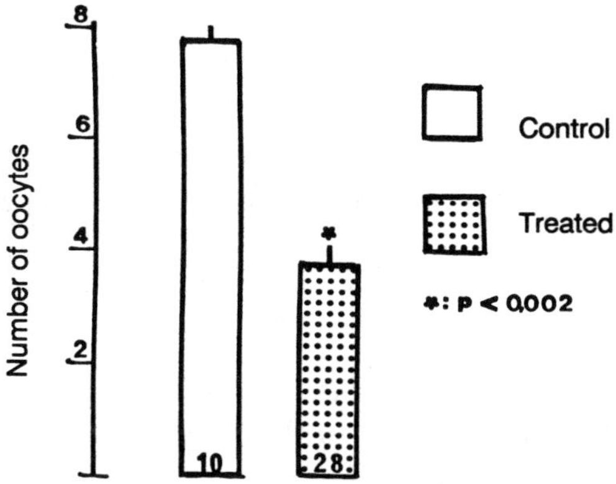

FIGURE 3. Number of oocytes on the day of the first estrus in THC-treated rats. See Figure 2 for details.

GENERAL CONCLUSIONS

It is evident that marijuana and its main psychoactive component, THC, can produce reversible and irreversible effects on the reproductive system. These effects can be observed in both sexes. Although direct effects on both gonadal and pituitary tissue may occur, the actions of THC are mediated primarily through the inhibition of hypothalamic regulatory centers by a direct action of THC or its metabolites within the CNS.[116]

It is unclear whether THC has direct toxic effects on the developing embryo and fetus. However, the effects on placental function and on delivery mechanisms may cause an increase of stillbirths and delay of pregnancy.

The possible effects of marijuana in adolescents have to be seriously taken into consideration. Although the lay public generally considers marijuana to be a nontoxic drug in the context of pubertal development, this is clearly an erroneous assumption. It is particularly important that the combined effect of marijuana and other drugs (including alcohol) on the developing reproductive system is even more dangerous. Further epidemiological studies are needed to assess the incidence of perinatal and pubertal reproductive, neurological, and behavioral disorders associated with marijuana exposure.

ACKNOWLEDGMENT

Work from Dr. Wenger's laboratory was supported by a grant provided by the Hungarian National Science Foundation (OTKA 1035).

REFERENCES

1. **Nahas, G. G.**, Hashish in Islam 9th to 18th century, *Bull. N.Y. Acad. Sci.*, 58, 814, 1982.
2. **Hollister, L. E.**, Cannabis 1988, *Acta Psychiatr. Scand.*, 345(Suppl.), 78, 108, 1988.
3. **Dogoloff, L. L.**, What's wrong with marijuana?, *Psychiatr. Lett.*, V/9, 44, 1987.
4. **Abel, E. L. and Dintcheff, B. A.**, Increased marijuana-induced fetotoxicity by a low dose of concomitant alcohol administration, *J. Stud. Alcohol*, 47, 440, 1986.
5. **Albert, M. and Solomon, J.**, Effect of neonatal administration of delta-9-THC and/or estradiol benzoate (EB) on reproductive development and function of the male rat, in *Cannabinoids: Chemistry, Pharmacology and Therapeutic Aspects*, Agurell, S., Dewey, W. L., and Willette, R., Eds., Academic Press, Orlando, FL, 1984, 339.
6. **Bain, P. A., Bartke, A., and Shrenker, P.**, Luteinizing hormone releasing hormone mediates effects of tetrahydrocannabinol on male sexual behavior, *Neuroendocrinol. Lett.*, 10, 147, 1988.
7. **Cooper, R. L., Goldman, J. M., and Rehnberg, G. L.**, Pituitary function following treatment with reproductive toxins, *Env. Health Behav.*, 70, 177, 1986.
8. **Doms, R. W., Nichols, S. F., and Harclerode, J.**, The effects of Δ^9-tetrahydrocannabinol and phencyclidine hydrochloride on the gonadotrophs of the rat anterior pituitary, *Fed. Proc.*, 40, 279, 1981.
9. **Maskarinec, M. P., Shipley, G., Novotny, M., Brown, D. J., and Forney, R. B.**, Endocrine effects of cannabis in male rats, *Toxicol. Appl. Pharmacol.*, 45, 617, 1978.
10. **Puder, M., Nir, I., Siegel, R. A., Weindenfeld, J., and Ayalon, D.**, The effect of Δ^1-tetrahydrocannabinol on luteinizing hormone release in castrated and hypothalamic deafferented male rats, *Exp. Brain. Res.*, 59, 213, 1985.
11. **Wright, P. L., Smith, S. H., Keplinger, M. L., Calandra, J. C., and Braude, M.**, Reproductive and teratologic studies with Δ^9-THC and crude marijuana extract, *Toxicol. Appl. Pharmacol.*, 38, 223, 1976.
12. **Zuckerman, B., Frank, D. A., Hingson, R., Amaro, H., Levenson, S., Kayne, H., Parker, S., Vinci, R., Aboagye, K., Frie, L. E., Cabral, H., Timperi, R., and Bauchner, H.**, Effects of maternal marijuana and cocaine use on fetal growth, *N. Engl. J. Med.*, 320, 762, 1989.
13. **Mechoulam, R.**, Marihuana chemistry, *Science*, 168, 1159, 1970.
14. **Bloch, E., Thysen, B., Morrill, G. A., Gardner, E., and Fujimoto, G.**, Effects of cannabinoids on reproduction and development, *Vitamins Hormones*, 36, 203, 1978.
15. **Hutchings, D. E., Brake, S. C., and Morgan, B.**, Animal studies of prenatal delta-9-tetrahydrocannabinol: female embryolethality and effects on somatic and brain growth, in *Prenatal Abuse of Licit and Illicit Drugs, Ann. N.Y. Acad. Sci.*, 562, 133, 1989.
16. **Nahas, G. and Frick, H. C.**, Developmental effects of cannabis, *Neurotoxicology*, 7, 381, 1986.

17. **Pace, H. B., Davis, W. M., and Borgen, L. A.**, Teratogenesis and marijuana, *Ann. N.Y. Acad. Sci.,* 191, 123, 1971.
18. **Smith, C. G.**, Effects of marijuana on neuroendocrine function, *NIDA Res. Monogr. Ser.,* 31, 120, 1980.
19. **Smith, C. G., Besch, N. F., and Asch, R. H.**, Effects of marijuana on the reproductive system, in *Advances in Sex Hormone Research,* Vol. 4, Thomas, J. A. and Singhal, R., Eds., Urban and Swarzenberg, Baltimore, 1980, 273.
20. **Smith, C. G. and Asch, R. H.**, Acute, short-term and chronic effects of marijuana on the female primate reproductive function, in *Marijuana Effects on the Endocrine and Reproductive Systems,* Braude, M. C. and Ludford, J. P., Eds., NIDA, Rockville, MD, 1984, 82.
21. **Tyrey, L.**, Endocrine aspects of cannabinoid action in female subprimates: search for sites of action, in *Marijuana Effects on the Endocrine and Reproductive Systems,* Braude, M. C., and Ludford, J. P., Eds., NIDA, Rockville, MD, 1984, 65.
22. **Abel, E. L. and Sokol, R. J.**, Marihuana and cocaine use during pregnancy, in *Drug Use and Pregnancy,* Niebyl, J.R., Ed., Lea and Fieberger, Philadelphia, 1986.
23. **Abel, E. L., Tan, S. E., and Subramanian, M.**, Effects of Δ^9-tetrahydrocannabinol, phenobarbital, and their combination on pregnancy and offspring in rats, *Teratology,* 36, 193, 1987.
24. **Bailey, J. R., Cunny, H. C., Paule, M. G., and Slikker, W.**, Fetal disposition of Δ^9-tetrahydrocannabinol (THC) during late pregnancy in the Rhesus monkey, *Toxicol. Appl. Pharmacol.,* 90, 315, 1987.
25. **Dalterio, S. and Bartke, A.,** Perinatal exposure to cannabinoids alters male reproductive function in mice, *Science,* 205, 1420, 1979.
26. **Dalterio, S., Steger, R., Mayfield, D., and Bartke, A.**, Early cannabinoid exposure influences neuroendocrine and reproductive functions in male mice. I. Prenatal exposure, *Pharmacol. Biochem. Behav.*, 20, 107, 1984.
27. **Dalterio, S., Steger, R., Mayfield, D., and Bartke, A.**, Early cannabinoid exposure influences neuroendocrine and reproductive functions in mice. II. Postnatal effects, *Pharmacol. Biochem. Behav.,* 20, 115, 1984.
28. **Evans, M. A. and Harbison, R. D.**, Cocaine, marijuana, LSD: pharmacological effects in the fetus and newborn, in *Drug Abuse in Pregnancy, Neonatal Effects,* Rementeria, J. L., Ed., Mosby, St. Louis, 1977, 195.
29. **Field, E. and Tyrey, L.**, Delayed sexual maturation in the female rat during chronic exposure to delta-9-tetrahydrocannabinol, *Life Sci.,* 35, 1725, 1984,
30. **Hutchings, D. E., Morgan, B., Brake, S. C., Shi, T., and Lasalle, E.**, Delta-9-tetrahydrocannabinol during pregnancy in the rat. I. Differential effects on maternal nutrition, embryotoxicity, and growth in the offspring, *Neurotoxicol. Teratol.,* 9, 39, 1987.
31. **Smith, C. G. and Asch, R.**, Drug abuse and reproduction, *Fertil. Steril.,* 48, 355, 1987.
32. **Wenger, T., Aranyi, Z., and Bereczky, M.**, The effect of marijuana (delta-9-tetrahydrocannabinol) on pregnancy in rat, *Neuroendocrinol. Lett.,* 10, 39, 1988.
33. **Wenger, T., Croix, D., Tramu, G., and Leonardelli, J.**, Prenatally administered delta-9-tetrahydrocannabinol temporary inhibits the developing hypothalamo-pituitary system in rats, paper presented at Marijuana '90 Symp., Kolympari, Greece, July 8 to 11, 1990.
34. **Almirez, R. G., Smith, C. G., and Asch, R.** The effects of marijuana extract and Δ^9-tetrahydrocannabinol on luteal function in the rhesus monkey, *Fertil. Steril.*, 39, 212, 1983.

35. **Asch, R. H. and Smith, C. G.**, Effects of Δ^9-THC, the principal psychoactive component of marijuana, during pregnancy in the Rhesus monkey, *J. Reprod. Med.*, 31, 1071, 1986.
36. **Borgen L. A., Davis, W. M., and Pace, H. B.**, Effects of synthetic Δ^9-tetrahydrocannabinol on pregnancy and offspring in the rat, *Toxicol. Appl. Pharmacol.*, 20, 480, 1971.
37. **Borgen, L. A. and Davis, W. M.**, Effects of chronic Δ^9-tetrahydrocannabinol on pregnancy in the rat, *The Pharmacologist,* Fall, 259, 1970.
38. **Chakravarty, I., Sheth, A. R., and Ghosh, J. J.**, Effect of acute Δ^9-tetrahydrocannabinol treatment on serum luteinizing hormone and prolactin levels in adult female rat, *Fertil. Steril.*, 26, 947, 1975.
39. **Dalterio, S., Bernard, S. A., and Esquivel, C. R.**, Acute Δ^9-tetrahydrocannabinol exposure alters Ca^{2+}ATPase activity in neuroendocrine and gonadal tissue in mice, *Eur. J. Pharmacol.*, 137, 91, 1987.
40. **Field, E. and Tyrey, L.**, Blockade of first ovulation in pubertal rats by delta-9-tetrahydrocannabinol: requirement for advanced treatment due to early initiation of the critical period, *Biol. Reprod.*, 34, 512, 1986.
41. **Walters, D. E. and Carr, L. A.**, Perinatal exposure to cannabinoids alters neurochemical development in rat brain, *Pharmacol. Biochem. Behav.*, 29, 213, 1988.
42. **Rosenkrantz, H. and Esber, H. J.**, Cannabinoid-induced hormone changes in monkeys and rats, *J. Toxicol. Environ. Health,* 6, 297, 1980.
43. **Hollister, L. E.**, Health aspects of cannabis, *Pharmacol. Rev.*, 38, 1, 1986.
44. **Lex, B. W., Griffin, M. L., Mello, N. K., and Mendelson, J.**, Alcohol, marijuana, and mood states in young women, *Int. J. Addict.*, 24, 405, 1989.
45. **Mendelson, J. H., Cristofaro, P., Ellingboe, J., Benedikt, R., and Mello, N. K.**, Acute effects of marijuana on luteinizing hormone in menopausal women, *Pharmacol. Biochem. Behav.*, 23, 765, 1985.
46. **Mendelson, J. H., Mello, N. K., and Ellingboe, J.**, Acute effects of marihuana in prolactin levels in human females, *J. Pharmacol. Exp. Ther.*, 232, 220, 1985.
47. **Hutchings, D. E., Martin, B. R., Gamagaris, Z., Miller, N., and Pico, T.**, Plasma concentrations of delta-9-tetrahydrocannabinol in dams and fetuses following acute or multiple prenatal dosing in rats, *Life Sci.*, 44, 697, 1989.
48. **Chakravarty, I., Shah, P. G., Sheth, A. R., and Ghosh, J. J.,** Mode of action of delta-9-tetrahydrocannabinol on hypothalamo-pituitary function in adult female rats, *J. Reprod. Fertil.*, 57, 113, 1979.
49. **Kumar, M. S. A. and Chen, C. L.** Effect of an acute dose of Δ^9-THC on hypothalamic luteinizing hormone releasing hormone and met-enkephalin content and serum levels of testosterone and corticosterone in rats, *Subst. Alcohol. Misuse*, 4, 37, 1983.
50. **Cordova, T., Ayalon, D., Naftali, L., Mechoulam, R., Nir, I., Puder, M., and Lindner, R.**, The ovulation effect of cannabinoids: structure-activity relationships, *Psycho-Neuroendocrinology*, 5, 53, 1980.
51. **Murphy, L. L. and Tyrey, L.**, Induction of luteinizing hormone release by electrochemical stimulation of the medial preoptic area in Δ^9-tetrahydrocannabinol-blocked proestrus rats, *Neuroendocrinology*, 43, 471, 1986.
52. **Steger, R. W., Silverman, A. Y., Siler-Khodr, T. M., and Asch, R. H.**, The effect of Δ^9-tetrahydrocannabinol on the positive and negative feedback control of luteinizing hormone release, *Life Sci.*, 27, 1911, 1980.
53. **Ayalon, D., Nir, I., Cordova, T., Bauminger, S., Puder, M., Naor, Z., Kashi, R., Zor, U., Harell, A., and Lindner, H. R.**, Acute effect of Δ^1-tetrahydrocannabinol on the hypothalamo-ovarian axis in the rat, *Neuroendocrinology,* 23, 31, 1977.

54. **Daley, J. D., Branda, L. A., Rosenfeld, J., and Younglai, E. V.**, Increase of serum prolactin in male rats by (–)-*trans*-Δ^9-tetrahydrocannabinol. *J. Endocrinol.,* 63, 415, 1974.
55. **Field, E. and Tyrey, L.**, Tolerance to the luteinizing hormone and prolactin suppressive effects of Δ^9-tetrahydrocannabinol develops during chronic prepuberal treatment of female rats, *J. Pharmacol. Exp. Ther.,* 238, 1034, 1986.
56. **Wenger, T., Croix, D., Tramu, G., and Leonardelli, J.**, Effects of prenatally administered delta-9-tetrahydrocannabinol on hypothalamic neuropeptide Y content in rat offsprings, *Neuroendocrinol. Lett.,* 13, 15, 1991.
57. **Wenger, T., Croix, D., and Tramu, G.**, Marihuana and pregnancy, in *Progress in Neuropeptide Research,* Dohler, K. and Pawlikowski, M., Eds., Birkhauser Verlag, Basel, 1989, 111.
58. **Wenger, T., Croix, D., and Tramu, G.**, The effect of chronic prepuberal administration of marijuana (delta-9-tetrahydrocannabinol) on the onset of puberty and the postpuberal reproductive functions in female rats, *Biol. Rep.,* 39, 540, 1989.
59. **Kumar, A. M., Solomon, J., Patel, V., Kream, R. M., Drieze, J. M., and Millard, W. J.**, Early exposure to Δ^9-tetrahydrocannabinol influences neuroendocrine and reproductive functions in female rats, *Neuroendocrinology,* 44, 260, 1986.
60. **Patel, V., Borysenko, M., and Kumar, M. S. A.**, Effect of Δ^9-THC on brain and plasma catecholamine levels as measured by HPLC, *Brain Res. Bull.,* 14, 85, 1985.
61. **Hughes, C. L., Everett, J. W., and Tyrey, L.**, Δ^9-tetrahydrocannabinol suppression of prolactin secretion in the rat: lack of direct pituitary effect, *Endocrinology*, 109, 876, 1981.
62. **Marks, B. H.**, Δ^1-Tetrahydrocannabinol and luteinizing hormone secretion, *Prog. Brain Res.,* 39, 331, 1973.
63. **Tyrey, L. and Murphy, L.**, Inhibition of suckling-induced milk ejections in the lactating rat by Δ^9-tetrahydrocannabinol, *Endocrinology,* 123, 469, 1988.
64. **Tyrey, L.**, Reversal of the delta-9-tetrahydrocannabinol inhibitory effect on prolactin secretion by rostral deafferentation of the medial basal hypothalamus, *Neuroendocrinology*, 44, 204, 1986.
65. **Morrill, G. A., Kostellow, A. B., Ziegler, D. H., and Fujimoto, G. I.**, Effects of cannabinoids on function of testis and secondary sex organs in the Fischer rat, *Pharmacology*, 26, 20, 1983.
66. **Murphy, L. L., Steger, R. W., Smith, S., and Bartke, A.**, Effects of delta-9-tetrahydrocannabinol, cannabinol and cannabidiol alone and in combinations, on luteinizing hormone and prolactin release and on hypothalamic neurotransmitters in the male rat, *Neuroendocrinology,* 52, 316, 1990.
67. **Brake, S. C., Hutchings, D. E., Morgan, B., Lasalle, E., and Shi, T.**, Delta-9-tetrahydrocannabinol during pregnancy in the rat. II. Effects on ontogeny of locomotor activity and nipple attachment in the offspring, *Neurotoxicol. Teratol.,* 9, 45, 1987.
68. **Wenger, T., Croix, D., West, M., and Tramu, G.**, Acute effect of intracerebroventricularly injected Δ^9-tetrahydrocannabinol on serum LH and FSH levels in adult male rats, *Neuroendocrinol. Lett.*, 6, 327, 1984.
69. **Rettori, V., Wenger, T., Dalterio, S., and McCann, S. M.**, Efecto de delta-9-tetrahydrocannabinol sobre la secretion de prolactina y hormona de crecimento en la rata, *Proc. Soc. Neuroendocrinologia, Buenos Aires,* 577, 1986.
70. **Collu, R.**, Endocrine effects of chronic intraventricular administration of Δ^9-tetrahydrocannabinol to prepuberal and adult male rat, *Life Sci.,* 18, 223, 1976.

71. **Wenger, T., Rettori, V., Snyder, G., Dalterio, S., and McCann, S. M.**, Effects of delta-9-tetrahydrocannabinol (THC) on the hypothalamic-pituitary control of LH and FSH secretion in adult male rats, *Neuroendocrinology,* 46, 488, 1987.
72. **Kumar, M. S. A. and Simpkins, J. W.,** Naloxone blocks the effects of Δ^9-tetrahydrocannabinol on serum luteinizing hormone and prolactin in rats, *Subst. Alcohol Misuse*, 4, 347, 1983.
73. **Harvey, D. J.**, In vivo metabolism of (+)-*trans*-delta-9-tetrahydrocannabinol in the mouse, *Biomed. Env. Mass Spectrom,* 15, 117, 1988.
74. **Dalterio, S., Steger, R., Peluso, J., and DePaolo, L.**, Acute Δ^9-tetrahydrocannabinol exposure: effects on hypothalamic-pituitary-testicular activity in mice, *Pharmacol. Biochem. Behav.,* 26, 533, 1987.
75. **Smith, C. G., Besch, N. F., Smith, R. G., and Besch, P. K.**, Effect of tetrahydrocannabinol on the hypothalamo-pituitary axis in the ovariectomized rhesus monkey, *Fertil. Steril.,* 31, 335, 1979.
76. **Dalterio, S. L.**, Perinatal or adult exposure to cannabinoids alters male reproductive functions in mice, *Pharmacol. Biochem. Behav.,* 12, 43, 1980.
77. **McDonald, J. K.**, NPY and related substances, *Crit. Rev., Neurobiol.,* 4, 97, 1988.
78. **Smith, C. G., Smith, M. T., Besch, N. F., Smith, R. G., and Asch, R. H.**, Effect of delta-9-tetrahydrocannabinol (THC) on female reproductive function, *Adv. Biosci.,* 22–23, 449, 1978.
79. **Smith, R. G., Besch, N. F., Besch, P. K., and Smith, C. G.**, Inhibition of gonadotropin by delta-9-tetrahydrocannabinol: mediation by steroid receptors, *Science,* 204, 325, 1979.
80. **Tyrey, L.**, Δ^9-tetrahydrocannabinol suppression of episodic luteinizing hormone secretion in the ovariectomized rat, *Endocrinology,* 102, 1808, 1978.
81. **Tyrey, L.**, Delta-9-tetrahydrocannabinol: a potent inhibitor of episodic luteinizing hormone secretion, *J. Pharmacol. Exp. Ther.,* 213, 306, 1980.
82. **Besch, N. F., Smith, C. G., Besch, P. K., and Kaufman, R. H.**, The effect of marihuana (delta-9-tetrahydrocannabinol) on the secretion of luteinizing hormone in the ovariectomized Rhesus monkey, *Am. J. Obstet. Gynecol.,* 128, 635, 1977.
83. **Besch, P. K., Stanley, R., Milwidsky, A., Smith, R. G., Besch, N. D., and Smith, C. G.**, The effect of Δ^9-tetrahydrocannabinol on steroid binding and aromatase activity in the human term placenta, paper presented at NIDA Conference on Genetic, Perinatal and Developmental Effects of Abused Substances, NIDA, 1979.
84. **Asch, R. H., Fernandez, E. O., Smith, C. G., and Pauerstein, C. J.**, Precoital single doses of Δ^9-tetrahydrocannabinol block ovulation in the rabbit, *Fertil. Steril.,* 31, 331, 1979.
85. **Nir, T., Ayalon, D., Tsafari, A., Cordova, T., and Lindner, H. R.**, Suppression of the cyclic surge of luteinizing hormone secretion in the rat by Δ^1-tetrahydrocannabinol, *Nature (London),* 243, 470, 1973.
86. **Wenger, T., Croix, D., Tramu, G., and Watanabe, K.**, Further evidence that delta-9-tetrahydrocannabinol decreases serum LH but not FSH in adult ovariectomized rats, *Neuroendocrinol. Lett.,* 8, 47, 1986.
87. **Rettori, V., Wenger, T., Snyder, G., Dalterio, S., and McCann, S. M.**, Hypothalamic action of delta-9-tetrahydrocannabinol (THC) to inhibit the release of prolactin (PRL) and growth hormone (GH) in the rat, *Neuroendocrinology,* 47, 498, 1988.
88. **Raine, J. M., Wing, D. R., and Paton, W. D.**, The effects of delta-1-tetrahydrocannabinol on mammary gland growth, enzyme activity and plasma prolactin levels in the mouse, *Eur. J. Pharmacol.,* 51, 11, 1978.

89. **Asch, R. H., Smith, C. G., Siler-Khodr, T. M., and Pauerstein, C. J.**, Acute decreases in serum prolactin concentrations caused by Δ^9-tetrahydrocannabinol in nonhuman primates, *Fertil. Steril.,* 32, 571, 1979.
90. **Kramer, J. and Ben-David, M.**, Prolactin suppression by (–)Δ^9-tetrahydrocannabinol (THC): involvement of serotonergic and dopaminergic pathways, *Endocrinology,* 103, 452, 1978.
91. **Barnett, G. and Chiang, C. W.**, Effects of marijuana on testosterone in male subjects, *J. Theor. Biol.,* 104, 685, 1983.
92. **Dixit, V. P., Arya, M., and Lohiya, N. K.**, The effect of chronically administered cannabis extract on the female genital tract of mice and rats, *Endocrinology,* 66, 365, 1975.
93. **Dalterio, S., Mayfield, D., Bartke, A., and Morgan, W.**, Effects of psychoactive and non-psychoactive cannabinoids on neuroendocrine and testicular responsiveness in mice, *Life Sci.,* 36, 1299, 1985.
94. **Dalterio, S., Bartke, A., Roberson, C., Watson, D., and Burstein, S.**, Direct and pituitary-mediated effects of delta^9-THC and cannabinol on the testis, *Pharmacol. Biochem. Behav.,* 8, 673, 1978.
95. **Murphy, L. L., Steger, R. W., and Bartke, A.**, Psychoactive and non-psychoactive cannabinoids and their effects on reproductive neuroendocrine parameters, in *Biochemistry and Physiology of Substance Abuse*, Vol. II, Watson, R. R., Ed., CRC Press, Boca Raton, FL, 1990, 73.
96. **Abel, E. L. and Tan, S. E.**, Effects of delta-9-tetrahydrocannabinol, chlordiazepoxide (Librium) and their combination on pregnancy and offspring in rat, *Reprod. Toxicol.,* 1, 37, 1987.
97. **King, T., Schenken, R. S., Kang, I. S., Javors, M. A., and Riehl, R. M.**, Cocaine disrupts estrous cyclicity and alters the reproductive neuroendocrine axis in the rat, *Neuroendocrinology,* 51, 15, 1990.
98. **Nathanielsz, P. W.**, Endocrine mechanisms of parturition, *Annu. Rev. Physiol.,* 40, 411, 1978.
99. **Blackard, C. and Tennes, K.,** Human placental transfer of cannabinoids, *N. Engl. J. Med.,* 311, 797, 1984.
100. **Harvey, D. J.**, Chemistry, metabolism and pharmacokinetics of the cannabinoids, in *Marihuana in Science and Medicine,* Nahas, G., Ed., Raven Press, New York, 1984, 37.
101. **Ryrfeldt, A., Ramsay, C. H., Nilsson, I. M., Widman, M., and Agurell, S.**, Metabolism of cannabis. XIII. Whole body autography of Δ^1-THC in mouse. Pharmacokinetic aspects of Δ^1-THC and its metabolites, *Acta Pharmacol. Suec.,* 10, 13, 1973.
102. **Freudenthal, R. I., Martin, J., and Vall, M. E.**, Distribution of Δ^9-tetrahydrocannabinol in the mouse, *Br. J. Pharmacol.,* 44, 244, 1972.
103. **Reichman, M., Nen, W., and Hokin, L. E.**, Effects of Δ^9-tetrahydrocannabinol on prostaglandin formation in brain, *Mol. Pharmacol.,* 32, 686, 1988.
104. **Howes, J. F. and Osgood, P. F.**, Cannabinoids and the inhibition of prostaglandin synthesis, in *Marihuana: Chemistry, Biochemistry and Cellular Effects,* Nahas, G., Paton, W. D. M., and Idanpaan-Heikkila, J. E., Eds., Springer, New York, 1976, 415.
105. **Persaud, T. V. N. and Ellington, A. C.**, Cannabis in early pregnancy, *Lancet,* II, 1306, 1967.
106. **Fried, P. A.**, Postnatal consequences of maternal marijuana use, in *Prenatal Abuse of Licit and Illicit Drugs, Ann. N.Y. Acad. Sci.,* 562, 123, 1989.
107. **Bromley, B. L., Rabii, J., Gordon, J. H., and Zimmerman, E.**, Δ^9-tetrahydrocannabinol inhibition of suckling-induced prolactin release in the lactating rat, *Endocrinol. Res. Commun.,* 5, 271, 1978.

108. **Chao, F. C., Green, D. E., Forrest, I. S., Kaplan, J. N., Winship-Ball, A., and Braude, M.**, The passage of C-14-delta-9-tetrahydrocannabinol into the milk of lactating squirrel monkeys, *Res. Commun. Chem. Pathol. Pharmacol.,* 14, 302, 1976.
109. **Ojeda, S. and Urbansky, H. F.**, Puberty in the rat, in *The Physiology of Reproduction,* Knobil, E. and Neill, J., Eds., Raven Press, New York, 1988, 1699.
110. **Field, E. and Tyrey, L.**, Delayed sexual maturation during prepuberal cannabinoid treatment: importance of the timing of treatment, *J. Pharmacol. Exp. Ther.,* 254, 171, 1990.
111. **Andrews, W. W. and Ojeda, S. R.**, A detailed analysis of the serum luteinizing hormone secretory profile in conscious free-moving rats during the time of puberty, *Endocrinology,* 109, 2032, 1981.
112. **Urbanski, H. F. and Ojeda, S. R.**, *In vitro* simulation of prepubertal changes in pulsatile luteinizing hormone release enhances progesterone and 17-estradiol secretion from immature rat ovaries, *Endocrinology,* 117, 638, 1985.
113. **Everett, J. W. and Sawyer, C. H.**, A 24-hour periodicity in the "LH release apparatus" of female rats, disclosed by barbiturate sedation, *Endocrinology,* 47, 198, 1950.
114. **Kalra, S. P. and Kalra P. S.**, Neural regulation of luteinizing hormone secretion in the rat, *Endocrinol. Rev.,* 4, 311, 1983.
115. **Devane, W. A., Dysarz, F. A., Johnson, M. R., Melvin, L. S., and Howlett, A. C.**, Determination and characterization of a cannabinoid receptor in rat brain, *Mol. Pharmacol.,* 34, 605, 1988.
116. **Ho, B. T., Estevez, V., Englert, L. F., and Isaac, W. M.**, Δ^9-tetrahydrocannabinol and its metabolites in monkey brain, *J. Pharm. Pharmacol.,* 24, 414, 1972.

INDEX

A

B

C

D

E

G

H

M

O

Q

R

S

U

V

W

X

Y